Customizing
AutoCAD® 2000

Sham Tickoo

Professor
Department of Manufacturing Engineering Technologies
And Supervision
Purdue University Calumet
Hammond, Indiana
U.S.A.

Contributing Author
Gregory Neff

Associate Professor
Department of Manufacturing Engineering Technologies
And Supervision
Purdue University Calumet

Autodesk.
Press

Thomson Learning™

Africa • Australia • Canada • Denmark • Japan • Mexico • New Zealand
Philippines • Puerto Rico • Singapore • Spain • United Kingdom • United States

NOTICE TO THE READER

Trademarks

Autodesk Press Staff
Executive Director: Alar Elken
Executive Editor: Sandy Clark
Development: John Fisher
Executive Marketing Manager: Maura Theriault
Executive Production Manager: Mary Ellen Black
Production Coordinator: Jennifer Gaines
Art and Design Coordinator: Mary Beth Vought
Marketing Coordinator: Paula Collins
Technology Project Manager: Tom Smith

Cover illustration by Brucie Rosch

Printed in the United States of America
1 2 3 4 5 6 7 8 9 10 XXX 04 03 02 01 00 99

For more information, contact
Autodesk Press
3 Columbia Circle, Box 15-015
Albany, New York USA 12212-15015;
or find us on the World Wide Web at http://www.autodeskpress.com

Library of Congress Cataloging-in-Publication Data

Tickoo, Sham.
   Customizing AutoCAD 2000 / Sham Tickoo.
     p. cm.
   ISBN 0-7668-1240-5
   1.   Computer Graphics. 2. AutoCAD. I. Title

   T385.T526 1999
   620'.0042'02855369—dc21

                                        99-047779
                                           CIP

Table of Contents

Chapter 3: Creating Linetypes and Hatch Patterns

Chapter 4: Customizing the ACAD.PGP File

Chapter 5: Pull-down, Shortcut, and Partial Menus and Customizing Toolbars

Chapter 6: Tablet Menus

Chapter 7: Image Tile Menus

Chapter 8: Image Tile Menus

Chapter 9: Screen Menus

Chapter 10: Customizing the Standard AutoCAD Menu

Chapter 11: Shapes and Text Fonts

Chapter 12: AutoLISP

Chapter 13: AutoLISP: Editing the Drawing Database

Chapter 14: Visual LISP

Chapter 15: Programmable Dialog Boxes Using Dialog Control Language

Chapter 16: DIESEL: A String Expression Language

Chapter 17: Visual Basic

Chapter 18: Accessing External Databases

Chapter 19: Defining Block Attributes

Chapter 20: AutoCAD on the Internet

Chapter 21: Inquiry Commands, Data Exchange, and Object Linking and Embedding

Chapter 22: Rendering

Appendices

Index

Preface

AutoCAD, developed by Autodesk Inc., is the most popular PC-CAD system available in the market. Nearly 2.1 million people in 80 countries around the world are using AutoCAD to generate various kinds of drawings. In 1999 the market share of AutoCAD grew to about 78 percent, making it the worldwide standard for generating drawings. Also, AutoCAD's open architecture has allowed third-party developers to write application software that has significantly added to its popularity. For example, the author of this book has developed a software package "SMLayout" for sheet metal products that generates flat layout of various geometrical shapes such as transitions, intersections, cones, elbows, and tank heads. Several companies in Canada and the United States are using this software package with AutoCAD to design and manufacture various products. AutoCAD has also provided facilities that allow users to customize AutoCAD to make it more efficient and therefore increase their productivity.

The purpose of this book is to unravel the customizing power of AutoCAD and explain it in a way that is easy to understand. Every customizing technique is thoroughly explained with examples and illustrations that make it easy to comprehend the customizing concepts of AutoCAD. When you are done reading this book, you will be able to generate a Template drawing, write script files, edit existing menus, write your own menus, write shape and text files, create new linetypes and hatch patterns, define new commands, write programs in the AutoLISP programming language, edit the existing drawing database, create your own dialog boxes using DCL, customize the status line using DIESEL, and edit the Program Parameter file (ACAD.PGP). In the process, you will discover some new applications of AutoCAD that are unique and might have a significant effect on your drawings. You will also get a better idea of why AutoCAD has become such a popular software package and an international standard in PC-CAD.

To use this book, you do not need to be an AutoCAD expert or a programmer. If you know the basic AutoCAD commands, you will have no problem in understanding the material presented in this book. The book contains a detailed description of various customizing techniques that you can use to customize your system. Every chapter has several examples that illustrate some possible applications of these customizing techniques. At the end of each chapter are some exercises that provide a challenge to the user to solve the problems on his/her own. In a class situation, these exercises can be assigned to students to test their understanding of the material explained in the chapter. The chapters on AutoLISP programming assume that the user has no programming background and therefore all commands have been thoroughly explained in a way that makes programming easy to understand and interesting to learn. All chapters, except Slide Shows and Editing the Drawing Database, are independent and can be read in any order and used without reading the rest of the book. The user needs only to read the chapters on Script Files before the chapter on Slide Shows and the chapter on AutoLISP before the chapter on Editing the Drawing Database. However, in order to get a good understanding of customizing techniques, it is recommended to start from Chapter 1 and then progress through the chapters. AutoCAD Release 14 features are indicated by asterisk (*) at the end of the feature. The following is a summary of each chapter.

Chapter 1: Template Drawings
This chapter explains how to create a Template drawing and how to standardize the information that is common to all drawings. It also describes how to create a Template drawing with paper space and predefined viewports.

Chapter 2: Script Files and Slide Shows
This chapter introduces the user to script files and how to utilize them to group AutoCAD commands in a predetermined sequence to perform a given operation. This chapter also explains how to use script files to create a slide show that can be used for product presentation.

Chapter 3: Creating Linetypes and Hatch Patterns
This chapter explains how to create a new linetype and how to edit the linetype file, ACAD.LIN. This chapter also describes the techniques of creating a new hatch pattern and the effect of hatch scale and hatch angle on hatch.

Chapter 4: Customizing the ACAD.PGP File
This chapter explains the use of AutoCAD's Program Parameter file (ACAD.PGP) to define aliases for the operating system commands and some of the AutoCAD commands.

Chapter 5: Pull-down, Shortcut and Partial Menus and Customizing Toolbars
This chapter explains how to write pull-down, shortcut, and partial menus. The chapter also explains how to customize the toolbars. Several examples have been given with thorough explination.

Chapter 6: Tablet Menus
This chapter explains how to write a tablet menu, and how to load other menus from the tablet menu. Advantages of the tablet menu, design of the tablet menu, and how AutoCAD assigns commands to different blocks of the tablet menu are also discussed.

Chapter 7: Image Tile Menus
This chapter explains the Image tile menus and how to write an Image tile menu. It also discusses submenus and how to make slides for the Image tile menu.

Chapter 8: Buttons and Auxiliary Menus
This chapter deals with buttons and auxiliary menus and how to assign AutoCAD commands to different buttons of a multi-button pointing device.

Chapter 9: Screen Menus
This chapter describes the procedure to write a screen menu with multiple submenus and how to load Image tile or pull-down menus from the screen menu.

Chapter 10: Customizing the Standard AutoCAD Menu

This chapter describes how to edit and change various menu sections of the standard AutoCAD menu, ACAD.MNU. It also contains information about submenus and how to load different submenus.

Chapter 11: Shapes and Text Fonts

This chapter explains what shapes are and how to create shape and text fonts. It also contains a detailed description of special codes and their application to creating shapes and text fonts.

Chapter 12: AutoLISP

This chapter explains different AutoLISP functions and how to use these functions to write a program. It also introduces the user to basic programming techniques and use of relational and conditional statements in a program.

Chapter 13: AutoLISP: Editing the Drawing Database

This chapter describes those AutoLISP functions that allow a user to edit the drawing database.

Chapter 14: Visual LISP

This chapter explains how to start Visual LISP in AutoCAD, use Visual LISP text editor, run Visual LISP programs, use Visual LISP console and formatter, and how to debug Visual LISP programs.

Chapter 15: Programmable Dialog Boxes Using Dialog Control Language

This chapter is an introduction to Dialog Control Language and its applications in customizing the existing dialog boxes and writing new dialog boxes. It also explains the use of AutoLISP in controlling the dialog boxes.

Chapter 16: DIESEL: A String Expression Language

This chapter describes the DIESEL string expression language and its application in customizing the status line by altering the value of AutoCAD system variable MODEMACRO.

Chapter 17: Visual Basic

This chapter describes how to install the AutoCAD preview VBA, load and run sample VBA projects, utilize the visual basic editor, use AutoCAD objects and object properties, and apply and use AutoCAD methods.

Chapter 18: Accessing External Database, AutoCAD SQL Extension (ASE)

This chapter describes how to use SQL to access and manipulate the data that is stored in the external database and link data from the database to objects in a drawing. The chapter also explains how to edit the database.

Chapter 19: Defining Block Attributes

This chapter describes how to extract the attributes that have been assigned to blocks by using DDATTEXT or ATTEXT command. The chapter also explains how to create an attribute definition by using the DDATTDEF or ATTDEF commands and how to edit attributes.

Chapter 20: AutoCAD on the Internet

This chapter explains how to use Internet to exchange digital information around the world by bringing together text, graphics, audio, and movies in an easy to use format. The chapter also describes the other uses of the Internet including FTP (file transfer protocol for effortless binary file transfer), Gopher (presents data in a structured, subdirectory-like format), and USENET.

Chapter 21: Inquiry Commands, Data Exchange, and Object Linking and Embedding

This chapter describes how to use various data exchange formats that enable transfer (translation) of data from one data processing software to another. The chapter also explains how to work with different Windows-based applications by transferring information between them by creating links between the different applications and then updating those links, which in turn updates or modifies the information in the corresponding applications. It also describes how to create, edit, and use multiline and how to digitize drawings.

Chapter 22: Rendering

This chapter describes how to render images that makes it easier to visualize the shape and size of a three-dimensional (3D) object, compared to a wireframe image or a shaded image. The chapter also explains how to control the appearance of the object by defining the surface material and reflective quality of the surface and by adding lights to get the desired effects.

In the end, it is the author's sincere hope that the material provided in this book will prove valuable in mastering the customizing techniques and advanced features of AutoCAD 2000.

Chapter 1

Template Drawings

Learning Objectives

After completing this chapter, you will be able to:
- *Create template drawings.*
- *Load template drawings using dialog boxes and the command line.*
- *Do initial drawing setup.*
- *Customize drawings with layers and dimensioning specifications.*
- *Customize drawings with viewports and paper space.*

CREATING TEMPLATE DRAWINGS

One way to customize AutoCAD is to create template drawings that contain initial drawing setup information and, if desired, visible objects and text. When the user starts a new drawing, the settings associated with the template drawing are automatically loaded. If you start a new drawing from scratch, AutoCAD loads default setup values. For example, the default limits are (0.0,0.0), (12.0,9.0) and the default layer is 0 with white color and continuous linetype. Generally, these default parameters need to be reset before generating a drawing on the computer using AutoCAD. A considerable amount of time is required to set up the layers, colors, linetypes, limits, snaps, units, text height, dimensioning variables, and other parameters. Sometimes, border lines and a title block may also be needed.

In production drawings, most of the drawing setup values remain the same. For example, the company title block, border, layers, linetypes, dimension variables, text height, LTSCALE, and other drawing setup values do not change. You will save considerable time if you save these values and reload them when starting a new drawing. You can do this by making template drawings, which can contain the initial drawing setup information, set according to company specifications. They can also contain a border, title block, tolerance table, block definitions, floating viewports in the paper space, and perhaps some notes and instructions that are common to all drawings.

THE STANDARD TEMPLATE DRAWINGS

The AutoCAD software package comes with standard template drawings like acad.dwt, acadiso.dwt, ansi_a.dwt, din_a.dwt, iso_a4.dwt, and jis_a3.dwt. The ansi, din, and iso template drawings are based on the drawing standards developed by ANSI (American National Standards Institute), DIN (German), and ISO (International Organization for Standardization). When you start a new drawing, AutoCAD displays the **Create New Drawing** dialog box on the screen. To load the template drawing, select the **Use a Template** button and AutoCAD will display the list of standard template drawings. From this list you can select any template drawing according to your requirements. If you want to start a drawing with default settings, select the **Start from Scratch** button in the **Create New Drawing** dialog box. The following are some of the system variables, with the default values that are assigned to a new drawing:

System Variable Name	Default Value
BASE	0.0000,0.0000,0.0000
BLIPMODE	Off
CHAMFERA	0.5000
CHAMFERB	0.5000
COLOR	Bylayer
DIMALT	Off
DIMALTD	2
DIMALTF	25.4
DIMPOST	None
DIMASO	On
DIMASZ	0.18
DRAGMODE	Auto
ELEVATION	0
FILLMODE	On
FILLETRAD	0.5000
GRID	0.5000
GRIDMODE	0
ISOPLANE	Left
LIMMIN	0.0000,0.0000
LIMMAX	12.0000,9.0000
LTSCALE	1.0
MIRRTEXT	1 (Text mirrored like other objects)
ORTHOMODE	0 (Off)
TILEMODE	1 (On)
TRACEWID	0.0500

Example 1

Create a template drawing with the following specifications. The name of the template drawing is **PROTO1**.

Limits	18.0,12.0
Snap	0.25
Grid	0.50

Text height	0.125
Units	Decimal
	2-digits to the right of decimal point
	Decimal degrees
	2-digits to the right of decimal point
	0 angle along positive X axis (east)
	Angle positive if measured counterclockwise

Start AutoCAD and select the **Start from Scratch** button in the **Start Up** dialog box. You can also invoke the **Create New Drawing** dialog box (Figure 1-1) by selecting New in the **File** menu or entering **NEW** at AutoCAD Command prompt.

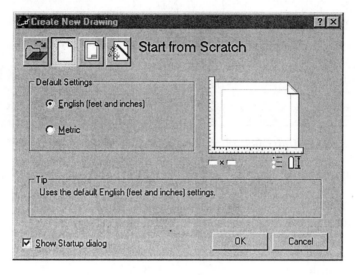

Figure 1-1 Create New Drawing dialog box

Once you are in the Drawing Window, use the following AutoCAD commands to set up the values as given in Example 1.

Command: **LIMITS**
Specify lower left corner or [ON/OFF] <0.00,0.00>: **0,0**
Specify upper right corner <12.0,9.0>: **18.0,12.0**

Command: **SNAP**
Specify snap spacing or [ON/OFF/Aspect/Rotate/Style/Type] <0.5000>: **0.25**

Command: **GRID**
Specify grid spacing(X) or [ON/OFF/Snap/Aspect] <0.5000>: **0.50**

Command: **SETVAR**
Enter variable name or [?]: **TEXTSIZE**
Enter new value for TEXTSIZE <0.2000>: **0.125**

Using the Dialog Box. You can use the **Drawing Units** dialog box (Figure 1-2) to set the units. To invoke the **Drawing Units** dialog box, enter **DDUNITS** or **UNITS** at AutoCAD Command prompt or select **Units** in the **Format** menu.

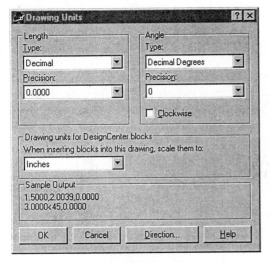

Figure 1-2 Drawing Units dialog box

*Figure 1-3 **Direction Control** dialog box*

Using UNITS Command. You can also use the **-UNITS** command to set the units.

Command: **-UNITS**

Report formats:	Examples:
1. Scientific	1.55E+01
2. Decimal	15.50
3. Engineering	1'-3.50"
4. Architectural	1'-3 1/2"
5. Fractional	15 1/2

With the exception of the Engineering and Architectural formats, these formats can be used with any basic units of measurement. For example, Decimal mode is perfect for metric units as well as decimal English units.

Enter choice, 1 to 5<2>: **2**
Number of digits to right of decimal point (0 to 8)<4>: **2**

Systems of angle measure:	Examples:
1. Decimal degrees	45.0000
2. Degrees/minutes/seconds	45d0'0"
3. Grads	50.0000g
4. Radians	0.7854r
5 Surveyor's units	N 45d0'0"

Enter choice, 1 to 5<1>: **1**

Number of fractional places for display of angles (0 to 8)<0>: **2**
Direction for angle 0.00:

East	3 o'clock	= 0.00
North	12 o'clock	= 90.00
West	9 o'clock	= 180.00
South	6 o'clock	= 270.00

Enter direction for angle 0.00<0.00>: **0**
Do you want angles measured clockwise?<N>: **N**

Now, save the drawing as **PROTO1.DWT** using AutoCAD's **SAVE** or **SAVEAS** command. You must select template (*DWT) from the list box in the dialog box. This drawing is now saved as PROTO1.DWT on the default drive. You can also save this drawing on a floppy diskette in drives A or B.

Command: **SAVE**
Save Drawing as <Drawing.dwg>: **A:PROTO1.DWT**

LOADING A TEMPLATE DRAWING
Using the Dialog Box

You can use the template drawing any time you want to start a new drawing. To use the preset values of the template drawing, start AutoCAD or enter the **NEW** command (Command: **NEW**) to start a new drawing. AutoCAD displays the **Create New Drawing** dialog box (Figure 1-4).

You can also start a new drawing by selecting the **File** menu and then selecting the **New** option from this menu. To load the template drawing, select the **Use a Template** button; AutoCAD will display the list of template drawings. Select **PROTO1** template drawing and then select the **OK**

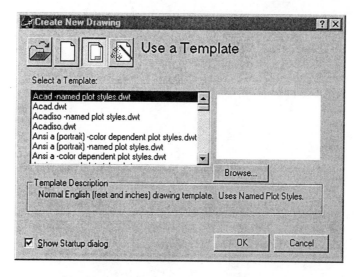

Figure 1-4 Create New Drawing dialog box

button to exit the dialog box. AutoCAD will start a new drawing that will have the same setup as that of template drawing, **PROTO1**.

You can have several template drawings, each with a different setup. For example, **PROTOB** for a 18" by 12" drawing, **PROTOC** for a 24" by 18" drawing, **PROTOD** for a 36" by 24" drawing, and **PROTOE** for a 48" by 36" drawing. Each template drawing can be created according to user-defined specifications. You can then load any of these template drawings as discussed previously.

Using the NEW Command

You can create a new drawing without using the dialog boxes by assigning a value of 0 to the AutoCAD system variable **FILEDIA**. To start a new drawing, enter **NEW** at the Command: prompt (Command: **NEW**) and AutoCAD will prompt you to enter the name of the drawing. The format for entering the name of the template drawing is:

> Command: **NEW**
> Enter template file name or [. (for none)] <current>: *Enter template drawing name.*

CUSTOMIZING DRAWINGS WITH LAYERS AND DIMENSIONING SPECIFICATIONS

Most production drawings need multiple layers for different groups of objects. In addition to layers, it is a good practice to assign different colors to different layers to control the line width at the time of plotting. You can generate a template drawing that contains the desired number of layers with linetypes and colors according to your company specifications. You can then use this template drawing to make a new drawing. The next example illustrates the procedure used for customizing a drawing with layers, linetypes, and colors.

Example 2

You want to create a template drawing (**PROTO2**) that has a border and the company's title block, as shown in Figure 1-5. In addition to this, you want the following initial drawing setup:

Limits	48.0,36.0
Snap	1.0
Grid	4.00
Text height	0.25
PLINE width	0.02
Ltscale	4.0

DIMENSIONS	
DIMSCALE	4.0
DIMTAD	1
DIMTIX	ON
DIMTOH	OFF
DIMTIH	OFF
DIMSCALE	25

(Use the **DIMSTYLE** command to save these values in the MYDIM1 dimension style file.)

LAYERS

Layer Names	Line Type	Color
0	Continuous	White
OBJ	Continuous	Red
CEN	Center	Yellow
HID	Hidden	Blue
DIM	Continuous	Green
BOR	Continuous	Magenta

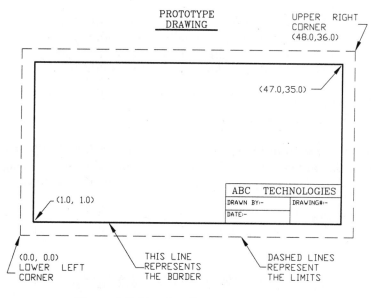

Figure 1-5 *Template drawing for Example 2*

Start a new drawing with default parameters. You can do this by selecting the **Start from Scratch** button in the **Create New Drawing** dialog box. Once you are in the drawing editor, use the AutoCAD commands to set up the values as given for this example. Also, draw a border and a title block as shown in Figure 1-5. In this figure the hidden lines indicate the drawing limits. The border lines are 0.5 units inside the drawing limits. For the border lines, use a polyline of width 0.02 units. Use the following command sequence to produce the prototype drawing for Example 2:

Command: **LIMITS**
Specify lower left corner or [ON/OFF] <0.00,0.00>: **0,0**
Specify upper right corner <12.0,9.0>: **48.0,36.0**

Command: **SNAP**
Specify snap spacing or [ON/OFF/Aspect/Rotate/Style/Type] <0.5000>: **1.0**

Command: **GRID**
Specify grid spacing(X) or [ON/OFF/Snap/Aspect] <1.0>: **4.00**

Command: **SETVAR**
Enter variable name or [?]: **TEXTSIZE**
Enter new value for TEXTSIZE <0.2000>: **0.25**

Command: **PLINEWID**
Enter new value for PLINEWID <0.0000>: **0.02**

Command: **PLINE**
Specify start point: **1.0,1.0**
Current line-width is **0.02**
Specify next point or [Arc/Close/Halfwidth/Length/Undo/Width]:**47,1**
Specify next point or [Arc/Close/Halfwidth/Length/Undo/Width]:**47,35**
Specify next point or [Arc/Close/Halfwidth/Length/Undo/Width]:**1,35**
Specify next point or [Arc/Close/Halfwidth/Length/Undo/Width]:**C**

Command: **LTSCALE**
Enter new linetype scale factor<Current>: **1.0**

Using the Dimension Style Dialog Box. You can use the **Dimension Style Manager** dialog box (Figure 1-6) to set the dimension variables. To invoke the **Dimension Style Manager** dialog box, enter **DDIM** at AutoCAD Command prompt or select **Style** in the **Dimension** menu.

Using the DIM Command. You can also use the **DIM** command to set the dimensions. The following is the command prompt sequence for setting the dimension variables and saving the dimension style file.

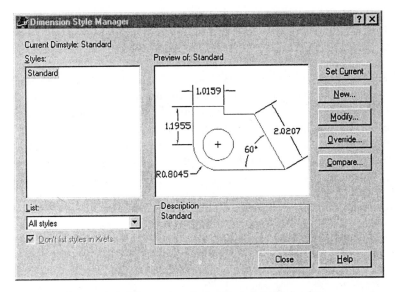

Figure 1-6 Dimension Style Manager dialog box

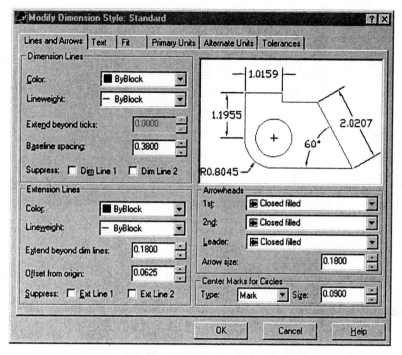

Figure 1-7 Modify Dimension Style dialog box

Command: **DIM**
Dim: **DIMSCALE**
Enter new value for dimension variable <Current>: **1.0**
Dim: **DIMTIX**
Enter new value for dimension variable <Off>: **ON**
Dim: **DIMTAD**
Enter new value for dimension variable <0>: **1**
Dim: **DIMTOH**
Enter new value for dimension variable <On>: **OFF**
Dim: **DIMTIH**
Enter new value for dimension variable <On>: **OFF**
Dim: **STYLE**
Enter new text style <STANDARD>: **MYDIM1**
Dim: **ESC**   (Press the Escape key)

Using the Layer Properties Manager Dialog Box. You can use the **Layer Properties Manager** dialog box (Figure 1-8) to set the layers and linetypes. To invoke this dialog box, enter **LAYER** at AutoCAD Command prompt or select **Layer** in the **Format** menu.

Using the LAYER command. You can also use the **-LAYER** command to set the layers and linetypes. The following is the command prompt sequence for setting the layers and linetypes.

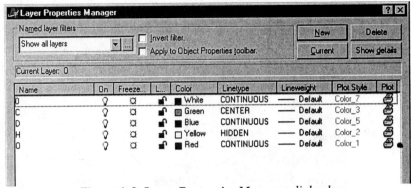

Figure 1-8 Layer Properties Manager dialog box

Command: -LAYER
Enter an option
[?/Make/Set/New/ON/OFF/Color/Ltype/LWeight/Plot/Freeze/Thaw/LOck/Unlock]: **N**
Enter name list for new layer(s): **OBJ,CEN,HID,DIM,BOR**

Enter an option
[?/Make/Set/New/ON/OFF/Color/Ltype/LWeight/Plot/Freeze/Thaw/LOck/Unlock]: **L**
Enter loaded linetype name or [?]<Continuous>: **HIDDEN**
Enter name list of layer(s) for linetype "HIDDEN" <0>: **HID**
Enter an option
[?/Make/Set/New/ON/OFF/Color/Ltype/LWeight/Plot/Freeze/Thaw/LOck/Unlock]: **L**
Enter loaded linetype name or [?]<Continuous>: **CENTER**
Enter name list of layer(s) for linetype "HIDDEN" <0>: **CEN**

Enter an option
[?/Make/Set/New/ON/OFF/Color/Ltype/LWeight/Plot/Freeze/Thaw/LOck/Unlock: **C**
Enter color name or number (1-255): **RED**
Enter name list of layer(s) for color 1 (red) <0>: **OBJ**
Enter an option
[?/Make/Set/New/ON/OFF/Color/Ltype/LWeight/Plot/Freeze/Thaw/LOck/Unlock: **C**
Enter color name or number (1-255): **Yellow**
Enter name list of layer(s) for color 2 (yellow) <0>: **CEN**

Enter an option
[?/Make/Set/New/ON/OFF/Color/Ltype/LWeight/Plot/Freeze/Thaw/LOck/Unlock: **C**
Enter color name or number (1-255): **Blue**
Enter name list of layer(s) for color 5 (blue) <0>: **HID**

Enter an option
[?/Make/Set/New/ON/OFF/Color/Ltype/LWeight/Plot/Freeze/Thaw/LOck/Unlock: **C**
Enter color name or number (1-255): **GREEN**
Enter name list of layer(s) for color 3 (green) <0>: **DIM**

Enter an option
[?/Make/Set/New/ON/OFF/Color/Ltype/LWeight/Plot/Freeze/Thaw/LOck/Unlock: **C**
Enter color name or number (1-255): **MAGENTA**
Enter name list of layer(s) for color 6 (magenta) <0>: **BOR**
Enter an option
[?/Make/Set/New/ON/OFF/Color/Ltype/LWeight/Plot/Freeze/Thaw/LOck/Unlock: (**RETURN**)

Next, add the title block and the text as shown in Figure 1-5. After completing the drawing, save it as PROTO2.DWT. You have created a template drawing (PROTO2) that contains all of the information given in Example 2.

CUSTOMIZING DRAWINGS ACCORDING TO PLOT SIZE AND DRAWING SCALE

You can generate a template drawing according to plot size and scale. For example, if the scale is 1/16" = 1' and the drawing is to be plotted on a 36" by 24" area, you can calculate drawing parameters like limits, **DIMSCALE**, and **LTSCALE** and save them in a template drawing. This will save considerable time in the initial drawing setup and provide uniformity in the drawings. The next example explains the procedure involved in customizing a drawing according to a certain plot size and scale. (Note, you can also use the paper space to specify the paper size and scale.)

Example 3

Generate a template drawing (**PROTO3**) with the following specifications:

Plotted sheet size	36" by 24" (Figure 1-9)
Scale	1/8" = 1'
Snap	3'
Grid	6'
Text height	1/4" on plotted drawing
Ltscale	Calculate
Dimscale	Calculate
Units	Architectural
	16-denominator of smallest fraction
	Angle in degrees/minutes/seconds
	4-number of fractional places for display of angles
	0 angle along positive X axis
	Angle positive if measured counterclockwise
Border	Border should be 1" inside the edges of the plotted drawing sheet, using PLINE 1/32" wide when plotted (Figure 1-9)

In this example, you need to calculate some values before you set the parameters. For example, the limits of the drawing depend on the plotted size of the drawing and the scale of the drawing. Similarly, **LTSCALE** and **DIMSCALE** depend on the limits of the drawing. The following calculations explain the procedure for finding the values of limits, ltscale, dimscale, and text height.

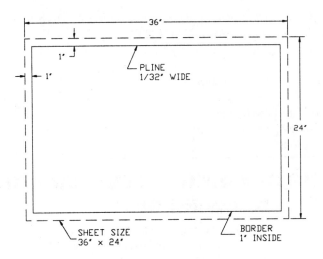

Figure 1-9 *Border of template drawing*

Limits

Given:
Sheet size   36" x 24"
Scale   1/8" = 1'
        or 1" = 8'

Calculate:
XLimit
YLimit
Since sheet size is 36" x 24" and scale is 1/8"=1'
Therefore, XLimit = 36 x 8' = 288'
        YLimit = 24 x 8' = 192'

Text height

Given:
Text height when plotted = 1/4"
Sheet size   36" x 24"
Scale   1/8" = 1'
Calculate:
Text height
Since scale is 1/8" = 1'
            or 1/8" = 12"
            or 1" = 96"
Therefore, scale factor = 96
        Text height = 1/4" x 96
                = 24" = 2'

Ltscale and Dimscale

Known:
Since scale is 1/8" = 1'
or 1/8" = 12"
or   1" = 96"

Calculate:
Ltscale and Dimscale
Since scale factor = 96
Therefore, LTSCALE = Scale factor = 96
Similarly, DIMSCALE = 96
(All dimension variables, like DIMTXT and DIMASZ, will be multiplied by 96.)

Pline Width

Given:
Scale is 1/8" = 1'
Calculate:
PLINE width
Since scale is 1/8" = 1'
or   1" = 8'
or   1" = 96"

Therefore,
PLINE width = 1/32 x 96
= 3"

After calculating the parameters, use the following AutoCAD commands to set up the drawing, then save the drawing as **PROTO3.DWT.**

Command: **-UNITS**

Report formats:	**Examples:**
1. Scientific	1.55E+01
2. Decimal	15.50
3. Engineering	1'-3.50"
4. Architectural	1'-3 1/2"
5. Fractional	15 1/2

With the exception of Engineering and Architectural formats, these formats can be used with any basic unit of measurement. For example, decimal mode is perfect for metric units as well as decimal English units.

Enter choice, 1 to 5<2>: **4**
Denominator of smallest fraction to display
1, 2, 4, 8, 16, 32, or 64 <16>: **16**

Systems of angle measure:		**Examples:**
1.	Decimal degrees	45.0000
2.	Degrees/minutes/seconds	45d0'0"
3.	Grads	50.0000g
4.	Radians	0.7854r
5.	Surveyor's units	N 45d0'0" E

Enter choice, 1 to 5<1>: **2**
Number of fractional places for display of angles (0 to 8)<0>: **4**

Direction for angle 0.00:

East	3 o'clock	= 0d0'0"
North	12 o'clock	= 90d0'0"
West	9 o'clock	= 180d0'0"
South	6 o'clock	= 270d0'0"

Enter direction for angle 0d0'0"<0d0'0">: **RETURN**
Do you want angles measured clockwise?<N>: **N**
Command: **LIMITS**
Specify lower left corner or [ON/OFF] <0'-0",0'-0">: **0,0**
Specify upper right corner <1'-0",0'-9">: **288',192'**

Command: **SNAP**
Specify snap spacing or [ON/OFF/Aspect/Rotate/Style/Type] <0'-1">: **3'**

Command: **GRID**
Specify grid spacing(X) or [ON/OFF/Snap/Aspect] <0'-0">: **6'**

Command: **SETVAR**
Enter variable name or [?] <current>: **TEXTSIZE**
Enter new value for TEXTSIZE <current>: **2'**

Command: **LTSCALE**
Enter new linetype scale factor <1.0000>: **96**

Command: **DIM**
Dim: **Dimscale**
Enter new value for dimension variable <1.0000> New value: **96**
Dim: **Style**
New text style <STANDARD>: **MYDIM2**

Command: **PLINE**
Specify start point: **8',8'**
Current line-width is **0.0000**
Specify next point or [Arc/Close/Halfwidth/Length/Undo/Width]:**W**
Specify starting width<0.00>: **3**
Specify ending width<0'-3">: **ENTER**

Specify next point or [Arc/Close/Halfwidth/Length/Undo/Width]: **280',8'**
Specify next point or [Arc/Close/Halfwidth/Length/Undo/Width]: **280',184'**
Specify next point or [Arc/Close/Halfwidth/Length/Undo/Width]: **8',184'**
Specify next point or [Arc/Close/Halfwidth/Length/Undo/Width]: **C**
Now, save the drawing as PROTO3.DWT

CUSTOMIZING DRAWINGS WITH VIEWPORTS

In certain applications you might need a standard viewport configuration to display different views of an object. It involves setting up the desired viewports and then changing the viewpoint for different viewports. You can generate a prototype drawing that contains a required number of viewports and the viewpoint information. Now, if you insert a 3D object in one of the viewports of the prototype drawing, you will automatically get different views of the object without setting viewports or viewpoints. The following example illustrates the procedure for creating a prototype

Example 4

drawing with a standard number (four) of viewports and viewpoints.
Generate a prototype drawing with four viewports, as shown in Figure 1-10. The viewports should have the following viewpoints (vpoints):

Viewports	Vpoint	View
Top right	1,-1,1	3D view
Top left	0,0,1	Top view
Lower right	1,0,0	Right side view
Lower left	0,-1,0	Front view

Start AutoCAD and create a new drawing, PROTO5. Use the following commands to set the viewports and vpoints.

Command: **-VPORTS**
Enter an option [Save/Restore/Delete/Join/SIngle/?/2/<3>/4]: **4**

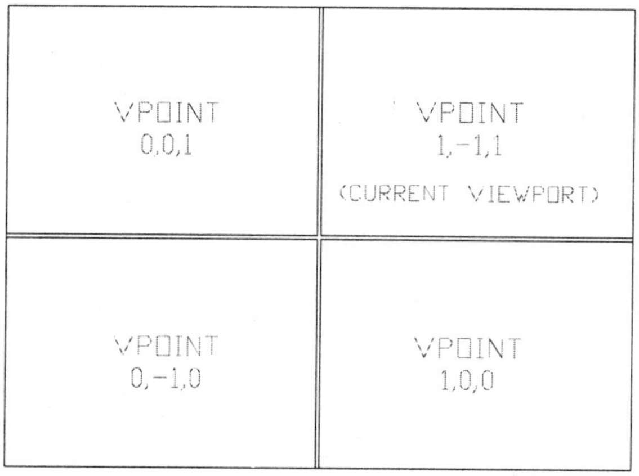

Figure 1-10 Viewports with different viewpoints

Regenerating model

Make the top right viewport current and use the following command to set up the vpoint:

Command: **VPOINT**
Current view direction: VIEWDIR=0.0000,0.0000,1.0000
Specify a view point or [Rotate] <display compass and tripod>: **1,-1,1**
Regenerating model

Make the top left viewport current and use the following command to set up the vpoint:

Command: **VPOINT**
Specify a view point or [Rotate] <display compass and tripod>: **0,0,1**
Regenerating drawing

Make the lower right viewport current and use the following command to set up the vpoint:

Command: **VPOINT**
Specify a view point or [Rotate] <display compass and tripod>: **1,0,0**
Regenerating drawing

Make the lower left viewport current and use the following command to set up the vpoint:

Command: **VPOINT**
Specify a view point or [Rotate] <display compass and tripod>: **0,-1,0**
Regenerating drawing

Command: **SAVE**
Save drawing as <Drawing.dwg>: **PROTO5.DWT**

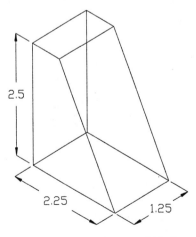

Figure 1-11 3D tapered block

The drawing will be saved under the filename, PROTO5. Now, start a new drawing, TBLOCK. Draw the 3D tapered block shown in Figure 1-11 (Page 1-16) and save it as TBLOCK. (Assume proportionate dimensions for the tapered block.)

Again, start a new drawing, TEST, using the prototype drawing PROTO5. Make the top right viewport current and insert the drawing TBLOCK. Four different views will be automatically displayed on the screen as shown in Figure 1-12.

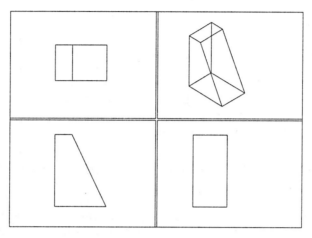

Figure 1-12 *Different views of 3D tapered block*

CUSTOMIZING A DRAWING WITH LAYOUT

The Layout (paper space) provides a convenient way to plot multiple views of a 3D drawing or multiple views of a regular 2D drawing. It takes quite some time to set up the viewports in model space with different vpoints and scale factors. You can create prototype drawings that contain predefined viewport settings, with vpoint and other desired information. Now if you create a new drawing, or insert a drawing, the views are automatically generated. The following example illustrates the procedure for generating a prototype drawing with paper space and model space viewports.

Example 5

Generate a prototype of the drawing in Figure 1-13 with four views in Layout3 (paper space) that display front, top, side, and 3D views of the object. The plot size is 10.5 by 8 inches. The plot scale is 0.5 or 1/2" = 1". The paper space viewports should have the following vpoint setting:

Viewports	Vpoint	View
Top right	1,-1,1	3D view
Top left	0,0,1	Top view
Lower right	1,0,0	Right side view
Lower left	0,-1,0	Front view

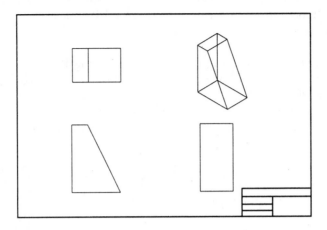

Figure 1-13 *Paper space with four viewports*

Start AutoCAD and create a new drawing. Use the following commands to set up various parameters.

The first step is to create a layout. To create a layout you can use the **LAYOUT** command. You can also select the **Layout1** tab or right-click on **Model** or any **Layout** tab to display the shortcut menu. From the shortcut menu select **New layout**.

Command: **LAYOUT**
Enter layout option [Copy/Delete/New/Template/Rename/SAveas/Set/?] <set>: **N**
Enter new Layout name <Layout3>: **Layout3**

The next step is to select the new layout (Layout3) tab. When you select this tab, AutoCAD displays the **Page Setup-Layout3** dialog box. Select the **Plot Device** tab and then select the printer or plotter that you want to use. In this example HP LaserJet4000 is used.

Next, select the **Layout Settings** tab and select the paper size that is supported by the selected plotting device. In this example the paper size is 8.5x11. Select the **OK** button to accept the settings and exit the dialog box. AutoCAD displays the new layout (Layout3) on the screen with default viewport. Use the ERASE command to erase this viewport.

The next step is to set up a layer (VIEW) for viewports and assign it a color (green). To accomplish this you can us the **Layer Properties Manager** dialog box or **-LAYER** command.

Command: **-LAYER**
Enter an option
[?/Make/Set/New/ON/OFF/Color/Ltype/LWeight/Plot/Freez/Thaw/LOck/Unlock]: **M**
Enter name for new layer (becomes the current layer) <0>: **VIEW**
Enter an option
[?/Make/Set/New/ON/OFF/Color/Ltype/LWeight/Plot/Freez/Thaw/LOck/Unlock]: **C**

Enter color name or number (1-255): **GREEN**
Enter name list of layer(s) for color 3 (green <VIEW>: *Press ENTER.*
[?/Make/Set/New/ON/OFF/Color/Ltype/LWeight/Plot/Freez/Thaw/LOck/Unlock]: *Press ENTER.*

Now, use AutoCAD's **MVIEW** command to set up a viewport, then switch to model space to zoom the display to half the size.

Command: **MVIEW**
Specify corner of viewport or
[ON/OFF/Fit/Hideplot/Lock/Object/Polygonal/Restore/2/3/4] <Fit>:**4**
Specify first corner or [Fit] <Fit>: **0.25,0.25**
Specify opposite corner: **10.25,7.75**

Command: **MSPACE**   (or **MS**)

Make the first viewport active by slecting a point in the viewport and then use the **ZOOM** command to specify the paper space scale factor.

Command: **ZOOM**
Specify corner of window, enter a scale factor (nX or nXP), 0r
[All/Center/Dynamic/Extents/Previous/Scale/Window] <real time>: **0.5XP**

Now, make the next viewport active and specify the scale factor. Do the same for the remaining viewports. The next step is to change the vpoints of different paper space viewports by using the **VPOINT** command. The vpoint values for different viewports are shown in Example 5. To set the view point for the lower-left viewport the command prompt sequence is as follows:

Command: **VPOINT**
Current view direction: VIEWDIR=0.0000,0.0000,1.0000
Specify a view point or [Rotate] <display compass and tripod>: **0,-1,0**

Use the **PSPACE** command to change to paper space and set a new layer, PBORDER, with the color yellow. Make the PBORDER layer current, draw a border, and if needed, a title block using the PLINE command.

Command: **PSPACE**

Command: **PLINE**
Specify start point: **0,0**
Current line-width is 0.0000
Specify next point or [Arc/Close/Halfwidth/Length/Undo/Width]: **8.0,0**
Specify next point or [Arc/Close/Halfwidth/Length/Undo/Width]: **8.0,10.5**
Specify next point or [Arc/Close/Halfwidth/Length/Undo/Width]: **0,10.5**
Specify next point or [Arc/Close/Halfwidth/Length/Undo/Width]: **C**

The last step is to select the **Model** tab (or change the **TILEMODE** to 1) and save the prototype drawing. To test the layout that you just created, insert the TBLOCK drawing that you created in

Example 5. If you switch to **Layout3** tab, you will find four different views of TBLOCK (Figure 1-14). You can freeze the VIEW layer so that the viewports do not appear on the drawing. Now you can plot this drawing from Layout3 with a plot scale factor of 1=1 and the size of the plot will be exactly as specified.

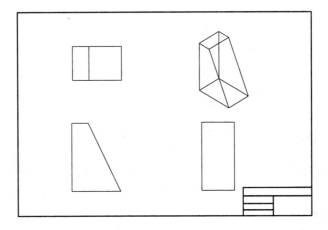

Figure 1-14 Four views of TBLOCK in paper space

Review Questions

1. The three standard template drawings that come with AutoCAD software are _____ , _____ , and _____ .

2. To use a template file, select the _____ button in the **Create New Drawing** dialog box.

3. To start a drawing with default setup, select the _____ button in the **Create New Drawing** dialog box.

4. The default value of DIMSCALE is _____ .

5. The default value for DIMTXT is _____ .

6. The default value for SNAP is _____ .

7. Architectural units can be selected by using AutoCAD's _____ or _____ commands.

8. If plot size is 36" x 24", and the scale is 1/2" = 1', then XLimit = _____ and YLimit = _____.

9. If the plot size is 24" x 18", and the scale is 1 = 20, the XLimit = _____ and YLimit = _____.

10. If the plot size is 200 x 150 and limits are (0.00,0.00) and (600.00,450.00), the **LTSCALE** factor = _____. (S.Factor=600/200=3 LTSCALE=3)

11. _____ provides a convenient way to plot multiple views of a 3D drawing or multiple views of a regular 2D drawing.

12. You can use AutoCAD's _____ command to set up a viewport in paper space.

13. You can use AutoCAD's _____ command to change to paper space.

14. You can use AutoCAD's _____ command to change to model space.

15. The values that can be assigned to TILEMODE are _____ and _____.

16. In the model space, if you want to reduce the display size by half, the scale factor you enter in **ZOOM**-Scale command is _____.

Exercises

Exercise 1   *General*

Generate a template drawing (**PROTOE1**) with the following specifications:

Limits	36.0,24.0
Snap	0.5
Grid	1.0
Text height	0.25
Units	Decimal
	2-number of digits to right of decimal point
	Decimal degrees
	0-number of fractional places for display of angles
	0-angle along positive X axis
	Angle positive if measured counterclockwise

Exercise 2   *General*

Generate a template drawing **PROTOE2** with the following specifications:

Limits	48.0,36.0
Snap	0.5
Grid	2.0
Text height	0.25
PLINE width	0.03
Ltscale	Calculate
Dimscale	Calculate
Plot size	10.5 x 8

LAYERS

Layer Names	Line Type	Color
0	Continuous	White
OBJECT	Continuous	Green
CENTER	Center	Magenta
HIDDEN	Hidden	Blue
DIM	Continuous	Red
BORDER	Continuous	Cyan

Exercise 3   *General*

Generate a template drawing (**PROTOE3**) with the following specifications:

Plotted sheet size	36" x 24" (Figure 1-16)
Scale	1/2" = 1.0'
Text height	1/4" on plotted drawing
Ltscale	Calculate
Dimscale	Calculate
Units	Architectural
	32-denominator of smallest fraction to display
	Angle in degrees/minutes/seconds
	4-number of fractional places for display of angles
	0d0'0"-direction for angle
	Angle positive if measured counterclockwise
Border	Border is 1-1/2" inside the edges of the plotted drawing sheet, using PLINE 1/32" wide when plotted (Figure 1-16)

Exercise 4   *General*

Generate a prototype drawing with the following specifications (the name of the drawing is PROTOE4):

Plotted sheet size	24" x 18" (Figure 1-17)

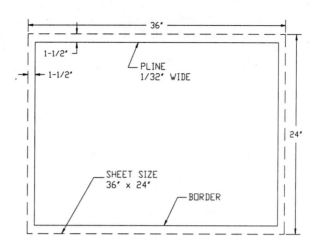

Figure 1-16 Drawing for Exercise 3

Scale	1 = 50
Border	The border is 1" inside the edges of the plotted drawing sheet, using PLINE 0.05" wide when plotted (Figure 1-17)
DIMTAD	ON
DIMTIX	ON
DIMTOH	OFF
DIMTIH	OFF
DIMSCALE	Calculate
DIMALT	ON
DIMASO	OFF
DIMTOFL	ON

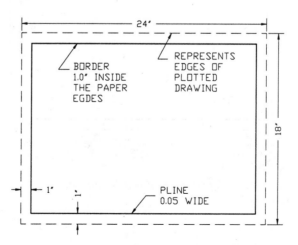

Figure 1-17 Prototype drawing

Exercise 5   *General*

You want to set up a prototype drawing with the following specifications (the name of the drawing is PROTOE5):

Limits	36.0,24,0
Border	35.0,23.0
Grid	1.0
Snap	0.5
Text height	0.15
Dimscale	2.0
Units	Decimal (up to 2 places)
Ltscale	3
Current layer	Object

LAYERS

Layer Name	Linetype	Color
0	Continuous	White
Object	Continuous	Red
Hidden	Hidden	Yellow
Center	Center	Green
Dim	Continuous	Blue
Border	Continuous	Magenta
Notes	Continuous	White

This prototype drawing should have a border line and title block as shown in Figure 1-18.

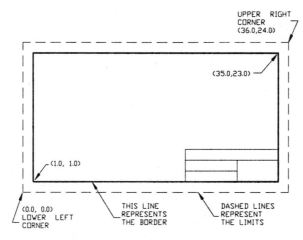

Figure 1-18 Prototype drawing

Chapter 2

Script Files and Slide Shows

Learning Objectives

After completing this chapter, you will be able to:
* *Write script files and use the **SCRIPT** command to run script files.*
* *Use the **RSCRIPT** and **DELAY** commands in script files.*
* *Invoke script files when loading AutoCAD.*
* *Create a slide show.*
* *Preload slides when running a slide show.*

WHAT ARE SCRIPT FILES?

AutoCAD has provided a facility called **script files** that allows you to combine different AutoCAD commands and execute them in a predetermined sequence. The commands can be written as a text file using any text editor like Notepad or AutoCAD's **EDIT** command (if the **ACAD.PGP** file is present and **EDIT** is defined in the file). These files, generally known as script files, have extension **.SCR** (example: **PLOT1.SCR**). A script file is executed with the AutoCAD **SCRIPT** command.

Script files can be used to generate a slide show, do the initial drawing setup, or plot a drawing to a predefined specification. They can also be used to automate certain command sequences that are used frequently in generating, editing, or viewing a drawing. Scripts cannot access dialog boxes or menus. When commands that open the file and plot dialog boxes are issued from a script file, AutoCAD runs the command line version of the command instead of opening the dialog box.

Example 1

Write a script file that will perform the following initial setup for a drawing (file name **SCRIPT1.SCR**). It is assumed that the drawing will be plotted on 12x9 size paper (Scale factor for plotting = 4).

Ortho	On	Zoom	All
Grid	2.0	Text height	0.125
Grid	Off	Ltscale	4.0
Snap	0.5	Dimscale	4.0
Limits	0,0 48.0,36.0		

Before writing a script file, you need to know the AutoCAD commands and the entries required in response to the command prompts. To find out the sequence of the prompt entries, you can type the command at the keyboard and then respond to different prompts. The following is a list of AutoCAD commands and prompt entries for Example 1:

Command: **ORTHO**
Enter mode [ON/OFF] <OFF>: **ON**

Command: **GRID**
Specify grid spacing(X) or [ON/OFF/Snap/Aspect] <1.0>: **2.0**

Command: **GRID**
Specify grid spacing(X) or [ON/OFF/Snap/Aspect] <1.0>: **OFF**

Command: **SNAP**
Specify snap spacing or [ON/OFF/Aspect/Rotate/Style/Type] <1.0>: **0.5**

Command: **LIMITS**
Reset Model space limits:
Specify lower left corner or [ON/OFF] <0.0,0.0>: **0,0**
Specify upper right corner <12.0,9.0>: **48.0,36.0**

Command: **ZOOM**
Specify corner of window, enter a scale factor (nX or nXP), or
[All/Center/Dynamic/Extents/Previous/Scale/Window] <real time>: **A**

Command: **SETVAR**
Enter variable name or [?]: **TEXTSIZE**
Enter new value for TEXTSIZE <0.02>: **0.125**

Command: **LTSCALE**
Enter new linetype scale factor <1.0000>: **4.0**

Command: **SETVAR**
Enter variable name or [?] <TEXTSIZE>: **DIMSCALE**
Enter new value for DIMSCALE <1.0000>: **4.0**

Once you know the AutoCAD commands and the required prompt entries, you can write the script file using the AutoCAD **EDIT** command or any text editor. The following file is a listing of the script file for Example 1:

```
ORTHO
ON
GRID
2.0
GRID
OFF
SNAP
0.5
LIMITS
0,0
48.0,36.0
ZOOM
ALL
SETVAR
TEXTSIZE
0.125
LTSCALE
4.0
SETVAR
DIMSCALE 4.0
```

Notice that the commands and the prompt entries in this file are in the same sequence as mentioned before. You can also combine several statements in one line, as shown in the following list:

```
ORTHO ON
GRID 2.0 GRID OFF
SNAP 0.5 SNAP ON
LIMITS 0,0 48.0,36.0 ZOOM ALL
SETVAR TEXTSIZE 0.125
LTSCALE 4.0
SETVAR DIMSCALE 4.0
```

Note

In the script file, a space is used to terminate a command or a prompt entry. Therefore, spaces are very important in these files. Make sure there are no extra spaces, unless they are required to press ENTER more than once.

*After you change the limits, it is a good practice to use the **ZOOM** command with the **All** option to display the new limits on the screen.*

AutoCAD ignores and does not process any lines that begin with a semicolon (;). This allows you to put comments in the file.

Chapter 2

SCRIPT COMMAND

The AutoCAD **SCRIPT** command allows you to run a script file while you are in the drawing editor. To execute the script file, type the **SCRIPT** command and press ENTER. AutoCAD will prompt you to enter the name of the script file. You can accept the default file name or enter a new file name. The default script file name is the same as the drawing name. If you want to enter a new file name, type the name of the script file **without** the file extension (**.SCR**). (The file extension is assumed and need not be included with the file name.)

To run the script file of Example 1, type the **SCRIPT** command and press ENTER; AutoCAD will display the **Select Script File** dialog box (Figure 2-1). In the dialog box, select the file SCRIPT1 and then choose the Open button. You will see the changes taking place on the screen as the script file commands are executed. The format of the **SCRIPT** command is:

Command: **SCRIPT**

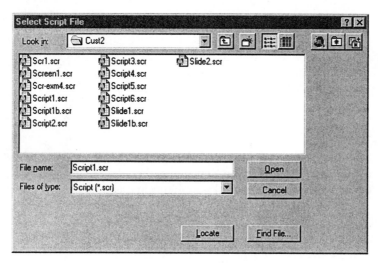

Figure 2-1 Select Script File dialog box

You can also enter the name of the script file at the Command prompt by setting **FILEDIA**=0. The format of the **SCRIPT** command is:

Command: **FILEDIA**
Enter new value for FILEDIA <1>: **0**
Command: **SCRIPT**
Enter script file name <default>: *Script file name.*

For example:
Command: **SCRIPT**
Script file <CUSTOM>: **SCRIPT1**
        Where   **<CUSTOM>** ------------- Default drawing file name
               SCRIPT1 ------------------ Name of the script file

Example 2

Write a script file that will set up the following layers with the given colors and linetypes (file name **SCRIPT2.SCR**).

Layer Names	Color	Linetype	Line Weight
Object	Red	Continuous	default
Center	Yellow	Center	default
Hidden	Blue	Hidden	default
Dimension	Green	Continuous	default
Border	Magenta	Continuous	default
Hatch	Cyan	Continuous	0.05

As mentioned earlier, you need to know the AutoCAD commands and the required prompt entries before writing a script file. For Example 2, you need the following commands to create the layers with the given colors and linetypes:

Command: **LAYER**
Enter an option
[?/Make/Set/New/ON/OFF/Color/Ltype/LWeight/Plot/Freeze/Thaw/LOck/Unlock]: **N**
Enter name list for new layer(s): **OBJECT,CENTER,HIDDEN,DIM,BORDER,HATCH**

Enter an option
[?/Make/Set/New/ON/OFF/Color/Ltype/LWeight/Plot/Freeze/Thaw/LOck/Unlock]: **L**
Enter loaded linetype name or [?] <Continuous>: **CENTER**
Enter name list of layer(s) for linetype "CENTER" <0>: **CENTER**

Enter an option
[?/Make/Set/New/ON/OFF/Color/Ltype/LWeight/Plot/Freeze/Thaw/LOck/Unlock]: **L**
Enter loaded linetype name or [?] <Continuous>: **HIDDEN**
Enter name list of layer(s) for linetype "HIDDEN" <0>: **HIDDEN**

Enter an option
[?/Make/Set/New/ON/OFF/Color/Ltype/LWeight/Plot/Freeze/Thaw/LOck/Unlock]: **C**
Enter color name or number (1-255): **RED**
Enter name list of layer(s) for color 1 (red) <0>: **OBJECT**

Enter an option
[?/Make/Set/New/ON/OFF/Color/Ltype/LWeight/Plot/Freeze/Thaw/LOck/Unlock]: **C**
Enter color name or number (1-255): **YELLOW**
Enter name list of layer(s) for color 2 (yellow) <0>: **CENTER**

Enter an option
[?/Make/Set/New/ON/OFF/Color/Ltype/LWeight/Plot/Freeze/Thaw/LOck/Unlock]: **C**
Enter color name or number (1-255): **BLUE**
Enter name list of layer(s) for color 5 (blue)<0>: **HIDDEN**

Enter an option
[?/Make/Set/New/ON/OFF/Color/Ltype/LWeight/Plot/Freeze/Thaw/LOck/Unlock]: **C**
Enter color name or number (1-255): **GREEN**
Enter name list of layer(s) for color 3 (green)<0>: **DIM**

Enter an option
[?/Make/Set/New/ON/OFF/Color/Ltype/LWeight/Plot/Freeze/Thaw/LOck/Unlock]: **C**
Enter color name or number (1-255): **MAGENTA**
Enter name list of layer(s) for color 6 (magenta)<0>: **BORDER**

Enter an option
[?/Make/Set/New/ON/OFF/Color/Ltype/LWeight/Plot/Freeze/Thaw/LOck/Unlock]: **C**
Enter color name or number (1-255): **CYAN**
Enter name list of layer(s) for color 4 (cyan)<0>: **HATCH**

Enter an option
[?/Make/Set/New/ON/OFF/Color/Ltype/LWeight/Plot/Freeze/Thaw/LOck/Unlock]: LW
Enter lineweight (0.0mm - 2.11mm): 0.05
Enter name list of layers(s) for lineweight 0.05mm <0>:HATCH
[?/Make/Set/New/ON/OFF/Color/Ltype/LWeight/Plot/Freeze/Thaw/LOck/Unlock]: **(RETURN)**

The following file is a listing of the script file that creates different layers and assigns the given colors and linetypes to these layers:

```
;This script file will create new layers and
;assign different colors and linetypes to layers
LAYER
NEW
OBJECT,CENTER,HIDDEN,DIM,BORDER,HATCH
L
CENTER
CENTER
L
HIDDEN
HIDDEN
C
RED
OBJECT
C
YELLOW
CENTER
C
BLUE
HIDDEN
C
GREEN
DIM
C
```

MAGENTA
BORDER
C
CYAN
HATCH
       (This is a blank line to terminate the **LAYER** command. End of script file.)

Example 3

Write a script file that will rotate the circle and the line, as shown in Figure 2-2, around the lower endpoint of the line through 45-degree increments. The script file should be able to produce a continuous rotation of the given objects with a delay of two seconds after every 45-degree rotation (file name **SCRIPT3.SCR**).

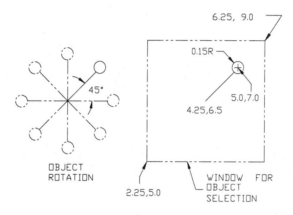

Figure 2-2 Line and circle rotated through 45-degree increments

Before writing the script file, enter the required commands and the prompt entries at the keyboard. Write down the exact sequence of the entries in which they have been entered to perform the given operations. The following is a listing of the AutoCAD command sequence needed to rotate the circle and the line around the lower endpoint of the line:

    Command: **ROTATE**   *(Enter **ROTATE** command.)*
    Current positive angle in UCS: ANGDIR=counterclockwise ANGBASE=0
    Select objects: **W**   *(Window option to select object)*
    Specify first corner: **2.25, 5.0**
    Specify opposite corner: **6.25, 9.0**
    Select objects: **<RETURN>**
    Specify base point: **4.25,6.5**
    Specify rotation angle or [Reference]: **45**

Once the AutoCAD commands, command options, and their sequences are known, you can write a script file. As mentioned earlier, you can use any text editor to write a script file. The following file is a listing of the script file that will create the required rotation of the circle and line of Example 3.

```
    ROTATE                                                        1
    W                                                             2
    2.25,5.0                                                      3
    6.25,9.0                                                      4
              (Blank line for Return.)                            5
    4.25,6.5                                                      6
    45                                                            7
```

Line 1
ROTATE
In this line, **ROTATE** is an AutoCAD command that rotates the objects.

Line 2
W
In this line, W is the Window option for selecting the objects that need to be edited.

Line 3
2.25,5.0
In this line, 2.25 defines the X coordinate and 5.0 defines the Y coordinate of the lower left corner of the object selection window.

Line 4
6.25,9.0
In this line, 6.25 defines the X coordinate and 9.0 defines the Y coordinate of the upper right corner of the object selection window.

Line 5
Line 5 is a blank line that terminates the object selection process.

Line 6
4.25,6.5
In this line, 4.25 defines the X coordinate and 6.5 defines the Y coordinate of the base point for rotation.

Line 7
45
In this line, 45 is the incremental angle for rotation.

Note

One of the limitations of the script files is that all the information has to be contained within the file. These files do not let you enter information. For instance, in Example 3, if you want to use the Window option to select the objects, the Window option (W) and the two points that define this window must be contained within the script file. The same is true for the base point and all other information that goes in a script file. There is no way that a script file can prompt you to enter a particular piece of information and then resume the script file, unless you embed AutoLISP commands to prompt for user input.

RSCRIPT COMMAND

The AutoCAD **RSCRIPT** command allows the user to execute the script file indefinitely until canceled. It is a very desirable feature when the user wants to run the same file continuously. For example, in the case of a slide show for a product demonstration, the **RSCRIPT** command can be used to run the script file again and again until it is terminated by pressing the ESC (Escape) key from the keyboard. Similarly, in Example 3, the rotation command needs to be repeated indefinitely to create a continuous rotation of the objects. This can be accomplished by adding **RSCRIPT** at the end of the file, as in the following file:

```
ROTATE
W
2.25,5.0
6.25,9.0
            (Blank line for Return.)
4.25,6.5
45
RSCRIPT
```

The **RSCRIPT** command on line 8 will repeat the commands from line 1 to line 7, and thus set the script file in an indefinite loop. The script file can be stopped by pressing the ESC or the BACK-SPACE key.

Note

You cannot provide conditional statements in a script file to terminate the file when a particular condition is satisfied, unless you use the AutoLISP functions in the script file.

DELAY COMMAND

In the script files, some of the operations happen very quickly and make it difficult to see the operations taking place on the screen. It might be necessary to intentionally introduce a pause between certain operations in a script file. For example, in a slide show for a product demonstration, there must be a time delay between different slides so that the audience has enough time to see them. This is accomplished by using the AutoCAD **DELAY** command, which introduces a delay before the next command is executed. The general format of the **DELAY** command is:

Command: DELAY Time
        Where   **Command** ------ AutoCAD command prompt
                DELAY ---------- **DELAY** command
                Time ------------- Time in milliseconds

The **DELAY** command is to be followed by the delay time in milliseconds. For example, a delay of 2,000 milliseconds means that AutoCAD will pause for approximately two seconds before executing the next command. It is approximately two seconds because computer processing speeds vary. The maximum time delay you can enter is 32,767 milliseconds (about 33 seconds). In Example 3, a two-second delay can be introduced by inserting a **DELAY** command line between line 7 and line 8, as in the following file listing:

Chapter 2

```
ROTATE
W
2.25,5.0
6.25,9.0
              (Blank line for Return.)
4.25,6.5
45
DELAY 2000
RSCRIPT
```

The first seven lines of this file rotate the objects through a 45-degree angle. Before the **RSCRIPT** command on line 8 is executed, there is a delay of 2,000 milliseconds (about two seconds). The **RSCRIPT** command will repeat the script file that rotates the objects through another 45-degree angle. Thus, a slide show is created with a time delay of two seconds after every 45-degree increment.

RESUME COMMAND

If you cancel a script file and then want to continue it, you can do so by using the AutoCAD **RESUME** command.

Command: **RESUME**

The **RESUME** command can also be used if the script file has encountered an error that causes it to be suspended. The **RESUME** command will skip the command that caused the error and continue with the rest of the script file. If the error occurred when the command was in progress, use a leading apostrophe with the **RESUME** command (**'RESUME**) to invoke the **RESUME** command in transparent mode.

Command: **'RESUME**

COMMAND LINE SWITCHES

The command line switches can be used as arguments to the acad.exe file that launches AutoCAD. You can also use the **Options** dialog box to set the environment or by adding a set of environment variables in the autoexec.bat file. The command line switches and environment variables override the values set in the **Options** dialog box for the current session only. These switches do not alter the system registry. The following is the list of the command line switches:

Switch	Function
/c	Controls where AutoCAD stores and searches for the hardware configuration file (acad2000.cfg)
/s	Specifies which directories to search for support files if they are not in the current directory
/d	Specifies which directories to search for ADI drivers
/b	Designates a script to run after AutoCAD starts
/t	Specifies a template to use when creating a new drawing
/nologo	Starts AutoCAD without first displaying the logo screen

/v	Designates a particular view of the drawing to be displayed upon start-up of AutoCAD
/r	Reconfigures AutoCAD with the default device configuration settings
/p	Specifies the profile to use on start-up

INVOKING A SCRIPT FILE WHEN LOADING AUTOCAD

The script files can also be run when loading AutoCAD, without getting into the drawing editor. The format of the command for running a script file when loading AutoCAD is:

Drive>AutoCAD2000 [existing-drawing] [/t template] [/v view] /b script-file

In the following example, AutoCAD will open the existing drawing (Mydwg1) and then run the script file (Setup).

Example
C:\AutoCAD2000>ACAD Mydwg1 /b Setup

Where **AutoCAD2000** - AutoCAD2000 subdirectory containing
AutoCAD system files
ACAD ------------ ACAD command to start AutoCAD
MyDwg1 -------- Existing drawing file name
Setup ------------- Name of the script file

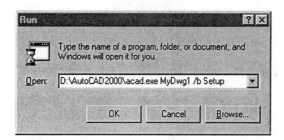

Figure 2-3 *Invoking script file when loading AutoCAD using the* **Run** *dialog box*

In the following example, AutoCAD will start a new drawing with the default name (Drawing), using the template file temp1, and then run the script file (Setup).

Example
C:\AutoCAD2000\ACAD /t temp1 /b Setup

Where **temp1** ------------ Existing template file name
Setup ------------- Name of the script file

or

C:\AutoCAD2000\ACAD /t c:\AutoCAD2000\mytemp\temp1 /b Setup

Where **c:\AutoCAD2000\mytemp** ------ Path name

In the following example, AutoCAD will start a new drawing with the default name (Drawing), and then run the script file (Setup).

Example

C:\AutoCAD2000\ACAD /b Setup

Where **Setup** ------------- Name of the script file

Here, it is assumed that the AutoCAD system files are loaded in the AutoCAD2000 directory.

 Note

For invoking a script file when loading AutoCAD, the drawing file or the template file specified in the command must exist in the search path. You cannot start a new drawing with a given name.

You should avoid abbreviations to prevent any confusion. For example, a C can be used as a close option when you are drawing lines. It can also be used as a command alias for drawing a circle. If you use both of these in a script file, it might be confusing.

Example 4

Write a script file that can be invoked when loading AutoCAD and create a drawing with the following setup (filename SCRIPT4.SCR):

Grid	3.0
Snap	0.5
Limits	0,0
	36.0,24.0
Zoom	All
Text height	0.25
Ltscale	3.0
Dimscale	3.0

Layers

Name	Color	Linetype
Obj	Red	Continuous
Cen	Yellow	Center
Hid	Blue	Hidden
Dim	Green	Continuous

First, write a script file and save the file under the name SCRIPT4.SCR. The following file is a listing of this script file that does the initial setup for a drawing:

```
GRID 2.0
SNAP 0.5
LIMITS 0,0 36.0,24.0 ZOOM ALL
SETVAR TEXTSIZE 0.25
LTSCALE 3
SETVAR DIMSCALE 3.0
LAYER NEW
OBJ,CEN,HID,DIM
L CENTER CEN
```

> L HIDDEN HID
> C RED OBJ
> C YELLOW CEN
> C BLUE HID
> C GREEN DIM
> *(Blank line for ENTER.)*

After you have written and saved the file, quit the drawing editor. To run the script file, SCRIPT4, select Start, Run, and then enter the following command line:

C:\AutoCAD2000\ACAD /t C:\EX4 D:\SCRIPT4
> Where **ACAD** ------------ ACAD to load AutoCAD
> **EX4** -------------- Drawing filename
> **SCRIPT4** ------- Name of the script file

Here it is assumed that the template file (EX4) is on C drive and the script file (SCRIPT4) is on D drive.

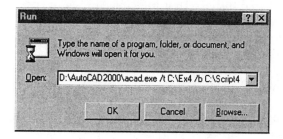

Figure 2-4 *Invoking script file when loading AutoCAD using the **Run** dialog box*

When you enter this line, AutoCAD is loaded and the file EX4.DWT is opened. The script file, SCRIPT4, is then automatically loaded and the commands defined in the file are executed.

In the following example, AutoCAD will start a new drawing with the default name (Drawing), and then run the script file (SCRIPT4).

Example
C:\AutoCAD2000\ACAD /b SCRIPT4
> Where **SCRIPT4** ------- Name of the script file

Here, it is assumed that the AutoCAD system files are loaded in the AutoCAD2000 directory.

Example 5

Write a script file that will plot a 36" by 24" to maximum plot size, using your system printer/plotter. Use the Window option to select the drawing to be plotted.

Before writing a script file to plot a drawing, find out the plotter specifications that must be entered in the script file to obtain the desired output. To determine the prompt entries and their sequences to set up the plotter specifications, enter the AutoCAD **-PLOT** command at the keyboard. Note the entries you make and their sequence (the entries for your printer or plotter will probably be different). The following is a listing of the plotter specification with the new entries:

Command: **-PLOT**
Detailed plot configuration? [Yes/No] <No>: **Yes**
Enter a layout name or [?] <Model>: Model
Enter an output device name or [?] <HP LaserJet 4000 Series PCL 6>: *Press ENTER.*
Enter paper size or [?] <Letter (8 1/2 x 11 in)>: *Press ENTER.*
Enter paper units [Inches/Millimeters] <Inches>: **I**
Enter drawing orientation [Portrait/Landscape] <Landscape>: *Press ENTER.*
Plot upside down? [Yes/No] <No>: **No**
Enter plot area [Display/Extents/Limits/View/Window] <Display>: **W**
Enter lower left corner of window <0.000000,0.000000>: **0,0**
Enter upper right corner of window <0.000000,0.000000>: **36,24**
Enter plot scale (Plotted Inches=Drawing Units) or [Fit] <Fit>: **F**
Enter plot offset (x,y) or [Center] <0.00,0.00>: **0,0**
Plot with plot styles? [Yes/No] <Yes>: **Yes**
Enter plot style table name or [?] (enter . for none) <>: .
Plot with lineweights? [Yes/No] <Yes>: **Yes**
Remove hidden lines? [Yes/No] <No>: **No**
Write the plot to a file [Yes/No] <N>: **No**
Save changes to model tab [Yes/No]? <N> **No**
Proceed with plot [Yes/No] <Y>: **Y**
Effective plotting area: 7.07 wide by 10.60 high
Plotting viewport 2.

Now you can write the script file by entering the responses to these prompts in the file. The following file is a listing of the script file that will plot a 36" by 24" drawing on 9" by 6" paper after making the necessary changes in the plot specifications. The comments on the right are not a part of the file.

Plot
y
                    (Blank line for ENTER, selects default layout.)
                    (Blank line for ENTER, selects default printer.)
                    (Blank line for ENTER, selects the default paper size.)
I
L
N
w
0,0
36,24
F
0,0

Y
.   *(Enter . for none)*
Y
N
N
N
Y

Note

You can use a blank line to accept the default value for a prompt. A blank line in the script file will cause a Return. However, you must not accept the default plot specifications because the file might have been altered by another user or by another script file. Therefore, always enter the actual values in the file so that when you run a script file, it does not take the default values.

Exercise 1

Write a script file that will plot a 288' by 192' drawing on a 36" x 24" sheet of paper. The drawing scale is 1/8" = 1'. (The filename is SCRIPT6.SCR. In this example assume that AutoCAD is configured for the HPGL plotter and the plotter description is HPGL-Plotter.)

WHAT IS A SLIDE SHOW?

AutoCAD provides a facility using script files to combine the slides in a text file and display them in a predetermined sequence. In this way, you can generate a slide show for a slide presentation. You can also introduce a time delay in the display so that the viewer has enough time to view a slide.

A drawing or parts of a drawing can also be displayed by using the AutoCAD display commands. For example, you can use **ZOOM, PAN,** or other commands to display the details you want to show. If the drawing is very complicated, it takes quite some time to display the desired information, and it may not be possible to get the desired views in the right sequence. However, with slide shows you can arrange the slides in any order and present them in a definite sequence. In addition to saving time, this will also help to minimize the distraction that might be caused by constantly changing the drawing display. Also, some drawings are confidential in nature and you may not want to display some portions or views of them. By making slides, you can restrict the information that is presented through them. You can send a slide show to a client without losing control of the drawings and the information that is contained in them.

WHAT ARE SLIDES?

A **slide** is the snapshot of a screen display; it is like taking a picture of a display with a camera. The slides do not contain any vector information like AutoCAD drawings, which means that the entities do not have any information associated with them. For example, the slides do not retain any information about the layers, colors, linetypes, start point, or endpoint of a line or viewpoint. Therefore, slides cannot be edited like drawings. If you want to make any changes in the slide, you need to edit the drawing and then make a new slide from the edited drawing.

MSLIDE COMMAND

Slides are created by using the AutoCAD **MSLIDE** command. If **FILEDIA** is set to 0, the command will prompt you to enter the slide file name.

Command: **MSLIDE**
Enter name of slide file to create <Default>: *Slide file name.*

Example
Command: **MSLIDE**
Slide File: <Drawing1> SLIDE1
     Where **Drawing1** ------- Default slide file name
         SLIDE1 --------- Slide file name

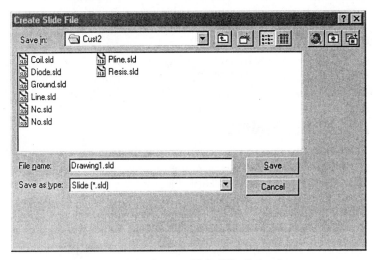

Figure 2-5 Create Slide File dialog box

In the preceding example, AutoCAD will save the slide file as **SLIDE1.SLD**. If **FILEDIA** is set to 1, the **MSLIDE** command displays the **Create Slide File** dialog box (Figure 2-5) on the screen. You can enter the slide file name in this dialog box.

Note

*In model space, you can use the **MSLIDE** command to make a slide of the existing display in the current viewport.*

If you are in the Layout (paper space) viewport, you can make a slide of the display in the Layout (paper space) that includes any floating viewports.

*When the viewports are not active, the **MSLIDE** command will make a slide of the current screen display.*

VSLIDE COMMAND

To view a slide, use the **VSLIDE** command. AutoCAD will then prompt you to enter the slide file name. Enter the name of the slide you want to view and press ENTER. Do not enter the extension after the slide file name. AutoCAD automatically assumes the extension **.SLD**.

Command: **VSLIDE**
Enter name of slide file to create<Default>: *Name*.

Example
Command: **VSLIDE**
Slide file <Drawing1>: SLIDE1
            Where   **Drawing1** ------- Default slide file name
                     SLIDE1 --------- Name of slide file

You can also use the **Select Slide File** dialog box (Figure 2-6) to view a slide.

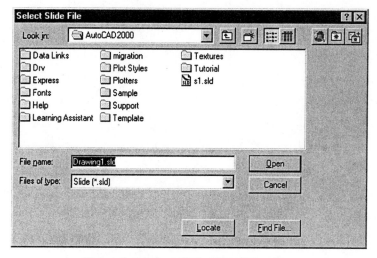

Figure 2-6 Select Slide Files dialog box

Note
*After viewing a slide, you can use the AutoCAD **REDRAW** command to remove the slide display and return to the existing drawing on the screen.*

*Any command that is automatically followed by a redraw will also display the existing drawing. For example, AutoCAD **GRID, ZOOM ALL**, and **REGEN** commands will automatically return to the existing drawing on the screen.*

You can view the slides on high-resolution or low-resolution monitors. Depending on the resolution of the monitor, AutoCAD automatically adjusts the image. However, if you are using a high-resolution monitor, it is better to make the slides on the same monitor to take full advantage of that monitor.

Example 6

Write a script file that will generate a slide show of the following slide files, with a time delay of 15 seconds after every slide:

SLIDE1, SLIDE2, SLIDE3, SLIDE4

The first step in a slide show is to create the slides. Figure 2-7 shows the drawings that have been saved as slide files **SLIDE1, SLIDE2, SLIDE3,** and **SLIDE4**. The second step is to find out the sequence in which you want these slides to be displayed, with the necessary time delay, if any, between slides. Then you can use any text editor or the AutoCAD **EDIT** command (provided the **ACAD.PGP** file is present and **EDIT** is defined in the file) to write the script file with the extension **.SCR**.

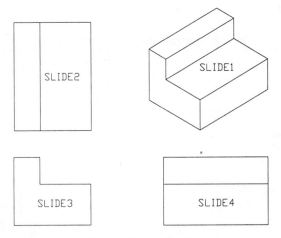

Figure 2-7 Slides for the slide show

The following file is a listing of the script file that will create a slide show of the slides in Figure 2-7. The name of the script file is **SLDSHOW1**.

```
VSLIDE SLIDE1
DELAY 15000
VSLIDE SLIDE2
DELAY 15000
VSLIDE SLIDE3
DELAY 15000
VSLIDE SLIDE4
DELAY 15000
```

To run this slide show, type SCRIPT in response to the AutoCAD Command prompt. Now, type the name of the script file **(SLDSHOW1)** and press ENTER. The slides will be displayed on the screen, with an approximate time delay of 15 seconds between them.

PRELOADING SLIDES

In the script file of Example 7, VSLIDE SLIDE1 in line 1 loads the slide file, **SLIDE1**, and displays it on screen. After a pause of 15,000 milliseconds, it starts loading the second slide file, **SLIDE2**. Depending on the computer and the disk access time, you will notice that it takes some time to load the second slide file; the same is true for the other slides. To avoid the delay in loading the slide files, AutoCAD has provided a facility to preload a slide while viewing the previous slide. This is accomplished by placing an asterisk (*) in front of the slide file name.

VSLIDE SLIDE1	*(View slide, SLIDE1.)*
VSLIDE *SLIDE2	*(Preload slide, SLIDE2.)*
DELAY 15000	*(Delay of 15 seconds.)*
VSLIDE	*(Display slide, SLIDE2.)*
VSLIDE *SLIDE3	*(Preload slide, SLIDE3.)*
DELAY 15000	*(Delay of 15 seconds.)*
VSLIDE	*(Display slide, SLIDE3.)*
VSLIDE *SLIDE4	
DELAY 15000	
VSLIDE	
DELAY 15000	
RSCRIPT	*(Restart the script file.)*

Example 7

Write a script file to generate a continuous slide show of the following slide files, with a time delay of two seconds between slides:

SLD1, SLD2, SLD3

The slide files are located in different subdirectories, as shown in Figure 2-8. The subdirectory **SUBDIR1** is the current subdirectory.

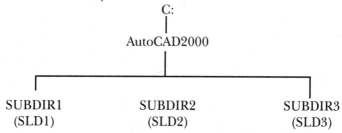

Figure 2-8 *Subdirectories of the C drive*

Where	**C:**	------------------- *Root directory.*
	AutoCAD2000	------------ *Subdirectory where the AutoCAD files are loaded.*
	SUBDIR1	---------------- *Drawing subdirectory.*
	SUBDIR2	---------------- *Drawing subdirectory.*
	SUBDIR3	---------------- *Drawing subdirectory.*

SLD1	------------------- *Slide file in SUBDIR1 subdirectory.*
SLD2	------------------- *Slide file in SUBDIR2 subdirectory.*
SLD3	------------------- *Slide file in SUBDIR3 subdirectory.*

The following file is the listing of the script files that will generate a slide show for the slides in Example 8:

```
VSLIDE SLD1
DELAY 2000
VSLIDE C:\AutoCAD2000\SUBDIR2\SLD2
DELAY 2000
VSLIDE C:\AutoCAD2000\SUBDIR3\SLD3
DELAY 2000
RSCRIPT
```

Line 1
VSLIDE SLD1
In this line, the AutoCAD command **VSLIDE** loads the slide file **SLD1**. Since, in this example, we are assuming you are in the subdirectory **SUBDIR1** and the first slide file, **SLD1**, is located in the same subdirectory, it does not require any path definition.

Line 2
DELAY 2000
This line uses the AutoCAD **DELAY** command to create a pause of approximately two seconds before the next slide is loaded.

Line 3
VSLIDE C:\AutoCAD2000\SUBDIR2\SLD2
In this line, the AutoCAD command **VSLIDE** loads the slide file **SLD2**, located in the subdirectory **SUBDIR2**. If the slide file is located in a different subdirectory, you need to define the path with the slide file.

Line 5
VSLIDE C:\AutoCAD2000\SUBDIR3\SLD3
In this line, the **VSLIDE** command loads the slide file SLD3, located in the subdirectory SUBDIR3.

Line 7
RSCRIPT
In this line, the **RSCRIPT** command executes the script file again and displays the slides on the screen. This process continues indefinitely until the script file is canceled by pressing the ESC key or the BACKSPACE key.

SLIDE LIBRARIES

AutoCAD provides a utility, SLIDELIB, which constructs a library of the slide files. The format of the SLIDELIB utility command is:

C:\>**SLIDELIB (Library filename)<(Slide list filename)**

Example
C:\>**SLIDELIB SLDLIB <SLDLIST**
        Where   **SLIDELIB** ------ AutoCAD's SLIDELIB utility
                SLDLIB --------- Slide library filename
                SLDLIST ------- List of slide filenames

The SLIDELIB utility is supplied with the AutoCAD software package. You can find this utility (SLIDELIB.EXE) in the support subdirectory. The slide file list is a list of the slide filenames that you want in a slide show. It is a text file that can be written by using any text editor or AutoCAD's **EDIT** command (provided ACAD.PGP file is present and **EDIT** is defined in the file). The slide files in the slide file list should not contain any file extension (.SLD). However, if you want to add a file extension it should be .SLD.

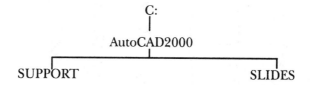

The slide file list can be also created by using the following command, if you have DOS version 5.0 or above:

    C:\AutoCAD2000\SLIDES>**DIR *.SLD/B>SLDLIST**

In this example assume that the name of the slide file list is **SLDLIST** and all slide files are in the SLIDES subdirectory. To use this command to create a slide file list, all slide files must be in the same directory.

When you use the SLIDELIB utility, it reads the slide filenames from the file that is specified in the slide list and the file is then written to the file specified by the library. In Example 8, the SLIDELIB utility reads the slide filenames from the file SLDLIST and writes them to the library file SLDLIB:

    C:\>**SLIDELIB SLDLIB<SLDLIST**

Note
*You **cannot** edit a slide library file. If you want to change anything, you have to create a new list of the slide files and then use the SLIDELIB utility to create a new slide library.*

*If you edit a slide while the slide is displayed on the screen, the slide is not edited. Instead, the current drawing that is behind the slide will be edited. Therefore, do not use any editing commands while you are viewing a slide. Use the **VSLIDE** and **DELAY** commands only when viewing a slide.*

The path name is not saved in the slide library; therefore, if you have more than one slide with the same name, even though they are in different subdirectories, only one slide will be saved in the slide library.

If you are using DOS, use a smaller library file. If the library file contains a large number of files, it might exceed the computer's memory. In that case, the SLIDELIB utility will display a warning message.

Example 8

Use AutoCAD's SLIDELIB utility to generate a continuous slide show of the following slide files with a time delay of 2.5 seconds between the slides. (The filenames are: SLDLIST for slide list file, SLDSHOW1 for slide library, SHOW1 for script file.)

FRONT, TOP, RSIDE, STAIRS, 3DVIEW, LROOM, FROOM, BROOM

The slide files are located in different subdirectories as shown in Figure 2-9.

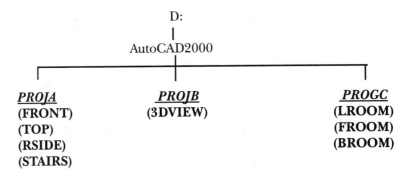

Figure 2-9 Subdirectories of D drive

Where   **D** ------------------ *(D drive.)*
        AutoCAD2000 - *(Subdirectory where the AutoCAD files are loaded.)*
        PROJA ---------- *(Drawing subdirectory.)*
        PROJB ---------- *(Drawing subdirectory.)*
        PROJC ---------- *(Drawing subdirectory.)*

Step 1

The first step is to create a list of the slide filenames with the drive and the directory information. Assume that you are in the ACAD subdirectory. You can use a text editor or AutoCAD's EDIT function to create a list of the slide files that you want to include in the slide show. These files do not need a file extension. However, if you choose to give them a file extension, it should be .SLD. The following file is a listing of the file SLDLIST for Example 8:

        D:\AutoCAD2000\PROJA\FRONT
        D:\AutoCAD2000\PROJA\TOP
        D:\AutoCAD2000\PROJA\RSIDE
        D:\AutoCAD2000\PROJA\STAIRS
        D:\AutoCAD2000\PROJB\3DVIEW
        D:\AutoCAD2000\PROJC\LROOM
        D:\AutoCAD2000\PROJC\FROOM

D:\AutoCAD2000\PROJC\BROOM

Step 2

The second step is to use AutoCAD's SLIDELIB utility program to create the slide library. The name of the slide library is assumed to be SLDSHOW1 for this example.

D:\AutoCAD2000\SUPPORT>SLIDELIB SLDSHOW1<SLDLIST
Where **SLIDELIB** ------ AutoCAD's SLIDELIB utility
SLDSHOW1 --- Slide library
SLDLIST ------- Slide list

Step 3

Now you can write a script file for the slide show that will use the slides in the slide library. The name of the script file for this example is assumed to be SHOW1.

```
VSLIDE SLDSHOW1(FRONT)
DELAY 2500
VSLIDE SLDSHOW1(TOP)
DELAY 2500
VSLIDE SLDSHOW1(RSIDE)
DELAY 2500
VSLIDE SLDSHOW1(STAIRS)
DELAY 2500
VSLIDE SLDSHOW1(3DVIEW)
DELAY 2500
VSLIDE SLDSHOW1(LROOM)
DELAY 2500
VSLIDE SLDSHOW1(RROOM)
DELAY 2500
VSLIDE SLDSHOW1(BROOM)
DELAY 2500
RSCRIPT
```

Step 4

Start AutoCAD, if you are not in AutoCAD. With AutoCAD's **SCRIPT** command run the script file, SHOW1, and you will see the slides displayed on the screen.

You can also invoke the **SCRIPT** command from the Command prompt after setting the system variable FILEDIA to 0.

Command: **SCRIPT**
Enter script file name<default>: **SHOW1**

Chapter 2

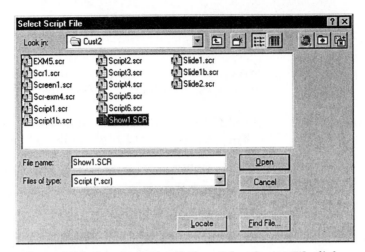

Figure 2-10 *Selecting script file from* **Select Script File** *dialog*

SLIDE SHOWS WITH RENDERED IMAGES

Slide shows can also contained rendered images. In fact, use of rendered images is highly recommended for a more dynamically pleasing, entertaining, and interesting show. AutoCAD provides the following AutoLISP function for using rendered images during a slide show presentation.

(C:REPLAY FILENAME TYPE [<XOFF> <YOFF> <XSIZE> <YSIZE>])

> Where:

> C:REPLAY ----------------- AutoLISP function calling AutoCAD's *Replay* command.

> FILENAME --------------- Name of rendered file, without the extension. The filename is to be enclosed in quotes. Include directory path and one forward slashe (/) for directory structures.

> TYPE   -------------------- The rendered file type (e.g., Bmp, Tif, Tga). The file type should also be enclosed in quotes.

> XOFF   -------------------- The rendered file's "X" coordinate offset from 0,0.

> YOFF   -------------------- The rendered file's "Y" coordinate offset from 0,0.

> XSIZE   -------------------- The size of the rendered file to be displayed in the "X" (horizontal) direction. This information can generally be obtained when creating a rendered file with AutoCAD's **SAVEIMG** command.

> YSIZE   -------------------- The size of the rendered file to be displayed in the "Y" (vertical) direction. This information can generally be obtained when creating a rendered file with AutoCAD's **SAVEIMG** command.

AutoCAD ships with a file call BIGLAKE.TGA as an example of a background texture map, Figure 2-11. To display this rendered image enter the following at the Command prompt. Note: The "biglake.tga" file must be in AutoCAD's support directory's search path.

(C:REPLAY "C:/PROGRAM FILES/AutoCAD2000//TEXTURES/BIGLAKE" "TGA" 150 50 944 564)

Where:

C:REPLAY:	----------------	Function for AutoCAD to perform - issue the *Replay* command.
"C:/PRO../BIGLAKE"	--	Name of the rendered image file, including drive and directory.
"TGA"	--------------------	The rendered image file type.
150	--------------------	The "X" offset for the image displayed.
50	--------------------	The "Y" offset for the image displayed.
944	--------------------	The "X" size (horizontal) for the image displayed.
564)	--------------------	The "Y" size (vertical) for the image displayed.

Combining Slides and Rendered Images

The following is an example of a script file (replay.scr) that combines the use of AutoCAD slides and rendered images (*.bmp files) into a seamless slide show presentation. The "F_" files are the wire FRAME images, while the "R_" files are the RENDERED images. It is assumed that the BMP image files are on A drive.

```
VSLIDE A:\SLIDES\F_ROLL
DELAY 3000
(C:REPLAY "A:/RENDERS/R_ROLL" "BMP" 0 0 944 564)
DELAY 3000
VSLIDE A:\SLIDES\F_BKCASE
DELAY 3000
(C:REPLAY "A:/RENDERS/R_BKCASE" "BMP" 0 0 944 564)
DELAY 3000
VSLIDE A:\SLIDES\F_MOUSE
DELAY 3000
(C:REPLAY "A:/RENDERS/R_MOUSE" "BMP" 0 0 944 564)
DELAY 3000
VSLIDE A:\SLIDES\F_TABLE2
DELAY 3000
(C:REPLAY "A:/RENDERS/R_TABLE2" "BMP" 0 0 944 564)
DELAY 3000
VSLIDE A:\SLIDES\F_MIC_3D
DELAY 3000
(C:REPLAY "A:/RENDERS/R_MIC_3D" "BMP" 0 0 944 564)
DELAY 3000
VSLIDE A:\SLIDES\F_BOX
DELAY 3000
(C:REPLAY "A:/RENDERS/R_BOX" "BMP" 0 0 944 564)
DELAY 3000 RSCRIPT
```

Chapter 2

Figure 2-11 Displaying a rendered image using a script file

Review Questions

SCRIPT FILES

1. AutoCAD has provided a facility of _____ that allows you to combine different AutoCAD commands and execute them in a predetermined sequence.

2. The _____ files can be used to generate a slide show, do the initial drawing setup, or plot a drawing to a predefined specification.

3. Before writing a script file, you need to know the AutoCAD _____ and the _____ required in response to the command prompts.

4. In a script file, you can _____ several statements in one line.

5. In a script file, the _____ are used to terminate a command or a prompt entry.

6. The AutoCAD _____ command is used to run a script file.

7. When you run a script file, the default script file name is the same as the _____ name.

8. When you run a script file, type the name of the script file without the file _____.

9. One of the limitations of script files is that all the information has to be contained _____ the file.

10. The AutoCAD _____ command allows you to re-execute a script file indefinitely until the command is canceled.

11. You cannot provide a _____ statement in a script file to terminate the file when a particular condition is satisfied.

12. The AutoCAD _____ command introduces a delay before the next command is executed.

13. The **DELAY** command is to be followed by _____ in milliseconds.

14. If the script file was canceled and you want to continue the script file, you can do so by using the AutoCAD _____ command.

SLIDE SHOWS

15. AutoCAD provides a facility through _____ files to combine the slides in a text file and display them in a predetermined sequence.

16. A _____ can also be introduced in the script file so that the viewer has enough time to view a slide.

17. Slides are the _____ of a screen display.

18. Slides do not contain any _____ information, which means that the entities do not have any information associated with them.

19. Slides _____ edited like a drawing.

20. Slides can be created using the AutoCAD _____ command.

21. Slide file names can be up to _____ characters long.

22. In model space, you can use the **MSLIDE** command to make a slide of the _____ display in the _____ viewport.

23. If you are in paper space, you can make a slide of the display in paper space that _____ any floating viewports.

24. To view a slide, use the AutoCAD _____ command.

25. If you want to make any change in the slide, you need to _____ the drawing, then make a new slide from the edited drawing.

Chapter 2

26. If the slide is in the slide library and you want to view it, the slide library name has to be _____ with the slide filename.

27. AutoCAD provides a utility that constructs a library of the slide files. This is done with AutoCAD's utility program called _____.

28. You cannot _____ a slide library file. If you want to change anything, you have to create a new list of the slide files and then use the _____ utility to create a new slide library.

29. The path name _____ be saved in the slide library. Therefore, if you have more than one slide with the same name, although with different subdirectories, only one slide will be saved in the slide library.

Exercises

SCRIPT FILES

Exercise 2   *General*

Write a script file that will do the following initial setup for a drawing:

Grid	2.0
Snap	0.5
Limits	0,0
	18.0,12.0
Zoom	All
Text height	0.25
Ltscale	2.0
Dimscale	2.0
Dimtix	On
Dimtoh	Off
Dimtih	Off
Dimtad	1
Dimcen	0.75

Exercise 3

General

Write a script file that will set up the following layers with the given colors and linetypes (filename SCRIPTE2.SCR):

Contour	Red	Continuous
SPipes	Yellow	Center
WPipes	Blue	Hidden
Power	Green	Continuous
Manholes	Magenta	Continuous
Trees	Cyan	Continuous

Exercise 4

General

Write a script file that will do the following initial setup for a new drawing:

Limits	0,0 24,18
Grid	1.0
Snap	0.25
Ortho	On
Snap	On
Zoom	All
Pline width	0.02
PLine	0,0 24,0 24,18 0,18 0,0
Units	Decimal units
	Number of decimal digits (2)
	Decimal degrees
	Number of decimal digits (2)
	Direction of 0 angle (3 o'clock)
	Angle measured counterclockwise
Ltscale	1.5

Layers

Name	**Color**	**Linetype**
Obj	Red	Continuous
Cen	Yellow	Center
Hid	Blue	Hidden
Dim	Green	Continuous

Exercise 5

General

Write a script file that will **PLOT** a given drawing according to the following specifications. (Use the plotter for which your system is configured and adjust the values accordingly.)

Plot, using the Window option
Window size (0,0 24,18)
Do not write the plot to file

Chapter 2

Size in Inch units
Plot origin (0.0,0.0)
Maximum plot size (8.5,11 or the smallest size available on your printer/plotter)
90 degree plot rotation
No removal of hidden lines
Plotting scale (Fit)

Exercise 6   *General*

Write a script file that will continuously rotate a line in 10-degree increments around its midpoint
(Figure 2-12). The time delay between increments is one second.

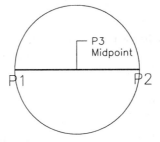

Figure 2-12 Drawing for Exercise 6

SLIDE SHOWS

Exercise 7   *General*

Make the slides shown in Figure 2-13 and write a script file for a continuous slide show. Provide a
time delay of 5 seconds after every slide. (You do not have to use the slides shown in Figure 2-13;
you can use any slides of your choice.)

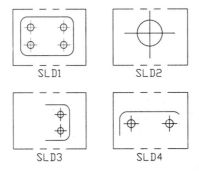

Figure 2-13 Slides for slide show

Exercise 8 *General*

From Exercise 6, list the slides in a file SLDLIST2 and create a slide library file SLDLIB2. Then write a script file SHOW2 using the slide library with a time delay of five seconds after every slide.

Exercise 9 *General*

Write a script file to generate a continuous slide show of the following slide files with a time delay of three seconds between the slides.

SLDX1, SLDX2, SLDX3
SLDY1, SLDY2, SLDY3
SLDZ1, SLDZ2, SLDZ3

Assume that the slide files are located in different subdirectories as shown in Figure 2-14 and ACAD is the current subdirectory.

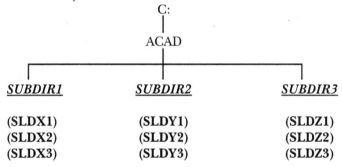

Figure 2-14 *Subdirectories of C drive*

Where
C: -------------------- *(C drive.)*
ACAD -------------------- *(Subdirectory where the AutoCAD files are loaded.)*
SUBDIR1 ---------------- *(Drawing subdirectory.)*
SUBDIR2 ---------------- *(Drawing subdirectory.)*
SUBDIR3 ---------------- *(Drawing subdirectory.)*

Chapter 3

Creating Linetypes and Hatch Patterns

![Learning Objectives]

After completing this chapter, you will be able to:

Create Linetypes:
- *Write linetype definitions.*
- *Create different linetypes.*
- *Create linetype files.*
- *Determine **LTSCALE** for plotting the drawing to given specifications.*
- *Define alternate linetypes and modify existing linetypes.*
- *Create string and shape complex linetypes.*

Create Hatch Patterns:
- *Understand hatch pattern definition.*
- *Create new hatch patterns.*
- *Determine the effect of angle and scale factor on hatch.*
- *Create hatch patterns with multiple descriptors.*
- *Save hatch patterns in a separate file.*
- *Define custom hatch pattern file.*
- *Add hatch pattern slides to AutoCAD slide library.*

STANDARD LINETYPES

The AutoCAD software package comes with a library of standard linetypes that has 38 different linetypes, including ISO linetypes. These linetypes are saved in the **ACAD.LIN** file. You can modify existing linetypes or create new ones.

LINETYPE DEFINITION

All linetype definitions consist of two parts: **header line and pattern line**.

Header Line

The **header line** consists of an asterisk (*) followed by the name of the linetype and the linetype description. The name and the linetype description should be separated by a comma. If there is no description, the comma that separates the linetype name and the description is not required.

The format of the header line is:

* **Linetype Name, Description**

Example
*HIDDENS,___ ___ ___ ___ ___ ___

Where   * ------------------- Asterisk sign
              HIDDENS ------ Linetype name
            , -------------------- Comma
        ___ ___ ___ ___ ------ Linetype description

All linetype definitions require a linetype name. When you want to load a linetype or assign a linetype to an object, AutoCAD recognizes the linetype by the name you have assigned to the linetype definition. The names of the linetype definition should be selected to help the user recognize the linetype by its name. For example, a linetype name LINEFCX does not give the user any idea about the type of line. However, a linetype name like DASHDOT gives a better idea about the type of line that a user can expect.

The linetype description is a textual representation of the line. This representation can be generated by using dashes, dots, and spaces at the keyboard. The graphic is used by AutoCAD when you want to display the linetypes on the screen by using the AutoCAD **LINETYPE** command with the ? option or by using the dialog box. The linetype description cannot exceed 47 characters.

Pattern Line

The **pattern line** contains the definition of the line pattern. The definition of the line pattern consists of the alignment field specification and the linetype specification. The alignment field specification and the linetype specification are separated by a comma.

The format of the pattern line is:

Alignment Field Specification, Linetype Specification

Example
A,.75,-.25,.75
Where   **A** ----------------- Alignment field specification
        , -------------------- Comma
      .75,-.25,.75 ----- Linetype specification

The letter used for alignment field specification is A. This is the only alignment field supported by AutoCAD; therefore, the pattern line will always start with the letter A. The linetype specification defines the configuration of the dash-dot pattern to generate a line. The maximum number for dash length specification in the linetype is 12, provided the linetype pattern definition fits on one 80-character line.

ELEMENTS OF LINETYPE SPECIFICATION

All linetypes are created by combining the basic elements in a desired configuration. There are three basic elements that can be used to define a linetype specification.

Dash   (Pen down)
Dot   (Pen down, 0 length)
Space   (Pen up)

Example

```
_____  .  _____  .  _____  .  _____
```

Where   . -------------------- Dot (pen down with 0 length)
Blank space --------------- Space (pen up)
_____ -------------------- Dash (pen down with specified length)

The dashes are generated by defining a positive number. For example, .5 will generate a dash 0.5 units long. Similarly, spaces are generated by defining a negative number. For example, -.2 will generate a space 0.2 units long. The dot is generated by defining a 0 length.

Example
A,.5,-.2,0,-.2,.5

Where **0** ------------------- Dot (zero length)
-.2 ---------------- Length of space (pen up, a negative number)
.5 ------------------ Length of dash (pen down, a positivr number)

CREATING LINETYPES

Before creating a linetype, you need to decide the type of line you want to generate. Draw the line on a piece of paper and measure the length of each element that constitutes the line. You need to define only one segment of the line, because the pattern is repeated when you draw a line. Linetypes can be created or modified by one of the following methods:

Using the AutoCAD LINETYPE command
Using a text editor (such as Notepad).

Consider the following example, which creates a new linetype, first using the AutoCAD **LINETYPE** command and then using a text editor.

Chapter 3

Example 1

Using the AutoCAD **LINETYPE** command, create linetype DASH3DOT (Figure 3-1) with the following specifications:

Length of the first dash 0.5
Blank space 0.125
Dot
Blank space 0.125
Dot
Blank space 0.125
Dot
Blank space 0.125

Using the AutoCAD Linetype Command

To create a linetype using the AutoCAD **LINETYPE** command, first make sure that you are in the drawing editor. Then enter the **LINETYPE** command and select the Create option to create a linetype.

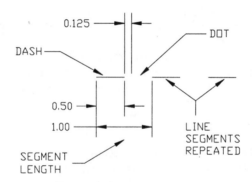

Figure 3-1 Linetype specifications of DASH3DOT

Command: -**LINETYPE**
Enter an option [?/Create/Load/Set]: **C**

Enter the name of the linetype and the name of the library file in which you want to store the definition of the new linetype.

Enter name of linetype to create: **DASH3DOT**

If **FILEDIA**=1, the **Create or Append Linetype File** dialog box (Figure 3-2) will appear on the screen. If **FILEDIA**=0, AutoCAD will prompt you to enter the name of the file.

Enter linetype file name for new linetype definition <default>: **Acad**

If the linetype already exists, the following message will be displayed on the screen:

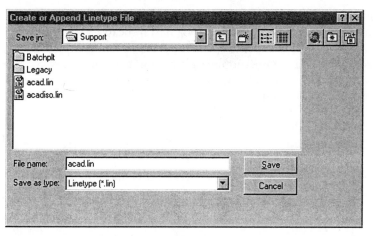

Figure 3-2 Create or Append Linetype File dialog box

Wait, checking if linetype already defined...
"Linetype" already exists in this file. Current definition is:
*DASH3DOT,____ ... ____...____
Overwrite?<N>

If you want to redefine the existing line style, enter Y; otherwise, type N or press RETURN to choose the default value of N. You can then repeat the process with a different name of the linetype.

After entering the name of the linetype and the library file name, AutoCAD will prompt you to enter the descriptive text and the pattern of the line.

Descriptive text: ***DASH3DOT,____ ... ____ ... ____**
Enter linetype pattern (on next line):
A,.5,-.125,0,-.125,0,-.125,0,-.125

Descriptive Text

   *DASH3DOT,____ ... ____ ... ____

For the descriptive text, you have to type an asterisk (*) followed by the name of the linetype. For Example 1, the name of the linetype is DASH3DOT. The name *DASH3DOT can be followed by the description of the linetype; the length of this description cannot exceed 47 characters. In this example, the description is dashes and dots, ____ ... ____. It could be any text or alphanumeric string. The description is displayed on the screen when you list the linetypes.

Pattern

   A,.5,-.125,0,-.125,0,-.125,0,-.125

The line pattern should start with an alignment definition. Currently, AutoCAD supports only one type of alignment—A. Therefore, it is automatically displayed on the screen when you select the

LINETYPE command with the Create option. After entering **A** for pattern alignment, you must define the pen position. A positive number (.5 or 0.5) indicates a "pen-down" position, and a negative number (-.25 or -0.25) indicates a "pen-up" position. The length of the dash or the space is designated by the magnitude of the number. For example, 0.5 will draw a dash 0.5 units long, and -0.25 will leave a blank space of 0.25 units. A dash length of 0 will draw a dot (.). Here are the pattern definition elements for Example 1:

.5	pen down	0.5 units long dash
-.125	pen up	.125 units blank space
0	pen down	dot
-.125	pen up	.125 units blank space
0	pen down	dot
-.125	pen up	.125 units blank space
0	pen down	dot
-.125	pen up	.125 units blank space

After you enter the pattern definition, the linetype (DASH3DOT) is automatically saved in the **ACAD.LIN** file. The linetype (DASH3DOT) can be loaded using the AutoCAD **LINETYPE** command and selecting the **Load** option.

Note
The name and the description must be separated by a comma (,). The description is optional. If you decide not to give one, omit the comma after the linetype name DASH3DOT.

Using a Text Editor

You can also use a text editor (like Notepad) to create a new linetype. Using the text editor, load the file and insert the lines that define the new linetype. The following file is a partial listing of the **ACAD.LIN** file after adding a new linetype to the file:

```
*BORDER,__ __ . __ __ . __ __ . __ __ . __ __ . __ __ .
A,.5,-.25,.5,-.25,0,-.25
*BORDER2,__._.__._.__._.__._.__._.__._.__._.__._.__._.__.
A,.25,-.125,.25,-.125,0,-.125
*BORDERX2,____ ____ . ____ ____ . ____ ____ .
A,1.0,-.5,1.0,-.5,0,-.5

*CENTER,____ _ ____ _ ____ _ ____ _ ____ _ ____
A,1.25,-.25,.25,-.25
*CENTER2,___ _ ___ _ ___ _ ___ _ ___ _ ___ _ ___
A,.75,-.125,.125,-.125
*CENTERX2,_____ __ _____ __ _____ __ __
A,2.5,-.5,.5,-.5
*DASHDOT,__ . __ . __ . __ . __ . __ . __ .
A,.5,-.25,0,-.25
*DOTX2,. . . . . . . . . . . . . . . .
```

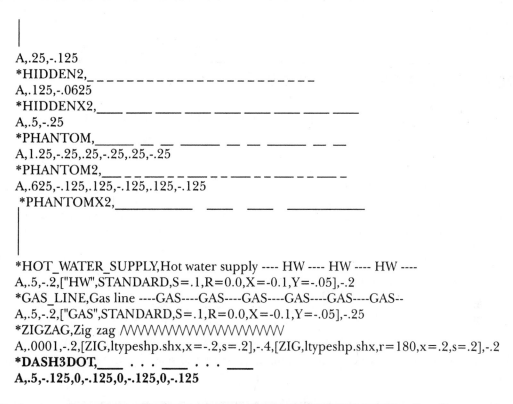

```
A,.25,-.125
*HIDDEN2,_ _ _ _ _ _ _ _ _ _ _ _ _ _ _ _ _ _ _ _ _ _
A,.125,-.0625
*HIDDENX2,___ ___ ___ ___ ___ ___ ___ ___ ___
A,.5,-.25
*PHANTOM,_____ __ __ _____ __ __ _____ __ __
A,1.25,-.25,.25,-.25,.25,-.25
*PHANTOM2,___ __ __ ___ __ __ ___ __ __ ___ __ __ ___ _
A,.625,-.125,.125,-.125,.125,-.125
 *PHANTOMX2,_____     ___   ___   _____
A,.5,-.25
```

```
*HOT_WATER_SUPPLY,Hot water supply ---- HW ---- HW ---- HW ----
A,.5,-.2,["HW",STANDARD,S=.1,R=0.0,X=-0.1,Y=-.05],-.2
*GAS_LINE,Gas line ----GAS----GAS----GAS----GAS----GAS----GAS--
A,.5,-.2,["GAS",STANDARD,S=.1,R=0.0,X=-0.1,Y=-.05],-.25
*ZIGZAG,Zig zag /\/\/\/\/\/\/\/\/\/\/\/\/\/\/\/\
A,.0001,-.2,[ZIG,ltypeshp.shx,x=-.2,s=.2],-.4,[ZIG,ltypeshp.shx,r=180,x=.2,s=.2],-.2
```

*DASH3DOT,___ . . . ___ . . . ___
A,.5,-.125,0,-.125,0,-.125,0,-.125

The last two lines of this file define the new linetype, DASH3DOT. The first line contains the name DASH3DOT and the description of the line (___ . . .___). The second line contains the alignment and the pattern definition. Save the file and then load the linetype using the AutoCAD **LINETYPE** command with the Load option. The lines and polylines that this linetype will generate are shown in Figure 3-3.

Figure 3-3 *Lines created by linetype DASH3DOT*

Note
If you change the LTSCALE factor, all lines in the drawing are affected by the new ratio.

CREATING LINETYPE FILES

You can start a new linetype file and then add the line definitions to this file. Use any text editor like Notepad to start a new file (NEWLT.LIN) and then add the following two lines to the file to define the DASH3DOT linetype.

*DASH3DOT,___ . . . ___ . . . ___
A,.5,-.125,0,-.125,0,-.125,0,-.125

You can load the DASH3DOT linetype from this file using AutoCAD's **LINETYPE** command and then selecting the LOAD option.

Command: **-LINETYPE**
Enter an option [?/Create/Load/Set]: **L**
Enter linetype(s) to load: **DASHDOT**
Enter name of linetype file to search <default>: **NEWLT**

You can also load the linetype by using **Linetype Manager** dialog box (Figure 3-4) that can be invoked by entering **LINETYPE** at the Command prompt and selecting **Linetype** from the **Format** menu.

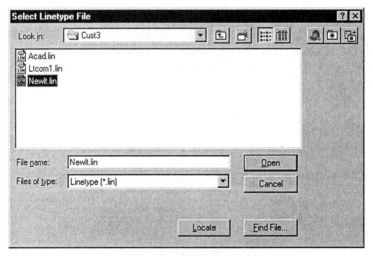

*Figure 3-4 Loading linetypes using **Linetype Properties Manager** dialog box*

ALIGNMENT SPECIFICATION

The alignment specifies the pattern alignment at the start and the end of the line, circle, or arc. In other words, the line would always start and end with the dash (___). The alignment definition "A" requires that the first element be a dash or dot (pen down), followed by a negative (pen up) segment. The minimum number of dash segments for alignment A is two. If there is not enough space for the line AutoCAD will draw a continuous line.

For example, in the linetype DASH3DOT of Example 1, the length of each line segment is 1.0 (.5 + .125 + .125 + .125 + .125 = 1.0). If the length of the line drawn is less than 1.00, the line will be drawn as a continuous line (Figure 3-5). If the length of the line is 1.00 or greater, the line will be drawn according to DASH3DOT linetype. AutoCAD automatically adjusts the length of the dashes and the line will always start and end with a dash. The length of the starting and ending dashes will be at least half the length of the dash as specified in the file. If the length of the dash as specified in the file is 0.5, the length of the starting and ending dashes will be at least 0.25. To fit a line that starts and ends with a dash, the length of these dashes can also increase as shown in Figure 3-5.

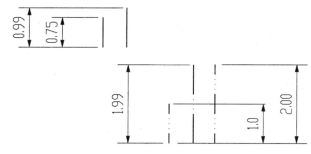

Figure 3-5 *Alignment of linetype DASH3DOT*

LTSCALE COMMAND

As we mentioned previously, the length of each line segment in the DASH3DOT linetype is 1.0 (.5 + .125 + .125 + .125 + .125 = 1.0). If you draw a line that is less than 1.0 units long, AutoCAD will draw a single dash that looks like a continuous line (Figure 3-6). This problem can be rectified by changing the linetype scale factor variable **LTSCALE** to a smaller value. This can be accomplished by using AutoCAD's **LTSCALE** command:

Command: **LTSCALE**
Enter new linetype scale factor <default>: *New value*.

The default value of the **LTSCALE** variable is 1.0. If the LTSCALE is changed to 0.75, the length of each segment is reduced by 0.75 (1.0 x 0.75 = 0.75). Then, if you draw a line 0.75 units or longer, it will be drawn according to the definition of DASH3DOT (__ . . . __) (Figures 3-7 and 3-8).

The appearance of the lines is also affected by the limits of the drawing. Most of the AutoCAD linetypes work fine for a drawing that has the limits 12,9. Figure 3-9 shows a line of linetype DASH3DOT that is four units long and the limits of the drawing are 12,9. If you increase the limits to 48,36 the lines will appear as continuous lines. If you want the line to appear the same as before **on the screen**, the LTSCALE should be changed. Since the limits of the drawing have increased four times, the LTSCALE should also increase by the same amount. If you change the

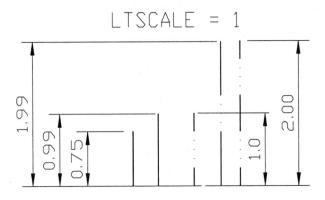

Figure 3-6 *Alignment when Ltscale = 1*

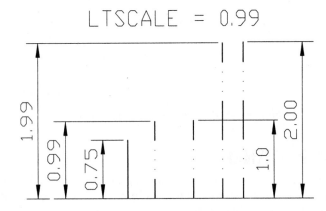

Figure 3-7 *Alignment when Ltscale = 0.99*

scale factor to four, the line segments will also increase by a factor of four. As shown in Figure 3-9, the length of the starting and the ending dash has increased to one unit.

In general, the approximate LTSCALE factor for **screen display** can be obtained by dividing the X-limit of the drawing by the default X-limit (12.00).

> **LTSCALE factor for SCREEN DISPLAY = X-limits of the drawing/12.00**

Example
> Drawing limits are 48,36
> LTSCALE factor for screen display = 48/12 = 4
>
> Drawing sheet size is 36,24 and scale is 1/4" = 1'
> LTSCALE factor for screen display = 12 x 4 x (36 / 12) = 144

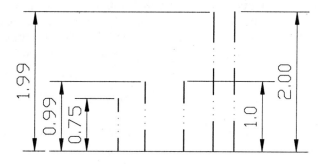

Figure 3-8 *Alignment when Ltscale = 0.75*

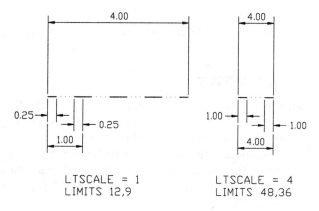

Figure 3-9 *Linetype DASH3DOT before and after changing the LTSCALE factor*

LTSCALE FACTOR FOR PLOTTING

The LTSCALE factor for plotting depends on the size of the sheet you are using to plot the drawing. For example, if the limits are 48 by 36, the drawing scale is 1:1, and you want to plot the drawing on a 48" by 36" size sheet, the LTSCALE factor is 1. If you check the specification of a hidden line in the ACAD.LIN file, the length of each dash is 0.25. Therefore, when you plot a drawing with 1:1 scale, the length of each dash in a hidden line is 0.25.

However, if the drawing scale is 1/8" = 1' and you want to plot the drawing on a 48" by 36" paper, the LTSCALE factor must be 96 (8 x 12 = 96). The length of each dash in the hidden line will increase by a factor of 96 because the LTSCALE factor is 96. Therefore, the length of each dash will be 24 units (0.25 x 96 = 24). At the time of plotting, the scale factor for plotting must be 1:96 to plot the 384' by 288' drawing on a 48" by 36" size paper. Each dash of the hidden line that was 24" long on the drawing will be 0.25 (24/96 = 0.25) inch long when plotted. Similarly, if the desired text size on the paper is 1/8", the text height in the drawing must be 12" (1/8 x 96 = 12").

Ltscale Factor for PLOTTING = Drawing Scale

Sometimes your plotter may not be able to plot a 48" by 36" drawing or you might like to decrease the size of the plot so that the drawing fits within a specified area. To get the correct dash lengths for hidden, center, or other lines, you must adjust the LTSCALE factor. For example, if you want to plot the previously mentioned drawing in a 45" by 34" area, the correction factor is:

Correction factor	= 48/45
	= 1.0666
New LTSCALE factor	= LTSCALE factor x Correction factor
	= 96 x 1.0666
	= 102.4

New Ltscale Factor for PLOTTING = Drawing Scale x Correction Factor

Note

If you change the LTSCALE factor, all lines in the drawing are affected by the new ratio.

ALTERNATE LINETYPES

One of the problems with the LTSCALE factor is that it affects all the lines in the drawing. As shown in Figure 3-10(a), the length of each segment in all DASH3DOT type lines is approximately equal, no matter how long the lines. You might want to have a small segment length if the lines are small and a longer segment length if the lines are long. You can accomplish this by using CELTSCALE (discussed later in this chapter) or by defining an alternate linetype with a different segment length. For example, you can define a linetype DASH3DOT and DASH3DOTX with different line pattern specifications.

```
*DASH3DOT,____ . . . ____ . . . ____ . . . ____
A,0.5,-.125,0,-.125,0,-.125,0,-.125
*DASH3DOTX,_____   . . .   _____
A,1.0,-.25,0,-.25,0,-.25,0,-.25
```

In DASH3DOT linetype the segment length is one unit, whereas in DASH3DOTX linetype the segment length is two units. You can have several alternate linetypes to produce the lines with different segment lengths. Figure 3-10(b) shows the lines generated by DASH3DOT and DASH3DOTX.

Note

Although you might have used different linetypes with different segment lengths, the lines will be affected equally when you change the LTSCALE factor. For example, if the LTSCALE factor is 0.5, the segment length of DASH3DOT line will be 0.5 and the segment length of DASH3DOTX will be 1.0 units.

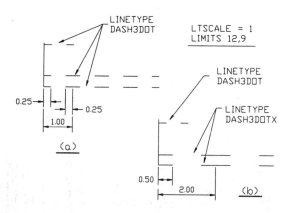

Figure 3-10 Linetypes generated by DASH3DOT and DASH3DOTX

MODIFYING LINETYPES

You can also modify the linetypes that are defined in the ACAD.LIN file. You need a text editor, such as Notepad, to modify the linetype. You can also use the EDIT function of DOS, or AutoCAD's **EDIT** command (provided the ACAD.PGP file is present and EDIT is defined in the file). For example, if you want to change the dash length of the border linetype from 0.5 to 0.75, load the file, then edit the pattern line of the border linetype. The following file is a partial listing of the ACAD.LIN file after changing the border and centerx2 linetypes.

```
;;  AutoCAD Linetype Definition file, Version 2.0
;;  Copyright 1991, 1992, 1993, 1994, 1996 by Autodesk, Inc.
;;
*BORDER,__ __ . __ __ . __ __ . __ __ . __ __ . __ __ . __ __ .
A,.75,-.25,.75,-.25,0,-.25
*BORDER2,Border (.5x) __ . __ . __ . __ . __ . __ . __ . __ .
A,.25,-.125,.25,-.125,0,-.125
*BORDERX2,Border (2x) ___ ___ . ___ ___ . ___
A,1.0,-.5,1.0,-.5,0,-.5

*CENTER,Center ____ _ ____ _ ____ _ ____ _ ____ _ ____
A,1.25,-.25,.25,-.25
*CENTER2,Center (.5x) __ _ __ _ __ _ __ _ __ _ __ _ __
A,.75,-.125,.125,-.125
*CENTERX2,Center (2x) _____ __ _____ __ _____
A,3.5,-.5,.5,-.5

*DASHDOT,Dash dot __ . __ . __ . __ . __ . __ . __
A,.5,-.25,0,-.25
*DASHDOT2,Dash dot (.5x) _._._._._._._._._._._.
A,.25,-.125,0,-.125
*DASHDOTX2,Dash dot (2x) ___ . ___ . ___ . ___
```

A,1.0,-.5,0,-.5

*DASHED,Dashed __ __ __ __ __ __ __ __ __ __ __ __ __
A,.5,-.25
*DASHED2,Dashed (.5x) _ _ _ _ _ _ _ _ _ _ _ _ _ _ _
A,.25,-.125
*DASHEDX2,Dashed (2x) ___ ___ ___ ___ ___ __
A,1.0,-.5

*DIVIDE,Divide ___ . . ___ . . ___ . . ___ . . ___
A,.5,-.25,0,-.25,0,-.25
*DIVIDE2,Divide (.5x) __.._.._.._.._.._..__.._.._.._.._.._..__
A,.25,-.125,0,-.125,0,-.125
*DIVIDEX2,Divide (2x) _____ . . _____ . . _
A,1.0,-.5,0,-.5,0,-.5

*DOT,Dot . . . .
A,0,-.25
*DOT2,Dot (.5x) . . . .
A,0,-.125
*DOTX2,Dot (2x) . . . .
A,0,-.5

*HIDDEN,Hidden __ __ __ __ __ __ __ __ __ __ __ __
A,.25,-.125
*HIDDEN2,Hidden (.5x) _ _ _ _ _ _ _ _ _ _ _ _ _ _ _ _
A,.125,-.0625
*HIDDENX2,Hidden (2x) ___ ___ ___ ___ ___ ___ ___
A,.5,-.25

*PHANTOM,Phantom _____ __ __ _____ __ __ _____
A,1.25,-.25,.25,-.25,.25,-.25
*PHANTOM2,Phantom (.5x) ___ _ _ ___ __ ___ __ __
A,.625,-.125,.125,-.125,.125,-.125
*PHANTOMX2,Phantom (2x) _____ ___ ___ _
A,2.5,-.5,.5,-.5,.5,-.5
;; ISO 128 (ISO/DIS 12011) linetypes
;; The size of the line segments for each defined ISO line, is
;; defined for an usage with a pen width of 1 mm. To use them with
;; the other ISO predefined pen widths, the line has to be scaled
;; with the appropriate value (e.g. pen width 0,5 mm -> ltscale 0.5).
;;
*ACAD_ISO02W100,ISO dash __ __ __ __ __ __ __ __ __ __ __ __ __
A,12,-3
*ACAD_ISO03W100,ISO dash space __ _ __ __ __ __ __
A,12,-18
*ACAD_ISO04W100,ISO long-dash dot ____ . ____ . ____ . ____ . _

A,24,-3,.5,-3
*ACAD_ISO05W100,ISO long-dash double-dot ____ .. ____ .. ____ .
A,24,-3,.5,-3,.5,-3
*ACAD_ISO06W100,ISO long-dash triple-dot ____ ... ____ ... ____
A,24,-3,.5,-3,.5,-3,.5,-3
*ACAD_ISO07W100,ISO dot . . . .
A,.5,-3
*ACAD_ISO08W100,ISO long-dash short-dash ____ __ ____ __ ____ _
A,24,-3,6,-3
*ACAD_ISO09W100,ISO long-dash double-short-dash ____ __ __ ____
A,24,-3,6,-3,6,-3
*ACAD_ISO10W100,ISO dash dot __ . __ . __ . __ . __ . __ .
A,12,-3,.5,-3
*ACAD_ISO11W100,ISO double-dash dot __ __ . __ __ . __ __ . __ _
A,12,-3,12,-3,.5,-3
*ACAD_ISO12W100,ISO dash double-dot __ .. __ .. __ .. __ ..
A,12,-3,.5,-3,.5,-3
*ACAD_ISO13W100,ISO double-dash double-dot __ __ .. __ __ .. _
A,12,-3,12,-3,.5,-3,.5,-3
*ACAD_ISO14W100,ISO dash triple-dot __ ... __ ... __ ... _
A,12,-3,.5,-3,.5,-3,.5,-3
*ACAD_ISO15W100,ISO double-dash triple-dot __ __ ... __ __ ..
A,12,-3,12,-3,.5,-3,.5,-3,.5,-3
;; Complex linetypes
;;
;; Complex linetypes have been added to this file.
;; These linetypes were defined in LTYPESHP.LIN in
;; Release 13, and are incorporated in ACAD.LIN in
;; Release 14.
;;
;; These linetype definitions use LTYPESHP.SHX.
;;
*FENCELINE1,Fenceline circle ——0——0——0——0——0——0—
A,.25,-.1,[CIRC1,ltypeshp.shx,x=-.1,s=.1],-.1,1
*FENCELINE2,Fenceline square ——[]——[]——[]——[]——[]—
A,.25,-.1,[BOX,ltypeshp.shx,x=-.1,s=.1],-.1,1
*TRACKS,Tracks -|-|-|-|-|-|-|-|-|-|-|-|-|-|-|-|-|-|-|-|-|-
A,.15,[TRACK1,ltypeshp.shx,s=.25],.15
*BATTING,Batting SSSSSSSSSSSSSSSSSSSSSSSSSSSSSSSSSSSSSSSSSSSSSS
A,.0001,-.1,[BAT,ltypeshp.shx,x=-.1,s=.1],-.2,[BAT,ltypeshp.shx,r=180,x=.1,s=.1],-.1
*HOT_WATER_SUPPLY,Hot water supply —— HW —— HW —— HW ——
A,.5,-.2,["HW",STANDARD,S=.1,R=0.0,X=-0.1,Y=-.05],-.2
*GAS_LINE,Gas line ——GAS——GAS——GAS——GAS——GAS——GAS—
A,.5,-.2,["GAS",STANDARD,S=.1,R=0.0,X=-0.1,Y=-.05],-.25
*ZIGZAG,Zig zag /\/\/\/\/\/\/\/\/\/\/\/\/\
A,.0001,-.2,[ZIG,ltypeshp.shx,x=-.2,s=.2],-.4,[ZIG,ltypeshp.shx,r=180,x=.2,s=.2],-.2

Example 2

Create a new file, **NEWLINET.LIN,** and define a linetype, VARDASH, with the following specifications:

> Length of first dash 1.0
> Blank space 0.25
> Length of second dash 0.75
> Blank space 0.25
> Length of third dash 0.5
> Blank space 0.25
> Dot
> Blank space 0.25
> Length of next dash 0.5
> Blank space 0.25
> Length of next dash 0.75

Use a text editor and insert the following lines that define the new linetype VARDASH. Save the file as NEWLINET.LIN.

> ***VARDASH,——— —— — . — —— ———**
> **A,1,-.25,.75,-.25,.5,-.25,0,-.25,.5,-.25,.75,-.25**

The type of lines that this linetype will generate are shown in Figure 3-11.

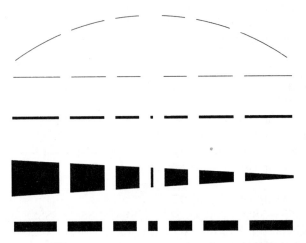

Figure 3-11 Lines generated by linetype VARDASH

CURRENT LINETYPE SCALING (CELTSCALE)

Like **LTSCALE,** the **CELTSCALE** system variable controls the linetype scaling. The difference is that **CELTSCALE** determines the current linetype scaling. For example, if you set the **CELTSCALE** to 0.5, all lines drawn after setting the new value for **CELTSCALE** will have the linetype scaling factor of 0.5. The value is retained in the **CELTSCALE** system variable. The first line (a) in Figure 3-12 is drawn with the **CELTSCALE** factor of 1, and the second line (b) is drawn with the **CELTSCALE** factor of 0.5. The length of the dashed line is reduced by a factor of 0.5 when the **CELTSCALE** is 0.5.

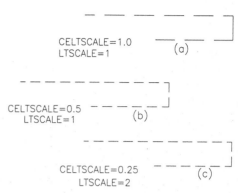

The **LTSCALE** system variable controls the global scale factor. For example, if **LTSCALE** is set to 2, all lines in the drawing will be affected by a factor of 2. The net scale factor is equal to the product of **CELTSCALE** and **LTSCALE**.

*Figure 3-12 Using **CELTSCALE** to control current linetype scaling*

Figure 3-12(c) shows a line that is drawn with **LTSCALE** of 2 and **CELTSCALE** of 0.25. The net scale factor is = **LTSCALE** x **CELTSCALE** = 2 x 0.25 = 0.5.

COMPLEX LINETYPES

AutoCAD has provided a facility to create complex linetypes. The complex linetypes can be classified into two groups: string complex linetype and shape complex linetype. The difference between the two is that the string complex linetype has a text string inserted in the line, whereas the shape complex linetype has a shape inserted in the line. The facility of creating complex linetypes increases the functionality of lines. For example, if you want to draw a line around a building that indicates the fence line, you can do it by defining a complex linetype that will automatically give you the desired line with the text string (Fence). Similarly, you can define a complex linetype that will insert a shape (symbol) at predefined distances along the line.

Creating a String Complex Linetype

When writing the definition of a string complex linetype, the actual text and its attributes must be included in the linetype definition. The format of the string complex linetype is:

   ["String", Text Style, Text Height, Rotation, X-Offset, Y-Offset]

String. It is the actual text that you want to insert along the line. The text string must be enclosed in quotation marks (" ").

Text Style. This is the name of the text style file that you want to use for generating the text string. The text style must be predefined.

Text Height. This is the actual height of the text, if the text height defined in the text style is 0. Otherwise, it acts as a scale factor for the text height specified in the text style. In Figure 3-13, the height of the text is 0.1 units.

Rotation. This is the rotation of the text string with respect to the positive X axis. The angle is always measured with respect to the positive X axis, no matter what AutoCAD's direction setting.

The angle can be specified in radians (r), grads (g), or degrees (d). The default is degrees.

X-Offset. This is the distance of the lower left corner of the text string from the endpoint of the line segment measured along the line. If the line is horizontal, then the X-Offset distance is measured along the X axis. In Figure 3-13, the X-Offset distance is 0.05.

Y-Offset. This is the distance of the lower left corner of the text string from the endpoint of the line segment measured perpendicular to the line. If the line is horizontal, then the Y-Offset distance is measured along the Y axis. In Figure 3-13, the Y-Offset distance is -0.05. The distance is negative because the start point of the text string is 0.05 units below the endpoint of the first line segment.

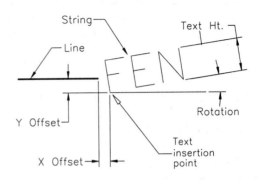

Figure 3-13   The attributes of a string complex linetype

Example 3   *General*

In the following example, you will write the definition of a string complex linetype that consists of the text string "Fence" and line segments. The length of each line segment is 0.75. The height of the text string is 0.1 units, and the space between the end of the text string and the following line segment is 0.05 (Figure 3-14).

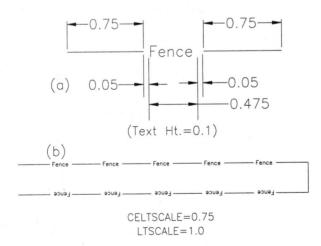

Figure 3-14   The attributes of a string complex linetype and line specifications for Example 3

Step 1
Before writing the definition of a new linetype, it is important to determine the line specification. One of the ways this can be done is to actually draw the lines and the text the way you want them to appear in the drawing. Once you have drawn the line and the text to your satisfaction, measure

the distances needed to define the string complex linetype. In this example, the values are given as follows:

Text string=	Fence
Text style=	Standard
Text height=	0.1
Text rotation=	0
X-Offset=	0.05
Y-Offset=	-0.05
Length of the first line segment=	0.75
Distance between the line segments=	0.575

Step 2

Use a text editor to write the definition of the string complex linetype. You can add the definition to the AutoCAD **ACAD.LIN** file or create a separate file. The extension of the file must be **.LIN**. The following file is the listing of the **FENCE.LIN** file for Example 3. The name of the linetype is NEWFence.

```
*NEWFence1,New fence boundary line
A,0.75,["Fence",Standard,S=0.1,A=0,X=0.05,Y=-0.05],-0.575
or
A,0.75,-0.05,["Fence",Standard,S=0.1,A=0,X=0,Y=-0.05],-0.525
```

Step 3

To test the linetype, load the linetype using the **LINETYPE** command with the Load option, and assign it to a layer. Draw a line or any object to check if the line is drawn to the given specifications. Notice that the text is always drawn along X axis. Also, when you draw a line at an angle, polyline, circle, or spline, the text string does not align with the object (Figure 3-15).

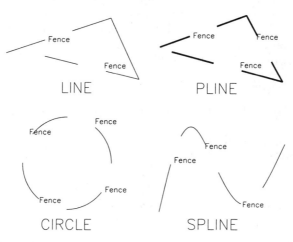

Figure 3-15 *Using string complex linetype with angle A=0*

Step 4

In the NEWFence linetype definition, the specified angle is 0 degrees (Absolute angle A = 0). Therefore, when you use the NEWFence linetype to draw a line, circle, polyline, or spline, the text string (Fence) will be at zero degrees. If you want the text string (Fence) to align with the polyline (Figure 3-16), spline, or circle, specify the angle as relative angle (R = 0) in the NEWFence linetype definition. The following is the linetype definition for NEWFence linetype with relative angle R = 0:

    *NEWFence2,New fence boundary line
    A,0.75,["Fence",Standard,S=0.1,R=0,X=0.05,Y=-0.05],-0.575

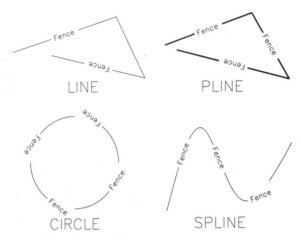

Figure 3-16 Using a string complex linetype with angle R = 0

Step 5

In Figure 3-16, you might notice that the text string is not properly aligned with the circumference of the circle. This is because AutoCAD draws the text string in a direction that is tangent to the circle at the text insertion point. To resolve this problem, you must define the middle point of the text string as the insertion point. Also, the line specifications should be measured accordingly. Figure 3-17 gives the measurements of the NEWFence linetype with the middle point of the text as the insertion point.

The following is the linetype definition for NEWFence linetype:

    *NEWFence3,New fence boundary line
    A,0.75,-0.287,["FENCE",Standard,S=0.1,X=-0.237,Y=-0.05],-0.287

Note
If no angle is defined in the line definition, it defaults to angle R = 0.

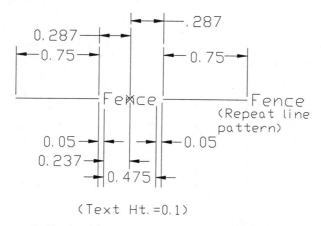

Figure 3-17 *Specifications of a string complex linetype with the middle point of the text string as the text insertion point*

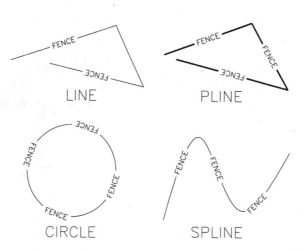

Figure 3-18 *Using a string complex linetype with the middle point of the text string as the text insertion point*

Creating a Shape Complex Linetype

As with the string complex linetype, when you write the definition of a shape complex linetype, the name of the shape, the name of the shape file, and other shape attributes, like rotation, scale, X-Offset, and Y-Offset, must be included in the linetype definition. The format of the shape complex linetype is:

[Shape Name, Shape File, Scale, Rotation, X-Offset, Y-Offset]

The following is the description of the attributes of Shape Complex Linetype (Figure 3-19).

Shape Name. This is the name of the shape that you want to insert along the line. The shape name must exist; otherwise, no shape will be generated along the line.

Shape File. This is the name of the compiled shape file (.SHX) that contains the definition of the shape being inserted in the line. The name of the subdirectory where the shape file is located must be in the ACAD search path. The shape files (.SHP) must be compiled before using the SHAPE command to load the shape.

Scale. This is the scale factor by which the defined shape size is to be scaled. If the scale is 1, the size of the shape will be the same as defined in the shape definition (.SHP file).

Rotation. This is the rotation of the shape with respect to the positive X axis.

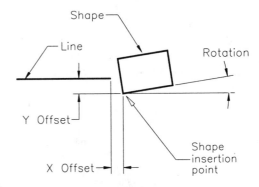

X-Offset. This is the distance of the shape insertion point from the endpoint of the line segment measured along the line. If the line is horizontal, then the X-Offset distance is measured along the X axis. In Figure 3-19, the X-Offset distance is 0.2.

Y-Offset. This is the distance of the shape insertion point from the endpoint of the line segment measured perpendicular to the line. If the line is horizontal, then the Y-Offset distance is measured along the Y axis. In Figure 3-19, the Y-Offset distance is 0.

Figure 3-19   The attributes of a shape complex linetype

Example 4

In the following example, you will write the definition of a shape complex linetype that consists of the shape (Manhole; the name of the shape is MH) and a line. The scale of the shape is 0.1, the length of each line segment is 0.75, and the space between line segments is 0.2.

Step 1
Before writing the definition of a new linetype, it is important to determine the line specifications. One of the ways this can be done is to actually draw the lines and the shape the way you want them to appear in the drawing (Figure 3-20). Once you have drawn the line and the shape to your satisfaction, measure the distances needed to define the shape complex linetype. In this example, the values are as follows:

```
Shape name   MH
Shape file name  MHOLE.SHX   (Name of the compiled shape file.)
Scale          0.1
Rotation       0
X-Offset       0.2
Y-Offset       0
```

Length of the first line segment = 0.75
Distance between the line segments = 0.2

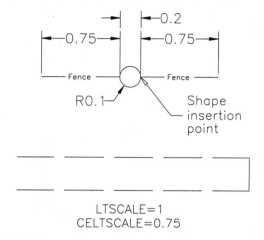

Figure 3-20 The attributes of the shape complex linetype and line specifications for Example 4

Step 2

Use a text editor to write the definition of the shape file. The extension of the file must be **.SHP**. The following file is the listing of the **MHOLE.SHP** file for Example 4. The name of the shape is MH. (For details, see Chapter 11, "Shape and Text Fonts".)

```
*215,9,MH
001,10,(1,007),
001,10,(1,071),0
```

Step 3

Use the **COMPILE** command to compile the shape file (**.SHP** file). When you use this command, AutoCAD will prompt you to enter the name of the shape file (Figure 3-21). For this example, the name is **MHOLE.SHP**. The following is the command sequence for compiling the shape file:

Command: **COMPILE**
Enter shape (.SHP) or PostScript font (.PFB) file name: **MHOLE**

You can also compile the shape from the command line by setting **FILEDIA**=0 and then using the **COMPILE** command.

Step 4

Use a text editor to write the definition of the shape complex linetype. You can add the definition to the AutoCAD ACAD.LIN file or create a separate file. The extension of the file must be .LIN. The following file is the listing of the MHOLE.LIN file for Example 4. The name of the linetype is MHOLE.

*MHOLE,Line with Manholes
A,0.75,[MH,MHOLE.SHX,S=0.10,X=0.2,Y=0],-0.2

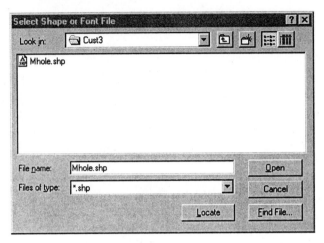

Figure 3-21 Select Shape or Font File dialog box

Step 5

To test the linetype, load the linetype using the **LINETYPE** command with the Load option and assign it to a layer. Draw a line or any object to check if the line is drawn to the given specifications. The shape is drawn upside down when you draw a line from right to left.

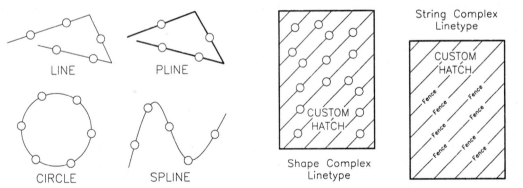

Figure 3-22 Using a shape complex linetype

Figure 3-23 Using shape and string complex linetypes to create custom hatch

HATCH PATTERN DEFINITION

The AutoCAD software comes with a hatch pattern library file, **ACAD.PAT,** that contains 67 hatch patterns. These hatch patterns are sufficient for general drafting work. However, if you need a different hatch pattern, AutoCAD lets you create your own. There is no limit to the number of hatch patterns you can define.

The hatch patterns you define can be added to the hatch pattern library file, **ACAD.PAT.** You can also create a new hatch pattern library file, provided the file contains only one hatch pattern definition, and the name of the hatch is the same as the name of the file. The hatch pattern definition consists of the following two parts: **header line and hatch descriptors**.

Header Line

The **header line** consists of an asterisk (*) followed by the name of the hatch pattern. The hatch name is the name used in the hatch command to hatch an area. After the name, you can give the hatch description, which is separated from the hatch name by a comma (,). The general format of the header line is:

*HATCH Name [, Hatch Description]
>   Where   ***** ----------------------------- Asterisk
>   **HATCH Name** ----------- Name of hatch pattern
>   **Hatch Description** ------ Description of hatch pattern

The description can be any text that describes the hatch pattern. It can also be omitted, in which case, a comma should not follow the hatch pattern name.

Example
*DASH45, Dashed lines at 45 degrees
>   Where **DASH45** --------- Hatch name
>   **Dashed lines at 45 degrees** ------ Hatch description

Hatch Descriptors

The **hatch descriptors** consist of one or more lines that contain the definition of the hatch lines. The general format of the hatch descriptor is:

Angle, X-origin, Y-origin, D1, D2 [,Dash Length.....]
>   Where   **Angle** ------------ Angle of hatch lines
>   **X-origin** --------- X coordinate of hatch line
>   **Y-origin** --------- Y coordinate of hatch line
>   **D1** ---------------- Displacement of second line (Delta-X)
>   **D2** ---------------- Distance between hatch lines (Delta-Y)
>   **Length** ----------- Length of dashes and spaces (Pattern line definition)

Example
45,0,0,0,0.5,0.5,-0.125,0,-0.125
>   Where **45** ----------------- Angle of hatch line
>   **0** ------------------- X-Origin
>   **0** ------------------- Y-Origin
>   **0** ------------------- Delta-X
>   **0.5** ---------------- Delta-Y
>   **0.5** ---------------- Dash (pen down)
>   **-0.125** ----------- Space (pen up)
>   **0** ------------------- Dot (pen down)
>   **-0.125** ----------- Space (pen up)
>   **0.5,-0.125,0,-0.125** Pattern line definition

Hatch Angle

X-origin and Y-origin. The hatch angle is the angle that the hatch lines make with the positive X axis. The angle is positive if measured counterclockwise (Figure 3-24), and negative if the angle is measured clockwise. When you draw a hatch pattern, the first hatch line starts from the point defined by X-origin and Y-origin. The remaining lines are generated by offsetting the first hatch line by a distance specified by delta-X and delta-Y. In Figure 3-25(a), the first hatch line starts from the point with the coordinates X = 0 and Y = 0. In Figure 3-25(b) the first line of hatch starts from a point with the coordinates X = 0 and Y = 0.25.

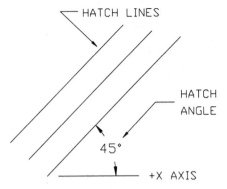

Figure 3-24 Hatch angle

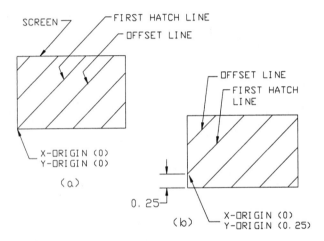

Figure 3-25 X-origin and Y-origin of hatch lines

Delta-X and Delta-Y. Delta-X is the displacement of the offset line in the direction in which the hatch lines are generated. For example, if the lines are drawn at a 0-degree angle and delta-X = 0.5, the offset line will be displaced by a distance delta-X (0.5) along the 0-angle direction. Similarly, if the hatch lines are drawn at a 45-degree angle, the offset line will be displaced by a distance delta-X (0.5) along a 45-degree direction (Figure 3-26). Delta-Y is the displacement of the offset lines measured perpendicular to the hatch lines. For example, if delta-Y = 1.0, the space between any two hatch lines will be 1.0 (Figure 3-26).

HOW HATCH WORKS

When you hatch an area, AutoCAD generates an infinite number of hatch lines of infinite length. The first hatch line always passes through the point specified by the X-origin and Y-origin. The remaining lines are generated by offsetting the first hatch line in both directions. The offset distance is determined by delta-X and delta-Y. All selected entities that form the boundary of the

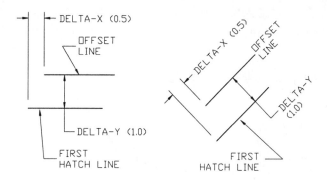

Figure 3-26 Delta-X and delta-Y of hatch lines

hatch area are then checked for intersection with these lines. Any hatch lines found within the defined hatch boundaries are turned on, and the hatch lines outside the hatch boundary are turned off, as shown in Figure 3-27. Since the hatch lines are generated by offsetting, the hatch lines in different areas of the drawing are automatically aligned relative to the drawing's snap origin. Figure 3-27(a) shows the hatch lines as computed by AutoCAD. These lines are not drawn on the screen; they are shown here for illustration only. Figure 3-27(b) shows the hatch lines generated in the circle that was defined as the hatch boundary.

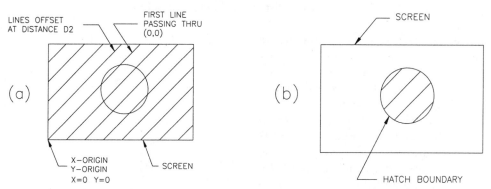

Figure 3-27 Hatch lines outside the hatch boundary are turned off

SIMPLE HATCH PATTERN

It is good practice to develop the hatch pattern specification before writing a hatch pattern definition. For simple hatch patterns it may not be that important, but for more complicated hatch patterns you should know the detailed specifications. Example 5 illustrates the procedure for developing a simple hatch pattern.

Example 5

Write a hatch pattern definition for the hatch pattern shown in Figure 3-28, with the following specifications:

Name of the hatch pattern =	HATCH1
X-Origin =	0
Y-Origin =	0
Distance between hatch lines =	0.5
Displacement of hatch lines =	0
Hatch line pattern =	Continuous

This hatch pattern definition can be added to the existing **ACAD.PAT** hatch file. You can use any text editor (like Notepad) to write the file. Load the **ACAD.PAT** file that is located in **AutoCAD2000\SUPPORT** directory and insert the following two lines at the end of the file.

```
*HATCH1,Hatch Pattern for Example 5
45,0,0,0,.5
```

Where **45** ----------------- Hatch angle
0 ------------------- X-origin
0 ------------------- Y-origin
0 ------------------- Displacement of second hatch line
.5 ----------------- Distance between hatch lines

The first field of hatch descriptors contains the angle of the hatch lines. That angle is 45 degrees with respect to the positive X axis. The second and third fields describe the X and Y coordinates of the first hatch line origin. The first line of the hatch pattern will pass through this point. If the values of the X-origin and Y-origin were 0.5 and 1.0, respectively, then the first line would pass through the point with the X coordinate of 0.5 and the Y coordinate of 1.0, with respect to the drawing origin 0,0. The remaining lines are generated by offsetting the first line, as shown in Figure 3-28.

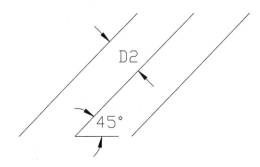

Figure 3-28 Hatch pattern angle and offset

EFFECT OF ANGLE AND SCALE FACTOR ON HATCH

When you hatch an area, you can alter the angle and displacement of hatch lines you have specified in the hatch pattern definition to get a desired hatch spacing. You can do this by entering an appropriate value for angle and scale factor in the AutoCAD **HATCH** command.

Command: **HATCH**
Enter a pattern name or [?/Solid/User defined] <default>: **Hatch1**
Specify a scale for the pattern <default>: **1**
Specify an angle for the pattern<default>: **0**

To understand how the angle and the displacement can be changed, hatch an area with the hatch pattern HATCH1 in Example 5. You will notice that the hatch lines have been generated according to the definition of hatch pattern HATCH1. Notice the effect of hatch angle and scale factor on the hatch. Figure 3-29(a) shows a hatch that is generated by the AutoCAD **HATCH** command with a 0-degree angle and a scale factor of 1.0. If the angle is 0, the hatch will be generated with the same angle as defined in the hatch pattern definition (45 degrees in Example 5). Similarly, if the scale factor is 1.0, the distance between the hatch lines will be the same as defined in the hatch pattern definition (0.5 in Example 5). Figure 3-29(b) shows a hatch that is generated when the hatch scale factor is 0.5. If you measure the distance between the successive hatch lines, it will be 0.5 x 0.5 = 0.25. Figures 3-29(c) and (d) show the hatch when the angle is 45 degrees and the scale factors are 1.0 and 0.5, respectively. You can enter any value in response to the **HATCH** command prompts to generate hatch lines at any angle and with any line spacing.

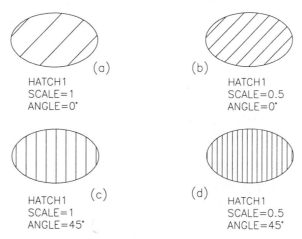

Figure 3-29 *Effect of angle and scale factor on hatch*

HATCH PATTERN WITH DASHES AND DOTS

The lines you can use in a hatch pattern definition are not restricted to continuous lines. You can define any line pattern to generate a hatch pattern. The lines can be a combination of dashes, dots, and spaces in any configuration. However, the maximum number of dashes you can specify in the line pattern definition of a hatch pattern is six. Example 6 uses a dash-dot line to create a hatch pattern.

Example 6

Write a hatch pattern definition for the hatch pattern shown in Figure 3-30, with the following specifications:

Name of the hatch pattern	HATCH2
Hatch angle =	0
X-origin =	0
Y-origin =	0
Displacement of lines (D1) =	0.25

Distance between lines (D2) =   0.25
Length of each dash =   0.5
Space between dashes and dots = 0.125
Space between dots =   0.125

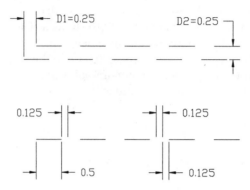

Figure 3-30 *Hatch lines made of dashes and dots*

You can use the AutoCAD **EDIT** command or any text editor (Notepad) to edit the **ACAD.PAT** file. The general format of the header line and the hatch descriptors is:

*HATCH NAME, Hatch Description
Angle, X-Origin, Y-Origin, D1, D2 [,Dash Length.....]

Substitute the values from Example 6 in the corresponding fields of header line and field descriptor:

*HATCH2,Hatch with dashes and dots
0,0,0,0.25,0.25,0.5,-0.125,0,-0.125,0,-0.125
                  Where **0** ------------------- Angle
                         0 ------------------- X-origin
                         0 ------------------- Y-origin
                         0.25 -------------- Delta-X
                         0.25 -------------- Delta-Y
                         0.5 ---------------- Length of dash
                         -0.125 ------------ Space (pen up)
                         0 ------------------- Dot (pen down)
                         -0.125 ------------ Space (pen up)
                         0 ------------------- Dot
                         -0.125 ------------ Space

The hatch pattern this hatch definition will generate is shown in Figure 3-31. Figure 3-31(a) shows the hatch with 0-degree angle and a scale factor of 1.0. Figure 3-31(b) shows the hatch with a 45-degree angle and a scale factor of 0.5.

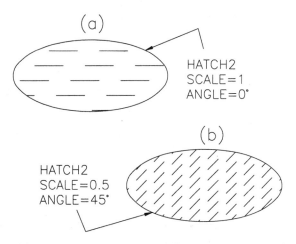

Figure 3-31 *Hatch pattern at different angles and scales*

HATCH WITH MULTIPLE DESCRIPTORS

Some hatch patterns require multiple lines to generate a shape. For example, if you want to create a hatch pattern of a brick wall, you need a hatch pattern that has four hatch descriptors to generate a rectangular shape. You can have any number of hatch descriptor lines in a hatch pattern definition. It is up to the user to combine them in any conceivable order. However, there are some shapes you cannot generate. A shape that has a nonlinear element, like an arc, cannot be generated by hatch pattern definition. However, you can simulate an arc by defining short line segments because you can use only straight lines to generate a hatch pattern. Example 7 uses three lines to define a triangular hatch pattern.

Example 7

Write a hatch pattern definition for the hatch pattern shown in Figure 3-32, with the following specifications:

Name of the hatch pattern =	HATCH3
Vertical height of the triangle =	0.5
Horizontal length of the triangle =	0.5
Vertical distance between the triangles =	0.5
Horizontal distance between the triangles =	0.5

Each triangle in this hatch pattern consists of the following three elements: a vertical line, a horizontal line, and a line inclined at 45 degrees.

Vertical Line. For the vertical line, the specifications are (Figure 3-33):

Hatch angle =	90 degrees
X-origin =	0
Y-origin =	0

Delta-X (D1) =   0
Delta-Y (D2) =   1.0
Dash length =   0.5
Space =   0.5

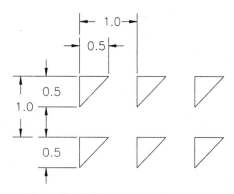

Figure 3-32 Triangle hatch pattern

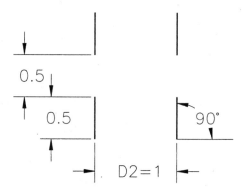

Figure 3-33 Vertical line

Substitute the values from the vertical line specification in various fields of the hatch descriptor to get the following line:

90,0,0,0,1,.5,-.5

Where	**90** ----------------	Hatch angle
	0 ------------------	X-origin
	0 ------------------	Y-origin
	0 ------------------	Delta-X
	1 ------------------	Delta-Y
	.5 -----------------	Dash (pen down)
	-.5 ----------------	Space (pen up)

Horizontal Line. For the horizontal line (Figure 3-34), the specifications are:

Hatch angle =   0 degrees
X-origin =   0
Y-origin =   0.5
Delta-X (D1) =   0
Delta-Y (D2) =   1.0
Dash length =   0.5
Space =   0.5

The only difference between the vertical line and the horizontal line is the angle. For the horizontal line, the angle is 0 degrees, whereas for the vertical line, the angle is 90 degrees. Substitute the values from the vertical line specification to obtain the following line:

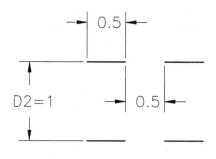

Figure 3-34 Horizontal line

0,0,0.5,0,1,.5,-.5

Where **0** ------------------ Hatch angle
0 ------------------ X-origin
0.5 ---------------- Y-origin
0 ------------------ Delta-X
1 ------------------ Delta-Y
.5 ----------------- Dash (pen down)
-.5 ---------------- Space (pen up)

Line Inclined at 45 Degrees. This line is at an angle; therefore, you need to calculate the distances delta-X (D1) and delta-Y (D2), the length of the dashed line, and the length of space. Figure 3-35 shows the calculations to find these values.

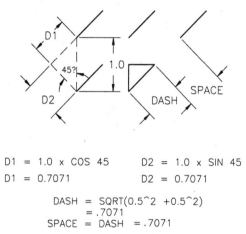

D1 = 1.0 x COS 45   D2 = 1.0 x SIN 45
D1 = 0.7071   D2 = 0.7071

DASH = SQRT(0.5^2 +0.5^2)
= .7071
SPACE = DASH = .7071

Figure 3-35 *Line inclined at 45 degrees*

Hatch angle =	45 degrees
X-Origin =	0
Y-Origin =	0
Delta-X (D1) =	0.7071
Delta-Y (D2) =	0.7071
Dash length =	0.7071
Space =	0.7071

After substituting the values in the general format of the hatch descriptor, you will obtain the following line:

45,0,0,.7071,.7071,.7071,-.7071

Where ·**45** ---------------- Hatch angle
0 ------------------ X-origin
0 ------------------ Y-origin
.7071 ------------ Delta-X
.7071 ------------ Delta-Y

.7071 ------------- Dash (pen down)
-.7071 ------------ Space (pen up)

Now you can combine these three lines and insert them at the end of the **ACAD.PAT** file. You can also use the AutoCAD **EDIT** command to edit the file and insert the lines.

Figure 3-36 shows the hatch pattern that will be generated by this hatch pattern (HATCH3). In Figure 3-36(a) the hatch pattern is at a 0-degree angle and the scale factor is 0.5. In Figure 3-36(b) the hatch pattern is at a -45-degree angle and the scale factor is 0.5.

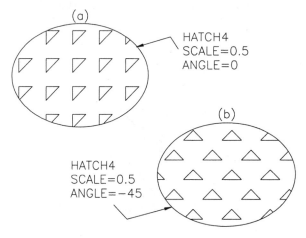

Figure 3-36   *Hatch generated by HATCH3 pattern*

The following file is a partial listing of the **ACAD.PAT** file, after adding the hatch pattern definitions from Examples 5, 6, and 7.

```
*SOLID, Solid fill
45, 0,0, 0,.125
*angle,Angle steel
0, 0,0, 0,.275, .2,-.075
90, 0,0, 0,.275, .2,-.075
*ansi31,ANSI Iron, Brick, Stone masonry
45, 0,0, 0,.125
*ansi32,ANSI Steel
45, 0,0, 0,.375
45, .176776695,0, 0,.375
*ansi33,ANSI Bronze, Brass, Copper
45, 0,0, 0,.25
45, .176776695,0, 0,.25, .125,-.0625
*ansi34,ANSI Plastic, Rubber
45, 0,0, 0,.75
45, .176776695,0, 0,.75
45, .353553391,0, 0,.75
```

45, .530330086,0, 0,.75
*ansi35,ANSI Fire brick, Refractory material
45, 0,0, 0,.25
45, .176776695,0, 0,.25, .3125,-.0625,0,-.0625
*ansi36,ANSI Marble, Slate, Glass
45, 0,0, .21875,.125, .3125,-.0625,0,-.0625

*swamp,Swampy area
0, 0,0, .5,.866025403, .125,-.875
90, .0625,0, .866025403,.5, .0625,-1.669550806
90, .078125,0, .866025403,.5, .05,-1.682050806
90, .046875,0, .866025403,.5, .05,-1.682050806
60, .09375,0, .5,.866025403, .04,-.96
120, .03125,0, .5,.866025403, .04,-.96
*trans,Heat transfer material
0, 0,0, 0,.25
0, 0,.125, 0,.25, .125,-.125
*triang,Equilateral triangles
60, 0,0, .1875,.324759526, .1875,-.1875
120, 0,0, .1875,.324759526, .1875,-.1875
0, -.09375,.162379763, .1875,.324759526, .1875,-.1875
*zigzag,Staircase effect
0, 0,0, .125,.125, .125,-.125
90, .125,0, .125,.125, .125,-.125
HATCH1,Hatch at 45 Degree Angle
45,0,0,0,.5
HATCH2,Hatch with Dashes & Dots:
0,0,0,.25,.25,0.5,-.125,0,-.125,0,-.125
HATCH3,Triangle Hatch:
90,0,0,0,1,.5,-.5
0,0,0.5,0,1,.5,-.5
45,0,0,.7071,.7071,.7071,-.7071

SAVING HATCH PATTERNS IN A SEPARATE FILE

When you load a certain hatch pattern, AutoCAD looks for that definition in the ACAD.PAT file; therefore, the hatch pattern definitions must be in that file. However, you can add the new pattern definition to a different file and then copy that file to ACAD.PAT. Be sure to make a copy of the original ACAD.PAT file so that you can copy that file back when needed. Assume the name of the file that contains your custom hatch pattern definitions is CUSTOMH.PAT.

1. Copy ACAD.PAT file to ACADORG.PAT
2. Copy CUSTOMH.PAT to ACAD.PAT

If you want to use the original hatch pattern file, copy the ACADORG.PAT file to ACAD.PAT

CUSTOM HATCH PATTERN FILE

As mentioned earlier, you can add the new hatch pattern definitions to the **ACAD.PAT** file. There is no limit to the number of hatch pattern definitions you can add to this file. However, if you have only one hatch pattern definition, you can define a separate file. It has the following three requirements:

1. The name of the file has to be the same as the hatch pattern name.
2. The file can contain only one hatch pattern definition.
3. The hatch pattern name—and, therefore, the hatch file name—should be unique.
4. If you want to save the hatch pattern on the A drive, then the drive letter (A:) should precede the hatch name. For example, if the hatch name is HATCH3, the header line will be ***A:HATCH3, Triangle Hatch:** and the file name **HATCH3.PAT**.

    *HATCH3,Triangle Hatch:
    90,0,0,0,1,.5,-.5
    0,0,0.5,0,1,.5,-.5
    45,0,0,.7071,.7071,.7071,-.7071

Note
*The hatch lines can be edited after exploding the hatch with the AutoCAD **EXPLODE** command. After exploding, each hatch line becomes a separate object.*

It is good practice not to explode a hatch because it increases the size of the drawing database. For example, if a hatch consists of 100 lines, save it as a single object. However, after you explode the hatch, every line becomes a separate object and you have 99 additional objects in the drawing. Keep the hatch lines in a separate layer to facilitate editing of the hatch lines.

Assign a unique color to hatch lines so that you can control the width of the hatch lines at the time of plotting.

You can also add hatch pattern slides to the AutoCAD slide library.

ADDING HATCH PATTERN SLIDES TO THE AUTOCAD SLIDE LIBRARY

The hatch patterns that you create can be added to AutoCAD's slide library so that the hatch patterns are displayed in the hatch icons. If you enter the **BHATCH** command, AutoCAD displays the **Boundary Hatch** dialogue box on the screen. If you select the **Pattern...** box, AutoCAD displays the hatch icons with the hatch names. Once you have created the new hatch patterns, you can make slides of these hatch patterns and add them to AutoCAD's slide library.

The first step is to make slides of the hatch patterns that you have created (refer to Chapter 2 "Script Files and Slide Shows"). **Make sure that the names of the slide files are same as the names of the hatch pattern.** For example, if the name of the hatch pattern is HATCH1, the name of the slide file must be HATCH1.SLD. When the slide name in the ACAD.SLB file matches with the hatch pattern name in the ACAD.PAT file, the slide is displayed in the **Hatch pattern palette** dialog box.

Before you make any changes in the standard slide library file (ACAD.SLB) supplied with AutoCAD, make a copy of this file. You can find this file in the support subdirectory. The original slide files that were used to create the slide file list are not supplied with the software. However, there are some programs available from third-party software developers that can be used to extract slides from the ACAD.SLB file.

Note

One such shareware program (Slide Manager, SLIDEMGR) that will allow you to extract the slides from AutoCAD's ACAD.SLB file is available from JMIcro, 335 Washington Street, Suite 178, Woburn, MA 01801.

After you make the new slides and extract the slides from the ACAD.SLB file, move the slide files to a separate directory on the hard disk of your computer. For example, you can make a subdirectory, SLIDES, which will contain all the slide files that you are using in your applications. Next, use any text editor to write a slide file list (ACADSLD.SLD). This file lists all the slide filenames that are used in the ACAD.SLB file. Add the slide filenames of your hatch patterns to this file.

The slide file list can be also created by using the following command, if you have DOS version 5.0 or above:

C:\AutoCAD 20000\SLIDES>**DIR *.SLD/B> ACADSLD.SLD**

In this example, assume that the name of the slide file list is **ACADSLD.SLD** (the file extension .SLD is optional) and all slide files are in the SLIDES subdirectory. (To use this command to create a slide file list, all slide files must be in the same directory.)

AutoCAD provides a SLIDELIB utility that constructs a library of the slide files. When you use the SLIDELIB utility, it reads the slide filenames from the slide file list (ACADSLD.SLD) and the file is then written to the slide library file (ACAD.SLB). Use the following command to create the slide library, ACAD.SLB:

C:\AutoCAD 2000\SLIDES> **SLIDELIB ACAD<ACADSLD.SLD**

The new hatch patterns that you create have been added to the slide library ACAD.SLB and these hatch patterns will be displayed in the hatch icons.

Note

The SLIDELIB utility that constructs the slide library file is in the SUPPORT subdirectory. If this subdirectory is not in the path, use the following command to create the slide library file, ACAD.SLB:

C:\AutoCAD 2000\SLIDES>C:\AutoCAD 2000\SUPPORT\SLIDELIB ACAD<ACADSLD.SLD

Review Questions

CREATING LINETYPES

1. The AutoCAD _____ command can be used to create a new linetype.

2. The AutoCAD _____ command can be used to load a linetype.

3. The AutoCAD _____ command can be used to change the linetype scale factor.

4. In AutoCAD, the linetypes are saved in the _____ file.

5. The linetype description should not be more than _____ characters long.

6. A positive number denotes a pen _____ segment.

7. The segment length _____ generates a dot.

8. AutoCAD supports only _____ alignment field specification.

9. A line pattern definition always starts with _____.

10. A header line definition always starts with _____.

CREATING HATCH PATTERNS

11. The **ACAD.PAT** file contains _____ number of hatch pattern definitions.

12. The header line consists of an asterisk, the pattern name, and _____.

13. The first hatch line passes through a point whose coordinates are specified by _____ and _____.

14. The perpendicular distance between the hatch lines in a hatch pattern definition is specified by _____.

15. The displacement of the second hatch line in a hatch pattern definition is specified by _____.

16. The maximum number of dash lengths that can be specified in the line pattern definition of a hatch pattern is _____.

17. The hatch lines in different areas of the drawing will automatically _____ since the hatch lines are generated by offsetting.

18. The hatch angle as defined in the hatch pattern definition can be changed further when you use the AutoCAD _____ command.

19. When you load a hatch pattern, AutoCAD looks for that hatch pattern in the _____ file.

20. The hatch lines can be edited after _____ the hatch by using the AutoCAD _____ command.

Exercises

Creating Linetypes

Exercise 1   *General*

Using the AutoCAD **LINETYPE** command, create a new linetype "DASH3DASH" with the following specifications:

    Length of the first dash 0.75
    Blank space 0.125
    Dash length 0.25
    Blank space 0.125
    Dash length 0.25
    Blank space 0.125
    Dash length 0.25
    Blank space 0.125

Exercise 2   *General*

Use a text editor to create a new file, **NEWLT2.LIN**, and a new linetype, DASH2DASH, with the following specifications:

    Length of the first dash 0.5
    Blank space 0.1
    Dash length 0.2
    Blank space 0.1
    Dash length 0.2
    Blank space 0.1

Exercise 3   *General*

Using AutoCAD's **LINETYPE** command, create a linetype DASH3DOT with the following specifications:

Chapter 3

Length of the first dash 0.75
Blank space 0.25
Dot
Blank space 0.25
Dot
Blank space 0.25
Dot
Blank space 0.25

Exercise 4   *General*

Using AutoCAD's **EDIT** command, create a new file, NEWLINET.LIN, and define a linetype
VARDASHX with the following specifications:

Length of first dash 1.5
Blank space 0.25
Length of second dash 0.75
Blank space 0.25
Dot
Blank space 0.25
Length of third dash 0.75

Exercise 5   *General*

a. Write the definition of a string complex linetype (Hot water line) as shown in Figure 3-37(a).

b. Write the definition of a string complex linetype (Gas line) as shown in Figure 3-37(b).

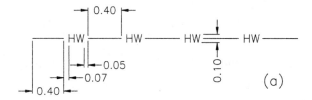

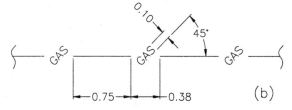

Figure 3-37 Specifications for a string complex linetype

Exercise 6   *General*

Write the shape file for the shape shown in Figure 3-38(a). Compile the shape and use it in defining the shape complex linetype so that you can draw a fence line as shown in Figure 3-38(b).

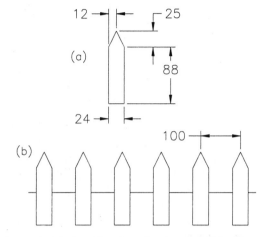

Figure 3-38 *Specifications for shape complex linetype*

Creating Hatch Patterns

Exercise 7   *General*

Determine the hatch specifications and write a hatch pattern definition for the hatch pattern in Figure 3-39.

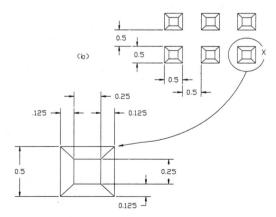

Figure 3-39 *Square hatch pattern*

Exercise 8   *General*

Determine the hatch pattern specifications and write a hatch pattern definition for the hatch pattern in Figure 3-40.

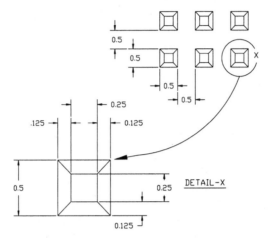

Figure 3-40 Hatch pattern for Exercise 8

Exercise 9   *General*

Determine the hatch specifications and write a hatch pattern definition for the hatch pattern as shown in Figure 3-41. Use this hatch to hatch a circle or rectangle.

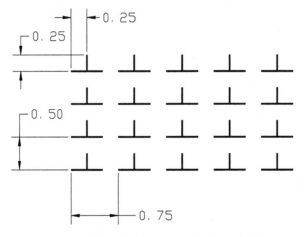

Figure 3-41 Hatch pattern for Exercise 9

Exercise 10 *General*

Determine the hatch specifications and write a hatch pattern definition for the hatch pattern as shown in Figure 3-42. Use this hatch to hatch a circle or rectangle.

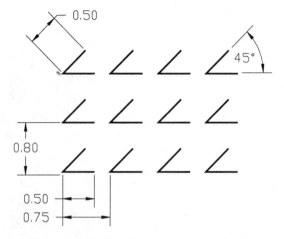

Figure 3-42 *Hatch pattern for Exercise 10*

Chapter 4

Customizing the
ACAD.PGP File

Learning Objectives

After completing this chapter, you will be able to:
- *Customize the **ACAD.PGP** file.*
- *Edit different sections of the **ACAD.PGP** file.*
- *Abbreviate commands by defining command aliases.*
- *Use the **REINIT** command to reinitialize the PGP file.*

WHAT IS THE ACAD.PGP FILE?

AutoCAD software comes with the program parameters file **ACAD.PGP,** which defines aliases for the operating system commands and some of the AutoCAD commands. When you install AutoCAD, this file is automatically copied on the **AutoCAD 2000\SUPPORT** subdirectory of the hard drive. The **ACAD.PGP** file lets you access the operating system commands from the drawing editor. For example, if you want to delete a file, all you need to do is enter DEL at the Command prompt **(Command: DEL)**, and then AutoCAD will prompt you to enter the name of the file you want to delete.

The file also contains command aliases of some frequently used AutoCAD commands. For example, the command alias for the **LINE** command is L. If you enter L at the Command prompt **(Command: L)**, AutoCAD will treat it as the **LINE** command. The **ACAD.PGP** file also contains comment lines that give you some information about different sections of the file.

The following file is a partial listing of the standard **ACAD.PGP** file. Some of the lines have been deleted to make the file shorter.

; AutoCAD Program Parameters File For AutoCAD 2000
; External Command and Command Alias Definitions
; Copyright (C) 1997-1999 by Autodesk, Inc.

; Each time you open a new or existing drawing, AutoCAD searches
; the support path and reads the first acad.pgp file that it finds.
; -- External Commands --
; While AutoCAD is running, you can invoke other programs or utilities
; such Windows system commands, utilities, and applications.
; You define external commands by specifying a command name to be used
; from the AutoCAD command prompt and an executable command string
; that is passed to the operating system.
; -- Command Aliases --
; You can abbreviate frequently used AutoCAD commands by defining
; aliases for them in the command alias section of acad.pgp.
; You can create a command alias for any AutoCAD command,
; device driver command, or external command.
; Recommendation: back up this file before editing it.
; External command format:
; <Command name>,[<DOS request>],<Bit flag>,[*]<Prompt>,
; The bits of the bit flag have the following meanings:
; Bit 1: if set, don't wait for the application to finish
; Bit 2: if set, run the application minimized
; Bit 4: if set, run the application "hidden"
; Bit 8: if set, put the argument string in quotes
; Fill the "bit flag" field with the sum of the desired bits.
; Bits 2 and 4 are mutually exclusive; if both are specified, only
; the 2 bit is used. The most useful values are likely to be 0
; (start the application and wait for it to finish), 1 (start the
; application and don't wait), 3 (minimize and don't wait), and 5
; (hide and don't wait). Values of 2 and 4 should normally be avoided,
; as they make AutoCAD unavailable until the application has completed.
; Bit 8 allows commands like DEL to work properly with filenames that
; have spaces such as "long filename.dwg". Note that this will interfere
; with passing space delimited lists of file names to these same commands.
; If you prefer multiplefile support to using long file names, turn off
; the "8" bit in those commands.
; Examples of external commands for command windows

```
CATALOG,   DIR /W,       8,File specification: ,
DEL,       DEL,          8,File to delete: ,
DIR,       DIR,          8,File specification: ,
EDIT,      START EDIT,   9,File to edit: ,
SH,        ,            1,*OS Command: ,
SHELL,     ,            1,*OS Command: ,
START,     START,        1,*Application to start: ,
TYPE,      TYPE,         8,File to list: ,
```

External
commands
area

; Examples of external commands for Windows
; See also the (STARTAPP) AutoLISP function for an alternative method.

EXPLORER, START EXPLORER, 1,,
NOTEPAD,   START NOTEPAD, 1,*File to edit: ,
PBRUSH,   START PBRUSH, 1,,

; Command alias format:
;   <Alias>,*<Full command name>

; The following are guidelines for creating new command aliases.
; 1. An alias should reduce a command by at least two characters.
;   Commands with a control key equivalent, status bar button,
;   or function key do not require a command alias.
;   Examples: Control N, O, P, and S for New, Open, Print, Save.
; 2. Try the first character of the command, then try the first two,
;   then the first three.
; 3. Once an alias is defined, add suffixes for related aliases:
;   Examples: R for Redraw, RA for Redrawall, L for Line, LT for
;   Linetype.
; 4. Use a hyphen to differentiate between command line and dialog
;   box commands.
;   Example: B for Block, -B for -Block.
;
; Exceptions to the rules include AA for Area, T for Mtext, X for Explode.

; -- Sample aliases for AutoCAD commands --
; These examples include most frequently used commands.

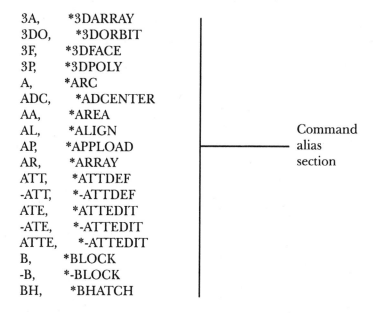

3A,	*3DARRAY	
3DO,	*3DORBIT	
3F,	*3DFACE	
3P,	*3DPOLY	
A,	*ARC	
ADC,	*ADCENTER	
AA,	*AREA	
AL,	*ALIGN	
AP,	*APPLOAD	Command
AR,	*ARRAY	alias
ATT,	*ATTDEF	section
-ATT,	*-ATTDEF	
ATE,	*ATTEDIT	
-ATE,	*-ATTEDIT	
ATTE,	*-ATTEDIT	
B,	*BLOCK	
-B,	*-BLOCK	
BH,	*BHATCH	

Chapter 4

```
BO,      *BOUNDARY
-BO,     *-BOUNDARY
BR,      *BREAK
C,       *CIRCLE
CH,      *PROPERTIES
-CH,     *CHANGE
CHA,     *CHAMFER
COL,     *COLOR
COLOUR,  *COLOR
CO,      *COPY
D,       *DIMSTYLE
DAL,     *DIMALIGNED
DAN,     *DIMANGULAR
DBA,     *DIMBASELINE
DBC,     *DBCONNECT
DCE,     *DIMCENTER
DCO,     *DIMCONTINUE
DDI,     *DIMDIAMETER
DED,     *DIMEDIT
DI,      *DIST
DIV,     *DIVIDE



L,       *LINE
LA,      *LAYER
-LA,     *-LAYER
LE,      *QLEADER
LEN,     *LENGTHEN
LI,      *LIST
LINEWEIGHT, *LWEIGHT
LO,      *-LAYOUT
LS,      *LIST
LT,      *LINETYPE
-LT,     *-LINETYPE
LTYPE,   *LINETYPE
-LTYPE,  *-LINETYPE
LTS,     *LTSCALE
LW,      *LWEIGHT
M,       *MOVE
MA,      *MATCHPROP
ME,      *MEASURE
MI,      *MIRROR
ML,      *MLINE
MO,      *PROPERTIES
MS,      *MSPACE
MT,      *MTEXT
```

```
MV,       *MVIEW
O,        *OFFSET
OP,       *OPTIONS
ORBIT,    *3DORBIT
OS,       *OSNAP
-OS,      *-OSNAP
P,        *PAN
-P,       *-PAN
PA,       *PASTESPEC
PARTIALOPEN, *-PARTIALOPEN
PE,       *PEDIT
PL,       *PLINE
PO,       *POINT
POL,      *POLYGON
PR,       *OPTIONS
PRCLOSE,  *PROPERTIESCLOSE
PROPS,    *PROPERTIES
PRE,      *PREVIEW
PRINT,    *PLOT
PS,       *PSPACE
PU,       *PURGE
R,        *REDRAW
RA,       *REDRAWALL
RE,       *REGEN
REA,      *REGENALL
REC,      *RECTANGLE
REG,      *REGION


SP,       *SPELL
SPL,      *SPLINE
SPE,      *SPLINEDIT
ST,       *STYLE
SU,       *SUBTRACT
T,        *MTEXT
-T,       *-MTEXT
TA,       *TABLET
TH,       *THICKNESS
TI,       *TILEMODE
TO,       *TOOLBAR
TOL,      *TOLERANCE
TOR,      *TORUS
TR,       *TRIM
UC,       *DDUCS
UCP,      *DDUCSP
```

; The following are alternative aliases and aliases as supplied
; in AutoCAD Release 13.

```
AV,       *DSVIEWER
CP,       *COPY
DIMALI,   *DIMALIGNED
DIMANG,   *DIMANGULAR
DIMBASE,  *DIMBASELINE
DIMCONT,  *DIMCONTINUE
DIMDIA,   *DIMDIAMETER
DIMED,    *DIMEDIT
DIMTED,   *DIMTEDIT
DIMLIN,   *DIMLINEAR
DIMORD,   *DIMORDINATE
DIMRAD,   *DIMRADIUS
DIMSTY,   *DIMSTYLE
DIMOVER,  *DIMOVERRIDE
LEAD,     *LEADER
TM,       *TILEMODE
```

; Aliases for Hyperlink/URL Release 14 compatibility
```
SAVEURL, *SAVE
OPENURL, *OPEN
INSERTURL, *INSERT
```

; Aliases for commands discontinued in AutoCAD 2000:
```
AAD,      *DBCONNECT
AEX,      *DBCONNECT
ALI,      *DBCONNECT
ASQ,      *DBCONNECT
ARO,      *DBCONNECT
ASE,      *DBCONNECT
DDATTDEF, *ATTDEF
DDATTEXT, *ATTEXT
DDCHPROP, *PROPERTIES
DDCOLOR,  *COLOR
DDLMODES, *LAYER
DDLTYPE,  *LINETYPE
DDMODIFY, *PROPERTIES
DDOSNAP,  *OSNAP
DDUCS,    *UCS
```

SECTIONS OF THE ACAD.PGP FILE

The contents of the AutoCAD program parameters file (**ACAD.PGP**) can be categorized into three sections. These sections merely classify the information that is defined in the **ACAD.PGP** file. They do not have to appear in any definite order in the file, and they have no section headings. For example, the comment lines can be entered anywhere in the file; the same is true with external commands and AutoCAD command aliases. The **ACAD.PGP** file can be divided into these three sections: **comments, external commands**, and **command aliases**.

Comments

The comments of ACAD.PGP file can contain any number of comment lines and can occur anywhere in the file. Every comment line must start with a semicolon (;) (This is a comment line). Any line that is preceded by a semicolon is ignored by AutoCAD. You should use the comment line to give some relevant information about the file that will help other AutoCAD users to understand, edit, or update the file.

External Command

In the external command section you can define any valid external command that is supported by your system. The information must be entered in the following format:

<Command name>, [OS Command name],<Bit flag>, [*]<Command prompt>,

Command Name. This is the name you want to use to activate the external command from the AutoCAD drawing editor. For example, you can use **goword** as a command name to load the word program (**Command: goword**). The command name must not be an AutoCAD command name or an AutoCAD system variable name. If the name is an AutoCAD command name, the command name in the **PGP** file will be ignored. Also, if the name is an AutoCAD system variable name, the system variable will be ignored. You should use the command names that reflect the expected result of the external commands. (For example, **hello** is not a good command name for a directory file.) The command names can be uppercase or lowercase.

OS Command Name. The OS Command name is the name of a valid system command that is supported by your operating system. For example, in DOS the command to delete files is DEL; therefore, the OS Command name used in the **ACAD.PGP** file must be DEL. The following is a list of the types of commands that can be used in the PGP file:

OS Commands (del, dir, type, copy, rename, edlin, etc.)
Commands for starting a word processor, or text editors (word, shell, etc.)
Name of the user-defined programs and batch files

Bit flag. This field must contain a number, preferably 8 or 1. The following are the bit flag values and their meaning:

Bit flag set to	Meaning
1	Do not wait for the application to finish
2	Runs the application minimized
4	Runs the application hidden
8	Puts the argument string in quotes

Chapter 4

Command Prompt. The command prompt field of the command line contains the prompt you want to display on the screen. It is an optional field that must be replaced by a comma if there is no prompt. If the operating system (OS) command that you want to use contains spaces, the prompt must be preceded by an asterisk (*). For example, the DOS command **EDIT NEW.PGP** contains a space between EDIT and NEW; therefore, the prompt used in this command line must be preceded by an asterisk. The command can be terminated by pressing ENTER. If the **OS** command consists of a single word (DIR, DEL, TYPE), the preceding asterisk must be omitted. In this case you can terminate the command by pressing the SPACEBAR or ENTER.

Command Aliases

It is time consuming to enter AutoCAD commands at the keyboard because it requires typing the complete command name before pressing ENTER. AutoCAD provides a facility that can be used to abbreviate the commands by defining aliases for the AutoCAD commands. This is made possible by the AutoCAD program parameters file (**ACAD.PGP** file). Each command alias line consists of two fields (**L, *LINE**). The first field (**L**) defines the alias of the command; the second field (***LINE**) consists of the AutoCAD command. The AutoCAD command must be preceded by an asterisk for AutoCAD to recognize the command line as a command alias. The two fields must be separated by a comma. The blank lines and the spaces between the two fields are ignored. In addition to AutoCAD commands, you can also use aliases for AutoLISP command names, provided the programs that contain the definition of these commands are loaded.

Example 1

Add the following external commands and AutoCAD command aliases to the AutoCAD program parameters file (**ACAD.PGP**).

External Commands

Abbreviation	Command Description
GOWORD	This command loads the word processor (winword) program from the C:\Program Files\Winword.
RN	This command executes the rename command of DOS.
COP	This command executes the copy command of DOS.

Command Aliases Section

Abbreviation	Command	Abbreviation	Command
EL	Ellipse	T	Trim
CO	Copy	CH	Chamfer
O	Offset	ST	Stretch
S	Scale	MI	Mirror

The **ACAD.PGP** file is an ASCII text file. To edit this file you can use the AutoCAD EDIT command (provided the EDIT command is defined in the **ACAD.PGP** file), or any text editor (Notepad or Wordpad). The following is a partial listing of the **ACAD.PGP** file after insertion of the lines for the command aliases of Example 5. **The line numbers are not a part of the file; they are shown here for reference only.** The lines that have been added to the file are highlighted in bold.

DEL,DEL,	8,File to delete: ,	1
DIR,DIR,	8,File specification ,	2
EDIT, START EDIT,	8,File to edit: ,	3
SH,,	1,*OS Command: ,	4
SHELL,,	1,*OS Command: ,	5
START,START,	1,Application to start: ,	6
		7
GOWORD, START WINWORD,1,,		8
RN, RENAME,8,File to rename:,		9
COP,COPY,8,File to copy:,		10
DIMLIN	*DIMLINEAR	11
DIMORD,	*DIMORDINATE	12
DIMRAD,	*DIMRADIUS	13
DIMSTY,	*DIMSTYLE	14
DIMOVER,	*DIMOVERRIDE	15
LEAD,	*LEADER	16
TM,	*TILEMODE	17
EL,	***ELLIPSE**	**18**
CO,	***COPY**	**19**
O,	***OFFSET**	**20**
S,	***SCALE**	**21**
MI,	***MIRROR**	**22**
ST,	***STRETCH**	**23**

Explanation

Line 8
GOWORD,   START WINWORD,1,,
In line 8, **goword** loads the word processor program (**winword**). The **winword.exe** program is located in the winword directory under Program Files.

Lines 9 and 10
RN, RENAME,8,File to rename:,
COP,COPY,8,File to copy:,
Line 9 defines the alias for the DOS command **RENAME,** and the next line defines the alias for the DOS command **COPY**. The 8 is a bit flag, and the command prompts **File to rename and File to copy** are automatically displayed to let you know the format and the type of information that is expected.

Lines 18 and 19
EL, *ELLIPSE
CO, *COPY
Line 18 defines the alias **(EL)** for the AutoCAD command **ELLIPSE,** and the next line defines the alias **(CO)** for the **COPY** command. The AutoCAD commands must be preceded by an asterisk. You can put any number of spaces between the alias abbreviation and the AutoCAD command.

Chapter 4

REINIT COMMAND

When you make any changes in the **ACAD.PGP** file, there are two ways to reinitialize the **ACAD.PGP** file. One is to quit AutoCAD and then reenter it. When you start AutoCAD, the **ACAD.PGP** file is automatically loaded.

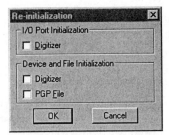

Figure 4-1 **Re-initialization** *dialog box*

You can also reinitialize the **ACAD.PGP** file by using the AutoCAD **REINIT** command. The **REINIT** command lets you reinitialize the I/O ports, digitizer, display, and AutoCAD program parameters file, **ACAD.PGP**. When you enter the **REINIT** command, AutoCAD will display a dialog box (Figure 4-1). To reinitialize the **ACAD.PGP** file, select the corresponding toggle box, and then choose **OK**. AutoCAD will reinitialize the program parameters file **(ACAD.PGP)**, and then you can use the command aliases defined in the file.

Review Questions

1. The comment section can contain any number of lines. (T/F)

2. AutoCAD ignores any line that is preceded by a semicolon. (T/F)

3. The command alias must not be an AutoCAD command. (T/F)

4. The bit flag filed must contain 8. (T/F)

5. In the command alias section, the command alias must be preceded by a semicolon. (T/F)

6. You cannot use aliases for AutoLISP commands. (T/F)

7. The ACAD.PGP file does not come with AutoCAD software. (T/F)

8. The ACAD.PGP file is an ASCII file. (T/F)

Exercises

Exercise 1 *General*

Add the following external commands and AutoCAD command aliases to the AutoCAD program parameters file (**ACAD.PGP**).

External Command Section

Abbreviation	Command Description
MYWORDPAD	This command loads the WORDPAD program that resides in the **Program Files\Accessories** directory.
MYEXCEL	This command loads the EXCEL program that resides in the **Program Fies\Microsoft Office** directory.
CD	This command executes the CHKDSK command of DOS.
FORMAT	This command executes the FORMAT command of DOS.

Command Aliases Section

Abbreviation	Command	Abbreviation	Command
BL	**BLOCK**	LT	**LTSCALE**
INS	**INSERT**	EX	**EXPLODE**
DIS	**DISTANCE**	G	**GRID**
T	**TIME**		

Chapter **5**

Pull-down, Shortcut, and Partial Menus and Customizing Toolbars

Learning Objectives

After completing this chapter, you will be able to:
- *Write pull-down menus.*
- *Load menus.*
- *Write cascading submenus in pull-down menus.*
- *Write cursor menus.*
- *Swap pull-down menus.*
- *Write partial menus.*
- *Define accelerator keys.*
- *Write toolbar definitions.*
- *Write menus to access online help.*
- *Customize toolbars.*

AUTOCAD MENU

The AutoCAD menu provides a powerful tool to customize AutoCAD. The AutoCAD software package comes with a standard menu file named **ACAD.MNU**. When you start AutoCAD, the menu file **ACAD.MNU** is automatically loaded. The AutoCAD menu file contains AutoCAD commands, separated under different headings for easy identification. For example, all draw commands are under **Draw** and all editing commands are under **Modify** menu. The headings are named and arranged to make it easier for you to locate and access the commands. However, there are some commands that you may never use. Also, some users might like to regroup and rearrange the commands so that it is easier to access those most frequently used.

AutoCAD lets the user eliminate rarely used commands from the menu file and define new ones. This is made possible by editing the existing **ACAD.MNU** file or writing a new menu file. There is no limit to the number of files you can write. You can have a separate menu file for each application. For example, you can have separate menu files for mechanical, electrical, and architectural drawings. You can load these menu files any time by using the AutoCAD **MENU** command. The menu files are text files with the extension **.MNU**. These files can be written by using any text editor like Wordpad or Notepad. The menu file can be divided into ten sections, each section identified by a.section label. AutoCAD uses the following labels to identify different sections of the AutoCAD menu file:

***SCREEN	
***TABLET(n)	**n** is from 1 to 4
***IMAGE	
***POP(n)	**n** is from 1 to 499 (For shortcut menus n=0 and 500 to 999)
***BUTTONS(n)	**n** is from 1 to 4
***AUX(n)	**n** is from 1 to 4
***MENUGROUP	
***TOOLBARS	
***HELPSTRING	
***ACCELERATORS	

The tablet menu can have up to four different sections. The **POP** menu (pull-down and cursor menu) can have up to 16 sections, and auxiliary and buttons menus can have up to four sections.

Tablet Menus	**Pull-Down and Cursor Menus**
***TABLET1	***POP0
***TABLET2	***POP1
***TABLET3	***POP2
***TABLET4	***POP3
	***POP4
BUTTONS Menus	***POP5
***BUTTONS1	***POP6
***BUTTONS2	***POP7
***BUTTONS3	***POP8
***BUTTONS4	
Auxiliary Menus	
***AUX1	
***AUX2	***POP498
***AUX3	***POP499
***AUX4	

STANDARD MENUS

The menu is a part of the AutoCAD standard menu file, **ACAD.MNU**. The **ACAD.MNU** file is automatically loaded when you start AutoCAD, provided the standard configuration of AutoCAD has not been changed.

The menus can be selected by moving the crosshairs to the top of the screen, into the menu bar area. If you move the pointing device sideways, different menu bar titles are highlighted and you can select the desired item by pressing the pick button on your pointing device. Once the item is selected, the corresponding menu is displayed directly under the title (Figure 5-1). The menu can have 499 sections, named as POP1, POP2, POP3, . . ., POP499.

WRITING A MENU

Before you write a menu, you need to design a menu and to know the exact sequence of AutoCAD commands and the prompts associated with particular commands. To design a menu, you should select and arrange the commands in a way that provides easy access to the most frequently used commands. A careful design will save a lot of time in the long run. Therefore, consider several possible designs with different command combinations, and then

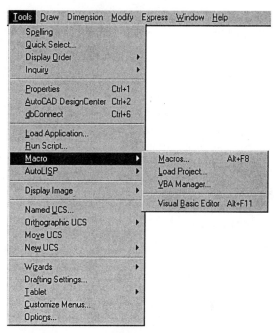

Figure 5-1 *Pull-down and cascading menus*

select the one best suited for the job. Suggestions from other CAD operators can prove very valuable.

The second important thing in developing a menu is to know the exact sequence of the commands and the prompts associated with each command. To better determine the prompt entries required in a command, you should enter all the commands and the prompt entries at the keyboard. The following is a description of some of the commands and the prompt entries required for Example 1.

    LINE Command
    Command: **LINE**

Notice the command and input sequence:

    LINE
    <RETURN>
    CIRCLE (C,R) Command
    Command: **CIRCLE**
    Specify center point for circle or [3P/2P/Ttr (tan tan radius)]: *Specify center point.*
    Specify radius of circle or [Diameter]: *Enter radius.*

Notice the command and input sequence:

    CIRCLE
    <RETURN>

Chapter 5

Center point
<RETURN>
Radius
<RETURN>

CIRCLE (C,D) Command
Command: **CIRCLE**
Specify center point for circle or [3P/2P/Ttr (tan tan radius)]: *Specify center point.*
Specify radius of circle or [Diameter]: **D**
Specify diameter of circle: *Enter diameter.*

Notice the command and input sequence:

CIRCLE
<RETURN>
Center Point
<RETURN>
D
<RETURN>
Diameter
<RETURN>

CIRCLE (2P) Command
Command: **CIRCLE**
Specify center point for circle or [3P/2P/Ttr (tan tan radius)]: **2P**
Specify first end point of circle's diameter: *Specify first point.*
Specify second end point of circle's diameter: *Specify second point.*

Notice the command and input sequence:

CIRCLE
<RETURN>
2P
<RETURN>
Select first point on diameter
<RETURN>
Select second point on diameter
<RETURN>

ERASE Command
Command: **ERASE**

Notice the command and input sequence:

ERASE
<RETURN>

MOVE Command
Command: **MOVE**

Notice the command and prompt entry sequence:

MOVE
<RETURN>

The difference between the Center-Radius and Center-Diameter options of the **CIRCLE** command is that in the first one the RADIUS is the default, whereas in the second one you need to enter D to use the diameter option. This difference, although minor, is very important when writing a menu file. Similarly, the 2P (two-point) option of the **CIRCLE** command is different from the other two options. Therefore, it is important to know both the correct sequence of the AutoCAD commands and the entries made in response to the prompts associated with those commands.

You can use any text editor (like Notepad) to write the file. You can also use AutoCAD's **EDIT** command to write the menu file. If you use the **EDIT** command, AutoCAD will prompt you to enter the file name you want to edit. The file name can be up to eight characters long, and the file extension must be **.MNU**. If the file name exists, it will be automatically loaded; otherwise a new file will be created. To understand the process of developing a pull-down menu, consider the following example.

Note

*If AutoCAD's **EDIT** command does not work, check the **ACAD.PGP** file and make sure that this command is defined in the file.*

Example 1

Write a menu for the following AutoCAD commands:

LINE	ERASE	REDRAW	SAVE
PLINE	MOVE	REGEN	QUIT
CIRCLE C,R	COPY	ZOOM ALL	PLOT
CIRCLE C,D	STRETCH	ZOOM WIN	
CIRCLE 2P	EXTEND	ZOOM PRE	
CIRCLE 3P	OFFSET		

The first step in writing any menu is to design the menu so that the commands are arranged in the desired configuration. Figure 5-2 shows one of the possible designs of this menu. This menu has four different groups of commands; therefore, it will have four sections: POP1, POP2, POP3, and POP4, and each section will have a section label. The following file is a listing of the pull-down menu file for Example 1. **The line numbers are not a part of the file; they are shown here for reference only.**

```
***POP1                                                          1
[DRAW]                                                           2
```

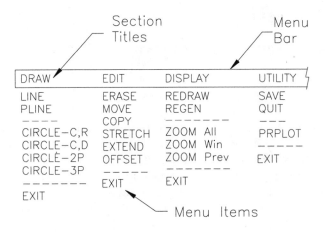

Figure 5-2 *Design of menu*

[LINE]*^C^CLINE	3
[PLINE]^C^CPLINE	4
[—]	5
[CIR-C,R]^C^CCIRCLE	6
[CIR-C,D]^C^CCIRCLE \D	7
[CIR-2P]^C^CCIRCLE 2P	8
[CIR-3P]^C^CCIRCLE 3P	9
[—]	10
[Exit]^C	11
***POP2	12
[EDIT]	13
[ERASE]*^C^CERASE	14
[MOVE]^C^CMOVE	15
[COPY]^C^CCOPY	16
[STRETCH]^C^CSTRETCH;C	17
[OFFSET]^C^COFFSET	18
[EXTEND]^C^CEXTEND	19
[—]	20
[Exit]^C	21
***POP3	22
[DISPLAY]	23
[REDRAW]'REDRAW	24
[REGEN]^C^CREGEN	25
[—]	26
[ZOOM-All]^C^C'ZOOM A	27
[ZOOM-Window]'ZOOM W	28
[ZOOM-Prev]'ZOOM PREV	29
[—]	30
[~Exit]^C	31

```
***POP4                                                    32
[UTILITY]                                                  33
[SAVE]^C^CSAVE;                                            34
[QUIT]^C^CQUIT                                             35
[——]                                                       36
[PLOT]^C^CPLOT                                             37
[—]                                                        38
[Exit]^C                                                   39
```

Explanation

Line 1
***POP1

POP1 is the section label for the first pull-down menu. All section labels in the AutoCAD menu begin with three asterisks (***), followed by the section label name, such as POP1.

Line 2
[DRAW]

In this menu item **DRAW** is the menu bar title displayed when the cursor is moved in the menu bar area. The title names should be chosen so you can identify the type of commands you expect in that particular pull-down menu. In this example, all the draw commands are under the title **DRAW** (Figure 5-3), all edit commands are under **EDIT**, and so on for other groups of items. The menu bar title can be of any length. However, it is recommended to keep them short to accommodate other menu items. Some of the display devices provide a maximum of 80 characters. To have 16 sections in a single row in the menu bar, the length of each title should not exceed five characters. If the length of the menu bar items

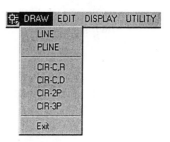

Figure 5-3 Draw menu

exceeds 80 characters, AutoCAD will wrap the items that cannot be accommodated in 80 characters space, and display them in the next line. This will result in two menu item lines in the menu bar.

If the first line in a menu section is blank, the title of that section is not displayed in the menu bar area. Since the menu bar title is not displayed, you **cannot** access that menu. This allows you to turn off the menu section. For example, if you replace [**DRAW**] with a blank line, the **DRAW** section (POP1) of the menu will be disabled; the second section (POP2) will be displayed in its place.

Example

```
    ***POP1                       Section label
                                  Blank line (turns off POP1)
    [LINE:]^CLINE                 Menu item
    [PLINE:]^CPLINE
    [CIRCLE:]^CCIRCLE
```

The menu bar titles are left-justified. If the first title is not displayed, the rest of the menu titles will be shifted to the left. In Example 1, if the **DRAW** title is not displayed in the menu bar area, then the **EDIT**, **DISPLAY**, and **FILE** sections of the menu will move to the left.

Line 3
*^C^CLINE
In this menu item, the command definition starts with an asterisk (*). This feature allows the command to be repeated automatically until it is canceled by pressing ESCAPE, entering CTRL C, or by selecting another menu command. ^C^C cancels the existing command twice; **LINE** is an AutoCAD command that generates lines.

*^C^CLINE
> Where   ***** ------------------- Repeats the menu item (command)
>   **^C^C** ---------- Cancels existing command twice
>   **LINE** ------------- AutoCAD's **LINE** command

Line 5
[--]
To separate two groups of commands in any section, you can use a menu item that consists of two or more hyphens (--). This line automatically expands to fill the **entire width** of the menu. You can use a blank line in a menu. If any section of a menu (**POP section) has a blank line, it is ignored.

Line 11
[Exit]^C
In this menu item, ^C command definition has been used to cancel the menu. This item provides you with one more option for canceling the menu. This is especially useful for new AutoCAD users who are not familiar with all AutoCAD features. The menu can also be canceled by any of the following actions:

1. Selecting a point.
2. Selecting an item in the screen menu area.
3. Selecting or typing another command.
4. Pressing ESC at the keyboard.
5. Selecting any menu title in the menu bar.

Line 28
[ZOOM-Window]'ZOOM W
In this menu item the single quote (') preceding the **ZOOM** command makes the **ZOOM** Window command transparent. When a command is transparent, the existing command is not canceled. After the **ZOOM** Window command (Figure 5-4), AutoCAD will automatically resume the current operation.

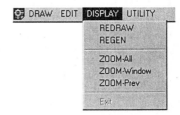

Figure 5-4 Display menu

[ZOOM-Window]'ZOOM W
> Where   **W** ----------------- Window option
>   **ZOOM** ----------- AutoCAD **ZOOM** command
>   **'** ------------------- Single quote makes **ZOOM** transparent

Line 31
[~Exit] ^ C
This menu item is similar to the menu item in line 11, except for the tilde (~). Since this menu item has a tilde (~), the menu item is not available (displayed grayed out), and if you select this item it will not cancel the menu. You can use this feature to disable a menu item or to indicate that the item is not a valid selection. If there is an instruction associated with the item, the instruction will not be executed when you select the item. For example, [~OSNAPS] ^ C ^ C$S = OSNAPS will not load the OSNAPS submenu on the screen.

Line 34
[SAVE] ^ C ^ CSAVE;
In this menu item, the semicolon (;) that follows the **SAVE** command enters RETURN. The semicolon is not required; the command will also work without a semicolon.

> **[SAVE] ^ C ^ CSAVE;**
> Where **SAVE** ------------- AutoCAD **SAVE** command
> ; ------------------ Semicolon enters RETURN

Line 36
[----]
This menu item has four hyphens. The line will extend to fill the width of the menu. If there is only one hyphen [-], AutoCAD gives a syntex error when you load the menu.

Note
For all menus, the menu items are displayed directly beneath the menu title and are left-justified. If any menu [for example, the rightmost menu (POP16)] does not have enough space to display the entire menu item, the menu will expand to the left to accommodate the entire length of the longest menu item.

You can use // (two forward slashes) for comment lines. AutoCAD ignores lines that start with //.

From this example it is clear that every statement in the menu is based on the AutoCAD commands and the information that is needed to complete these commands. This forms the basis for creating a menu file and should be given consideration. Following is a summary of the AutoCAD commands used in Example 1 and their equivalents in the menu file.

<u>AutoCAD Commands</u>	<u>Menu File</u>
Command: **LINE**	[LINE] ^ C ^ CLINE
Command: **CIRCLE** Specify center point for circle or [3P/2P/Ttr (tan tan radius)]: Specify radius of circle or [Diameter]:	[CIR-C,R] ^ C ^ CCIRCLE
Command: **CIRCLE** Specify center point for circle or [3P/2P/Ttr (tan tan radius)]: Specify radius of circle or [Diameter]: D	

Specify diameter of circle: **[CIR-C,D] ^ C ^ CCIRCLE;\D**

Command: **CIRCLE**
Specify center point for circle or [3P/2P/Ttr (tan tan radius)]: **2P**
Specify first end point of circle's diameter:
Specify second end point of circle's diameter: **[CIR- 2P] ^ C ^ CCIRCLE;2P**

Command: **ERASE**   **[ERASE   ] ^ C ^ CERASE**

Command: **MOVE**   **[MOVE   ] ^ C ^ CMOVE**

LOADING MENUS

AutoCAD automatically loads the **ACAD.MNU** file when you get into the AutoCAD drawing editor. However, you can also load a different menu file by using the AutoCAD **MENU** command.

Command: **MENU**

When you enter the **MENU** command, AutoCAD displays the **Select Menu File** dialog box (Figure 5-5) on the screen. Select the menu file that you want to load and then choose the **Open** button. You can also load the menu file from the command line.

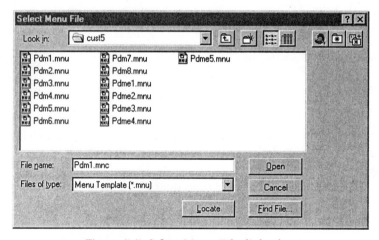

Figure 5-5 Select Menu File dialog box

Command: **FILEDIA**
Enter new value for FILEDIA <1>: **0**
Command: **MENU**

Enter menu file name or [. (for none)] <ACAD>: PDM1
          Where   **PDM1** ------------ Name of menu file
                  <ACAD> ------- Default menu file

After you enter the **MENU** command, AutoCAD will prompt for the file name. Enter the name of the menu file without the file extension (**.MNU**), since AutoCAD assumes the extension .MNU. AutoCAD will automatically compile the menu file into **MNC** and **MNR** files. When you load a menu file in windows, AutoCAD creates the following files:

.mnc and **.mnr** files   When you load a menu file (.mnu), AutoCAD compiles the menu file and creates **.mnc** and **.mnr** files. The .mnc file is a compiled menu file. The **.mnr** file contains the bitmaps used by the menu.

.mns file   When you load the menu file, AutoCAD also creates an **.mns** file. This is an ASCII file that is same as the **.mnu** file when you initially load the menu file. Each time you make a change in the contents of the file, AutoCAD changes the **.mns** file.

Note

*After you load the new menu, you cannot use the screen menu, buttons menu, or digitizer because the original menu, **ACAD.MNU**, is not present and the new menu does not contain these menu areas.*

*To activate the original menu again, load the menu file by using the **MENU** command:*

*Command: **MENU***
*Enter menu file name or [. (for none)] <SM1>: **ACAD***

If you need to use input from a keyboard or a pointing device, use the backslash (\). The system will pause for you to enter data.

There should be no space after the backslash (\).

The menu items, menu labels, and command definition can be uppercase, lowercase, or mixed.

You can introduce spaces between the menu items to improve the readability of the menu file. The blank lines are ignored and are not displayed on the screen.

If there are more items in the menu than the number of spaces available, the excess items are not displayed on the screen. For example, if the display device limits the number of items to 21, items in excess of 21 will not be displayed on the screen and are therefore inaccessible.

If you are using a high-resolution graphics board, you can increase the number of lines that can be displayed on the screen. On some devices this is 80 lines.

RESTRICTIONS

The menus are easy to use and provide a quick access to frequently used AutoCAD commands. However, the menu bar and the menus are disabled during the following commands:

DTEXT Command
After you assign the text height and the rotation angle to a **DTEXT** command, the menu is automatically disabled.

SKETCH Command

The menus are disabled after you set the record increment in the **SKETCH** command.

VPOINT Command

The menus are disabled while the axis tripod and compass are displayed on the screen.

ZOOM and DVIEW Commands

The menus are disabled when the dynamic zoom or dynamic view is in progress.

Exercise 1   *General*

Write a menu for the following AutoCAD commands.

DRAW	EDIT	DISP/TEXT	UTILITY
LINE	FILLET0	DTEXT,C	SAVE
PLINE	FILLET	DTEXT,L	QUIT
ELLIPSE	CHAMFER	DTEXT,R	END
POLYGON	STRETCH	ZOOM WIN	DIR
DONUT	EXTEND	ZOOM PRE	PLOT
	OFFSET		

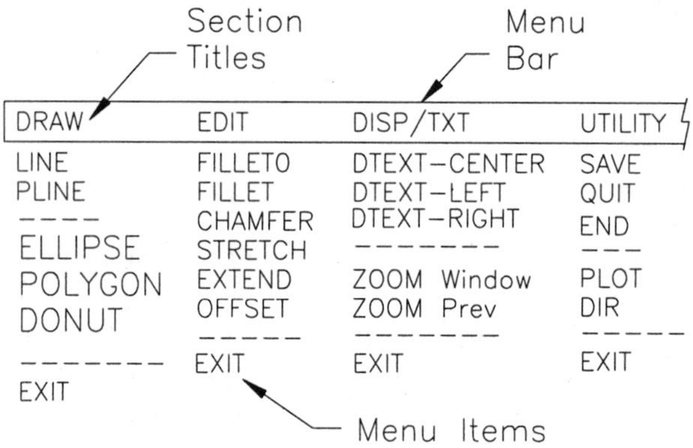

Figure 5-6 Menu display for Exercise 1

CASCADING SUBMENUS IN MENUS

The number of items in a menu or cursor menu can be very large, and sometimes they cannot all be accommodated on one screen. For example, the maximum number of items that can be displayed on some display devices is 21. If the menu or the cursor menu has more items than can be displayed, the excess menu items are not displayed on the screen and cannot be accessed. You can overcome this problem by using cascading menus that let you define smaller groups of items within a menu section. When an item is selected, it loads the cascading menu and displays the items, defined in the cascading menu, on the screen.

The cascading feature of AutoCAD allows pull-down and cursor menus to be displayed in a hierarchical order that makes it easier to select submenus. To use the cascading feature in pull-down and cursor menus, AutoCAD has provided some special characters. For example, -> defines a cascaded submenu and <- designates the last item in the menu. The following table lists some of the characters that can be used with the pull-down or cursor menus.

Character	**Character Description**
--	The item label consisting of two hyphens automatically expands to fill the entire width of the menu. Example: [--]
+	Used to continue the menu item to the next line. This character has to be the last character of the menu item. Example: [Triang:]^C^CLine;1,1;+3,1;2,2;
->	This label character defines a cascaded submenu; it must precede the name of the submenu. Example: [->**Draw**]
<-	This label character designates the last item of the cascaded pull-down or cursor menu. The character must precede the label item. Example: [**<-CIRCLE 3P**]^C^CCIRCLE;3P
<-<-...	This label character designates the last item of the pull-down or cursor menu and also terminates the parent menu. The character must precede the label item. Example: [**<-<-Center Mark**]^C^C_dim;_center
$(	This label character can be used with the pull-down and cursor menus to evaluate a DIESEL expression. The character must precede the label item. Example: $(if,$(getvar,orthomode),Ortho)
~	This item indicates the label item is not available (displayed grayed out); the character must precede the item. Example: [~Application not available]
!.	When used as a prefix, it displays the item with a check mark.
&	When placed directly before a character, the character is displayed underscored. For Example, [W&Block] is displayed as W<u>B</u>lock. It also specifies the character as a menu accelerator key in the pull-down or shortcut menu.
/c	When placed directly before an item with a character, the character is displayed underscored. For example, [/BWBlock] is displayed as

WBlock. It also specifies that the item has a menu accelarator key in the pull-down or shortcut menu.

\t   The label text to the right of \t is displayed to the right side of the menu.

Some display devices provide space for a maximum of 80 characters. Therefore, if there are 10 menus, the length of each menu title should average eight characters. If the combined length of all menu bar titles exceeds 80 characters, AutoCAD automatically wraps the excess menu items and displays them on the next line in the menu bar. The following is a list of some additional features of the menu.

1. The section labels of the menus are ***POP1 through ***POP16. The menu bar titles are displayed in the menu bar.

2. The menus can be accessed by selecting the menu title from the menu bar at the top of the screen.

3. A maximum of 999 menu items can be defined in the menu. This includes the items that are defined in the submenus. The menu items in excess of 999 are ignored.

4. The number of menu items that can be displayed depends on the display device you are using. If the cursor or the menu contains more items than can be accommodated on the screen, the excess items are truncated. For example, if your system can display only 21 menu items, the menu items in excess of 21 are automatically truncated.

Example 2

Write a menu for the commands shown in Figure 5-7. The menu must use the AutoCAD cascading feature.

The following file is a listing of the menu for Example 2. **The line numbers are not a part of the menu; they are shown here for reference only.**

```
***POP1                                              1
[DRAW]                                               2
[LINE]^C^CLINE                                        3
[PLINE]^C^CPLINE                                      4
[->ARC]                                               5
  [ARC]^C^CARC                                        6
  [ARC,3P]^C^CARC;\\DRAG                              7
  [ARC,SCE]^C^CARC;\C;\DRAG                           8
  [ARC,SCA]^C^CARC;\C;\A;DRAG                         9
  [ARC,CSE]^C^CARC;C;\\DRAG                          10
  [ARC,CSA]^C^CARC;C;\\A;DRAG                        11
  [<-ARC,CSL]^C^CARC;C;\\L;DRAG                      12
```

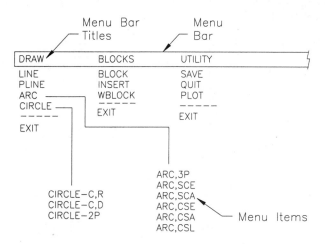

Figure 5-7 *Menu structure for Example 2*

[->CIRCLE]	13
[CIRCLE C,R]^C^CCIRCLE	14
[CIRCLE C,D]^C^CCIRCLE;\D	15
[CIRCLE 2P]^C^CCIRCLE;2P	16
[<-CIRCLE 3P]^C^CCIRCLE;3P	17
[—]	18
[Exit]^C	19
***POP2	20
[BLOCKS]	21
[BLOCK]^C^CBLOCK	22
[INSERT]*^C^CINSERT	23
[WBLOCK]^C^CWBLOCK	24
[—]	25
[Exit]^C	26
***POP3	27
[UTILITY]	28
[SAVE]^C^CSAVE	29
[QUIT]^C^CQUIT	30
[PLOT]^C^CPLOT	31
[—]	32
[Exit]^C	33

Explanation

Line 5
[->ARC]
In this menu item, **ARC** is the menu item label that is preceded by the special label character **->**. This special character indicates that the menu item has a submenu. The menu items that follow it (Lines 6-12) are the submenu items.

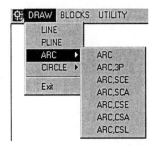

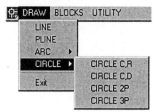

Figure 5-9 *Draw menu*
with CIRCLE submenu

Figure 5-8 *Draw menu*

Line 12
[<-ARC,CSL]^C^CARC;C;\\L;DRAG
In this line, the menu item label **ARC,CSL** is preceded by another special label character, **<-**, which indicates the end of the submenu. The item that contains this character must be the last menu item of the submenu.

Lines 13 and 17
[->CIRCLE]
[<-CIRCLE 3P]^C^CCIRCLE;3P
The special character -> in front of **CIRCLE** indicates that the menu item has a submenu; the character <- in front of **CIRCLE 3P** indicates that this item is the last menu item in the submenu. When you select the menu item **CIRCLE** from the menu, it will automatically display the submenu on the side.

Example 3

Write a menu that has the cascading submenus for the commands shown in Figure 5-10.

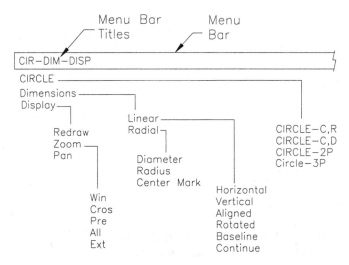

Figure 5-10 *Menu structure for Example 3*

The following file is a listing of the menu for Example 3. **The line numbers are not a part of the menu; they are shown here for reference only.**

```
***POP1                                              1
[DRAW]                                               2
[->CIRCLE]                                           3
  [CIRCLE C,R]^C^C_CIRCLE                            4
  [CIRCLE C,D]^C^C_CIRCLE;\_D                        5
  [CIRCLE 2P]^C^C_CIRCLE;_2P                         6
  [<-CIRCLE 3P]^C^C_CIRCLE;_3P                       7
[->Dimensions]                                       8
  [->Linear]                                         9
   [Horizontal]^C^C_dim;_horizontal                10
   [Vertical]^C^C_dim;_vertical                     11
   [Aligned]^C^C_dim;_aligned                       12
   [Rotated]^C^C_dim;_rotated                       13
   [Baseline]^C^C_dim;_baseline                     14
   [<-Continue]^C^C_dim;_continue                   15
  [->Radial]                                        16
   [Diameter]^C^C_dim;_diameter                     17
   [Radius]^C^C_dim;_radius                         18
   [<-<-Center Mark]^C^C_dim;_center                19
[->DISPLAY]                                          20
  [REDRAW]^C^CREDRAW                                 21
  [->ZOOM]                                           22
   [...Win]^C^C_ZOOM;_W                             23
   [...Cros]^C^C_ZOOM;_C                            24
   [...Pre]^C^C_ZOOM;_P                             25
   [...All]^C^C_ZOOM;_A                             26
   [<-...Ext]^C^C_ZOOM;_E                           27
  [<-PAN]^C^C_Pan                                    28
```

Explanation

Lines 8 and 9
[->Dimensions]
[->Linear]

The special label character **->** in front of the menu item **Dimensions** indicates that it has a submenu, and the character -> in front of **Linear** indicates that there is another submenu. The sec-

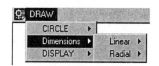

Figure 5-11 Draw menu with cascading menus

ond submenu **Linear** is within the first submenu Dimensions (Figure 5-11). The menu items on lines 10 to 15 are defined in the **Linear** submenu, and the menu items Linear and **Radial** are defined in the submenu Dimensions.

Line 16
[->Radial]

This menu item defines another submenu; the menu items on line numbers 17, 18, and 19 are part of this submenu.

Line 19

[<-<-Center Mark] ^ C ^ C_dim;_center

In this menu item the special label character **<-<-** terminates the **Radial** and **Dimensions** (parent submenu)
submenus.

Lines 27 and 28

[<-...Ext] ^ C ^ C_ZOOM;_E

[<-PAN] ^ C ^ C_Pan

The special character **<-** in front of the menu item **...Ext**
terminates the **ZOOM** submenu (Figure 5-12); the special
character in front of the menu item **PAN** terminates the
DISPLAY submenu.

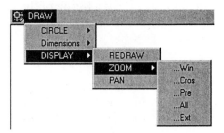

*Figure 5-12 Draw pull-down menu
with cascading menus*

SHORTCUT AND CONTEXT MENUS

The shortcut menus are similar to the pull-down menus, except the shortcut menu can contain
only 499 menu items compared with 999 items in the pull-down menu. The section label of the
shortcut menu can be ***POP0 and ***POP500 to ***POP 999. The shortcut menus are displayed near or at the cursor location. Therefore, they can be used to provide convenient and quick
access to some of the frequently used commands. The shortcut menus that are in the upper range
are also referred as context menus. The following is a list of some of the features of the shortcut
menu.

1. The section label of the shortcut menu are ***POP0 and ***POP500 to ***POP999. The
    menu bar title defined under this section label is not displayed in the menu bar.

2. On most systems, the menu bar title is not displayed at the top of the shortcut menu. However,
    for compatibility reasons you should give a dummy menu bar title.

3. The POP0 menu can be accessed through the **$P0=*** menu command. The shortcut menus
    POP500 through POP999 must be referenced by their alias names. The reserved alias names
    for AutoCAD use are GRIPS, CMDEFAULT, CMEDIT, and CMCOMMAND. For example, to
    reference POP500 for grips, use **GRIPS command line under POP5000. This commands
    can be issued by a menu item in another menu, such as the button menu, auxiliary menu, or
    the screen menu. The command can also be issued from an AutoLISP or ADS program.

4. A maximum of 499 menu items can be defined in the shortcut menu. This includes the items
    that are defined in the shortcut submenus. The menu items in excess of 499 are ignored.

5. The number of menu items that can be displayed on the screen depends on the system you are
    using. If the shortcut or pull-down menu contains more items than your screen can accommo-
    date, the excess items are truncated. For example, if your system displays 21 menu items, the
    menu items in excess of 21 are automatically truncated.

6. The system variable **SHORTCUTMENU** controls the availability of Default, Edit, and Com-
    mand mode shortcut menus. If the value is 0, it restores R14 legacy behavior and disables the
    Default, Edit, and Command mode shortcut menus. The default value of this variable is 11.

Example 4

Write a shortcut menu for the following AutoCAD commands using cascading submenus. The menu should be compatible with foreign language versions of AutoCAD. Use the third button of the **BUTTONS** menu to display the shortcut menu.

Osnaps	Draw	DISPLAY
Center	Line	REDRAW
Endpoint	PLINE	ZOOM
Intersection	CIR C,R	...Win
Midpoint	CIR 2P	...Cen
Nearest	ARC SCE	...Prev
Perpendicular	ARC CSE	...All
Quadrant		...Ext
Tangent		PAN
None		

The following file is a listing of the menu file for Example 4. **The line numbers are not a part of the file; they are for reference only.**

```
***AUX2                                    1
;                                          2
$P0=*                                      3
***POP0                                    4
[Osnaps]                                   5
[Center]_Center                            6
[End point]_Endp                           7
[Intersection]_Int                         8
[Midpoint]_Mid                             9
[Nearest]_Nea                             10
[Perpendicular]_Per                       11
[Quadrant]_Qua                            12
[Tangent]_Tan                             13
[None]_Non                                14
[—]                                       15
[->Draw]                                  16
  [Line]^C^C_Line                         17
  [PLINE]^C^C_Pline                       18
  [CIR C,R]^C^C_Circle                    19
  [CIR 2P]^C^C_Circle;_2P                 20
  [ARC SCE]^C^C_ARC;\C                    21
  [<-ARC CSE]^C^C_Arc;C                   22
[—]                                       23
[->DISPLAY]                               24
  [REDRAW]^C^_REDRAW                      25
  [->ZOOM]                                26
    [...Win]^C^C_ZOOM;_W                  27
```

```
      [...Cen]^C^C_ZOOM;_C                                    28
      [...Prev]^C^C_ZOOM;_P                                   29
      [...All]^C^C_ZOOM;_A                                    30
      [<-...Ext]^C^C_ZOOM;_E                                  31
    [<-PAN]^C^C_Pan                                           32
 ***POP1                                                      33
 [SHORTCUTMENU]                                               34
 [SHORTCUTMENU=0]^C^CSHORTCUTMENU;0                           35
 [SHORTCUTMENU=11]^C^CSHORTCUTMENU;11                         36
```

Explanation

Line 1
***AUX2
AUX2 is the section label for the second auxiliary menu; *** designates the menu section. The menu items that follow it, until the second section label, are a part of this buttons menu.

Lines 2 and 3
;
$P0=*
The semicolon (;) is assigned to the second button of the pointing device (the first button of the pointing device is the pick button); the special command **$P0=*** is assigned to the third button of the pointing device.

Lines 4 and 5
***POP0
[Osnaps]
The menu label **POP0** is the menu section label for the shortcut menu; Osnaps is the menu bar title. The menu bar title is not displayed, but is required. Otherwise, the first item will be interpreted as a title and will be disabled.

Line 6
[Center]_Center
In this menu item, **_Center** is the center object Snap mode. The menu files can be used with foreign language versions of AutoCAD, if AutoCAD commands and the command options are preceded by the underscore (_) character.

After loading the menu, if you press the third button of your pointing device, the shortcut menu (Figure 5-13) will be displayed at the cursor (screen crosshairs) location. If the cursor is close to the edges of the screen, the shortcut menu will be displayed at a location that is closest to the cursor position. When you select a submenu, the items contained in the submenu will be displayed, even if the shortcut menu is touching the edges of the screen display area.

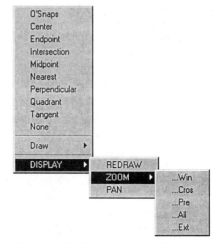

Figure 5-13 *Shortcut menu for Example 4*

Lines 33 and 34
***POP1
[Draw]
***POP1 defines the first pull-down menu. If no POPn sections are defined or the status line is turned off, the shortcut menu is automatically disabled.

Exercise 2	*General*

Write a menu for the following AutoCAD commands. Use a cascading menu for the **LINE** command options in the menu. (The layout of the menu is shown in Figure 5-14.)

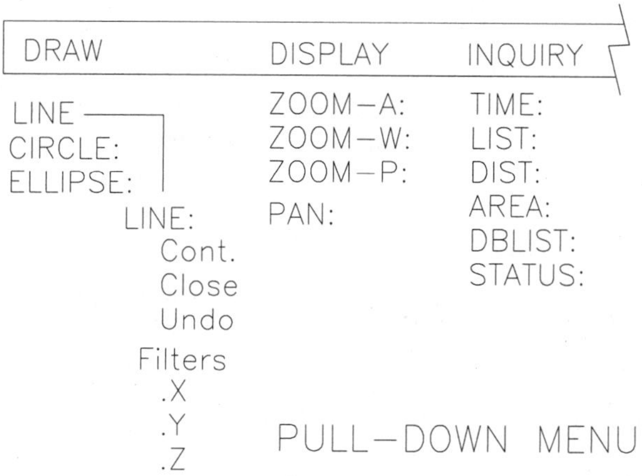

Figure 5-14 *Design of menu for Exercise 2*

LINE	**ZOOM All**	**TIME**
Continue	**ZOOM Win**	**LIST**
Close	**ZOOM Pre**	**DISTANCE**
Undo	**PAN**	**AREA**
.X	**DBLIST**	
.Y	**STATUS**	
.Z		
CIRCLE		
ELLIPSE		

SUBMENUS

The number of items in a pull-down menu or shortcut menu can be very large and sometimes they cannot all be accommodated on one screen. For example, the maximum number of items that can be displayed on most of the display devices is 21. If the menu or the shortcut menu has more than 21 items, the menu excess items are not displayed on the screen and cannot be accessed. You can overcome this problem by using submenus that let you define smaller groups of items within a menu section. When a submenu is selected, it loads the submenu items and displays them on the screen.

Chapter 5

The menus that use AutoCAD's cascading feature are the most efficient and easy to write. The submenus follow a logical pattern that are easy to load and use without causing any confusion. It is strongly recommended to use the cascading menus whenever you need to write the pull-down or the shortcut menus. However, AutoCAD provides the option to swap the submenus in the menus. These menus can sometimes cause distraction because the original menu is completely replaced by the submenu when swapping the menus.

Submenu Definition

A submenu definition consists of two asterisk signs (**) followed by the name of the submenu. A menu can have any number of submenus and each submenu should have a unique name. The items that follow a submenu, up to the next section label or submenu label, belong to that submenu. Following is the format of a submenu label:

**Name

        Where ** ----------------- Two asterisk signs (**) designate a submenu
                    Name ------------ Name of the submenu

Note

The submenu name can be up to 31 characters long.

The submenu name can consist of letters, digits, and special characters like: $ (dollar), - (hyphen), and _ (underscore).

The submenu name cannot have any embedded blanks.

The submenu names should be unique in a menu file.

Submenu Reference

The submenu reference is used to reference or load a submenu. It consists of a "$" sign followed by a letter that specifies the menu section. The letter that specifies a menu section is Pn, where n designates the number of the menu section. The menu section is followed by "=" sign and the name of the submenu that the user wants to activate. The submenu name should be without "**". Following is the format of a submenu reference:

$Section=Submenu

        Where **$** ------------------- "$" sign
                    Section ---------- Menu section specifier
                    = ------------------ "=" sign
                    Submenu ------- Name of submenu

Example
$P1=P1A

        Where **$P1** --------------- P1 Specifies pull-down menu section 1
                    P1A -------------- Name of submenu

Displaying a Submenu

When you load a submenu in a menu, the submenu items are not automatically displayed on the screen. For example, when you load a submenu P1A that has DRAW-ARC as the first item, the current title of POP1 will be replaced by the DRAW-ARC. But the items that are defined under DRAW-ARC are not displayed on the screen. To force the display of the new items on the screen, AutoCAD uses a special command $Pn=*.

   $Pn=*

             Where   **P** ------------------ P for menu
                        Pn ----------------- Menu section number (1 to 10)
                        * ------------------ Asterisk sign (*)

LOADING MENUS

From the menu, you can load any menu that is defined in the screen or image tile menu sections by using the appropriate load commands. It may not be needed in most of the applications, but if you want to, you can load the menus that are defined in other menu sections.

Loading Screen Menus

You can load any menu that is defined in the screen menu section from the menu by using the following load command:

   $S=X $S=LINE

             Where   **S** ------------------- S specifies screen menu
                        X ------------------ Submenu name defined in screen menu section
                        LINE ------------- Submenu name defined in screen menu section

The first load command ($S=X) loads the submenu X that has been defined in the screen menu section of the menu file. The X submenu can contain 21 blank lines, so that when it is loaded it clears the screen menu. The second load command ($S=LINE) loads the submenu **LINE** that has also been defined in the screen menu section of the menu file.

Loading an Image Tile Menu

You can also load an image tile menu from the menu by using the following load command:

   $I=IMAGE1 $I=*

             Where   **$I=Image1** ----- Load the submenu IMAGE1
                        $I=* ------------ To display the dialog box

This menu item consists of two load commands. The first load command $I=IMAGE1 loads the image tile submenu IMAGE1 that has been defined in the image tile menu section of the file. The second load command $I=* displays the new dialog box on the screen.

Example 5

Write a menu for the following AutoCAD commands. Use submenus for the **ARC** and **CIRCLE** commands.

LINE	**BLOCK**	**QUIT**
PLINE	**INSERT**	**SAVE**
ARC	**WBLOCK**	——
ARC 3P	**PLOT**	
ARC SCE		
ARC SCA		
ARC CSE		
ARC CSA		
ARC CSL		
CIRCLE		
CIRCLE C,R		
CIRCLE C,D		
CIRCLE 2P		

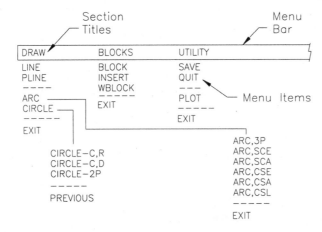

Figure 5-15 Design of menu for Example 5

The layout shown in Figure 5-15 is one of the possible designs for this menu. The **ARC** and **CIRCLE** commands are in separate groups that will be defined as submenus in the menu file.

The following file is a listing of the menu for Example 5. **The line numbers are not a part of the menu file; they are given here for reference only**.

```
***POP1                                                  1
**P1A                                                    2
[DRAW]                                                   3
[LINE]^C^CLINE                                           4
[PLINE]^C^CPLINE                                         5
```

```
[—]                                                                    6
[ARC]^C^C$P1=P1B $P1=*                                                  7
[CIRCLE]^C^C$P1=P1C $P1=*                                               8
[—]                                                                    9
[Exit]^C                                                               10
                                                                       11
**P1B                                                                  12
[ARC]                                                                  13
[ARC,3P]^C^CARC \\DRAG                                                 14
[ARC,SCE]^C^CARC \C \DRAG                                              15
[ARC,SCA]^C^CARC \C \A DRAG                                            16
[ARC,CSE]^C^CARC C \\DRAG                                              17
[ARC,CSA]^C^CARC C \\A DRAG                                            18
[ARC,CSL]^C^CARC C \\L DRAG                                            19
[—]                                                                   20
[Exit]^C                                                               21
                                                                       22
**P1C                                                                  23
[CIRCLE]                                                               24
[CIRCLE C,R]^C^CCIRCLE                                                 25
[CIRCLE C,D]^C^CCIRCLE \D                                              26
[CIRCLE 2P]^C^CCIRCLE 2P                                               27
[—]                                                                   28
[PREVIOUS]$P1=P1A $P1=*                                                29
                                                                       30
***POP2                                                                31
[BLOCKS]                                                               32
[BLOCK]^C^CBLOCK                                                       33
[INSERT]*^C^CINSERT                                                    34
[WBLOCK]^C^CWBLOCK                                                     35
[—]                                                                   36
[Exit]$P1=P1A $P1=*                                                    37
                                                                       38
***POP3                                                                39
[UTILITY]                                                              40
[SAVE]^C^CSAVE                                                         41
[QUIT]^C^CQUIT                                                         42
[~—]                                                                  43
[PLOT]^C^CPLOT                                                         44
[~—]                                                                  45
[Exit]^C                                                               46
                                                                       47
```

Line 2
P1A

**P1A defines the submenu P1A. All the submenus have two asterisk signs (**) followed by the name of the submenu. The submenu can have any valid name. In this example, P1A has been chosen because it is easy to identify the location of the submenu. P indicates that it is a menu, 1 indicates that it is in the first menu (POP1), and A indicates that it is the first submenu in that section.

Line 6
[--]
The two hyphens enclosed in the brackets will automatically expand to fill the entire width of the menu. This menu item cannot be used to define a command. If it does contain a command definition, the command is ignored. For example, if the menu item is [--]^C^CLINE, the command ^C^CLINE will be ignored.

Line 7
[ARC]^C^C$P1=P1B $P1=*
In this menu item, $P1=P1B loads the submenu P1B and assigns it to the first menu section (POP1), but the new menu is not displayed on the screen. $P1=* forces the display of the new menu on the screen.

For example, if you select **CIRCLE** from the first menu (POP1), the menu bar title **DRAW** will be replaced by **CIRCLE**, but the new menu is not displayed on the screen. Now, if you select **CIRCLE** from the menu bar, the command defined under the **CIRCLE** submenu will be displayed in the menu. To force the display of the menu that is currently assigned to POP1, you can use AutoCAD's special command **$P1=***. If you select **CIRCLE** from the first menu (POP1), the **CIRCLE** submenu will be loaded and automatically displayed on the screen.

Line 21
[EXIT]^C
When you select this menu item the current menu will be canceled. It will not return to the previous submenu (**DRAW**). If you check the menu bar, it will display **ARC** as the section title, not **DRAW**. Therefore, it is not a good practice to cancel a submenu. It is better to return to the first menu before canceling it or define a command that automatically loads the previous menu and then cancels the menu.

Figure 5-16 Arc pull-down menu replaces Draw menu

Example
[EXIT]$P1=P1A $P1=* ^C^C

Line 29
[PREVIOUS]$P1=P1A $P1=*
In this menu item $P1=P1A loads the submenu P1A, which happens to be the previous menu in this case. You can also use $P1= to load the previous menu. $P1=* forces the display of the submenu P1A.

PARTIAL MENUS

AutoCAD has provided a facility that allows users to write their own menus and then load them in the menu bar. For example, in Windows you can write partial menus, toolbars, and definitions for accelerator keys. After you write the menu, AutoCAD lets you load the menu and use it with the standard menu. For example, you could load a partial menu and use it like a menu. You can also unload the menus that you do not want to use. These features make it convenient to use the menus that have been developed by AutoCAD users and developers.

Menu Section Labels

The following is a list of the additional menu section labels for Windows.

Section label	Description
***MENUGROUP	Menu file group name
***TOOLBARS	Toolbar definition
***HELPSTRING	Online help
***ACCELERATORS	Accelerator key definitions

Writing Partial Menus

The following example illustrates the procedure for writing a partial menu for Windows:

Example 6

In this example you will write a partial menu for Windows. The menu file has two menus, POP1 (MyDraw) and POP2 (MyEdit), as shown in Figure 5-17.

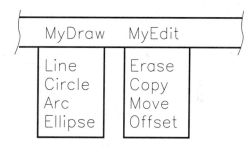

Figure 5-17 *Pull-down menus for Example 6*

Step 1

Use a text editor to write the following menu file. The name of the file is assumed to be **MYMENU1.MNU**. The following is the listing of the menu file for this example:

```
***MENUGROUP=Menu1                              1
***POP1                                         2
[/MMyDraw]                                      3
[/LLine]^C^CLine                                4
[/CCircle]^C^CCircle                            5
[/AArc]^C^CArc                                  6
[/EEllipse]^C^CEllipse                          7
***POP2                                         8
[/EMyEdit]                                      9
[/EErase]^C^CErase                             10
[/CCopy]^C^CCopy                               11
[/MMove]^C^CMove                               12
[/OOffset]^C^COffset                           13
```

Chapter 5

Explanation
Line 1
***MENUGROUP=Menu1
MENUGROUP is the section label, and Menu1 is the name tag for the menu group. The MENUGROUP label must precede all menu section definitions. The name of the MENUGROUP (Menu1) can be up to 32 characters long (alphanumeric), excluding spaces and punctuation marks. There is only one MENUGROUP in a menu file. All section labels must be preceded by *** (***MENUGROUP).

Line 2
***POP1
POP1 is the menu section label. The items on line numbers 3 through 7 belong to this section. Similarly, the items on line numbers 9 through 13 belong to the menu section **POP2**.

Line 3
[/MMyDraw]
/M defines the mnemonic key you can use to activate the menu item. For example, /M will display an underline under the letter M in the text string that follows it. If you enter the letter M, AutoCAD will execute the command defined in that menu item. MyDraw is the menu item label. The text string inside the brackets [ ] , except /M, has no function. It is used for displaying the function name so that the user can recognize the command that will be executed by selecting that item.

Line 4
[/LLine] ^ C ^ CLine
In this line, the /L defines the mnemonic key, and the Line that is inside the brackets is the menu item label. ^ C ^ C cancels the command twice, and the Line is the AutoCAD **LINE** command. The part of the menu item statement that is outside the brackets is executed when you select an item from the menu. When you select Line 4, AutoCAD will execute the **LINE** command.

Step 2
Save the file and then load the partial menu file using the AutoCAD **MENULOAD** command.

> Command: **MENULOAD**

When you enter the **MENULOAD** command, AutoCAD displays the **Menu Customization** dialog box (Figure 5-18). To load the menu file, enter the name of the menu file, **MYMENU1.MNU**, in the **File Name:** edit box. You can also use the **Browse...** option to invoke the **Select Menu File** dialog box. Select the name of the file, and then use the **OK** button to return to the **Menu Customization** dialog box. To load the selected menu file, choose the **LOAD** button. The name of the menu group (MENU1) will be displayed in the **Menu Groups** list box.
You can also load the menu file from the command line, as follows:

> Command: **FILEDIA**
> Enter new value for FILEDIA <1>: **0**   (*Disables the file dialog boxes.*)
> Command: **MENULOAD**
> Enter name of menu file to load: **MYMENU1.MNU**

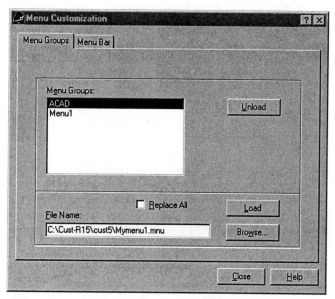

Figure 5-18 Menu Customization dialog box (Menu Groups tab)

You can also use the AutoLISP functions to set the **FILEDIA** system variable to 0 and then load the partial menu.

> Command: **(SETVAR "FILEDIA" 0)**
> Command: **(Command "MENULOAD" "MYMENU1")**

Step 3

In the **Menu Customization** dialog box, select the **Menu Bar** tab to display the menu bar options (Figure 5-19). In the **Menu Groups** list box select Menu1; the menus defined in the menu group (Menu1) will be displayed in the **Menus** list box. In the **Menus** list box select the menu (MyDraw) that you want to insert in the menu bar. In the **Menu Bar** list box select the position where you want to insert the new menu. For example, if you want to insert the new menu in front of Format, select the **Format** menu in the **Menu Bar** list box. Choose the **Insert** button to insert the selected menu (MyDraw) in the menu bar. The menu (MyDraw1) is displayed in the menu bar located at the top of your screen.

You can also load the menu from the Command line. Once the menu is loaded, use the **MENUCMD** command (AutoLISP function) to display the partial menus.

> Command: **(MENUCMD "P5 = +Menu1.POP1")**
> Command: **(MENUCMD "P6 = +Menu1.POP2")**

After you enter these commands, AutoCAD will display the menu titles in the menu bar, as shown in Figure 5-20. If you select MyDraw, the corresponding menu as defined in the menu file will be displayed on the screen. Similarly, selecting MyEdit will display the corresponding edit menu.

Chapter 5

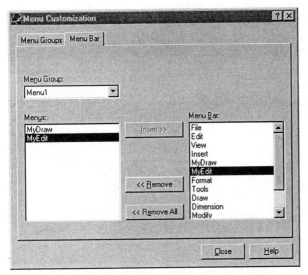

Figure 5-19 Menu Customization dialog box (Menu Bar tab)

MENUCMD is an AutoLISP function, and P5 determines where the POP1 menu will be displayed. In this example the POP1 (MyDraw) menu will be displayed as the fifth menu. Menu1 is the MENUGROUP name as defined in the menu file, and POP1 is the menu section label. The MENUGROUP name and the menu section label must be separated by a period (.).

Figure 5-20 MENUCMD command places the menu titles in the menu bar

Step 4
If you want to unload the menu (MyDraw), select the Menu Bar tab of the **Menu Customization** dialog box. In the **Menu Bar** list box select the menu item that you want to unload. Choose the **Remove** button to remove the selected item.

Step 5
You can also unload the menu groups (Menu1) using the **Menu Customization** dialog box. Enter the **MENULOAD** or **MENUUNLOAD** command to invoke the **Menu Customization** dialog box (Figure 5-18). AutoCAD will display the names of the menu files in the **Menu Groups:** list box. Select the **Menu1** menu, and then choose the **Unload** button. AutoCAD will unload the menu group. Choose the **Close** button to exit the dialog box. You can also unload the menu file from the command line, as follows:

Command: **FILEDIA**
Enter new value for FILEDIA <1>: **0**
Command: **MENUUNLOAD**
Enter the name of a MENUGROUP to unload: **MENU1.MNU**

The **MENUUNLOAD** command unloads the entire menu group. You can also unload an individual menu without unloading the entire menu group by using the following command:

Command: (MENUCMD "P5=-)

This command will unload the menu at position five (P5, MyDraw menu). The menu group is still loaded, but the P5 menu is not visible. The menu can also be reinitialized by using the **MENU** command to load the base menu **ACAD.MNU**. This will remove all partial menus and the tag definitions associated with the partial menus.

Accelerator Keys

AutoCAD for Windows also supports user-defined accelerator keys. For example, if you enter **C** at the Command prompt, AutoCAD draws a circle because it is a command alias for a circle as defined in the ACAD.PGP file. You cannot use the **C** key to enter the **COPY** command. To use the **C** key for entering the **COPY** command, you can define the accelerator keys. You can combine the SHIFT key with C in the menu file so that when you hold down the SHIFT key and then press the C key, AutoCAD will execute the **COPY** command. The following example illustrates the use of accelerator keys.

Example 7

In this example you will add the following accelerator keys to the partial menu of Example 6.

CONTROL+"E"	to draw an ellipse (**ELLIPSE** command)
SHIFT+"C"	to copy (**COPY** command)
[CONTROL'Q"]	to quit (**QUIT** command)

The following file is the listing of the partial menu file that uses the accelerator keys of Example 7.

```
***MENUGROUP=Menu1
***POP1
**Alias
[/MMyDraw]
[/LLine]^C^CLine
[/CCircle]^C^CCircle
[/AArc]^C^CArc
ID_Ellipse [/EEllipse]^C^CEllipse

***POP2
[/EMyEdit]
[/EErase]^C^CErase
ID_Copy [/CCopy]^C^CCopy
[/OOffset]^C^COffset
[/VMove]^C^CMov

***ACCELERATORS
ID_Ellipse [CONTROL+"E"]
ID_Copy [SHIFT+"C"]
[CONTROL'Q"]^C^CQuit
```

This menu file defines three accelerator keys. The **ID_Copy [SHIFT+"C"]** accelerator key consists of two parts. The ID_Copy is the name tag, which must be the same as used earlier in the menu item definition. The SHIFT+"C" is the label that contains the modifier (SHIFT) and the keyname (C). The keyname or the string, such as "ESCAPE", must be enclosed in quotation marks. After you load the file, SHIFT+C will enter the **COPY** command and CTRL+E will draw an ellipse. Similarly, Ctrl+Q will cancel the existing command and enter the **QUIT** command.

The accelerator keys can be defined in two ways. One way is to give the name tag followed by the label containing the modifier. The modifier is followed by a single character or a special virtual key enclosed in quotation marks [CONTROL+"E"] or ["ESCAPE"]. You can also use the plus sign (+) to concatenate the modifiers [SHIFT + CONTROL + "L"]. The other way of defining an accelerator key is to give the modifier and the key string, followed by a command sequence [CONTROL "Q"] ^ C ^ CQuit.

Special Virtual Keys

The following are the special virtual keys. These keys must be enclosed in quotation marks when used in the menu file.

String	Description	String	Description
"F1"	F1 key	"NUMBERPAD0"	0 key
"F2"	F2 key	"NUMBERPAD1"	1 key
"F3"	F3 key	"NUMBERPAD2"	2 key
"F4"	F4 key	"NUMBERPAD3"	3 key
"F5"	F5 key	"NUMBERPAD4"	4 key
"F6"	F6 key	"NUMBERPAD5"	5 key
"F7"	F7 key	"NUMBERPAD6"	6 key
"F8"	F8 key	"NUMBERPAD7"	7 key
"F9"	F9 key	"NUMBERPAD8"	8 key
"F10"	F10 key	"NUMBERPAD9"	9 key
"F11"	F11 key	"UP"	UP-ARROW key
"F12"	F12 key	"DOWN"	DOWN-ARROW key
"HOME"	HOME key	"LEFT"	LEFT-ARROW key
"END"	END key	"RIGHT"	RIGHT-ARROW key
"INSERT"	INS key	"ESCAPE"	ESC key
"DELETE"	DEL key		

Valid Modifiers. The following are the valid modifiers:

String	Description
CONTROL	The CTRL key on the keyboard
SHIFT	The SHIFT key (left or right)
COMMAND	The Apple key on Macintosh keyboards
META	The meta key on UNIX keyboards

Toolbars

The contents of the toolbar and its default layout can be specified in the Toolbar section (***TOOLBARS) of the menu file. Each toolbar must be defined in a separate submenu.

Toolbar Definition. The following is the general format of the toolbar definition:

```
***TOOLBARS
**MYTOOLS1
TAG1 [Toolbar ("tbarname", orient, visible, xval, yval, rows)]
TAG2 [Button ("btnname", id_small, id_large)]macro
TAG3 [Flyout ("flyname", id_small, id_large, icon, alias)]macro
TAG4 [control (element)]
[—]
```

***TOOLBARS is the section label of the toolbar, and **MYTOOLS1** is the name of the submenu that contains the definition of a toolbar. Each toolbar can have five distinct items that control different elements of the toolbar: TAG1, TAG2, TAG3, TAG4, and separator ([—]). The first line of the toolbar (TAG1) defines the characteristics of the toolbar. In this line, **Toolbar** is the keyword, and it is followed by a series of options enclosed in parentheses. The following describes the available options.

tbarname	This is a text string that names the toolbar. The tbarname text string must consist of alphanumeric characters with no punctuation other than a dash (-) or an underscore (_).
orient	This determines the orientation of the toolbar. The acceptable values are Floating, Top, Bottom, Left, and Right. These values are not case sensitive.
visible	This determines the visibility of the toolbar. The acceptable values are Show and Hide. These values are not case sensitive.
xval	This is a numeric value that specifies the X ordinate in pixels. The X ordinate is measured from the left edge of the screen to the right side of the toolbar.
yval	This is a numeric value that specifies the Y ordinate in pixels. The Y ordinate is measured from the top edge of the screen to the top of the toolbar.
rows	This is a numeric value that specifies the number of rows.

The second line of the toolbar (TAG2) defines the button. In this line the **Button** is the key word and it is followed by a series of options enclosed in parentheses. The following is the description of the available options.

btnname	This is a text string that names the button. The text string must consist of alphanumeric characters with no punctuation other than a dash (-) or an underscore (_). This text string is displayed as ToolTip when you place the cursor over the button.
id_small	This is a text string that names the ID string of the small image resource (16 by 16 bitmap). The text string must consist of alphanumeric characters with no punctuation other than a dash (-) or an underscore (_). The id_small text

	string can also specify a user-defined bitmap (Example: ICON_16_CIRCLE).
id_big	This is a text string that names the ID string of the large image resource (32 by 32 bitmap). The text string must consist of alphanumeric characters with no punctuation other than a dash (-) or an underscore (_). The id_big text string can also specify a user-defined bitmap (Example: ICON_32_CIRCLE).
macro	The second line (TAG2), which defines a button, is followed by a command string (macro). For example, the macro can consist of ^C^CLine. It follows the same syntax as that of any standard menu item definition.

The third line of the toolbar (TAG3) defines the flyout control. In this line the **Flyout** is the key word, and it is followed by a series of options enclosed in parentheses. The following describes the available options.

flyname	This is a text string that names the flyout. The text string must consist of alphanumeric characters with no punctuation other than a dash (-) or an underscore (_). This text string is displayed as ToolTip when you place the cursor over the flyout button.
id_small	This is a text string that names the ID string of the small image resource (16 by 16 bitmap). The text string must consist of alphanumeric characters with no punctuation other than a dash (-) or an underscore (_). The id_small text string can also specify a user-defined bitmap.
id_big	This is a text string that names the ID string of the large image resource (32 by 32 bitmap). The text string must consist of alphanumeric characters with no punctuation other than a dash (-) or an underscore (_). The id_big text string can also specify a user-defined bitmap.
icon	This is a Boolean key word that determines whether the button displays its own icon or the last icon selected. The acceptable values are **ownicon** and **othericon**. These values are not case-sensitive.
alias	The alias specifies the name of the toolbar submenu that is defined with the standard ****aliasname** syntax.
macro	The third line (TAG3), which defines a flyout control, is followed by a command string (macro). For example, the macro can consist of ^C^CCircle. It follows the same syntax as that of any standard menu item definition.

The fourth line of the toolbar (TAG4) defines a special control element. In this line the Control is the key word, and it is followed by the type of control element enclosed in parentheses. The following describes the available control element types.

element	This parameter can have one of the following three values: Layer: This specifies the layer control element.

Linetype: This specifies the linetype control element.
Color: This specifies the color control element.

The fifth line ([--]) defines a separator.

Example 8

In this example you will write a menu file for a toolbar for the **LINE**, **PLINE**, **CIRCLE**, **ELLIPSE**, and **ARC** commands. The name of the toolbar is MyDraw1 (Figure 5-21).

The following is the listing of the menu file:
MENUGROUP=M1
***TOOLBARS
**TB_MyDraw1
ID_MyDraw1[_Toolbar("MyDraw1", _Floating, _Hide, 10, 200, 1)]
ID_Line [_Button("Line", ICON_16_LINE, ICON_32_LINE)]^C^C_line
ID_Pline [_Button("Pline", ICON_16_PLine, ICON_32_PLine)]^C^C_PLine
ID_Circle[_Button("Circle", ICON_16_CirRAD, ICON_32_CirRAD)]^C^C_Circle
ID_ELLIPSE[_Button("Ellipse", ICON_16_EllCEN, ICON_32_EllCEN)]^C^C_ELLIPSE
ID_Arc[_Button("Arc 3Point", ICON_16_Arc3Pt, ICON_32_Arc3Pt)]^C^C_Arc

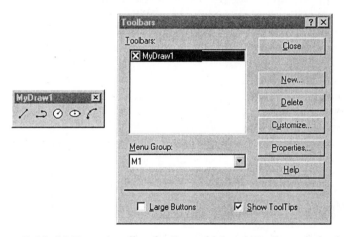

*Figure 5-21 **MyDraw1** toolbar for Example 8 and **Toolbars** dialog box*

Use the **MENULOAD** command to load the MyDraw1 menu group, as discussed earlier. To display the new toolbar (MyDraw1) on the screen, select Toolbars from the View menu and then select MyDraw1 in the **Menu Groups** list box. Turn the MyDraw1 toolbar on in the **Toolbars** dialog box; MyDraw1 toolbar is displayed on the screen.

You can also load the new toolbar from the command line. After using the **MENULOAD** command to load the MyDraw1 menu group, use the **-TOOLBAR** command to display the MyDraw1 toolbar.

Chapter 5

Command: -**TOOLBAR**
Enter toolbar name or [ALL]: **MYDRAW1**
Enter an option [Show/Hide/Left/Right/Top/Bottom/Float] <Show>: **S**

Example 9

In this example you will write a menu file for a toolbar with a flyout. The name of the toolbar is
MyDraw2 (Figure 5-22), and it contains two buttons, Circle and Arc. When you select the Circle
button, it should display a flyout with radius, diameter, 2P, and 3P buttons (Figure 5-23). Similarly,
when you select the Arc button, it should display the 3Point, SCE, and SCA buttons.

The following is the listing of the menu file:
***Menugroup=M2
***TOOLBARS
**TB_MyDraw2
ID_MyDraw2[_Toolbar("MyDraw2", _Floating, _Show, 10, 100, 1)]
ID_TbCircle[_Flyout("Circle", ICON_16_Circle, ICON_32_Circle, _OtherIcon, M2.TB_Circle)]
ID_TbArc[_Flyout("Arc", ICON_16_Arc, ICON_32_Arc, _OtherIcon, M2.TB_Arc)]
**TB_Circle
ID_TbCircle[_Toolbar("Circle", _Floating, _Hide, 10, 150, 1)]
ID_CirRAD[_Button("Circle C,R", ICON_16_CirRAD, ICON_32_CirRAD)]^C^C_Circle
ID_CirDIA[_Button("Circle C,D", ICON_16_CirDIA, ICON_32_CirDIA)]^C^C_Circle;\D
ID_Cir2Pt[_Button("Circle 2Pts", ICON_16_Cir2Pt, ICON_32_Cir2Pt)]^C^C_Circle;2P
ID_Cir3Pt[_Button("Circle 3Pts", ICON_16_Cir3Pt, ICON_32_Cir3Pt)]^C^C_Circle;3P

**TB_Arc
ID_TbArc[_Toolbar("Arc", _Floating, _Hide, 10, 150, 1)]
ID_Arc3PT[_Button("Arc,3Pts", ICON_16_Arc3PT, ICON_32_Arc3PT)]^C^C_Arc
ID_ArcSCE[_Button("Arc,SCE", ICON_16_ArcSCE, ICON_32_ArcSCE)]^C^C_Arc;\C
ID_ArcSCA[_Button("Arc,SCA", ICON_16_ArcSCA, ICON_32_ArcSCA)]^C^C_Arc;\C;\A

Explanation
ID_TbCircle[_Flyout("Circle", ICON_16_Circle, ICON_32_Circle, _OtherIcon,
M2.TB_Circle)]

In this line M2 is the MENUGROUP name (***MENUGROUP=M2) and TB_Circle is the name
of the toolbar submenu. **M2.TB_Circle** will load the submenu TB_Circle that has been defined in
the M2 menugroup. If M2 is missing, AutoCAD will not display the flyout when you select the
Circle button.

ID_CirDIA[_Button("Circle C,D", ICON_16_**CirDIA**, ICON_32_**CirDIA**)]^C^C_Circle;\D

CirDIA is a user-defined bitmap that displays the Circle-diameter button. If you use any other
name, AutoCAD will not display the desired button.

To load the toolbar, use the **MENULOAD** command to load the menu file. The **MyDraw1** toolbar
will be displayed on the screen.

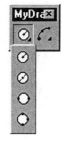

Figure 5-22 Toolbar for Example 9

Figure 5-23 Toolbar with flyout

Menu-Specific Help

AutoCAD for Windows allows access to online help. For example, if you want to define a helpstring for the **CIRCLE** and **ARC** commands, the syntax is as follows:

***HELPSTRING
ID_Copy *(This command will copy the selected object.)*
ID_Ellipse *(This command will draw an ellipse.)*

The ***HELPSTRING** is the section label for the helpstring menu section. The lines defined in this section start with a name tag (ID_Copy) and are followed by the label enclosed in square brackets. If you press the F1 key when the menu item is highlighted, the help engine gets activated and AutoCAD displays the helpstring defined with the name tag.

CUSTOMIZING THE TOOLBARS

AutoCAD has provided several toolbars that should be sufficient for general use. However, sometimes you may need to customize the toolbars so that the commands that you use frequently are grouped in one toolbar. This saves time in selecting commands. It also saves the drawing space because you do not need to have several toolbars on the screen. The following example explains the process involved in creating and editing the toolbars.

Example 10

In this example you will create a new toolbar (MyToolbar1) that has Line, Polyline, Circle (Center, Radius option), Arc (Center, Start, End option), Spline, and Paragraph Text (**MTEXT**) commands. You will also change the image and tooltip of Line button and perform other operations like deleting toolbars and buttons and copying buttons between toolbars.

1. Select the toolbars in the **View** menu. The **Toolbars** dialog box will appear on the screen. You can also invoke this dialog box by entering **TBCONFIG** at AutoCAD's Command prompt.

2. Choose **New...** button in the **Toolbars** dialog box; AutoCAD will display the **New Toolbar** dialog box at the top of the existing (Toolbars) dialog box.

3. Enter the name of the toolbar in the **Toolbar Name** edit box (i.e., MyToolbar1) and then chose the **OK** button to exit the dialog box. The name of the new toolbar (MyToolbar1) is displayed in the **Toolbars** dialog box (Figure 5-24).

4. Select the new toolbar, if it is not already selected, by selecting the box that is located just to the left of the toolbar name (MyToolbar1). The new toolbar (MyToolbar1) appears on the screen.

5. Choose the **Customize...** button in the **Toolbars** dialog box (Figure 5-25). The **Customize Toolbars** dialog box is displayed on the screen.

6. In the **Categories** edit box select the down arrow to display the list of toolbars and select the draw toolbar. (You can also type D to display the toolbars that start with the letter D). AutoCAD displays the draw tool buttons.

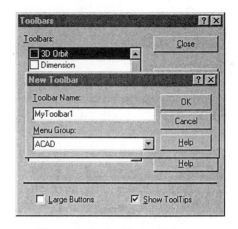

Figure 5-24 Toolbars dialog box

7. Choose and drag the Line button and position it in the MyToolbar1 toolbox. Repeat the same for Polyline, Circle (Center, Radius option), Arc (Center, Start, End option), Spline, and Text (Paragraph text) buttons.

8. Now, select the Dimensioning category and then select and drag some of the dimensioning commands to MyToolbar1.

9. Similarly, you can open any toolbar category and add commands to the new toolbar.

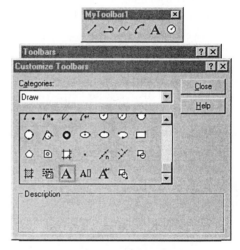

10. When you are done defining the commands for the new toolbar, choose the **Close** button in the **Customize Toolbars** dialog box to return to the **Toolbars** dialog box. Choose the **Close** button again to return to the AutoCAD screen.

Figure 5-25 MyToolbar1 toolbar and
Customize Toolbars dialog box

11. Test the buttons in the new toolbar (MyToolbar1).

Creating a New Image and Tooltip for a Button

1. Make sure the button with the image you want to edit is displayed on the screen. In this example we want to edit the image of the **Line** button of MyToolbar1.

2. Right-click on the **Line** button of MyToolbar1; the **Toolbars** dialog box appears on the screen. Again, right-click on the **Line** button; the **Button Properties** dialog box is displayed with the image of the existing **Line** button (Figure 5-26).

3. To edit the shape of the image, choose the **edit** button to access the **Button Editor**. Select the Grid to display the grid lines.

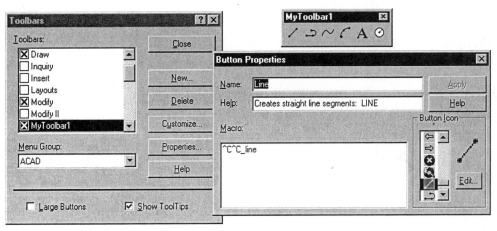

Figure 5-26 MyToolbar1 toolbar, Toolbars and **Button Properties** dialog box

You can edit the shape by using different tools in the **Button Editor**. You can draw a Line by choosing the **Line** button and specifying two points. You can draw a circle or an ellipse by using the **Circle** button. The **Erase** button can be used to erase the image.

4. In this example, we want to change the color of Line image. To accomplish this, erase the existing line, and then select the color and draw a line. Also, create the shape L in the lower right corner (Figure 5-27).

5. Choose the **SaveAs** button and save the image as **MyLine** in the Tutorial directory. Choose the **Close** button to exit the **Button Editor**.

6. In the **Button Properties** dialog box, enter MyLine in the **Name** edit box. This changes the tooltip of this button. Now, choose the **Apply** button to apply the changes to the image. Close the dialog boxes to return to the AutoCAD screen. Notice the change in the button image and tooltip (Figure 5-28).

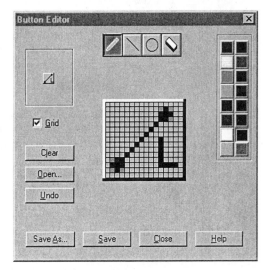

Figure 5-27 Button Editor dialog box

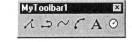

Figure 5-28 MyToollbar1 Toolbar

Deleting the Button from a Toolbar

1. Right-click on the button that you want to remove from the toolbox. The **Toolbars** dialog box is displayed. In this example, we want to remove the **Spline** button from the MyToolbar1 toolbox.

2. Choose the **Customize...** button in the **Toolbars** dialog box. The **Customize Toolbars** dialog box is displayed.

3. Click and drag out the **Spline** button from the MyToolbar1 toolbox. The button will be deleted from the toolbox.

4. Repeat the above step to delete other buttons, if needed.

5. Close the dialog boxes to return to the screen.

Deleting a Toolbar

1. Select Toolbars from the **View** menu.

2. Select the Toolbar that you want to delete and then choose the **Delete** button. The toolbar you selected is deleted.

Copying a Tool Button

In this example we want to copy the ordinate **dimensioning** button from the **Dimensioning** toolbar to MyToolbar1.

1. Select the Toolbars from the **View** menu or right-click on any button in any toolbar.

2. In the **Toolbars** dialog box select the check box for Dimensioning and MyToolbar1. The Dimensioning and MyToolbar1 toolbars are displayed on the screen.

3. In the **Toolbars** dialog box choose the **Customizing...** button; the **Customize Toolbars** dialog box appears.

4. Hold the CTRL key down and then choose and drag the **Ordinate dimensioning** button from the **Dimensioning** toolbar to MyToolbar1. The Ordinate dimensioning toolbar is copied to MyToolbar1.

 Note

*If you do not hold down the CTRL key, the button you selected will be moved from the **Dimensioning** toolbar to MyToolbar1.*

Any changes made in the toolbars are saved in the ACAD.MNS and ACAD.MNR files. The following is the partial listing of ACAD.MNS file:

```
**MYTOOLBAR1
ID_MyToolbar1_0 [_Toolbar("MyToolbar1", _Floating, _Show, 512, 177, 1)]
ID_Line_0     [_Button("MyLine", "ICON.bmp", "ICON_24_LINE")]^C^C_line
ID_CircleCenterRadius_0 [_Button("Circle Center Radius", "ICON_16_CIRRAD",
"ICON_24_CIRRAD")]^C^C_circle
ID_Polyline_0 [_Button("Polyline", "ICON_16_PLINE", "ICON_24_PLINE")]^C^C_pline
ID_ArcCenterStartEnd_0 [_Button("Arc Center Start End", "ICON_16_ARCCSE",
"ICON_24_ARCCSE")]^C^C_arc _c
```

Creating Custom Toolbars with Flyout Icons

In this example you will create a custom toolbar with flyout buttons.

1. Right-click on any toolbar; the **Toolbars** dialog box is displayed on the screen.

2. In the **Toolbars** dialog box, choose the **New...** button to display the **New Toolbars** dialog box. Enter the name of the new toolbar (for example, MYToolbar2) and then choose the **OK** button to exit the box.

3. In the **Toolbars** dialog box, select the new toolbar name (MYToolbar2) and then choose the **Customize...** button to display the **Customize Toolbars** dialog box. From the drop-down list select custom and then drag the **flyout** button to the new toolbar.

4. Right-click twice on the **flyout** button of the new toolbar. The **Flyout Properties** dialog box (Figure 5-29) is displayed on the screen. To associate it with a predefined toolbar, select the toolbar name (for example, ACAD.Draw) from the list.

5. Choose the **Apply** button and then close the dialog boxes. If you click on the **flyout** button of the new toolbar, the corresponding **flyout** buttons are displayed.

6. Now you can add new buttons to the toolbar or delete the ones you do not want in this toolbar.

7. If you select Show This Button's Icon, the selected icon will be displayed on the screen.

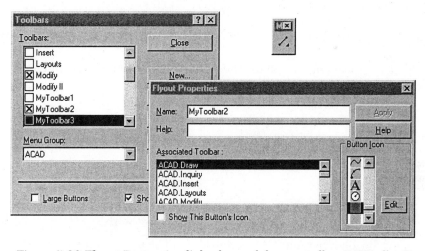

Figure 5-29 Flyout Properties dialog box and the new toolbar (MYToolbar2)

Review Questions

1. A pull-down menu can have _____ sections.

2. The length of the section title should not exceed _____ characters.

3. The section titles in a pull-down menu are _____ justified.

4. In a pull-down menu, a line consisting of two hyphens ([--]) _____ automatically to fill the _____ of the menu.

5. If the menu item begins with a tilde (~), the items will be _____

6. The pull-down menus are _____ when the dynamic zoom is in progress.

7. Every cascading menu in the menu file should have a _____ name.

8. The cascading menu name can be _____ characters long.

9. The cascading menu names should not have any _____ blanks.

10. In Windows you can write partial menus, toolbars, and accelerator key definitions. (T/F)

11. A menu file can contain only one MENUGROUP. (T/F)

12. You can load the partial menu file by using the AutoCAD _____ command.

13. Once the menu is loaded, you can use the _____ (AutoLISP function) to display the partial menus.

Exercises

Exercise 3   *General*

Write a menu for the following AutoCAD commands. (The layout of the menu is shown in Figure 5-30.)

LINE	DIM HORZ	DTEXT LEFT
CIRCLE C,R	DIM VERT	DTEXT RIGHT
CIRCLE C,D	DIM RADIUS	DTEXT CENTER
ARC 3P	DIM DIAMETER	DTEXT ALIGNED
ARC SCE	DIM ANGULAR	DTEXT MIDDLE
ARC CSE	DIM LEADER	DTEXT FIT

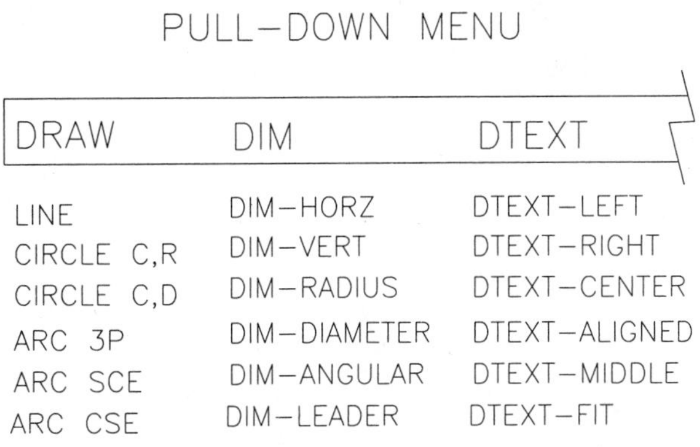

PULL—DOWN MENU

DRAW	DIM	DTEXT
LINE	DIM—HORZ	DTEXT—LEFT
CIRCLE C,R	DIM—VERT	DTEXT—RIGHT
CIRCLE C,D	DIM—RADIUS	DTEXT—CENTER
ARC 3P	DIM—DIAMETER	DTEXT—ALIGNED
ARC SCE	DIM—ANGULAR	DTEXT—MIDDLE
ARC CSE	DIM—LEADER	DTEXT—FIT

Figure 5-30 Layout of menu

Exercise 4   *General*

Write a menu for the following AutoCAD commands.

LINE	**BLOCK**
PLINE	**WBLOCK**
CIRCLE C,R	**INSERT**
CIRCLE C,D	**BLOCK LIST**
ELLIPSE AXIS ENDPOINT	**ATTDEF**
ELLIPSE CENTER	**ATTEDIT**

Exercise 5   *General*

Write a partial menu for Windows. The menu file should have two menus, POP1 (MyArc) and POP2 (MyDraw). The MyArc menu should contain all Arc options and must be displayed at the sixth position. Similarly, the MyDraw menu should contain **LINE**, **CIRCLE**, **PLINE**, **TRACE**, **DTEXT**, and **MTEXT** commands and should occupy the ninth position.

Exercise 6   *General*

Write a menu file for a toolbar with a flyout. The name of the toolbar is MyDrawX1, and it contains two buttons, **Polygon** and **Ellipse**. When you select the **Polygon** button, it should display a flyout with **Rectangle** and **Polygon** button. Similarly, when you choose the **Ellipse** button, it should display the **Ellipse-Center** Option and **Ellipse-Edge** Option button.

Chapter 5

Exercise 7

Write a menu for the following AutoCAD commands. (The layouts of the menu is shown in Figure 5-31.)

LAYER NEW	**SNAP** 0.25	**UCS WORLD**
LAYER MAKE	**SNAP** 0.5	**UCS PREVIOUS**
LAYER SET	**GRID** 1.0	**VPORTS** 2
LAYER LIST	**DRID** 10.0	**VPORTS** 4
LAYER ON	**APERTURE** 5	**VPORTS SING.**
LAYER OFF	**PICKBOX** 5	

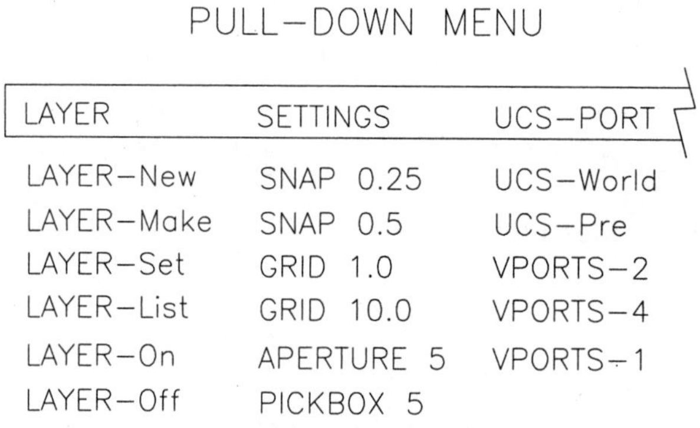

PULL—DOWN MENU

LAYER	SETTINGS	UCS—PORT
LAYER—New	SNAP 0.25	UCS—World
LAYER—Make	SNAP 0.5	UCS—Pre
LAYER—Set	GRID 1.0	VPORTS—2
LAYER—List	GRID 10.0	VPORTS—4
LAYER—On	APERTURE 5	VPORTS—1
LAYER—Off	PICKBOX 5	

Figure 5-31 Design of menu for Exercise 7

Chapter 6

Tablet Menus

Learning Objectives

After completing this chapter, you will be able to:

- *Understand the advantages of tablet menus.*
- *Write and customize tablet menus.*
- *Load menus and configure tablet menus.*
- *Write tablet menus with different block sizes.*
- *Assign commands to tablet overlays.*

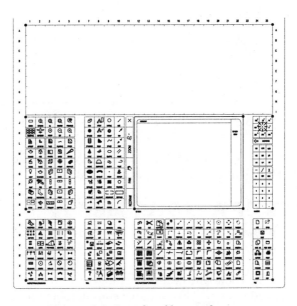

Figure 6-1 *Sample tablet template*

STANDARD TABLET MENU

The tablet menu provides a powerful alternative for entering commands. In the tablet menu, the commands are selected from the template that is secured on the surface of a digitizing tablet. To use the tablet menu you need a digitizing tablet and a pointing device. You also need a tablet template (Figure 6-1) that contains AutoCAD commands arranged in various groups for easy identification.

The standard AutoCAD menu file has four tablet menu sections: TABLET1, TABLET2, TABLET3, and TABLET4. When you start AutoCAD and get into the drawing editor, the tablet menu sections TABLET1, TABLET2, TABLET3, and TABLET4 are automatically loaded. The commands defined in these four sections are then assigned to different blocks of the template.

The first tablet menu section (TABLET1) has 225 blank blocks that can be used to assign up to 225 menu items. The remaining tablet menu sections contain AutoCAD commands, arranged in functional groups that make it easier to identify and access the commands. The commands contained in the TABLET2 section include RENDER, SOLID MODELING, DISPLAY, INQUIRY, DRAW, ZOOM, and PAPER SPACE commands. The TABLET3 section contains numbers, fractions, and angles. The commands contained in the TABLET4 section include **TEXT**, **DIMENSIONING**, **OBJECT SNAPS**, **EDIT**, **UTILITY**, **XREF**, and **SETTINGS**.

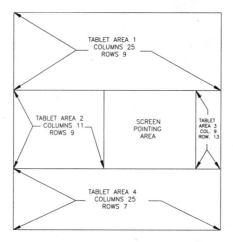

Figure 6-2 *Four tablet areas of the AutoCAD tablet template*

The AutoCAD tablet template has four tablet areas (Figure 6-2), which correspond to four tablet menu sections, TABLET1, TABLET2, TABLET3, and TABLET4.

ADVANTAGES OF A TABLET MENU

A tablet menu has the following advantages over the screen menu, pull-down menu, image menu, or keyboard.

1. In the tablet menu the commands can be arranged so that the most frequently used commands can be accessed directly. This can save considerable time in entering AutoCAD com-

mands. In the screen menu or the pull-down menu some of the commands cannot be accessed directly. For example, to generate a horizontal dimension you have to go through several steps. You first select Dimension from the root menu and then Linear. In the tablet menu you can select the Linear dimensioning command directly from the digitizer. This saves time and eliminates the distraction that takes place as you page through different screens.

2. You can have the graphical symbols of the AutoCAD commands drawn on the tablet template. This makes it much easier to recognize and select commands. For example, if you are not an expert in AutoCAD dimensioning, you may find Baseline and Continue dimensioning confusing. But if the command is supported by the graphical symbol illustrating what a command does, the chances of selecting a wrong command are minimized.

3. You can assign any number of commands to the tablet overlay. The number of commands you can assign to a tablet is limited only by the size of the digitizer and the size of the rectangular blocks.

CUSTOMIZING A TABLET MENU

As with a screen menu, you can customize the AutoCAD tablet menu. It is a powerful customizing tool to make AutoCAD more efficient.

The tablet menu can contain a maximum of four sections: TABLET1, TABLET2, TABLET3, and TABLET4. Each section represents a rectangular area on the digitizing tablet. These rectangular areas can be further divided into any number of rectangular blocks. The size of each block depends on the number of commands that are assigned to the tablet area. Also, the rectangular tablet areas can be located anywhere on the digitizer and can be arranged in any order. The AutoCAD **TABLET** command configures the tablet. The **MENU** command loads and assigns the commands to the rectangular blocks on the tablet template.

Before writing a tablet menu file, it is very important to design the layout of the design of tablet template. A well-thought-out design can save a lot of time in the long run. The following points should be considered when designing a tablet template:

1. Learn the AutoCAD commands that you use in your profession.

2. Group the commands based on their function, use, or relationship with other commands.

3. Draw a rectangle representing a template so that it is easy for you to move the pointing device around. The size of this area should be appropriate to your application. It should not be too large or too small. Also, the size of the template depends on the active area of the digitizer.

4. Divide the remaining area into four different rectangular tablet areas for TABLET1, TABLET2, TABLET3, and TABLET4. It is not necessary to use all four areas; you can have fewer tablet areas, but four is the maximum.

5. Determine the number of commands you need to assign to a particular tablet area; then determine the number of rows and columns you need to generate in each area. The size of the blocks does not need to be the same in every tablet area.

6. Use the **TEXT** command to print the commands on the tablet overlay, and draw the symbols of the command, if possible.

7. Plot the tablet overlay on good-quality paper or a sheet of Mylar. If you want the plotted side of the template to face the digitizer board, you can create a mirror image of the tablet overlay and then plot the mirror image.

WRITING A TABLET MENU

When writing a tablet menu you must understand the AutoCAD commands and the prompt entries required for each command. Equally important is the design of the tablet template and the placement of various commands on it. Give considerable thought to the design and layout of the template, and, if possible, invite suggestions from AutoCAD users in your trade. To understand the process involved in developing and writing a tablet menu, consider Example 1.

Example 1

Write a tablet menu for the following AutoCAD commands. The commands are to be arranged as shown in Figure 6-3. Make a tablet menu template for configuration and command selection (filename **TM1.MNU**).

LINE	**CIRCLE**	**PLINE**
CIRCLE C,D	**ERASE**	**CIRCLE 2P**

Figure 6-3 represents one of the possible template designs, where the AutoCAD commands are in one row at the top of the template, and the screen pointing area is in the center. There is only one area in this template; therefore, you can place all these commands under the section label TABLET1. To write a menu file you can use any text editor like Notepad.

The name of the file is TM1, and the extension of the file is **.MNU. The line numbers are not part of the file. They are shown here for reference only.**

***TABLET1	1
^C^CLINE	2
^C^CPLINE	3
^C^CCIRCLE	4
^C^CCIRCLE;\D	5
^C^CCIRCLE;2P	6
^C^CERASE	7

Explanation

Line 1
***TABLET1
TABLET1 is the section label of the first tablet area. All the section labels are preceded by three asterisks (***).

***TABLET1

Where *** ---------------- Three asterisks designate section label.
          TABLET1 ------ Section label for TABLET1

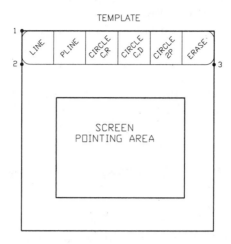

Figure 6-3 *Design of tablet template*

Line 2
^C^CLINE
^C^C cancels the existing command twice; **LINE** is an AutoCAD command. There is no space between the second ^C and **LINE**.

      ^C^C LINE
                Where **^C^C** ---------- Cancels the existing command twice
                       Line ------------- AutoCAD **LINE** command

Line 3
^C^CPLINE
^C^C cancels the existing command twice; **PLINE** is an AutoCAD command.

Line 4
^C^CCIRCLE
^C^C cancels the existing command twice; **CIRCLE** is an AutoCAD command. The default input for the **CIRCLE** command is the center and the radius of the circle; therefore, no additional input is required for this line.

Line 5
^C^CCIRCLE;\D
^C^C cancels the existing command twice; **CIRCLE** is an AutoCAD command like the previous line. However, this command definition requires the diameter option of the circle command. This is accomplished by using \D in the command definition. There should be no space between the backslash (\) and the D, but there should always be a space **before** the backslash (\). The backslash (\) lets the user enter a point, and in this case it is the center point of the circle. After you enter the center point, the diameter option is selected by the letter D, which follows the backslash (\).

^ C ^ CCIRCLE;\D

Where **CIRCLE** --------- **CIRCLE** command
      ; ------------------ Space for RETURN
      **** -------------------- Pause for input
      D ------------------ Diameter option

Line 6

^ C ^ CCIRCLE;2P

^ C ^ C cancels the existing command twice; **CIRCLE** is an AutoCAD command. Semicolon is for ENTER or RETURN. The 2P selects the two-point option of the **CIRCLE** command.

Line 7

^ C ^ CERASE

^ C ^ C cancels the existing command twice; **ERASE** is an AutoCAD command that erases the selected objects.

Note

In the tablet menu, the part of the menu item that is enclosed in the brackets is used for screen display only. For example, in the following menu item, T1-6 will be ignored and will have no effect on the command definition.

[T1-6] ^ C ^ CCIRCLE;2P

Where **[T1-6]** ------------ For reference only and has no effect on the command
                        definition
      T1 ----------------- Tablet area 1
      6 ------------------- Item number 6

The reference information can be used to designate the tablet area and the line number.

Before you can use the commands from the new tablet menu, you need to configure the tablet and load the tablet menu.

TABLET CONFIGURATION

To use the new template to select the commands, you need to configure the tablet so that AutoCAD knows the location of the tablet template and the position of the commands assigned to each block. This is accomplished by using the AutoCAD **TABLET** command. Secure the tablet template (Figure 6-3) on the digitizer with the edges of the overlay approximately parallel to the edges of the digitizer. Enter the AutoCAD **TABLET** command, select the **Configure option**, and respond to the following prompts. Figure 6-4 shows the points you need to select to configure the tablet.

Command: **TABLET**
Enter an option [ON/OFF/CAL/CFG]: **CFG**
Enter number of tablet menus desired (0–4) <current>: **1**
Do you want to realign tablet menus? [Yes/No] <N>: **Y**
Digitize upper left corner of menu area 1; **P1**
Digitize lower left corner of menu area 1; **P2**
Digitize lower right corner of menu area 1; **P3**

Enter the number of columns for menu area 1: **6**
Enter the number of rows for menu area 1: **1**
Do you want to respecify the Fixed Screen Pointing Area? [Yes/No]<N>: **Y**
Digitize lower left corner of Fixed Screen pointing area: **P4**
Digitize upper right corner of Fixed Screen pointing area: **P5**
Do you want to specify the Floating Screen Pointing Area? [Yes/No]<N>: **N**

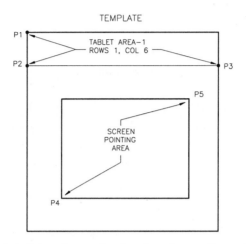

Figure 6-4 *Points that need to be selected to configure the tablet*

Note
*The three points P1, P2, and P3 should form a 90-degree angle. If the selected points do not form a
90-degree angle, AutoCAD will prompt you to enter the points again until they do.
The tablet areas should not overlap the screen pointing area.*

*The screen pointing area can be any size and located anywhere on the tablet as long as it is within the
active area of the digitizer. The screen pointing area should not overlap other tablet areas. The screen
pointing area you select will correspond to the monitor screen area. Therefore, the length-to-width
ratio of the screen pointing area should be the same as that of the monitor, unless you are using the
screen pointing area to digitize a drawing.*

LOADING MENUS

AutoCAD automatically loads the **ACAD.MNU** file when you get into the AutoCAD drawing editor. However, you can also load a different menu file by using the AutoCAD **MENU** command.

Command: **MENU**

When you enter the **MENU** command, AutoCAD displays the **Select Menu File** dialog box on the
screen (Figure 6-5). Select the menu file that you want to load and then choose the **Open** button.

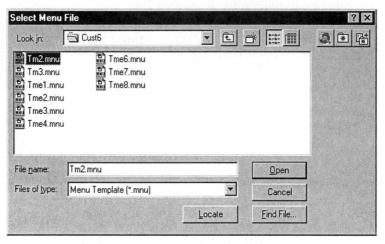

Figure 6-5 Select Menu File dialog box

You can also load the menu file from the command line.

    Command: **FILEDIA**
    Enter new value for FILEDIA <1>: **0**
    Command: **MENU**
    Menu filename <ACAD>: PDM1
                Where **ACAD** ------------ Default menu file
                       PDM1 ------------ Name of menu file

After you enter the **MENU** command, AutoCAD will prompt for the file name. Enter the name of the menu file without the file extension (**.MNU**), since AutoCAD assumes the extension **.MNU**. AutoCAD will automatically compile the menu file into **MNC** and **MNR** files.

Exercise 1   *General*

Write a tablet menu for the following AutoCAD commands. Make a tablet menu template for configuration and command selection (filename **TME1.MNU**).

LINE	**TEXT-Center**
CIRCLE C,R	**TEXT-Left**
ARC C.S.E	**TEXT-Right**
ELLIPSE	**TEXT-Aligned**
DONUT	

Use the template in Figure 6-6 to arrange the commands. The draw and text commands should be placed in two separate tablet areas.

TEMPLATE

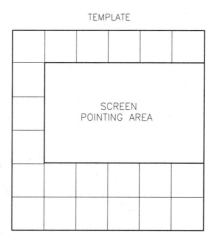

SCREEN
POINTING AREA

Figure 6-6 *Template for Exercise 1*

TABLET MENUS WITH DIFFERENT BLOCK SIZES

As mentioned earlier, the size of each tablet area can be different. The size of the blocks in these tablet areas can also be different. But the size of every block in a particular tablet area must be the same. This provides you with a lot of flexibility in designing a template. For example, you may prefer to have smaller blocks for numbers, fractions, or letters, and larger blocks for draw commands. You can also arrange these tablet areas to design a template layout with different shapes, such as L-shape and T-shape.

Tablet Area-1   ### Tablet Area-2

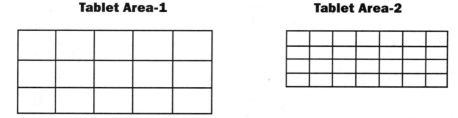

The following example illustrates the use of multiple tablet areas with different block sizes.

Example 2

Write a tablet menu for the tablet overlay shown in Figure 6-7(a). Figure 6-7(b) shows the number of rows and columns in different tablet areas (filename **TM2.MNU**).

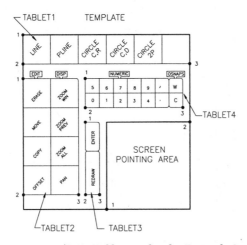

Figure 6-7(a)   *Tablet overlay for Example 2*

Notice that this tablet template has four different sections in addition to the screen pointing area. Therefore, this menu will have four section labels: TABLET1, TABLET2, TABLET3, and TABLET4. You can use any text editor (Notepad) to write the file. The following file is a listing of the tablet menu of Example 2.

```
***TABLET1                                                  1
^C^CLINE                                                    2
^C^CPLINE                                                   3
^C^CCIRCLE                                                  4
^C^CCIRCLE;\D                                               5
^C^CCIRCLE;2P                                               6
***TABLET2                                                  7
^C^CERASE                                                   8
^C^CZOOM;W                                                  9
^C^CMOVE                                                   10
^C^CZOOM;P                                                 11
^C^CCCOPY                                                  12
^C^CZOOM;A                                                 13
^C^COFFSET                                                 14
^C^CPAN                                                    15
***TABLET3                                                 16
;   .                                                      17
;                                                          18
'REDRAW                                                    19
'REDRAW                                                    20
'REDRAW                                                    21
***TABLET4                                                 22
5\                                                         23
6\                                                         24
```

7\	25
8\	26
9\	27
,\	28
WINDOW	29
0\	30
1\	31
2\	32
3\	33
4\	34
.\	35
CROSSING	36

Explanation

Lines 1-6
The first six lines are identical to the first six lines of the tablet menu in Example 1.

Line 9
^C^CZOOM;W
ZOOM is an AutoCAD command; W is the window option of the **ZOOM** command.

> **^C^CZOOM;W**
> Where **ZOOM** ----------- AutoCAD **ZOOM** command
> **;** ------------------ Space for Return
> **W** ---------------- Window option of **ZOOM** command

This menu item could also be written as:

> **^C^CZOOM W**
> Space between ZOOM nad W causes a RETURN

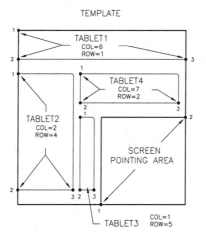

Figure 6-7(b) Number of rows and columns in different tablet

Lines 17 and 18
The semicolon (;) is for RETURN. It has the same effect as entering RETURN at the keyboard.

Lines 19-21
'REDRAW
REDRAW is an AutoCAD command that redraws the screen. Notice that there is no $^\wedge$C$^\wedge$C in front of the **REDRAW** command. If it had $^\wedge$C$^\wedge$C, the existing command would be canceled before redrawing the screen. This may not be desirable in most applications because you might want to redraw the screen without canceling the existing command. The single quotation (') in front of REDRAW makes the **REDRAW** command transparent.

Line 23
5
The backslash (\\) is used to introduce a pause for user input. Without the backslash you cannot enter another number or a character, because after you select the digit **5** it will automatically be followed by RETURN. For example, without the backslash (\\), you will not be able to enter a number like **5.6**. Therefore, you need the backslash to enable you to enter decimal numbers or any characters. To terminate the input, enter RETURN at the keyboard or select RETURN from the digitizer.

    5

                           Where \\ --------- Backslash for user input

ASSIGNING COMMANDS TO A TABLET

After loading the menu by means of the AutoCAD **MENU** command, you must configure the tablet. At the time of configuration AutoCAD actually generates and stores the information about the rectangular blocks on the tablet template. When you load the menu, the commands defined in the tablet menu are assigned to various blocks. For example, when you select the three points for tablet area 4, Figure 6-7(a), and enter the number of rows and columns, AutoCAD generates a grid of seven columns and two rows, as shown in the following diagram.

After Configuration

When you load the new menu, AutoCAD takes the commands under the section label TABLET4 and starts filling the blocks from left to right. That means "5", "6", "7", "8", "9", ",", and "Window" will be placed in the top row. The next seven commands will be assigned to the next row, starting from the left, as shown in the diagram on page 6-13.

After Loading Tablet Menu

5	6	7	8	9	.	Window
0	1	2	3	4	.	Cross

Similarly, tablet area 3 has been divided into five rows and one column. At first, it appears that this tablet area has only two rows and one column, as shown in Figure 6-7(a). When you configure this tablet area by specifying the three points and entering the number of rows and columns, AutoCAD divides the area into one column and five rows, as shown in the following diagrams.

After loading the menu, AutoCAD takes the commands in the TABLET3 section of the tablet menu and assigns them to the blocks. The first command (;) is placed in the first block. Since there are no more blocks in the first row, the next command (;) is placed in the second row. Similarly, the three **REDRAW** commands are placed in the next three rows. If you pick a point in the first two rows, you will select the **ENTER** command. Similarly, if you pick a point in the next three blocks, you will select the **REDRAW** command.

After Configuration   **After Loading Menu**

;
;
Redraw
Redraw
Redraw

This process is carried out for all the of tablet areas, and the information is stored in the AutoCAD configuration file, **ACAD.CFG**. If for any reason the configuration is not right, the tablet menu may not perform the desired function.

AUTOMATIC MENU SWAPPING

The screen menus can be automatically swapped by using the system variable **MENUCTL**. When the system variable **MENUCTL** is set to 1, AutoCAD automatically issues the **$S=CMDNAME** command, where **CMDNAME** is the name of the command that loads the submenu. For example, if you select the **LINE** command from the digitizer, menu, or enter the **LINE** command from the keyboard, the **CMDNAME** command will load the **LINE** submenu and display it in the screen menu area. To utilize this feature of AutoCAD, the command name and the submenu name must be the same. For example, if the name of the arc submenu is **ARC** and you select the **ARC** command, the arc submenu will be automatically loaded on the screen. However, if the submenu name is different (for example MYARC), then AutoCAD will not load the arc submenu on the screen. The default value of **MENUCTL** system variable is 1. If you set the **MENUCTL** variable to 0, AutoCAD will not utilize the **$S=CMDNAME** command feature to load the submenus.

Review Questions

1. The maximum number of tablet menu sections is _____.

2. A tablet menu area is _____ in shape.

3. The blocks in any tablet menu area are _____ in shape.

4. A tablet menu area can have _____ number of rectangular blocks.

5. You _____ assign the same command to more than one block on the tablet menu template.

6. The AutoCAD _____ command is used to configure the tablet menu template.

7. The AutoCAD _____ command is used to load a new menu.

Exercises

Exercise 2   *General*

Design the template and write a tablet menu to insert the following user-defined blocks:

BX1	BX5	BX9
BX2	BX6	BX10
BX3	BX7	BX11
BX4	BX8	BX12

Exercise 3   *General*

Design the template and the screen menu for the following AutoCAD commands:

LINE	**ZOOM-Win**	**DIM-Horz**
PLINE	**ZOOM-Dyn**	**DIM-Vert**
ARC	**ZOOM-All**	**DIM-Alig**

CIRCLE	ZOOM-Pre	DIM-Ang
ELLIPSE	ZOOM-Ext	DIM-Rad
POLYGON	ZOOM-Scl	DIM-Cen

Exercise 4 *General*

Write a tablet menu for the commands shown in the tablet menu template of Figure 6-8. Make a tablet menu template for configuration and command selection.

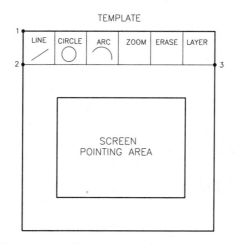

Figure 6-8 *Tablet menu template for Exercise 4*

Exercise 5 *General*

Write a tablet menu for the AutoCAD commands shown in Figure 6-9. Configure the tablet and then load the new menu. Make a tablet menu template for configuration and command selection.

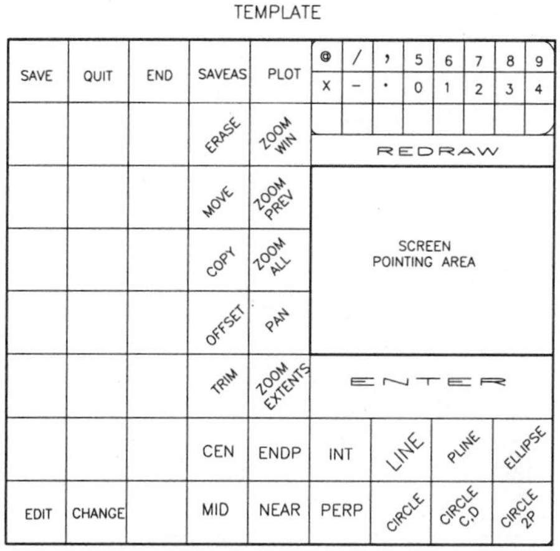

Figure 6-9 *Tablet overlay for Exercise 5*

Exercise 6
General

Write a tablet menu file for the AutoCAD commands shown in the template of Figure 6-10. Make a tablet menu template for configuration and command selection.

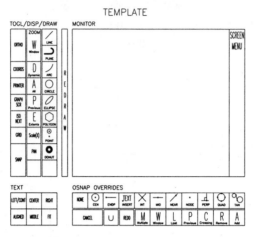

Figure 6-10 *Tablet menu template for Exercise 6*

Exercise 7
General

Write a combined pull-down and tablet menu file for the commands shown in the tablet menu template in Figure 6-11. Make a tablet menu template for configuration and command selection.

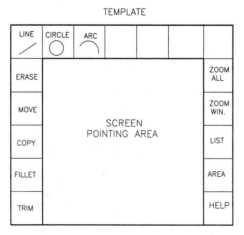

Figure 6-11 *Tablet menu template for Exercise 7*

Template for Example 1

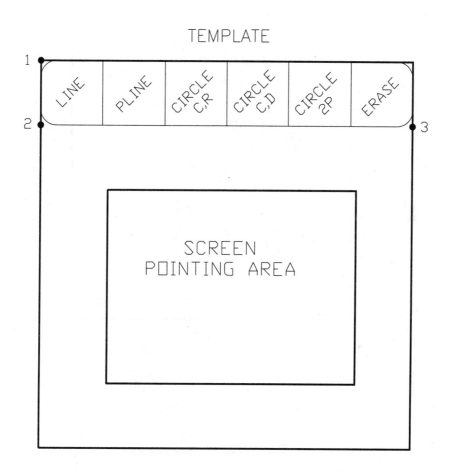

 Note

This template is for tablet configuration. You may make a copy of this page and then secure it on the digitizer surface for configuration.

Template for Exercise 1

TEMPLATE

SCREEN
POINTING AREA

 Note

This template is for tablet configuration. You may make a copy of this page and then secure it on the digitizer surface for configuration.

Chapter 6

Template for Example 2

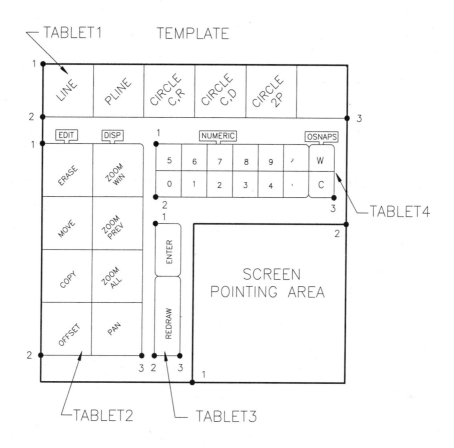

Note

This template is for tablet configuration. You may make a copy of this page and then secure it on the digitizer surface for configuration.

Template for Example 2

TEMPLATE

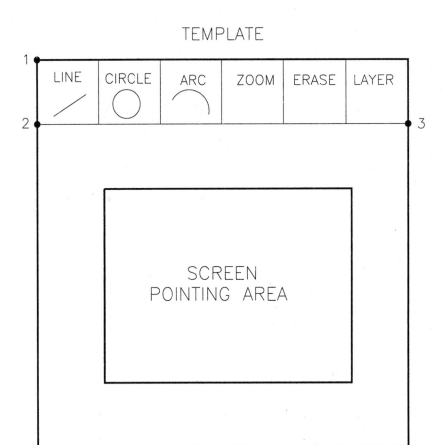

Note

This template is for tablet configuration. You may make a copy of this page and then secure it on the digitizer surface for configuration.

Template for Exercise 5

TEMPLATE

SAVE	QUIT	END	SAVEAS	PLOT	@ / , 5 6 7 8 9								
					X − • 0 1 2 3 4								
			ERASE	ZOOM WIN	REDRAW								
			MOVE	ZOOM PREV	SCREEN POINTING AREA								
			COPY	ZOOM ALL									
			OFFSET	PAN									
			TRIM	ZOOM EXTENTS	ENTER								
			CEN	ENDP	INT	LINE		PLINE		ELLIPSE			
EDIT	CHANGE		MID	NEAR	PERP	CIRCLE		CIRCLE C,D		CIRCLE 2P			

Note

This template is for tablet configuration. You may make a copy of this page and then secure it on the digitizer surface for configuration.

Template for Exercise 6

TEMPLATE

TOGL/DISP/DRAW   MONITOR

ORTHO	ZOOM W Window	LINE
COORDS	Dynamic	PLINE
PRINTER	All	ARC
GRAPH SCR	Previous	CIRCLE
ISO NEXT	Extents	ELLIPSE
GRID	Scale(X)	POLYGON
SNAP	PAN	POINT
		DONUT

R
E
D
R
A
W

SCREEN MENU

TEXT

| LEFT/CONT | CENTER | RIGHT |
| ALIGNED | MIDDLE | FIT |

OSNAP OVERRIDES

| NONE | CEN | ENDP | .TEXT INSERT | INT | MID | NEAR | NODE | PERP | QUAD | TAN |
| CANCEL | U | REDO | M Multiple | W Window | L Last | P Previous | C Crossing | R Remove | A Add |

Note

This template is for tablet configuration. You may make a copy of this page and then secure it on the digitizer surface for configuration.

Template for Exercise 7

TEMPLATE

LINE	CIRCLE	ARC				
ERASE						ZOOM ALL
MOVE		SCREEN POINTING AREA				ZOOM WIN.
COPY						LIST
FILLET						AREA
TRIM						HELP

Note

This template is for tablet configuration. You may make a copy of this page and then secure it on the digitizer surface for configuration.

Chapter 7

Image Tile Menus

Learning Objectives

After completing this chapter, you will be able to:
- *Write image tile menus.*
- *Reference and display submenus.*
- *Make slides for image tile menus.*

IMAGE TILE MENUS

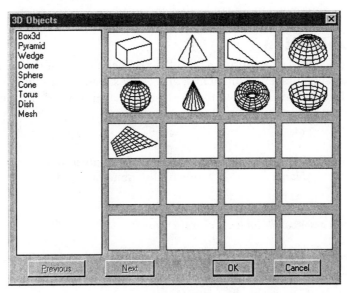

Figure 7-1 *Sample image tile menu display*

The image tile menus, also known as **icon menus**, are extremely useful for inserting a block, selecting a text font, or drawing a 3D object. You can also use the image tile menus to load an AutoLISP routine or a predefined macro. Thus, the image tile menu is a powerful tool for customizing AutoCAD.

The image tile menus can be accessed from the pull-down, tablet, button, or screen menu. However, the image tile menus **cannot** be loaded by entering the command from the keyboard. When you select an image tile, a dialog box that contains **20 image tiles** is displayed on the screen (Figure 7-1). The names of the slide files associated with image tiles appear on the left side of the dialog box with a scrolling bar that can be used to scroll the file names. The title of the image tile menu is displayed at the top of the dialog box (Figure 7-1). When you activate the image tile menu, an arrow that can be moved to select any image tile appears on the screen. You can select an image tile by selecting the slide file name from the dialog box and then choosing the **OK** button from the dialog box or double-clicking on the slide file name.

When you select the slide file, AutoCAD highlights the corresponding image tile by drawing a rectangle around the image tile. You can also select an image tile by moving the arrow to the desired image tile, and then pressing the PICK button of the pointing device. The corresponding slide file name will be automatically highlighted. And if you choose the **OK** button or double-click on the image tile, the command associated with that menu item will be executed. You can cancel an image tile menu by pressing ESCAPE on the keyboard, or selecting an image tile from the dialog box.

SUBMENUS

You can define an unlimited number of menu items in the image tile menu, but only 20 image tiles will be displayed at a time. If the number of items exceeds 20, you can use the **Next** and **Previous** buttons of the dialog box to page through different pages of image tiles. You can also define submenus that let you define smaller groups of items within an image tile menu section. When you select a submenu, the submenu items are loaded and displayed on the screen.

Submenu Definition

A submenu label consists of two asterisks (**) followed by the name of the submenu. The image tile menu can have any number of submenus, and every submenu should have a unique name. The items that follow a submenu, up to the next section label or submenu label, belong to that submenu. The format of a submenu label is:

    **Name

          Where ****** ----------------- Two asterisks designate a submenu
                   Name ------------ Name of the submenu

Note
The submenu name can be up to 31 characters long.

The submenu name can consist of letters, digits, and special characters, like $ (dollar), - (hyphen), and _ (underscore).

The submenu name should not have any embedded blanks.

Submenu names should be unique in a menu file.

Submenu Reference

The submenu reference is used to reference or load a submenu. It consists of a $ sign followed by a letter that specifies the menu section. The letter that specifies an image tile menu section is I. The menu section is followed by an equal sign (=) and the name of the submenu you want to activate. The submenu name should be without the **. Following is the format of a submenu reference:

$Section=Submenu
Where   **$** ------------------- "$" sign
        Section ---------- Menu section specifier
        = ----------------- "=" sign
        Submenu ------- Name of submenu

$I=IMAGE1
Where   **I** ------------------ I specifies image tile menu section
        IMAGE1 -------- Name of submenu

Displaying a Submenu

When you load a submenu, the new dialog box and the image tiles are not automatically displayed. For example, if you load submenu IMAGE1, the items contained in this submenu will not be displayed. To force the display of the new image tile menu on the screen, AutoCAD uses the special command $I=*:

$I=*
Where   **I** ------------------ I for image tile menu
        ***** ------------------ Asterisk (*)

WRITING AN IMAGE TILE MENU

The image tile menu consists of the section label ***IMAGE followed by image tiles or image tile submenus. The menu file can contain only one image tile menu section (***IMAGE); therefore, all image tiles must be defined in this section.

***IMAGE
Where   ******* --------------- Three asterisks designate a section label
        IMAGE ---------- Section label for an image tile

You can define any number of submenus in the image tile menu. All submenus have two asterisks followed by the name of the submenu (**PARTS or **IMAGE1):

**IMAGE1
Where   ****** ---------------- Two asterisks designate a submenu
        IMAGE1 -------- Name of submenu

The first item in the image tile menu is the title of the image tile menu, which is also displayed at the top of the dialog box. The image tile dialog box title has to be enclosed in brackets ([PLC-SYMBOLS]) and should not contain any command definition. If it does contain a command definition, AutoCAD ignores the definition. The remaining items in the image tile menu file contain slide names in the brackets and the command definition outside the brackets.

***IMAGE	Image tile menu section
**BOLTS	Image tile submenu (BOLTS)
[HEX-HEAD BOLTS]	Image tile title
[BOLT1]^C^CINSERT;B1	BOLT1 is slide file name;
	B1 is block name

Example 1

Write an image tile menu that will enable you to insert the block shapes from Figure 7-2 in a drawing by selecting the corresponding image tile from the dialog box. Use the menu to load the image tile menu.

PLC SYMBOLS   **ELECTRIC SYMBOLS**
NO (NORMALLY OPEN)   RESIS (RESISTANCE)
NC (NORMALLY CLOSED)   DIODE
COIL   GROUND

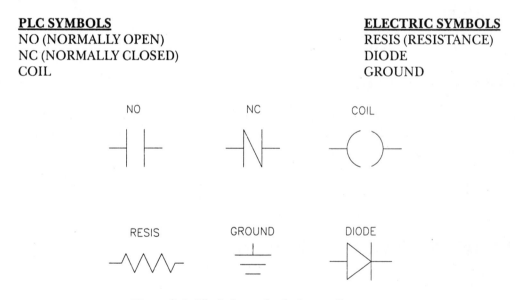

Figure 7-2 *Block shapes for the image tile menu*

As mentioned in earlier chapters, the first step in writing a menu is to design the menu so that the commands are arranged in a desired configuration. Figure 7-3 shows one possible design for the menu and the image tile menu for Example 1.

You can use the AutoCAD **EDIT** command to write the file. **The line numbers in the following file are for reference only and are not a part of the menu file.**

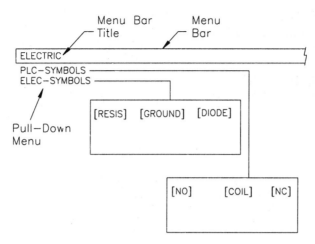

Figure 7-3 *Design of the menu and image tile menu for Example*

***POP1	1
[ELECTRIC]	2
[PLC-SYMBOLS]$I=IMAGE1 $I=*	3
[ELEC-SYMBOLS]$I=IMAGE2 $I=*	4
***IMAGE	5
**IMAGE1	6
[PLC-SYMBOLS]	7
[NO]^C^CINSERT;NO;\1.0;1.0;0	8
[NC]^C^CINSERT;NC;\1.0;1.0;0;	9
[COIL]^C^CINSERT;COIL	10
[No-Image]	11
[blank]	12
**IMAGE2	13
[ELECTRICAL SYMBOLS]	14
[RESIS]^C^CINSERT;RESIS;\\\\	15
[DIODE]^C^CINSERT;DIODE;\1.0;1.0;\	16
[GROUND]^C^CINSERT;GRD;\1.5;1.5;0;;	17

Explanation

Line 1
***POP1
In this menu item, ***POP1 is the section label and defines the first section of the menu.

Line 2
[ELECTRIC]
In this menu item [ELECTRIC] is the menu bar label for the POP1 menu. It will be displayed in the menu bar.

Line 3
[PLC-SYMBOLS]$I=IMAGE1 $I=*
In this menu item, $I=IMAGE1 loads the submenu IMAGE1; $I=* displays the current image tile menu on the screen.

> **[PLC-SYMBOLS]$I=IMAGE1 $I=***
> Where **Image1** ---------- Loads submenu IMAGE1
> $ ------------------- Forces display of current menu

Line 5
***IMAGE
In this menu item, ***IMAGE is the section label of the image tile menu. All the image tile menus have to be defined within this section; otherwise, AutoCAD cannot locate them.

Line 6
**IMAGE1
In this menu item, **IMAGE1 is the name of the image tile submenu.

Line 7
[PLC-SYMBOLS]
When you select line 3 ([PLC-SYMBOLS]$I=IMAGE1 $I=*), AutoCAD loads the submenu IMAGE1 and displays the title of the image tile at the top of the dialog box (Figure 7-4). This title is defined in Line 7. If this line is missing, the next line will be displayed at the top of the dialog box. Image tile titles can be any length, as long as they fit the length of the dialog box.

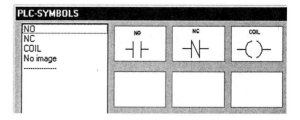

Figure 7-4 Image tile box for PLC-SYMBOLS

Line 8
[NO]^C^CINSERT;NO;\1.0;1.0;0
In this menu item, the first NO is the name of the slide and has to be enclosed within brackets. The name should not have any trailing or leading blank spaces. If the slides are not present, AutoCAD will not display any graphical symbols in the image tiles. However, the menu items will be loaded, and if you select this item, the command associated with the image tile will be executed. The sec-

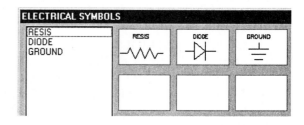

Figure 7-5 Image tile box for ELECTRICAL-SYMBOLS

ond NO is the name of the block that is to be inserted. The backslash (\) pauses for user input; in this case it is the block insertion point. The first 1.0 defines the Xscale factor. The second 1.0 defines the Yscale factor, and the following 0 defines the rotation.

[NO]^C^CINSERT;NO;\1.0;1.0;0

Where **NO** --------------- Block name

\------------------- Pause for block insertion point

1.0 --------------- Xscale factor

1.0 --------------- Yscale factor

0 ------------------ Rotation angle

When you select this item, it will automatically enter all the prompts of the **INSERT** command and insert the NO block at the given location. The only input you need to enter is the insertion point of the block.

Line 10

[COIL]^C^CINSERT;COIL

In this menu item the block name is given, but you need to define other parameters when inserting this block.

Line 11

[No-Image]

Notice the **blank space before No-Image**. If there is a space following the open bracket, AutoCAD does not look for a slide. AutoCAD instead displays the text, enclosed within the brackets, in the slide file list box of the dialog box.

Line 12

[blank]

Line 12 consists of **blank**; this displays a separator line in the list box and a blank image (no image) in the image box.

Line 15

[RESIS]^C^CINSERT;RESIS;

This menu item inserts the block RESIS. The first backslash (\) is for the block insertion point. The second and third backslashes are for the Xscale and Yscale factors. The fourth backslash is for the rotation angle. This menu item could also be written as:

[RESIS]^C^CINSERT;RESIS;

or

[RESIS]^C^CINSERT;RESIS

Line 16

[DIODE]^C^CINSERT;DIODE;\1.0;1.0;

If you select this menu item, AutoCAD will prompt you to enter the block insertion point and the rotation angle. The first backslash is for the block insertion point; the second backslash is for the rotation angle.

[DIODE]^C^CINSERT;DIODE;\1.0;1.0;

Where \------------------- Pause for insertion point

\------------------- Pause for rotation angle

Line 17

[GROUND]^C^CINSERT;GRD;\1.5;1.5;0;;

This menu item has two semicolons (;) at the end. The first semicolon after 0 is for RETURN and completes the block insertion process. The second semicolon enters a RETURN and repeats the **INSERT** command. However, when the command is repeated you will have to respond to all of the prompts. It does not accept the values defined in the menu item.

Note

The menu item repetition feature cannot be used with the image tile menus. For example, if the command definition starts with an asterisk ([GROUND]^C^CINSERT;GRD;\1.5;1.5;0;;), the command is not automatically repeated, as is the case with a pull-down menu.*

A blank line in an image tile menu terminates the menu and clears the image tiles.

The menu command $I=, which displays the current menu, cannot be entered at the keyboard.*

If you want to cancel or exit an image tile menu, press the ESC (Escape) key on the keyboard. AutoCAD ignores all other entries from the keyboard.

You can define any number of image tile menus and submenus in the image tile menu section of the menu file.

SLIDES FOR IMAGE TILE MENUS

The idea behind creating slides for the image tile menus is to display graphical symbols in the image tiles. This symbol makes it easier for you to identify the operation that the image tile will perform. Any slide can be used for the image tile. However, the following guidelines should be kept in mind when creating slides for the image tile menu:

1. When you make a slide for an image tile menu, draw the object so that it fills the entire screen. The **MSLIDE** command makes a slide of the existing screen display. If the object is small, the picture in the image tile menu will be small. Use **ZOOM** Extents or **ZOOM** Window to display the object before making a slide.

2. When you use the image tile menu, it takes some time to load the slides for display in the image tiles. The more complex the slides, the more time it will take to load them. Therefore, the slides should be kept as simple as possible and at the same time give enough information about the object.

3. Do not fill the object, because it takes a long time to load and display a solid object. If there is a solid area in the slide, AutoCAD does not display the solid area in the image tile.

4. If the objects are too long or too wide, it is better to center the image with the AutoCAD **PAN** command before making a slide.

5. The space available on the screen for image tile display is limited. Make the best use of this small area by giving only the relevant information in the form of a slide.

6. The image tiles that are displayed in the dialog box have the length-to-width ratio (aspect ratio) of 1.5:1. For example, if the length of the image tile is 1.5 units, the width is 1 unit. If the drawing area of your screen has an aspect ratio of 1.5 and the slide drawing is centered in the drawing area, the slide in the image tile will also be centered.

LOADING MENUS

AutoCAD automatically loads the **ACAD.MNU** file when you get into the AutoCAD drawing editor. However, you can also load a different menu file by using the AutoCAD **MENU** command.

Command: **MENU**

When you enter the **MENU** command, AutoCAD displays the **Select Menu File** dialog box (Figure 7-6) on the screen. Select the menu file that you want to load and then choose the **Open** button.

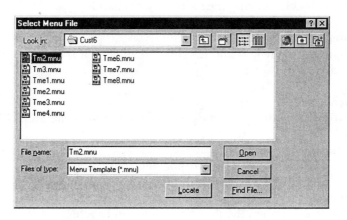

Figure 7-6 Select Menu File dialog box

You can also load the menu file from the command line.

Command: **FILEDIA**
Enter new value for FILEDIA <1>: **0**
Command: **MENU**
Menu filename <ACAD>: PDM1
        Where **PDM1** ------------ Name of menu file
               ACAD ------------ Default menu file

After you enter the **MENU** command, AutoCAD will prompt for the file name. Enter the name of the menu file without the file extension (**.MNU**), since AutoCAD assumes the extension **.MNU**. AutoCAD will automatically compile the menu file into **MNS** and **MNR** files.

Note

When you load the image tile menu, some of the commands will be displayed on the screen menu area. This happens because the screen menu area is empty. If the menu file contains a screen menu also, the pull-down menu items will not be displayed in the screen menu area.

When you load a new menu, the original menu you had on the screen prior to loading the new menu is disabled. You cannot select commands from the screen menu, pull-down menu, pointing device, or the digitizer, unless these sections are defined in the new menu file.

To load the original menu, use the **MENU** command again and enter the name of the menu file.

RESTRICTIONS

The pull-down and image tile menus are very easy to use and provide quick access to some of the frequently used AutoCAD commands. However, the menu bar, the pull-down menus, and the image tile menus are disabled during the following commands:

DTEXT Command
After you assign the text height and the rotation angle to a **DTEXT** command, the menu is automatically disabled.

SKETCH Command
The menus are disabled after you set the record increment in the **SKETCH** command.

VPOINT Command
The menus are disabled while the axis tripod and compass are displayed on the screen.

ZOOM Command
The menus are disabled when the dynamic zoom is in progress.

DVIEW
The menus are disabled when the dynamic view is in progress.

Exercise 1	*General*

Write an image tile menu for inserting the blocks shown in Figure 7-7. Arrange the blocks in two groups so that you have two submenus in the image tile menu.

PIPE FITTINGS	**ELECTRIC SYMBOLS**
GLOBE-P	BATTERY
GLOBE	CAPACITOR
REDUCER	COUPLER
CHECK	BREAKER

IMAGE TILE MENU ITEM LABELS

As with screen and menus, you can use menu item labels in the image menus. However, the menu item labels in the image tile menus use different formats, and each format performs a particular

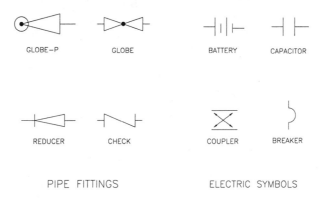

Figure 7-7 Block shapes for Exercise 1

function in the image tile menu. The menu item labels appear in the slide list box of the dialog box. The maximum number of characters that can be displayed in this box is 23. The characters in excess of 23 are not displayed in the list box. However, this does not affect the command that is defined with the menu item.

Menu Item Label Formats

[slidename]. In this menu item label format, **slidename** is the name of the slide displayed in the image tile. This name (slidename) is also displayed in the list box of the corresponding dialog box.

[slidename,label]. In this menu item label format, **slidename** is the name of the slide displayed in the image tile. However, unlike the previous format, the **slidename** is **not** displayed in the list box. The **label** text is displayed in the list box. For example, if the menu item label is **[BOLT1,1/ 2-24UNC-3LG]**, **BOLT1** is the name of the slide and **1/2-24UNC-3LG** is the label that will be displayed in the list box.

[slidelib(slidename)]. In this menu item label format, **slidename** is the name of the slide in the slide library file **slidelib**. The slide (slidename) is displayed in the image tile, and the slide file name (slidename) is also displayed in the list box of the corresponding dialog box.

[slidelib(slidename,label)]. In this menu item label format, **slidename** is the name of the slide in the slide library file **slidelib**. The slide (slidename) is displayed in the image tile, and the label text is displayed in the list box of the corresponding dialog box.

[blank]. This menu item will draw a line that extends through the width of the list box. It also displays a blank image tile in the dialog box.

[label]. If the **label** text is preceded by a space, AutoCAD does not look for a slide. The label text is displayed in the list box only. For example, if the menu item label is [EXIT]^C, the label text (EXIT) will be displayed in the list box. If you select this item, the cancel command (^C) defined with the item will be executed. The **label** text is **not** displayed in the image tile of the dialog box.

Example 2

Write the pull-down and image tile menus for inserting the following commands. B1 to B15 are the block names.

BLOCK
WBLOCK
ATTDEF
LIST
INSERT

BL1	BL6	BL11
BL2	BL7	BL12
BL3	BL8	BL13
BL4	BL9	BL14
BL5	BL10	BL15

The first step in writing a menu is to design the menu. Figure 7-8 shows the design of the menu. If you select Insert from the menu, the image tiles and block names will be displayed in the dialog box.

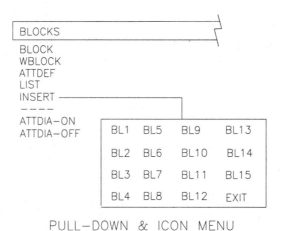

PULL—DOWN & ICON MENU

Figure 7-8 Design of screen and menus for Example 2

The following file is a listing of the menu file for Example 2. The file contains the screen, tablet, pull-down, and image tile menu sections. **The line numbers are not a part of the menu file; they are given for reference only.**

***POP1	1
[INSERT]	2
[BLOCK]^C^CBLOCK	3
[WBLOCK]^C^CWBLOCK	4
[ATTRIBUTE DEFINITION]^C^CATTDEF	5
[LIST BLOCK NAMES]^C^CINSERT;?	6

```
[INSERT]^C^C$I=IMAGE1 $I=*                                              7
[—]                                                                     8
[ATTDIA-ON]^C^CSETVAR ATTDIA 1                                          9
[ATTDIA-OFF]^C^CSETVAR ATTDIA 0                                        10
                                                                       11
***IMAGE                                                               12
**IMAGE1                                                               13
[BLOCK INSERTION FOR EXAMPLE-2]                                        14
[BL1]^C^C$S=INSERT1 INSERT;BL1;\1.0;1.0;\                              15
[BL2]^C^C$S=INSERT1 INSERT;BL2;\1.0;1.0;0                              16
[BL3]^C^C$S=INSERT1 INSERT;BL3;\;;\                                    17
[BL4]^C^C$S=INSERT1 INSERT;BL4;\;;;                                    18
[BL5]^C^C$S=INSERT1 INSERT;*BL5;\1.75                                  19
[BL6]^C^C$S=INSERT1 INSERT;BL6;\XYZ                                    20
[BL7]^C^C$S=INSERT1 INSERT;BL7;\XYZ;;;\0                               21
[BL8]^C^C$S=INSERT1 INSERT;BL8;\XYZ;;;;;                               22
[BL9]^C^C$S=INSERT1 INSERT;BL9;\XYZ;;;;\                               23
[BL10]^C^C$S=INSERT1 INSERT;*BL10;\XYZ;\                               24
[BL11]^C^C$S=INSERT2 INSERT;BL11;\XYZ;1;1.5;2;45                       25
[BL12]^C^C$S=INSERT2 INSERT;BL12;\XYZ;\\;;                             26
[BL13]^C^C$S=INSERT2 INSERT;*BL13;\\45                                 27
[BL14]^C^C$S=INSERT2 INSERT;BL14;\C;@1.0,1.0;0                         28
[BL15]^C^C$S=INSERT2 INSERT;BL15;\C;@1.0,2.0;\                         29
```

Explanation

Line 1
***POP1
This is the section label of the first menu. The menu items defined on lines 2 through 10 are defined in this section.

Line 12
***IMAGE
This is the section label of the image tile menu.

Line 13
**IMAGE1
IMAGE1 is the name of the submenu; the items on lines 14 through 29 are defined in this submenu.

Line 15
[BL1]^C^C$S=INSERT1 INSERT;BL1;\1.0;1.0;
In this menu item, BL1 is the name of the slide and $S=INSERT1 loads the submenu INSERT1 on the screen.

[BL1]^C^C$S=INSERT1 INSERT;BL1;\1.0;1.0;
        Where **INSERT** --------- AutoCAD **INSERT** command
                  INSERT1 ------- Loads submenu INSERT1
                  BL1 -------------- Slide file name

Exercise 2   *General*

Write a tablet and image tile menu for inserting the following blocks. (The template of the tablet menu is shown in Figure 7-9.)

B1	B4	B7
B2	B5	B8
B3	B6	B9

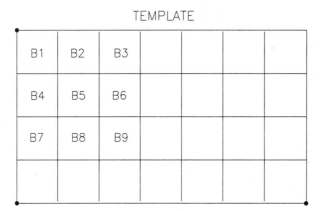

Figure 7-9 Tablet template for Exercise 2

Review Questions

1. The image tiles are displayed in the _____ box.

2. An **image tile** menu can be canceled by entering _____ at the keyboard.

3. The **image title** dialog box can contain a maximum of _____ image tiles.

4. A blank line in an image tile menu _____ the image tile menu.

5. The menu item repetition feature _____ be used with the image tile menu.

6. The drawing for a slide should be _____ on the entire screen before making a slide.

7. You _____ fill a solid area in a slide for an image tile menu.

8. An image tile menu _____ be accessed from a tablet menu.

Exercises

Exercise 3 *General*

Write an image tile menu for the following commands. Make the slides that will graphically illustrate the function of the command.

LINE	**CIRCLE C,R**
PLINE	**CIRCLE C,D**
	CIRCLE 2P

Exercise 4 *General*

Write a tablet and image tile menu for inserting the following blocks. The layout of the template for the tablet menu is shown in Figure 7-10.

B1	B2	B3	C1	C2	C3
B4	B5	B6	C4	C5	C6
B7	B8	B9	C7	C8	C9

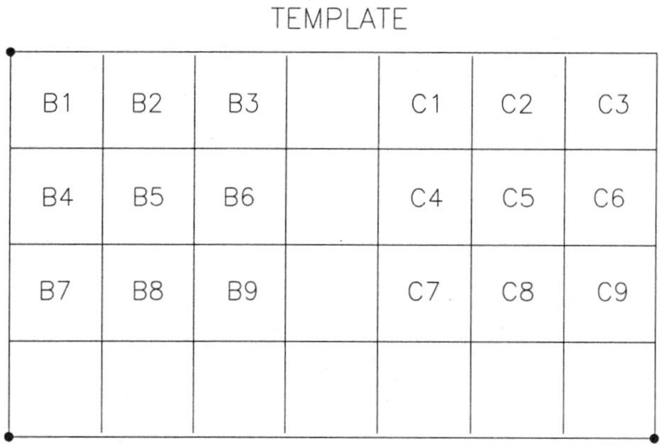

TEMPLATE

Figure 7-10 *Tablet template for Exercise 4*

Chapter 8

Button and Auxiliary Menus

Learning Objectives

After completing this chapter, you will be able to:
• *Write button menus.*
• *Learn special handling for button menus.*
• *Define and load submenus.*

BUTTON MENUS

You can use a multibutton pointing device to specify points, select objects, and execute commands. These pointing devices come with different numbers of buttons, but four-button and twelve-button pointing devices are very common. In addition to selecting points and objects, the multibutton pointing devices can be used to provide easy access to frequently used AutoCAD commands. The commands are selected by pressing the desired button; AutoCAD automatically executes the command or the macro that is assigned to that button. Figure 8-1 shows one such pointing device with 12 buttons.

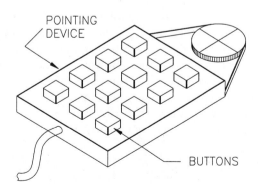

Figure 8-1 Pointing device with 12 buttons

The AutoCAD software package comes with a standard button menu that is part of the **ACAD.MNU** file. The **standard** menu is automatically loaded when you start AutoCAD and enter the drawing editor. You can write your own button menu and assign the desired commands or macros to various buttons of the pointing device.

AUXILIARY MENUS

In a menu file, you can have up to four auxiliary menu sections (AUX1 through AUX4). The auxiliary menu sections (***AUXn) are identical to the button menu sections. The difference is in the hardware. If your computer uses a system mouse, it will automatically use the auxiliary menu.

WRITING BUTTON AND AUXILIARY MENUS

In a menu file, you can have up to four button menus (BUTTONS1 through BUTTONS4) and four auxiliary menus (AUX1 through AUX4). The buttons and the auxiliary menus are identical. However, they are OS dependent. For example, in AutoCAD 2000 the system mouse uses auxiliary menus. If your system has a pointing device (digitizer puck), AutoCAD automatically assigns the commands defined in the BUTTONS sections of the menu file to the buttons of the pointing device. When you load the menu file, the commands defined in the BUTTONS1 section of the menu file are assigned to the pointing device (digitizer puck) and if your computer has a system mouse, the mouse will use the auxiliary menus. You can also access other button menus (BUTTONS2 through BUTTONS4) by using the following keyboard-and-button (buttons of the pointing device-digitizer puck) combinations.

Aux Menu	Buttons Menu	Keyboard + Button Sequence
AUX1	BUTTONS1	Press the button of the pointing device.
AUX2	BUTTONS2	Hold down the SHIFT key and press the button of the pointing device.
AUX3	BUTTONS3	Hold down the CTRL key and press the button of the pointing device.
AUX4	BUTTONS4	Hold down the SHIFT and CTRL keys and press the button of the pointing device.

One of the buttons, generally the first, is used as a pick button to specify the coordinates of the screen crosshairs and send that information to AutoCAD. This button can also be used to select commands from various other menus, such as tablet menu, screen menu, menu, and image tile menu. This button cannot be used to enter a command, but AutoCAD commands can be assigned to other buttons of the pointing device. Before writing a button menu, you should decide the commands and options you want to assign to different buttons, and know the prompts associated with those commands. The following example illustrates the working of the button menu and the procedure for assigning commands to different buttons.

Example 1

Write a button menu for the following AutoCAD commands. The pointing device has 12 buttons (Figure 8-2), and button number 1 is used as a pick button (filename **BM1.MNU**).

Button	Function	Button	Function
2	RETURN	3	CANCEL

4	CURSOR MENU	5	SNAP
6	ORTHO	7	AUTO
8	INT,END	9	LINE
10	CIRCLE	11	ZOOM Win
12	ZOOM Prev		

You can use the AutoCAD **EDIT** command or any other text editor to write the menu file. The following file is a listing of the button menu for Example 1. **The line numbers are for reference only and are not a part of the menu file.**

```
***BUTTONS1                                                    1
;                                                             2
^C^C                                                          3
$P0=*                                                         4
^B                                                            5
^O                                                            6
AUTO                                                          7
INT,ENDP                                                      8
^C^CLINE                                                      9
^C^CCIRCLE                                                   10
'ZOOM;Win                                                    11
'ZOOM;Prev                                                   12
```

Explanation

Line 1
***BUTTONS
***BUTTONS1 is the section label for the first button menu. When the menu is loaded, AutoCAD compiles the menu file and assigns the commands to the buttons of the pointing device.

Line 2
;
This menu item assigns a semicolon (;) to button number 2. When you specify the second button on the pointing device, it enters a Return. It is like entering Return at the keyboard or the digitizer.

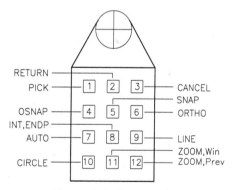

Figure 8-2 Pointing device

Line 3
^C^C
This menu item cancels the existing command twice (^C^C). This command is assigned to button number 3 of the pointing device. When you pick the third button on the pointing device, it cancels the existing command twice.

Line 4
$P0=*

This menu item loads and displays the cursor menu POP0, which contains various object snap modes. It is assumed that the POP0 menu has been defined in the menu file. This command is assigned to button number 4 of the pointing device. If you press this button, it will load and display the shortcut menu on the screen near the crosshairs location.

Line 5
^B
This menu item changes the Snap mode; it is assigned to button number 5 of the pointing device. When you pick the fifth button on the pointing device, it turns the Snap mode on or off. It is like holding down the CTRL key and then pressing the B key.

Line 6
^O
This menu item changes the ORTHO mode; it is assigned to button number 6. When you pick the sixth button on the pointing device, it turns the ORTHO mode on or off.

Line 7
AUTO
This menu item selects the AUTO option for creating a selection set; this command is assigned to button number 7 on the pointing device.

Line 8
INT,ENDP
In this menu item, INT is for the Intersection Osnap, and ENDP is for the Endpoint Osnap. This command is assigned to button number 8 on the pointing device. When you pick this button, AutoCAD looks for the intersection point. If it cannot find an intersection point, it then starts looking for the endpoint of the object that is within the pick box.

INT,ENDP
Where   **INT** --------------- Intersection object snap
        ENDP ------------ Endpoint object snap

Line 9
^C^CLINE
This menu item defines the **LINE** command; it is assigned to button number 9. When you select this button, AutoCAD cancels the existing command, and then selects the **LINE** command.

Line 10
^C^CCIRCLE
This menu item defines the **CIRCLE** command; it is assigned to button number 10. When you pick this button, AutoCAD automatically selects the **CIRCLE** command and prompts for the user input.

Line 11
'ZOOM;Win
This menu item defines a transparent **ZOOM** command with Window option; it is assigned to button number 11 of the pointing device.

'ZOOM;Win

Where '-------------------- Single quote makes **ZOOM** command transparent

          ZOOM ---------- AutoCAD **ZOOM** command

          ; ------------------- Semicolon for RETURN

          Win -------------- Window option of **ZOOM** command

Line 12

'ZOOM;Prev

This menu item defines a transparent **ZOOM** command with **previous** option; it is assigned to button number 12 of the pointing device.

Note

If the button menu has more menu items than the number of buttons on the pointing device, the menu items in excess of the number of buttons are ignored. This does not include the pick button. For example, if a pointing device has three buttons in addition to the pick button, the first three menu items will be assigned to the three buttons (buttons 2, 3, and 4). The remaining lines of the button menu are ignored.

The commands are assigned to the buttons in the same order in which they appear in the file. For example, the menu item that is defined on line 3 will automatically be assigned to the fourth button of the pointing device. Similarly, the menu item that is on line 4 will be assigned to the fifth button of the pointing device. The same is true of other menu items and buttons.

SPECIAL HANDLING FOR BUTTON MENUS

When you press any button on the multibutton pointing device (Figure 8-3), AutoCAD receives the following information:

1. **The button number**
2. **The coordinates of the screen crosshairs**

You can write a button menu that uses one or both pieces of information. The following example uses only the button number and ignores the coordinates of the screen crosshairs:

Example

^C^CLINE

If this command is assigned to the second button of the pointing device and you select this button, AutoCAD will receive the button number and the coordinates of the screen cross-hairs. AutoCAD will execute the command that is assigned to the second button, but it will ignore the coordinates of the crosshairs. The following example uses both button number and the coordinates of the screen cross hairs:

Example

^C^CLINE;\

In this menu item the **LINE** command is followed by a semicolon (;) and a backslash (\). The semicolon inputs an ENTER and the backslash normally causes a pause for user input. However, in the buttons menu AutoCAD will not pause for the user input. The backslash (\) in this menu item will use the coordinates of the screen crosshairs supplied by the pointing device as the starting point (From point) of the line. AutoCAD will then prompt for the second point of the line (To point).

Example 2

Write a button menu for the following AutoCAD commands. The menu items should use the information about the coordinate points of the screen crosshairs, where applicable (filename BM2.MNU).

Button	**Function**
2	Enter
3	ERASE (with SI and NEAR options)
4	INT,END
5	LINE
6	PLINE
7	CIRCLE

The following file is a listing of the button menu for Example 2. The line numbers are not a part of the file; they are for reference only.

```
***BUTTONS1                                          1
;                                                    2
^C^CERASE;SI;NEAR;\                                  3
INT,ENDP;\                                           4
LINE;\                                               5
PLINE;\                                              6
CIRCLE;\                                             7
```

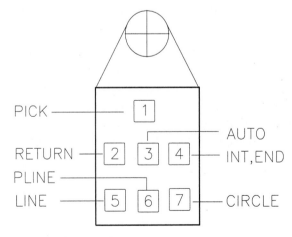

Figure 8-3 *Pointing device with seven buttons*

Line 3

^C^CERASE;SI;NEAR;

This menu item defines an **ERASE** command with single object selection option (SI) and near object snap (NEAR). The backslash is used to accept the coordinates supplied by the screen cross-hairs. This command is assigned to button number 3 of the pointing device. If you point to an object and press the third button on the pointing device, the object will be erased and AutoCAD will automatically return to the command prompt.

^C^CERASE;SI;NEAR;\

    Where **SI** ----------------- SIngle object selection mode
          NEAR ------------ NEAR object snap
          \-------------------- Accepts coordinates of screen cross-hairs

Line 4

INT,ENDP;

In this menu item INT is for the intersection snap and ENDP is for the endpoint snap. This command is assigned to button number 8 on the pointing device. When you pick this button, AutoCAD looks for the intersection point. If it cannot find an intersection point, it starts looking for the endpoint of the object that is within the pick box. The backslash (\) is used here to accept the coordinates of the screen crosshairs. If the crosshairs are near an object within the pick box AutoCAD will snap to the intersection's point, or the endpoint of the object when you press button number 4 on the pointing device.

INT,ENDP;\

    Where **INT** -------------- Intersection object snap
          EBDP ------------ Endpoint object snap
          \-------------------- Accepts the coordinates of the screen crosshairs

Line 7

CIRCLE;

This menu item generates a circle. The backslash (\) is used here to accept the coordinates of the screen crosshairs as the center of the circle. If you press button number 7 of the pointing device to draw a circle, you do not need to enter the center of the circle because the current position of the screen crosshairs automatically becomes the center of the circle. You only have to define the radius to generate the circle.

Note

The coordinate information associated with the button can be used with the first backslash only.

If a menu item in a button menu has more than one backslash (\), the remaining backslashes are ignored. For example, in the following menu item the first backslash uses the coordinates of screen crosshairs as the insertion point and the remaining backslashes are ignored and do not cause a pause as in other menus.

Example
   INSERT;B1\\\0

$$
\begin{array}{ll}
\text{Where} \quad \backslash \text{--------------------} & \text{Insertion point} \\
\backslash \text{--------------------} & \text{X-scale factor} \\
\backslash \text{--------------------} & \text{Y-scale factor} \\
\mathbf{0} \text{------------------} & \text{Rotation}
\end{array}
$$

SUBMENUS

The facility to define submenus is not limited to screen, pull-down, and image menus only. You can also define submenus in the button menu.

Submenu Definition

A submenu label consists of two asterisk signs (**) followed by the name of the submenu. The button menu can have any number of submenus, and every submenu should have a unique name. The items that follow a submenu, up to the next section label or submenu label, belong to that submenu. The format of a submenu label is:

   **Name

$$
\begin{array}{ll}
\text{Where} \quad \textbf{**} \text{-----------------} & \text{Two asterisk signs (**) designate a submenu} \\
\textbf{Name} \text{------------} & \text{Name of the submenu}
\end{array}
$$

Note
The submenu name can be up to 31 characters long.

The submenu name can consist of letters, digits, and the special characters like: $ (dollar), - (hyphen), and _ (underscore).

The submenu name should not have any embedded blanks.

The submenu names should be unique in a menu file.

Submenu Reference

The submenu reference is used to reference or load a submenu. It consists of a "$" sign followed by a letter that specifies the menu section. The letter that specifies a button menu section is B. The menu section is followed by "=" sign and the name of the submenu that the user wants to activate. The submenu name should be without "**". Following is the format of a submenu reference:

   $Section=Submenu

$$
\begin{array}{ll}
\text{Where} \quad \textbf{\$} \text{--------------------} & \text{"\$" sign} \\
\textbf{Section} \text{----------} & \text{Menu section specifier} \\
\textbf{=} \text{------------------} & \text{"=" sign} \\
\textbf{Submenu} \text{-------} & \text{Name of submenu}
\end{array}
$$

Example
        $B=BUTTON1

                Where **B** ----------------- B specifies buttons menu section
                       BUTTON1 ----- Name of submenu

LOADING MENUS

From the buttons menu, you can load any menu that is defined in the screen, pull-down, or image menu sections by using the appropriate load commands. It may not be needed in most of the applications, but if you want to, you can load the menus that are defined in other menu sections.

Loading Screen Menus

You can load any menu that is defined in the screen menu section from the button menu by using the following load commands:

        $S=X $S=INSERT

                Where **S** ------------------- S specifies screen menu
                       X ------------------ Submenu name defined in screen menu section
                       INSERT --------- Submenu name defined in screen menu section

The first load command ($S=X) loads the submenu X that has been defined in the screen menu section of the menu file. The X submenu can contain 21 blank lines, so that when it is loaded it clears the screen menu. The second load command ($S=INSERT) loads the submenu **INSERT** that has been defined in the screen menu section of the menu file.

Loading a Menu

You can load a menu from the button menu by using the following command:

        $P1=P1A $P1=*

                Where **$P1=P1A** -------- Loads the submenu P1A
                       $P1=* ----------- Forces the display of new menu items

The first load command $P1=P1A loads the submenu P1A that has been defined in the POP1 section of the menu file. The second load command $P1=* forces the new menu items to be displayed on the screen.

Loading an Image Menu

You can also load an image menu from the button menu by using the following load command:

        $I=IMAGE1 $I=*

                Where **$I=IMAGE1** --- Load the submenu IMAGE1
                       $I=* ------------ To display the dialog box

This menu item consists of two load commands. The first load command $I=IMAGE1 loads the image submenu IMAGE1 that has been defined in the image menu section of the file. The second load command $I=* displays the new dialog box on the screen.

Example 3

Write a button menu for a pointing device that has six buttons. The functions assigned to different buttons are shown in the following table (filename BM3.MNU):

SUBMENU B1	SUBMENU B2
1. PICK	1. PICK
2. Enter	2. Enter
3. LOAD OSNAPS SUBMENU	3. LOAD IMAGE1 SUBMENU
4. LOAD ZOOM1 SUBMENU	4. EXPLODE
5. LOAD BUTTON SUBMENU B1	5. LOAD BUTTON SUBMENU B1
6. LOAD BUTTON SUBMENU B2	6. LOAD BUTTON SUBMENU B2

Note

OSNAPS submenu is defined in the POP1 section of the menu.

ZOOM1 submenu is defined in the screen section of the menu file.

IMAGE1 submenu is defined in the IMAGE section of the menu file. The IMAGE1 submenu contains four images for inserting the blocks.

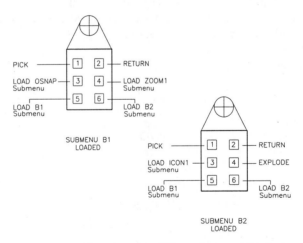

Figure 8-4 Commands assigned to different buttons of pointing device

In this example there are two submenus. The submenus are loaded by picking button 5 for submenu B1 and button 6 for submenu B2. When submenu B1 is loaded, AutoCAD assigns the commands that are defined under submenu B1 to the buttons of the pointing device. Similarly, when submenu B2 is loaded, AutoCAD assigns the commands that are defined under submenu B2 to the buttons of the pointing device. Figure 8-4 shows the commands that are assigned to the buttons after loading submenu B1, and submenu B2.

You can use AutoCAD's **EDIT** command to write the file. The following file is the listing of the button menu for Example 3. The line numbers are not a part of the file. They are shown here for reference only.

```
***BUTTON                                           1
**B1                                                2
;                                                   3
$P1=*                                               4
$S=X $S=ZOOM1 'ZOOM;Win                             5
$B=B1                                               6
$B=B2                                               7
**B2                                                8
;                                                   9
^C^C$I=IMAGE1 $I=*                                 10
EXPLODE;\                                           11
$B=B1                                              12
$B=B2                                              13
```

Line 2
**B1
This menu item defines a submenu. The name of the submenu is B1.

Line 4
$P1=*
This menu item loads and displays the pull-down menu that has been defined in the POP1 section of the menu.

Line 5
$S=X $S=ZOOM1 'ZOOM;Win
This menu item will load and display the ZOOM1 submenu on the screen and enter a transparent **ZOOM Window** command. $S=X loads the submenu X that has been defined in the screen menu section of the menu file. $S=ZOOM1 loads the submenu ZOOM1 that has also been defined in the screen menu section. **'ZOOM;Win** is a transparent zoom command with the window option.

$S=X $S=ZOOM1 'ZOOM;Win
        Where **$S=X** ----------- Loads X submenu that clears the screen menu
                     $S=ZOOM1 --- Loads ZOOM1 submenu
                     'Zoom;Win ----- Transparent ZOOM Window command

Line 8
**B2
This menu item defines the submenu B2.

Line 10
^C^C$I=IMAGE1 $I=*
This menu item cancels the existing command twice and then loads the submenu IMAGE1 that has been defined in the IMAGE menu. $I=* displays the current dialog box on the screen.

$\wedge$ C $\wedge$ C$I=IMAGE1 $I=*

    Where $\wedge$ **C** $\wedge$ **$** ----------- Cancels the existing command twice

                  I=IMAGE1 ---- Loads IMAGE1 submenu

                  $I=* ------------ Displays the dialog box with images

Line 11

EXPLODE;

This menu item will explode an object. It utilizes the special feature of the pointing device buttons that supply the coordinates of the screen crosshair. When you select this button, it will explode the object where the screen crosshair is located.

    EXPLODE;

           Where   **Explode** --------- AutoCAD's **EXPLODE** command

                     ; ------------------- Semicolon (;) for ENTER

                     **** -------------------- Utilizes the coordinates of screen

                                 crosshair as input

Line 12

$B=B1

This menu item loads submenu B1 and assigns the functions defined under this submenu to different buttons of the pointing device.

Line 13

$B=B2

This menu item loads submenu B2 and assigns the functions defined under this submenu to different buttons of the pointing device.

Review Questions

1. A multibutton pointing device can be used to specify _____ , or select _____, or enter AutoCAD _____.

2. AutoCAD receives the button _____ and _____ of screen crosshairs when a button is activated on the pointing device.

3. If the number of menu items in the button menu is more than the number of buttons on the pointing device, the excess lines are _____.

4. Commands are assigned to the buttons of the pointing device in the _____ order in which they appear in the buttons menu.

5. The format of referencing or loading a submenu that has been defined in the image menu is _____.

6. The format of displaying a loaded submenu that has been defined in the image menu is _____.

7. The format of the **LOAD** command for loading a submenu that has been defined in the image menu is _____.

Exercises

Exercise 1 *General*

Write a button menu for the following AutoCAD commands. The pointing device has 10 buttons (Figure 8-5), and button number 1 is used for specifying the points. The blocks are to be inserted with a scale factor of 1.00 and a rotation of 0 degrees (file name **BME1.MNU**).

1. PICK BUTTON
2. RETURN
3. CANCEL
4. OSNAPS
5. INSERT B1
6. INSERT B2
7. INSERT B3
8. OOM Window
9. ZOOM All
10. ZOOM Previous

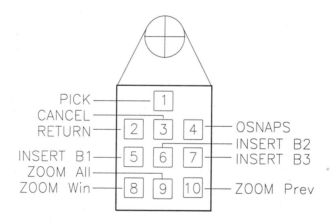

Figure 8-5 Pointing device with 10 buttons

1. B1, B2, B3 are the names of the blocks or Wblocks that have already been created.
2. Assume that the Osnap submenu has already been defined in the screen menu section of the menu file.
3. Use the transparent **ZOOM** command for **ZOOM Previous** and **ZOOM Window.**

Exercise 2 *General*

Write a button menu for a pointing device that has 10 buttons. The functions assigned to different buttons are shown in the following table and in Figure 8-6 (filename BME2.MNU):

SUBMENU B1		**SUBMENU B2**	
1.	PICK	1.	PICK
2.	Enter	2.	Enter
3.	LINE	3.	LOAD IMAGE1
4.	CIRCLE	4.	LOAD IMAGE2
5.	LOAD OSNAPS	5.	LOAD P2 (Pull-down)
6.	ZOOM Win	6.	LOAD P3 (Pull-down)
7.	ZOOM Prev	7.	INSERT
8.	ERASE	8.	EXPLODE
9.	LOAD BUTTON MENU B1	9.	LOAD BUTTON MENU B1
10.	LOAD BUTTON MENU B2	10.	LOAD BUTTON MENU B2

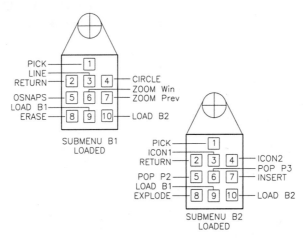

Figure 8-6 Commands assigned to different buttons of pointing device

Assume that:
1. OSNAPS submenu is defined in the POP1 or POP0 (Cursor menu) section of the menu.
2. Menus P2 and P3 are defined in the POP2 and POP3 sections of the menu.
3. IMAGE1 and IMAGE2 submenus are defined in the IMAGE section of the menu file. The IMAGE1 and IMAGE2 submenus contain four images each for inserting the blocks.

Chapter 9

Screen Menus

Learning Objectives

After completing this chapter, you will be able to:
* *Write screen menus.*
* *Load screen menus.*
* *Write submenus and referencing submenus.*
* *Write menu files with multiple submenus.*
* *Use menu command repetition in menus.*
* *Write menus for foreign languages.*
* *Use control and special characters in menu items.*
* *Use command definition without return or space.*
* *Use menu items with single object selection mode.*
* *Use AutoLISP and DIESEL expressions in menus.*

SCREEN MENU

When you are in the AutoCAD drawing editor the screen menu is displayed on the right of the screen. The AutoCAD screen menu displays AutoCAD at the top, followed by asterisk signs (* * * *) and a list of commands (Figure 9-1).

Depending on the scope of the menu, the size of the menu file can vary from a few lines to several hundred. A menu file consists of section labels, submenus, and menu items. A menu item consists of an item label and a command definition. The menu item label is enclosed in brackets, and the command definition (menu macro) is outside the brackets.

The menu item label that is enclosed in the brackets is displayed in the screen menu area of the monitor and is not a part of the command definition. The command definition, the part of the menu item outside the bracket, is the executable part of the menu item. To understand the process of developing and writing a screen menu, consider the following example.

[LINE   ] $S=X $S=LINE ^C^CLINE
            Where   **[LINE   ] $S=X $S=LINE ^C^CLINE** -------- Menu item
                    [LINE   ] ------- Item label or menu command name
                    $S=X $S=LINE ^C^CLINE - Command definition

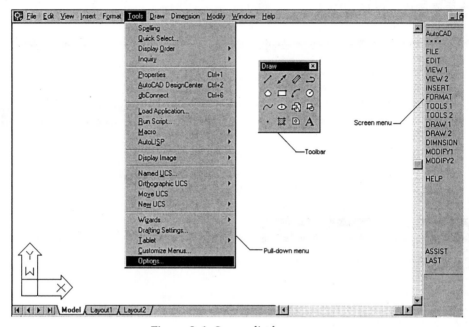

Figure 9-1 Screen display

Example 1

Write a screen menu for the following AutoCAD commands (file name **SM1.MNU**):

LINE
CIRCLE C,R
CIRCLE C,D
CIRCLE 2P
ERASE
MOVE

The layout of these commands is shown in Figure 9-2(a). This menu is named MENU-1 and it should be displayed at the top of the screen menu. It lets you know the menu you are using. You can use any text editor to write the file. The file name can be up to eight characters long, and the file extension must be **.MNU**. For Example 1, **SM1.MNU. SM1** is the name of the screen menu file, and **.MNU** is the extension of this file. All menu files have the extension **.MNU**.

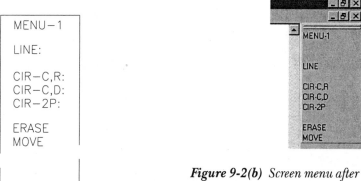

Figure 9-2(b) *Screen menu after loading the menu file*

Figure 9-2(a) *Layout of screen menu*

The following file is a listing of the screen menu for Example 1. **The line numbers on the right are not a part of the file. They are shown here for reference only.**

```
***SCREEN                                          1
[ MENU-1        ]                                  2
[              ]                                   3
[              ]                                   4
[LINE          ]^C^CLINE                           5
[              ]                                   6
[CIR-C,R       ]^C^CCIRCLE                         7
[CIR-C,D       ]^C^CCIRCLE;\D                      8
[CIR- 2P       ]^C^CCIRCLE;2P                      9
[              ]                                  10
[ERASE         ]^C^CERASE                         11
[MOVE          ]^C^CMOVE                          12
```

Explanation

Line 1
***SCREEN
***SCREEN is the section label for the screen menu. The lines that follow the screen menu are treated as a part of this menu. The screen menu definition will be terminated by another section label, such as ***TABLET1 or ***POP1.

Line 2
[MENU-1]
This menu item displays MENU-1 on the screen. Anything that is inside the brackets is for display only and has no effect on the command. The maximum number of characters or spaces that can be displayed inside these brackets is eight, because the width of the screen menu column on the screen is eight characters. If the number of characters is more than eight, the remaining characters are not displayed on the screen and can be used for comments. The part of the menu item that is outside the brackets is executed even if the number of characters inside the bracket is more than eight.

Chapter 9

Lines 3 and 4

[   ]

These menu items print a blank line on the screen menu. There are eight blank spaces inside the bracket. You could also use the brackets with no spaces between the brackets ([]). When the blank spaces are printed on the screen menu, it displays a blank line. This line does not contain anything outside the bracket; therefore, no command is executed. To provide the space in the menu you can also leave a blank space in the menu file or have two brackets ([ ]). The next line, line 4, also prints a blank line.

Line 5

[LINE   ]^C^CLINE

This menu item displays LINE on the screen. The first ^C (caret C) cancels the existing command, and the second ^C cancels the command again. The two **CANCEL** (^C^C) commands are required to make sure that the existing commands are canceled before a new command is executed. Most AutoCAD commands can be canceled with just one **CANCEL** command. However, some commands, like dimensioning and pedit, need to be canceled twice to get out of the command. **LINE** will prompt the user to enter the points to draw a line. Since there is nothing after the LINE, the spaces after the line automatically enters a Return.

    [LINE   ]^C^CLINE
            Where **[LINE   ]** --------- For screen display only
                        ^C --------------- ^C is the first **CANCEL** command
                        ^C --------------- ^C is the second **CANCEL** command
                        LINE ------------- AutoCAD's **LINE** command

Line 7

[CIR-C,R]^C^CCIRCLE

The part of the menu item that is enclosed within the brackets is for screen display only. The part of the menu item that is outside the brackets is executed when this line is selected. ^C^C (caret C) cancels the existing command twice. **CIRCLE** is an AutoCAD command that generates a circle. **A space after CIRCLE automatically causes a RETURN. The space acts like pressing the SPACEBAR on the keyboard.**

    [CIR-C,R]^C^CCIRCLE
            Where **[CIR-C,R]** ----- For screen display only
                        ^C --------------- ^C for first **CANCEL** command
                        ^C --------------- ^C for second **CANCEL** command
                        CIRCLE --------- AutoCAD's **CIRCLE** command

Line 8

[CIR-C,D]^C^CCIRCLE;\D

The part of the menu item that is enclosed in the brackets is for screen display only, and the part that is outside the brackets is the executable part. ^C^C will cancel the existing command twice.

    [CIR-C,D]^C^CCIRCLE;\D
            Where   **^C^C** ---------- ^C^C cancels the existing command
                        CIRCLE --------- **CIRCLE** command

       ; ------------------ Semicolon (;) for RETURN
       \ -------------------- AutoCAD pauses for user input
       D ----------------- Diameter option

The **CIRCLE** command is followed by a semicolon, a backslash (\), and a D for diameter option. **The semicolon (;) after the CIRCLE command causes a RETURN, which has the same effect as entering RETURN at the keyboard. The backslash (\) pauses for user input.** In this case it is the center point of the circle. D is for the diameter option and it is automatically followed by RE-TURN. The semicolon in this example can also be replaced by a blank space, as shown in the following line. However, the semicolon is easier to spot.

    [CIR-C,D]^C^CCIRCLE \D
          Where   Space between CIRCLE and \ --- Blank space for RETURN

Line 9
[CIR- 2P]^C^CCIRCLE;2P
In this menu item ^C^C cancels the existing command twice. The semicolon after **CIRCLE** enters a RETURN. 2P is for the two-point option, followed by the blank space that causes a RE-TURN. You will notice that the sequence of the command and input is the same as discussed earlier. Therefore, it is essential to know the exact sequence of the command and input; otherwise, the screen menu is not going to work.

    [CIR- 2P]^C^CCIRCLE;2P
          Where **^C^C** ------------ ^C^C Cancels existing command twice
                   CIRCLE --------- **CIRCLE** command
                   ; ------------------ Semicolon (;) for RETURN
                   2P ----------------- 2-Point option for **CIRCLE**

The semicolon after the **CIRCLE** command can be replaced by a blank space, as shown in the following line. The blank space causes a RETURN, like a semicolon (;).

    [CIR- 2P]^C^CCIRCLE 2P
          Where Space between CIRCLE and 2P -- Blank space or semicolon for RETURN

Line 11
[ERASE]^C^CERASE
In this menu item ^C^C cancels the existing command twice, and **ERASE** is an AutoCAD command that erases the selected objects.

    [ERASE]^C^CERASE
          Where **ERASE** ---------- AutoCAD **ERASE** command

Line 12
[MOVE   ]^C^CMOVE
In this menu item ^C^C cancels the existing command twice, and the **MOVE** command will move the selected objects.

[MOVE   ]^C^CMOVE
        Where   **Move** ------------ AutoCAD **MOVE** command

LOADING MENUS

AutoCAD automatically loads the **ACAD.MNS** file when you get into the AutoCAD drawing editor. However, you can also load a different menu file by using the AutoCAD **MENU** command.

Command: **MENU**

When you enter the **MENU** command, AutoCAD displays the **Select Menu File** dialog box (Figure 9-3) on the screen. Select the menu file that you want to load and then choose the **Open** button.

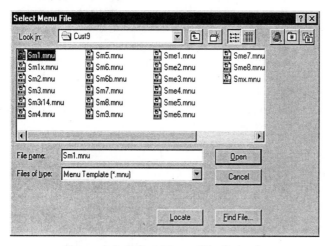

Figure 9-3 Select Menu File dialog box

You can also load the menu file from the command line.

Command: **FILEDIA**
Enter new value for FILEDIA <1>: **0**
Command: **MENU**
Enter menu file name or [. (for none)] <current>: PDM1.MNU
        Where **<current>** ------ Default menu file
               PDM1.MNU --- Name of menu file

After you enter the **MENU** command, AutoCAD will prompt for the file name. Enter the name of the menu file with the file extension (**.MNU**).

Note

*1. After you load the new menu, you cannot use the screen menu, buttons menu, or digitizer because the original menu, **ACAD.MNU**, is not present and the new menu does not contain these menu areas.*

2. *To activate the original menu again, load the menu file by using the* **MENU** *command:*

Command: **MENU**
Enter menu file name or [. (for none)] <current>: **ACAD.MNU**

3. *If you need to use input from a keyboard or a pointing device, in a menu macro use the backslash (\). The system will pause for you to enter data.*

4. *There should be no space after the backslash (\).*

5. *The menu items, menu labels, and command definition can be uppercase, lowercase, or mixed.*

6. *You can introduce spaces between the menu items to improve the readability of the menu file.*

7. *If there are more items in the menu than the number of spaces available, the excess items are not displayed on the screen. For example, if the display device limits the number of items to 21, items in excess of 21 will not be displayed on the screen and are therefore inaccessible.*

8. *If you are using a high-resolution graphics board, you can increase the number of lines that can be displayed on the screen. On some devices this is 80 lines.*

Exercise 1 *General*

Design and write a screen menu for the following AutoCAD commands (file name **SME1.MNU**):

PLINE
ELLIPSE (Center)
ELLIPSE (Axis endpoint)
ROTATE
OFFSET
SCALE

SUBMENUS

The screen menu file can be so large that all items cannot be accommodated on one screen. For example, the maximum number of items that can be displayed on most screens is 21. If the screen menu has more than 21 items, the menu items in excess of 21 are not displayed on the screen and therefore cannot be accessed. You can overcome this problem by using submenus that enable the user to define smaller groups of items within a menu section. When a submenu is selected, it loads the submenu items and displays them on screen. However, depending on the resolution of the monitor and the graphics card, the number of items you can display can be higher and you may not need submenus.

Submenu Definition

A submenu definition consists of two asterisks (**) followed by the name of the submenu. A menu can have any number of submenus, and every submenu should have a unique name. The items that follow a submenu, up to the next section label, or submenu label belong to that submenu. The format of a submenu definition is:

**Name

      Where ****** ---------------- Two asterisks (**) designate a submenu
                 Name ------------ Name of the submenu

**DRAW1

      Where **DRAW1** --------- Name of the submenu
                 ** ---------------- ** designates that DRAW1 is a submenu

Note

1. The submenu name can be up to 31 characters long.

2. The submenu name can consist of letters, digits, and such special characters as $ (dollar sign), - (hyphen), and _ (underscore).

3. The submenu name should not have any embedded blanks (spaces).

4. The submenu names should be unique in a menu file.

Submenu Reference

The submenu reference is used in a menu item to reference or load a submenu. It consists of a "$" sign followed by a letter that specifies the menu section. The letter that specifies a screen menu section is S. The section is followed by an "=" sign and the name of the submenu the user wants to activate. The submenu name should be without the **. The following is the format of a submenu reference:

$Section=Submenu

      Where **$** -------------------- "$" sign
                 Section ---------- Menu section specifier
                 = ------------------ "=" sign
                 Submenu ------- Name of submenu

$S=EDIT

      Where **S** -------------------- S specifies screen menu section
                 Edit -------------- Name of submenu

Note

$ *A special character code used to load a submenu in a menu file.*
$M=   *This is used to load a DIESEL macro from a menu item.*
The following are the section specifiers:
S   *Specifies the **SCREEN** menu.*
P0 - P16   *Specifies the POP menus, POP0 through POP16.*
I   *Specifies the Image Tile menu.*
B1 - B4   *Specifies the BUTTONS menu, B1 through B4.*
T1 - T4   *Specifies the TABLET menus, T1 through T4.*
A1 - A4   *Specifies the AUX menus A1 through A4.*

Nested Submenus

When a submenu is activated, the current menu is copied to a stack. If you select another submenu, the submenu that was current will be copied or pushed to the top of the stack. The maximum number of menus that can be stacked is eight. If the stack size increases to more than eight, the menu at the bottom of the stack is removed and forgotten. You can call the previous submenu by using the nested submenu call. The format of this call is:

$S=

Where **$** ------------------ "$" sign
       S ------------------ Screen menu specifier
       = ------------------ "=" sign

The maximum number of nested submenu calls is eight. Each time you call a submenu (issue $S=), this pops the last item from the stack and reactivates it.

Example 2

Design a menu layout and write a screen menu for the following commands.

LINE	**ERASE**
PLINE	**MOVE**
ELLIPSE-C	**ROTATE**
ELLIPSE-E	**OFFSET**
CIR-C,R	**COPY**
CIR-C,D	**SCALE**
CIR- 2P	

As mentioned earlier, the first and the most important part of writing a menu is the design of the menu and knowing the commands and the prompts associated with those commands. You should know the way you want the menu to look and the way you want to arrange the commands for maximum efficiency. Write out the menu on a piece of paper, and check it thoroughly to make sure you have arranged the commands the way you want them. Use submenus to group the commands based on their use, function, and relationship with other submenus. Make provisions to access other frequently used commands without going through the root menu.

Figure 9-4 shows one of the possible arrangements of the commands and the design of the screen menu. It has one main menu and two submenus. One of the submenus is for draw commands, the other is for edit commands. The colon (:) at the end of the commands is not required. It is used here to distinguish the commands from those items that are not used as commands. For example, DRAW in the root menu is not a command; therefore, it has no colon at the end. On the other hand, if you select **ERASE** from the **EDIT** menu, it executes the **ERASE** command, so it has a colon (:) at the end of the command.

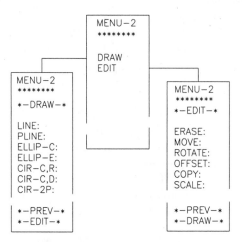

Figure 9-4 *Screen menu design*

The following file is a listing of the menu file of Example 1. **The line numbers on the right are not a part of the file; they are given here for reference only.**

```
***SCREEN                                                              1
[ MENU-2       ]                                                       2
[********      ]                                                       3
[             ]                                                        4
[             ]                                                        5
[             ]                                                        6
[             ]                                                        7
[DRAW         ]^C^C$S=DRAW                                             8
[EDIT         ]^C^C$S=EDIT                                             9
                                                                      10
**DRAW                                                                11
[ MENU-2       ]^C^C$S=SCREEN                                          12
[********      ]                                                       13
[                                                                     14
[*-DRAW-*      ]                                                       15
[             ]                                                        16
[LINE:        ]^C^CLINE                                                17
[PLINE:       ]^C^CPLINE;\W;0.1;0.1                                    18
[ELLIP-C:     ]^C^CELLIPSE;C                                           19
[ELLIP-E:     ]^C^CELLIPSE                                             20
[CIR-C,R:     ]^C^CCIRCLE                                              21
[CIR-C,D:     ]^C^CCIRCLE;\D                                           22
[CIR-2P:      ]^C^CCIRCLE;2P                                           23
[             ]                                                        24
[             ]                                                        25
[             ]                                                        26
```

```
[                ]                                    27
[                ]                                    28
[                ]                                    29
[*-PREV-*        ]^C^C$S=                             30
[*-EDIT-*        ]^C^C$S=EDIT                         31
                                                     32
**EDIT                                                33
[ MENU-2         ]^C^C$S=SCREEN                       34
[********        ]                                    35
[                ]                                    36
[*-EDIT-*        ]                                    37
[                ]                                    38
[ERASE:          ]^C^CERASE                           39
[MOVE:           ]^C^CMOVE                            40
[ROTATE:         ]^C^CROTATE                          41
[OFFSET:         ]^C^COFFSET                          42
[COPY:           ]^C^CCOPY                            43
[SCALE:          ]^C^CSCALE                           44
[                ]                                    45
[                ]                                    46
[                ]                                    47
[                ]                                    48
[                ]                                    49
[                ]                                    50
[                ]                                    51
[*-PREV-*        ]^C^C$S=                             52
[*-DRAW-*        ]^C^C$S=DRAW                         53
```

Explanation

Line 1
***SCREEN
***SCREEN is the section label for the screen menu.

Line 2
[MENU-2]
This menu item displays MENU-2 at the top of the screen menu.

Line 3
[********]
This menu item prints eight asterisk signs (********) on the screen menu.

Lines 4-7
[   ]
These menu items print four blank lines on the screen menu. The brackets are not required. They could be just four blank lines without brackets.

Line 8

[DRAW   ]^C^C$S=DRAW

[DRAW   ] displays DRAW on the screen, letting you know that by selecting this function you can access the draw commands. ^C^C cancels the existing command, and $S=DRAW loads the DRAW submenu on the screen.

[DRAW   ]^C^C$S=DRAW

Where   **^C^C** ---------- Cancels the existing command twice

= ----------------- Loads submenu DRAW

DRAW ---------- Name of submenu

Line 9

[EDIT   ]^C^C$S=EDIT

[EDIT   ] displays EDIT on the screen. ^C^C cancels the current command, and $S=EDIT loads the submenu **EDIT**.

Line 10

The blank lines between the submenus or menu items are not required. It just makes it easier to read the file.

Line 11

**DRAW

**DRAW is the name of the submenu; lines 12 through 31 are defined under this submenu.

Line 15

[*-DRAW-*]

This prints *-DRAW-* as a heading on the screen menu to let the user know that the commands listed on the menu are draw commands.

Line 18

[PLINE:]^C^CPLINE;\W;0.1;0.1

[PLINE:] displays PLINE: on the screen. ^C^C cancels the command, and **PLINE** is the AutoCAD polyline command. The semicolons are for RETURN. The semicolons can be replaced by a blank space. The backslash (\\) is for user input. In this case it is the start point of the polyline. W selects the width option of the polyline. The first 0.1 is the starting width of the polyline, and the second 0.1 is the ending width. This command will draw a polyline of 0.1 width.

[PLINE:]^C^CPLINE;\W;0.1;0.1

Where **PLINE** ------------ **PLINE** (Polyline) command

; ------------------- RETURN

\\-------------------- Pause for input (start pt.)

W ----------------- Width

; ------------------- RETURN

0.1 ---------------- Starting width

; ------------------- RETURN

0.1 ---------------- Ending width

Line 19
[ELLIP-C:] ^ C ^ CELLIPSE;C
[ELLIP-C:] displays ELLIP-C: on the screen menu. ^ C ^ C cancels the existing command, and **ELLIPSE** is an AutoCAD command to generate an ellipse. The semicolon is for RETURN, and C selects the center option of the **ELLIPSE** command.

> **[ELLIP-C:] ^ C ^ CELLIPSE;C**
> Where **CELLIPSE** ------ **ELLIPSE** command
> ; ------------------ RETURN
> C ----------------- Center option for ellipse

Line 20
[ELLIP-E:] ^ C ^ CELLIPSE
[ELLIP-E:] displays ELLIP-E: on the screen menu. ^ C ^ C cancels the existing command, and **ELLIPSE** is an AutoCAD command. Here the **ELLIPSE** command uses the default option, axis endpoint, instead of center.

Line 30
[*-PREV-*] ^ C ^ C$S=
[*-PREV-*] displays *-PREV-* on the screen menu, and ^ C ^ C cancels the existing command. $S= restores the previous menu that was displayed on the screen before loading the current menu.

> **[*-PREV-*] ^ C ^ C$S=**
> Where **$S** ----------------- Restores the previous screen menu

Line 31
[*-EDIT-*] ^ C ^ C$S=EDIT
[*-EDIT-*] displays *-EDIT-* on the screen menu, and ^ C ^ C cancels the existing command. $S=EDIT loads the submenu EDIT on the screen. This lets you access the **EDIT** commands without going back to the root menu and selecting **EDIT** from there.

> **[*-EDIT-*] ^ C ^ C$S=EDIT**
> Where **$S** ----------------- Loads the submenu EDIT
> **EDIT** ------------ Name of the submenu

Line 33
**EDIT
**EDIT is the name of the submenu, and lines 34 to 53 are defined under this submenu.

Line 34
[MENU-2] ^ C ^ C$S=SCREEN
[MENU-2] displays MENU-2 on the screen menu, and ^ C ^ C cancels the existing command. $S=SCREEN loads the root menu SCREEN on the screen menu.

Line 39
[ERASE:] ^ C ^ CERASE
[ERASE:] displays ERASE: on the screen menu, and ^ C ^ C cancels the existing command.

Chapter 9

ERASE is an AutoCAD command for erasing the selected objects.

Line 53
[*-DRAW-*] ^ C ^ C\$S=DRAW
$S=DRAW loads the DRAW submenu on the screen. It lets you load the DRAW menu without going through the root menu. The root menu is the first menu that appears on the screen when you load a menu or start AutoCAD.

When you select DRAW from the root menu, the submenu DRAW is loaded on the screen. The menu items in the draw submenu completely replace the menu items of the root menu. If you select MENU-2 from the screen menu now, the root menu will be loaded on the screen, but some of the items are not cleared from the screen menu (Figure 9-5). This is because the root menu does not have enough menu items to completely replace the menu items of the DRAW submenu.

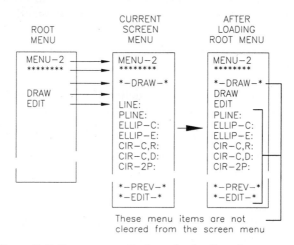

Figure 9-5 Screen menu display after loading the root menu

One of the ways to clear the screen is to define a submenu that has 21 blank lines. When this submenu is loaded it will clear the screen. If you then load another submenu, there will be no overlap, because the screen menu has already been cleared and there are no menu items left on the screen (Figure 9-6). Example 3 illustrates the use of such a submenu.

Another way to avoid overlapping the menu items is to define every submenu so that they all have the same number of menu items. The disadvantage with this approach is that the menu file will be long, because every submenu will have 21 lines.

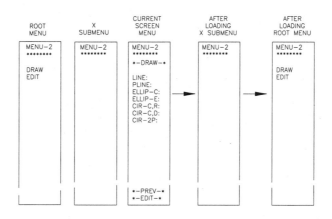

Figure 9-6 *Screen menu display after loading the root menu*

Exercise 2   *General*

Write a screen menu file for the following AutoCAD commands (file name **SM2.MNU**).

ARC	MIRROR
-3P	BREAK-F
-SCE	BREAK-@
-SCA	EXTEND
-SCL	STRETCH
-SEA	FILLET-0
POLYGON-C	FILLET
POLYGON-E	CHAMFER

MULTIPLE SUBMENUS

A menu file can have any number of submenus. All the submenu names have two asterisks (**) in front of them, even if there is a submenu within a submenu. Also, several submenus can be loaded in a single statement. If there is more than one submenu in a menu item, they should be separated from one another by a space. Example 3 illustrates the use of multiple submenus.

Example 3

Design the layout of the menu, and then write the screen menu for the following AutoCAD commands.

Draw	ARC	Edit	Display
LINE	3Point	EXTEND	ZOOM
Continue	SCE	STRETCH	REGEN
Close	SCA	FILLET	SCALE
Undo	CSE		PAN

Chapter 9

```
.X              CSA
.Y              CSL
.Z
.XY
.XZ
.YZ
```

There are several different ways of arranging the commands, depending on your requirements. The following layout is one of the possible screen menu designs for the given AutoCAD commands (Figure 9-7).

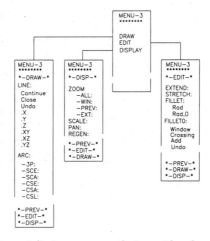

Figure 9-7 *Screen menu design with submenus*

The name of the following menu file is **SM3.MNU**. You can use a text editor to write the file. The following file is the listing of the menu file **SM3.MNU. The line numbers on the right are not a part of the file. They are shown here for reference only.**

```
***SCREEN                                                          1
**S                                                                2
[ MENU-3        ]^C^C$S=X $S=S                                      3
[********       ]$S=OSNAP                                           4
[               ]                                                   5
[DRAW           ]^C^C$S=X $S=DRAW                                   6
[EDIT           ]^C^C$S=X $S=EDIT                                   7
[DISPLAY        ]^C^C$S=X $S=DISP                                   8
**DRAW 3                                                            9
[*-DRAW-*       ]                                                  10
[               ]                                                  11
[LINE:          ]^C^CLINE                                          12
[ Continue      ]^C^CLINE;;                                        13
[ Close         ]CLOSE                                             14
[ Undo          ]UNDO                                              15
```

[.X	].X	16
[.Y	].Y	17
[.Z	].Z	18
[.XY	].XY	19
[.XZ	].XZ	20
[.YZ	].YZ	21
[	]	22
[ARC:	]^C^CArc	23
[-3P:	]^C^CARC \\DRAG	24
[-SCE:	]^C^CARC \C \DRAG	25
[-SCA:	]^C^CARC \C \A DRAG	26
[-CSE:	]^C^CARC C \\DRAG	27
[-CSA:	]^C^CARC C \\A DRAG	28
[-CSL:	]^C^CARC C \\L DRAG	29
[	]	30
[*-PREV-*	]$S= $S=	31
[*-EDIT-*	]^C^C$S=X $S=EDIT	32
[*-DISP-*	]$S=X $S=DISP	33
**EDIT 3		34
[*-EDIT-*	]	35
[	]	36
[EXTEND:	]^C^CEXTEND	37
[STRETCH:	]^C^CSTRETCH;C	38
[FILLET:	]^C^CFILLET	39
[Rad	]R;\Fillet	40
[Rad 0	]R;0;Fillet	41
[FILLET0:	]^C^CFillet;R;0::	42
Win.		43
Cross.		44
Add		45
Undo		46
[	]	47
[*-PREV-*	]$S= $S=	48
[*-DRAW-*	]^C^C$S=X $S=DRAW	49
[*-DISP-*	]$S=X $S=DISP	50
**DISP 3		51
[*-DISP-*	]	52
[	]	53
[ZOOM:	]'ZOOM	54
[-ALL	]A	55
[-WIN	]W	56
[-PREV	]P	57
[-EXT	]E	58
[	]	59
[SCALE:	]'ZOOM	60
[PAN:	]'PAN	61
[REGEN:	]^C^CREGEN	62

Chapter 9

```
[                        ]                                    63
[*-PREV-*               ]$S= $S=                              64
[*-EDIT-*               ]^C^C$S=X $S=EDIT                     65
[*-DRAW-*               ]$S=X $S=DRAW                         66
**X 3                                                         67
[                        ]                                    68
[                        ]                                    69
[                        ]                                    70
[                        ]                                    71
[                        ]                                    72
[                        ]                                    73
[                        ]                                    74
[                        ]                                    75
[                        ]                                    76
[                        ]                                    77
[                        ]                                    78
[                        ]                                    79
[                        ]                                    80
[                        ]                                    81
[                        ]                                    82
[                        ]                                    83
[                        ]                                    84
[                        ]                                    85
[                        ]                                    86
**OSNAP 2                                                     87
[-OSNAPS-               ]                                     88
[                        ]                                    89
[Center                 ]CEN $S=                             90
[Endpoint               ]END $S=                             91
[Insert                 ]INS $S=                             92
[Intersec               ]INT $S=                             93
[Midpoint               ]MID $S=                             94
[Nearest                ]NEA $S=                             95
[Node                   ]NOD $S=                             96
[Perpend                ]PER $S=                             97
[Quadrant               ]QUA $S=                             98
[Tangent                ]TAN $S=                             99
[None                   ]NONE $S=                            100
[                        ]                                    101
[                        ]                                    102
[                        ]                                    103
[                        ]                                    104
[                        ]                                    105
[*-PREV-*               ]$S=                                 106
```

Explanation
Line 3
[MENU-3] ^ C ^ C$S=X $S=S
In this menu item, [MENU-3] displays MENU-3 at the top of the screen menu. ^ C ^ C cancels the command, and $S=X loads submenu X. Similarly, $S=S loads submenu S. Submenu X, defined on line 171, consists of 18 blank lines. Therefore, when this submenu is loaded it prints blank lines on the screen and clears the screen menu area. After loading submenu X, the system loads submenu S, and the items that are defined in this submenu are displayed on the screen menu.

> **[MENU-3] ^ C ^ C$S=X $S=S**
> Where   **^ C ^ C** ---------- Cancels the existing command
>   **$S=X** ------------ Loads submenu X (clears screen)
>   **$S=S** ------------ Loads submenu S

Line 4
[********]$S=OSNAP
This menu item prints eight asterisks on the screen menu, and $S=OSNAP loads submenu OSNAP. The OSNAP submenu, defined on line 191, consists of object snap modes.

Line 6
[DRAW   ]^ C ^ C$S=X $S=DRAW
This menu item displays DRAW on the screen menu area, cancels the existing command, and then loads submenu X and submenu DRAW. Submenu X clears the screen menu area, and submenu DRAW, as defined on line 11, loads the items defined under the DRAW submenu.

Line 9
**DRAW 3
This line is the submenu label for DRAW; 3 indicates that the first line of the DRAW submenu will be printed on line 3. Nothing will be printed on line 1 or line 2. The first line of the DRAW submenu will be on line 3, followed by the rest of the menu. All the submenus except submenus S and OSNAP have a 3 at the end of the line. Therefore, the first two lines (MENU-3) and (********) will never be cleared and will be displayed on the screen all the time. If you select MENU-3 in any menu, it will load submenu S. If you select ********, it will load submenu OSNAP.

> ****DRAW 3**
> Where   **DRAW** ----------- Submenu name
>   **3** ------------------ Leaves two blank lines and prints from third line

Line 12
[LINE:]^ C ^ CLINE
^ C ^ C cancels the existing command, and **LINE** is the AutoCAD **LINE** command.

Line 13
[Continue]^ C ^ CLINE;;
In this menu item, the **LINE** command is followed by two semicolons that continue the **LINE** command. To understand it, look at the **LINE** command to see how a continue option is used in this command:

Command: **LINE**
Specify first point: **RETURN (Continue)**
Specify next point or [Undo]:

Following is the command and prompt entry sequence for the **LINE** command, if you want to start a line from the last point:

LINE
RETURN
RETURN
SELECT A POINT

Therefore, in the screen menu file the **LINE** command has to be followed by two Returns to continue from the previous point.

Line 16
[.X   ].X
In this menu item, .X extracts the X coordinate from a point. The bracket, .X, and spaces inside the brackets ([.X   ]) are not needed. The line could consist of .X only. The same is true for lines 39 through 43.

Example
[.X   ].X   can also be written (without brackets) as   .X
[.Y   ].Y   .Y
[.Z   ].Z   .Z

Line 31
[*-PREV-*]$S= $S=
This menu item recalls the previous menu twice. AutoCAD keeps track of the submenus that were loaded on the screen. The first **$S=** will load the previously loaded menu, and the second **$S=** will load the submenu that was loaded before that. For example, in line 16 ([ARC:]$S=X $S=ARC), two submenus have been loaded: first X, and then ARC. Submenu X is stacked before loading submenu ARC, which is current and is not stacked yet. The first **$S=** will recall the previous menu, in this case submenu X. The second **$S=** will load the menu that was on the screen before selecting the item in line 16.

          [*-PREV-*]$S= $S=
                    Where **$S=** --------------- Loads last menu
                           $S= -------------- Loads next to last menu

Line 33
[*-DISP-*]$S=X $S=DISP
In this menu item, $S=X loads submenu X, and $S=DISP loads submenu DISP. Notice that there is no CANCEL (^ C ^ C) command in the line. There are some menus you might like to load without canceling the existing command. For example, if you are drawing a line, you might want to zoom without canceling the existing command. You can select [*-DISP-*], select the appropriate zoom option, and then continue with the line command. However, if the line has a CANCEL

command ([*-DISP-*] ^ C ^ C$S=X $S=DISP), you cannot continue with the **LINE** command, because the existing command will be canceled when you select *-DISP-* from the screen. In line 28 ([*-EDIT-*] ^ C ^ C$S=X $S=EDIT), ^ C ^ C cancels the existing command, because you cannot use any editing command unless you have canceled the existing command.

Line 38

[STRETCH:] ^ C ^ CSTRETCH;C

^ C ^ C cancels the existing command, and **STRETCH** is an AutoCAD command. The C is for the crossing option that prompts you to enter two points to select the objects.

[STRETCH:] ^ C ^ CSTRETCH;C

> Where **STRETCH** ----- **STRETCH** command
> ; ------------------- RETURN
> C ------------------ Center O'SNAP

Line 40

[RAD]R;\FILLET

This menu item selects the radius option of the **FILLET** command, and then waits for you to enter the radius. After entering the radius, execute the **FILLET** command again to generate the desired fillet between the two selected objects.

[RAD]R;\FILLET

> Where **R** ------------------ Radius option of **FILLET** command
> ; ------------------ Space or semicolon for RETURN
> \-------------------- Pause for user input
> **FILLET** --------- AutoCAD **FILLET** command

Line 41

[RAD 0]R;0;FILLET

This menu item selects the radius option of the **FILLET** command and assigns it a 0 value. Then it executes the **FILLET** command to generate a 0 radius fillet between the two selected objects.

[RAD 0]R;0;FILLET

> Where **R** ------------------ Radius option of **FILLET** command
> ; ------------------ Space or semicolon for RETURN
> **0** ------------------ 0 radius
> ; ------------------ Space or semicolon for RETURN
> **FILLET** --------- AutoCAD **FILLET** command

Line 42

[FILLET0:] ^ C ^ CFILLET;R 0;;

This menu item defines a **FILLET** command with 0 radius, and then generates 0 fillet between the two selected objects.

[FILLET0:] ^ C ^ CFILLET;R 0;;

> Where **FILLET** --------- **FILLET** command
> ; ------------------ Semicolon or space for RETURN

R ------------------ R for radius option
Space ------------- Semicolon or space for RETURN
0 ------------------- 0 radius
; ------------------- Semicolon for RETURN
; ------------------- Repeats **FILLET** command

Line 54
[ZOOM:]'ZOOM
This menu item defines a transparent **ZOOM** command.

     [ZOOM:]'ZOOM
          Where ‘-------------------- Single quote (‘) makes **ZOOM** transparent
          ZOOM ----------- AutoCAD **ZOOM** command

Line 90
[Center]CEN $X=
In this menu item CEN is for center object snap, and $X= automatically recalls the previous screen menu after selecting the object.

     [Center]CEN $S=
          Where **CEN** -------------- Center object snap
          $S= -------------- Loads previous menu

Note
If any menu item on a screen menu includes more than one load command, the commands must be separated by a space.

Example
[LINE:]$S=X $S=LINE
    Note that there is a space between X and $S.

Similarly, if a menu item includes both a load command and an AutoCAD command, they should be separated by a space.

Example
[LINE:]$S=LINE ^C^CLINE

Exercise 3   *General*

Write a screen menu for the AutoCAD commands shown in Figure 9-8 (file name **SME3.MNU**). The ******** are to access the object snap submenu.

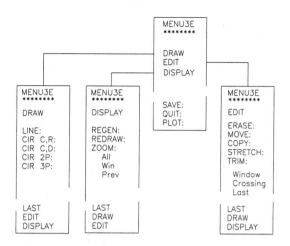

Figure 9-8 *Screen menu design with submenus*

Example 4

Write a combined screen and tablet menu for the commands shown in the tablet menu template shown in Figure 9-9(a) and the screen menu shown in Figure 9-9(b). When the user selects a command from the digitizer template, it should automatically load the corresponding screen menu on the screen (filename TM3.MNU).

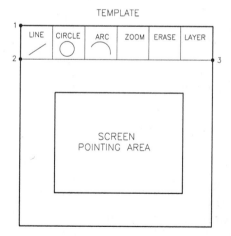

Figure 9-9(a) *Tablet menu template for Example 4*

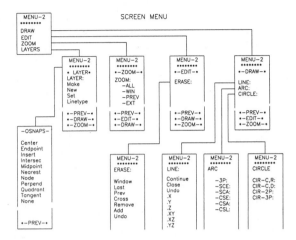

Figure 9-9(b) Screen menu displays

The following file is a listing of the combined menu for Example 4. **The line numbers are not a part of the file; they are shown here for reference only.**

```
***SCREEN                                                     1
**S                                                           2
[ MENU-3   ]^C^C$S=X $S=S                                     3
[********   ]$S=OSNAP                                         4
[          ]                                                  5
[          ]                                                  6
[DRAW      ]^C^C$S=X $S=DRAW                                  7
[EDIT      ]^C^C$S=X $S=EDIT                                  8
[ZOOM      ]^C^C$S=X $S=ZOOM                                  9
[LAYER     ]^C^C$S=X $S=LAYER                                10
                                                             11
**DRAW 3                                                     12
[*-DRAW-* ]                                                  13
[          ]                                                 14
[          ]                                                 15
[LINE:     ]$S=X $S=LINE ^C^CLINE                            16
[ARC:      ]$S=X $S=ARC                                      17
[CIRCLE:   ]$S=X $S=CIRCLE ^C^CCIRCLE                        18
[          ]                                                 19
[          ]                                                 20
[          ]                                                 21
[          ]                                                 22
[          ]                                                 23
[          ]                                                 24
[          ]                                                 25
[          ]                                                 26
```

```
[                    ]                                    27
[*-PREV-*    ]$S= $S=                                     28
[*-EDIT-*    ]^C^C$S=X $S=EDIT                            29
[*-ZOOM-*  ]$S=X $S=ZOOM                                  30
                                                         31
**LINE 3                                                 32
[LINE:       ]^C^CLINE                                   33
[                    ]                                    34
[                    ]                                    35
[Continue    ]^C^CLINE;;                                 36
[Close          ]CLOSE                                   37
[Undo          ]UNDO                                     38
[.X               ].X                                    39
[.Y               ].Y                                    40
[.Z               ].Z                                    41
[.XY             ].XY                                    42
[.XZ             ].XZ                                    43
[.YZ             ].YZ                                    44
[                    ]                                    45
[                    ]                                    46
[                    ]                                    47
[*-PREV-*    ]$S= $S=                                     48
[*-EDIT-*    ]^C^C$S=X $S=EDIT                            49
[*-ZOOM-*  ]$S=X $S=ZOOM                                  50
                                                         51
**ARC 3                                                  52
[ARC          ]                                          53
[                    ]                                    54
[  -3P:        ]^C^CARC \\DRAG                           55
[  -SCE:      ]^C^CARC \C \DRAG                          56
[  -SCA:     ]^C^CARC \C \A DRAG                         57
[  -CSE:      ]^C^CARC C \\DRAG                          58
[  -CSA:     ]^C^CARC C \\A DRAG                         59
[  -CSL:      ]^C^CARC C \\L DRAG                        60
[                    ]                                    61
[                    ]                                    62
[                    ]                                    63
[                    ]                                    64
[                    ]                                    65
[                    ]                                    66
[                    ]                                    67
[*-PREV-*    ]$S= $S=                                     68
[*-EDIT-*    ]^C^C$S=X $S=EDIT                            69
[*-ZOOM-*  ]$S=X $S=ZOOM                                  70
                                                         71
**CIRCLE 3                                               72
[CIRCLE:   ]                                             73
```

```
[              ]                                                           74
[  -C,R:    ]^C^CCIRCLE                                                   75
[  -C,D:    ]^C^CCIRCLE \D                                                76
[  -2P:     ]^C^CCIRCLE 2P                                                77
[  -3P:     ]^C^CCIRCLE 3P                                                78
[              ]                                                           79
[              ]                                                           80
[              ]                                                           81
[              ]                                                           82
[              ]                                                           83
[              ]                                                           84
[              ]                                                           85
[              ]                                                           86
[              ]                                                           87
[*-PREV-*  ]$S= $S=                                                       88
[*-EDIT-*  ]^C^C$S=X $S=EDIT                                              89
[*-ZOOM-* ]$S=X $S=ZOOM                                                   90
                                                                          91
                                                                          92

**EDIT 3                                                                  93
[*-EDIT-*  ]                                                              94
[              ]                                                           95
[              ]                                                           96
[ERASE:    ]$S=X $S=ERASE ^C^CERASE                                       97
[              ]                                                           98
[              ]                                                           99
[              ]                                                          100
[              ]                                                          101
[              ]                                                          102
[              ]                                                          103
[              ]                                                          104
[              ]                                                          105
[              ]                                                          106
[              ]                                                          107
[              ]                                                          108
[*-PREV-*  ]$S= $S=                                                      109
[*-DRAW-* ]^C^C$S=X $S=DRAW                                              110
[*-ZOOM-* ]$S=X $S=ZOOM                                                  111
                                                                         112

**ERASE 3                                                                113
[ERASE:    ]^C^CERASE                                                    114
[              ]                                                          115
Window                                                                   116
Last                                                                     117
Prev                                                                     118
Cross                                                                    119
Remove                                                                   120
```

```
        Add                                               121
        Undo                                              122
        [              ]                                  123
        [              ]                                  124
        [              ]                                  125
        [              ]                                  126
        [              ]                                  127
        [              ]                                  128
        [*-PREV-*   ]$S= $S=                              129
        [*-DRAW-*   ]^C^C$S=X $S=DRAW                     130
        [*-ZOOM-*  ]$S=ZOOM                               131
                                                          132
        **ZOOM 3                                          133
        [*-ZOOM-* ]                                       134
        [              ]                                  135
        [              ]                                  136
        [ZOOM:      ]'ZOOM                                137
        [ -ALL      ]A                                    138
        [ -WIN      ]W                                    139
        [ -PREV     ]P                                    140
        [ -EXT      ]E                                    141
        [              ]                                  142
        [              ]                                  143
        [              ]                                  144
        [              ]                                  145
        [              ]                                  146
        [              ]                                  147
        [              ]                                  148
        [*-PREV-*   ]$S= $S=                              149
        [*-EDIT-*   ]^C^C$S=X $S=EDIT                     150
        [*-DRAW-*  ]$S=X $S=DRAW                          151
                                                          152
        ***LAYER 3                                        153
        [*-LAYER* ]                                       154
        [              ]                                  155
        [              ]                                  156
        [LAYER:     ]^C^CLAYER                            157
        Make                                              158
        New                                               159
        Set                                               160
        Linetype                                          161
        Color                                             162
        [List          ]?;;                               163
        [              ]                                  164
        [              ]                                  165
        [              ]                                  166
        [              ]                                  167
```

Chapter 9

```
[                    ]                                          168
[*-PREV-*   ]$S= $S=                                           169
[*-EDIT-*   ]^C^C$S=X $S=EDIT                                  170
[*-DRAW-*   ]$S=X $S=DRAW                                      171
                                                               172
**X 3                                                          173
[                    ]                                          174
[                    ]                                          175
[                    ]                                          176
[                    ]                                          177
[                    ]                                          178
[                    ]                                          179
[                    ]                                          180
[                    ]                                          181
[                    ]                                          182
[                    ]                                          183
[                    ]                                          184
[                    ]                                          185
[                    ]                                          186
[                    ]                                          187
[                    ]                                          188
[                    ]                                          189
[                    ]                                          190
[                    ]                                          191
                                                               192
**OSNAP 2                                                      193
[-OSNAPS- ]                                                    194
[                    ]                                          195
[Center      ]CEN $S=                                          196
[Endpoint    ]END $S=                                          197
[Insert      ]INS $S=                                          198
[Intersec    ]INT $S=                                          199
[Midpoint    ]MID $S=                                          200
[Nearest     ]NEA $S=                                          201
[Node        ]NOD $S=                                          202
[Perpend     ]PER $S=                                          203
[Quadrant    ]QUA $S=                                          204
[Tangent     ]TAN $S=                                          205
[None        ]NONE $S=                                         206
[                    ]                                          207
[                    ]                                          208
[                    ]                                          209
[                    ]                                          210
[                    ]                                          211
[*-PREV-*   ]$S=                                               212
                                                               213
***TABLET1                                                     214
```

$S=X $S=LINE ^C^CLINE	215
$S=X $S=CIRCLE ^C^CCIRCLE	216
$S=X $S=ARC ^C^CARC	217
$S=X $S=ZOOM ^C^CZOOM	218
$S=X $S=ERASE ^C^CERASE	219
$S=X $S=LAYER ^C^CLAYER	220

Line 1
***SCREEN
This is the section label for the screen menu. The line numbers 1 through 213 are defined in this section of the menu file.

Line 3
[MENU-3]^C^C$S=X $S=S
In this menu item $S=X loads the submenu X and $S=S loads the submenu S.

Line 7
[DRAW   ]^C^C$S=X $S=DRAW
In this menu item $S=DRAW loads the submenu DRAW that is defined in the menu file.

Line 12
**DRAW 3
In this menu item DRAW is the name of the submenu. The 3 forces the submenu to be displayed on line 3. Since nothing is printed on the first two lines, [MENU-3   ] and [********] will be displayed on the screen all the time. If you select [MENU-3   ] from any menu, it will load the submenu S and display it on the screen. Similarly, if you pick [********] from any menu, it will load and display the OSNAP submenu.

Line 28
[*-PREV-*]$S= $S=
In this menu item $S= $S= loads the previous two submenus. One of them is the submenu X and the second submenu is the one that was displayed on the screen before the current menu.

Line 214
***TABLET1
TABLET1 is the section label for the tablet area 1. The line numbers 215 through 220 are defined in this section.

Line 215
$S=X $S=LINE ^C^CLINE
$S=X, defined in the screen menu section, loads the submenu X on the screen menu area. The purpose of loading the submenu X is to clear the screen so that there is no overlapping of the screen menu items. $S=LINE, defined in the screen menu section, loads the **LINE** submenu on the screen menu. ^C^CLINE cancels the existing command twice and then executes AutoCAD's **LINE** command.

When you select the LINE block from the digitizer, AutoCAD automatically clears the screen menu,

Chapter 9

loads the LINE submenu, and enters a **LINE** command. This makes it easier for the user to select the command options from the screen menu, since they are not on the digitizer template.

> **$S=X $S=LINE ^C^CLINE**
> Where   **$S=X** ------------ Loads submenu X
>   **LINE** ------------- Loads submenu LINE
>   **^C^C** ---------- Cancels the existing command
>   **LINE** ------------ AutoCAD's **LINE** command

You can also use the automatic menu swapping feature of AutoCAD to load the corresponding screen menus. If you want to utilize this feature, the TABLET1 section must be written as:

***TABLET1	214
$S=X ^C^CLINE	215
$S=X ^C^CCIRCLE	216
$S=X ^C^CARC	217
$S=X ^C^CZOOM	218
$S=X ^C^CERASE	219
$S=X ^C^CLAYER	220

LONG MENU DEFINITIONS

You can put any number of commands in one screen menu line. There is no limit to the number of commands and the order in which they appear in the line, as long as they satisfy all the requirements of the command and the sequence of prompt entries. Again, you need to know the AutoCAD commands, the options in the command, the command prompts, and the entries for various prompts. If the statement cannot fit on one line, you can put a plus (+) sign at the end of the first line and continue with a second line. The command definition that involves several commands put together in one line is also called a macro. The following example illustrates a long menu item or a macro.

Example 5

Write a screen menu command definition that performs the following functions (filename SM4.MNU):

Draw a border using polyline

Width =	0.01
Point-1	0,0
Point-2	12,0
Point-3	12,9
Point-4	0,9
Point-5	0,0

Initial drawing setup

Snap	0.25
Grid	0.5
Limits	12,9

              Zoom   All

Before writing a menu you should know the commands, options, prompts, and the prompt entries required for the commands. Therefore, first study the commands involved in the previously mentioned drawing setup.

POLYLINE
Command: **PLINE**
Specify start point: **0,0**
Specify next point or [Arc/Close/Halfwidth/Length/Undo/Width]: **W**
Specify starting width <0.0000>: **0.01**
Specify ending width <0.01>: *Press ENTER.*
Specify next point or [Arc/Close/Halfwidth/Length/Undo/Width]: **12,0**
Specify next point or [Arc/Close/Halfwidth/Length/Undo/Width]: **12,9**
Specify next point or [Arc/Close/Halfwidth/Length/Undo/Width]: **0,9**
Specify next point or [Arc/Close/Halfwidth/Length/Undo/Width]: **C**

Screen menu command definition for **PLINE**
PLINE;0,0;W;0.01;;12,0;12,9;0,9;C

SNAP
Command: **SNAP**
Specify snap spacing or [ON/OFF/Aspect/Rotate/Style/Type] <default>: **0.25**

Screen menu command definition for SNAP
SNAP;0.25

GRID
Command: **GRID**
Specify grid spacing(X) or [ON/OFF/Snap/Aspect] <default>: **0.5**

Screen menu command definition
GRID;0.5

LIMITS
Command: **LIMITS**
Specify lower left corner or [ON/OFF] <current co-ordinates>: **0,0**
Specify upper right corner <current co-ordinates>: **12,9**

Screen menu command definition
LIMITS;0,0;12,9

ZOOM
Command: **ZOOM**
Specify corner of window, enter a scale factor (nX or nXP), or
[All/Center/Dynamic/Extents/Previous/Scale/Window] <real time>: **A**

Screen menu command definition   ·
ZOOM;A

Now you can combine these individual command definitions to form a single screen menu command definition that will perform all the functions when you select NSETUP from the new screen menu.

   Combined screen menu line

   [-NSETUP-]PLINE;0,0;W;0.01;;12,0;12,9;0,9;C;+
   SNAP;0.25;GRID;0.5;LIMITS;+
   0,0;12,9;ZOOM;A

Exercise 4   *General*

Write a screen menu item that will set up the following parameters for the **UNITS** command (filename SME4.MNU):

   System of units: Scientific
   Number of digits to right of decimal point: 2
   System of angle measure: Decimal
   Number of fractional places for display of angles: 2
   Direction for angle: 0
   Angles measured counterclockwise

MENU COMMAND REPETITION

AutoCAD has made provision for repeating a menu item until it is canceled by the user by pressing the CTRL and C keys from the keyboard or selecting another menu item. It is particularly useful when you are editing a drawing and using the same command several times. A command can be repeated if the menu command definition starts with an asterisk sign (*).

   [ERASE,W:]*^C^CERASE W
              Where   * ------------------ Asterisk for command repetition
                      ^C^C ---------- Cancels existing command
                      Erase ------------- AutoCAD's **ERASE** command
                      W ----------------- Window option

If you select this menu item, AutoCAD will prompt you to enter two points to select the objects because the window option for the **ERASE** command requires two points. When you press ENTER the objects you selected will be erased and the command will be repeated.

Example 6

Write a screen menu for the following AutoCAD commands. Provide for automatically repeating the commands (filename SM7.MNU).

LINE	**LIST**
ERASE	**INSERT**
TRIM	**DIST**

The following file is a listing of the screen menu for Example 6 that will repeat the selected command until canceled by pressing the CTRL and C keys.

```
[-REPEAT-      ]
[              ]
[              ]
[LINE:         ]*^C^CLINE
[ERASE:        ]*^C^CERASE
[TRIM:         ]*^C^CTRIM
[LIST:         ]*^C^CLIST
[INSERT:       ]*^C^CINSERT
[DIST:         ]*^C^CDIST
```

Note

If the command definition is incorrect and you happen to select that item, the screen prompt display will be repeated indefinitely. To get out of this infinite loop, you must reboot the system. Pressing the ESCAPE key or selecting another command does not stop display of the prompt. In Windows 95, press CTRL+ALT+DEL to close the program.

*One of the major drawbacks of the menu command repetition is that it does not let you select a different command option. In the following menu item, the **ERASE** command uses the **C (Crossing)** option to select the objects. When the command is repeated, it does not allow you to use a different option for object selection.*

[ERASE,C:]*^C^CERASE C

AUTOMATIC MENU SWAPPING

Screen menus can be automatically swapped by using the system variable **MENUCTL**. When the system variable **MENUCTL** is set to 1, AutoCAD automatically issues the **$S=CMDNAME** command, where **CMDNAME** is the command that loads the submenu. For example, if you select the **LINE** command from the digitizer or pull-down menu, or enter the **LINE** command at the keyboard, the **CMDNAME** command will load the **LINE** submenu and display it in the screen menu area. To use this feature, the command name and the submenu name must be the same. For example, if the name of the arc submenu is ARC and you select the **ARC** command, the arc submenu will be automatically loaded on the screen. However, if the submenu name is different (for example, MYARC), AutoCAD will not load the arc submenu on the screen. The default value of the **MENUCTL** system variable is 1. If you set the **MENUCTL** variable to 0, AutoCAD will not use the **$S=CMDNAME** command feature to load the submenus.

MENUECHO SYSTEM VARIABLE

If the system variable **MENUECHO** is set to 0, the commands that you select from the digitizer, screen menu, pull-down menu, or buttons menu will be displayed in the command prompt area.

For example, if you select the **CIRCLE** command from the menu, AutoCAD will display **_circle Specify center point for circle or [3P/2P/Ttr (tan tan radius)]:.** If you set the value of the **MENUECHO** variable to 1, AutoCAD will suppress the echo of the menu item and display **Specify center point for circle or [3P/2P/Ttr (tan tan radius)]:** only. You will notice that the **_circle** is not displayed when **MENUECHO** is set to 1. You can use **^P** in the menu item to turn the echo on or off. The **MENUECHO** system variable can also be assigned a value of 2, 4, or 8, which controls the suppression of system prompts, disabling the ^P toggle, and debugging aid for DIESEL macros, respectively.

MENUS FOR FOREIGN LANGUAGES

Besides English, AutoCAD has several foreign language versions. If you want to write a menu that is compatible with other foreign language versions of AutoCAD, you must precede each command and keyword in the menu file with the underscore (_) character.

Examples
[New   ]^C^C_New
[Open]^C^C_Open
[Line   ]^C^C_Line
[Arc-SCA]^C^C_Arc;_C;_A

Any command or keyword that starts with the underscore character will be automatically translated. If you check the ACAD.MNU file, you will find that AutoCAD has made extensive use of this feature.

Example 7

Rewrite the screen menu of Example 1 so that the menu file is compatible with other foreign language versions of AutoCAD.

The following file is the listing of the menu file for Example 7:

```
***SCREEN
[ MENU-1      ]
[             ]
[             ]
[LINE         ]^C^C_LINE
[             ]
[CIR-C,R      ]^C^C_CIRCLE
[CIR-C,D      ]^C^C_CIRCLE;\_D
[CIR- 2P      ]^C^C_CIRCLE;_2P
[             ]
[ERASE        ]^C^C_ERASE
[MOVE         ]^C^C_MOVE
```

USE OF CONTROL CHARACTERS IN MENU ITEMS

You can use ASCII control characters in the command definition by using the caret sign (^)

followed by the control character. For example, if you want to write a menu item that will toggle the SNAP off or on, you can use a caret sign followed by the control character B as shown in the following example.

[SNAP-TOG]^B

Where **SNAP-TOG** ----- Command label for SNAP toggle

   ^ ---------------- Caret sign (^)

   B ----------------- Control character for SNAP

Examples

^C	(Cancel)
^G	(Grid on/off)
^H	(Backspace)
^O	(Ortho on/off)
^T	(Tablet on/off)
^E	(Isoplane top/left/right)

^B is equivalent to pressing the CTRL and B keys from the keyboard that toggles the SNAP mode. SNAP-TOG is the menu item label that will be displayed on the screen menu. You can use any ASCII control character in the command definition. Following are some of the ASCII control characters:

^@	(ASCII code 0)
^[	(ASCII code 27)
^\	(ASCII code 28)
^]	(ASCII code 29)
^^	(ASCII code 30)
^-	(ASCII code 31)

SPECIAL CHARACTERS

The following is a list of the special characters that you can use in the AutoCAD menus:

Character	Description
***	Three asterisks indicate a section title
**	Two asterisks indicate a submenu
[ ]	Used to enclose a menu item label
;	Causes a ENTER
Space	Space is equivalent to pressing the spacebar
\	AutoCAD pauses for user input
–	(Underscore character) translates the AutoCAD commands and keyword that follow it into English.
+	Used to continue the menu item definition to next line
=*	Forces the display of pull-down, cursor or image tile menus
*	Menu item repetition
$M=	Special character to load a DIESEL macro expression
$S=CMDNAME	Special command that loads a screen submenu

Chapter 9

^B	Toggles snap (On/Off)
^C	Cancels the existing command
^D	Toggles coordinate dial (On/Off)
^E	Changes the isometric plane (Left/Right/Top)
^G	Toggles grid (On/Off)
^H	Causes a backspace
^O	Toggles ortho mode (On/Off)
^P	Toggles MENUECHO on or off
^Q	Echoes all prompts to the printer
^T	Toggles tablet mode (On/Off)
^V	Change current viewport
^Z	Suppresses the automatic addition of spacebar at the end of a menu item

Example 8

Write a screen menu for the following toggle functions (filename SM5.MNU):

ORTHO	SNAP
GRID	COORDINATE DIAL
TABLET	ISOPLANE
PRINTER	

Before writing a screen menu for these toggle functions, it is important to know the control characters that AutoCAD uses to turn these functions on or off. The following is a list of the control characters for the functions in this example:

ORTHO	CTRL O
SNAP	CTRL B
GRID	CTRL G
COORDINATE DIAL	CTRL D
TABLET	CTRL T
ISOPLANE	CTRL E
CURRENT VIEWPORT	CTRL V

The following file is a listing of the screen menu that toggle the given functions. You can use AutoCAD's **EDIT** command to write this file.

```
[-TOGGLE-    ]
[            ]
[            ]
[            ]
[ORTHO       ]^O          ————————    turns ORTHO on/off
[SNAP        ]^B          ————————    turns SNAP on/off
[GRID        ]^G          ————————    turns GRID on/off
[CO-ORDS     ]^D          ————————    turns COORDINATE dial on/off
[TABLET      ]^T          ————————    turns TABLET on/off
```

```
[                    ]
[ISOPLANE       ]^E          ————————    turns ISOPLANE on/off
[CURRENT VIEWPORT]^V          ————————    Changes the current viewport
```

COMMAND DEFINITION WITHOUT ENTER OR SPACE

All command definitions that have been discussed so far use a semicolon (;) or a space for Enter. However, sometimes a user might want to define a menu item that is not followed by an Enter or a space. This is accomplished by using the backspace ASCII control character (^{H}). It is especially useful when you want to write a screen menu for a numeric keypad. ^{H} control character can be used with any character or a group of characters. The following is the format of menu item definition without an ENTER or space:

[9]9X^H

```
           Where   9 ------------------- Menu item label
                   9 ------------------ Character returned by selecting this item
                   X ------------------ X is erased because of ^H (backspace)
                                        (X could be substituted by any character)
                  ^H -------------- ^H for backspace
```

^{H} is the ASCII control character for backspace. When you select this item, ^{H} erases the previous character X and therefore returns 9 only. It is not necessary to have X preceding ^{H}. X is a dummy character; it could be any character. You can continue selecting more characters, and when you are done and you want to press ENTER, use the keyboard or the digitizer.

Example 9

Write a screen menu for the following characters (filename SM6.MNU):

```
        0           5           .
        1           6           ,
        2           7           X
        3           8           Y
        4           9           Z
```

The following file is a listing of the screen menu of Example 9 that will enable the user to pick the characters without an ENTER.

```
[-KEYPAD-    ]
[            ]
[            ]
[            ]
[0]0Y^H                 ————————    returns 0
[1]1Y^H                 ————————    returns 1
[2]2Y^H                 ————————    returns 2
[3]3Y^H
[4]4Y^H
[5]5Y^H
```

Chapter 9

```
[6]6Y^H
[7]7Y^H
[8]8Y^H
[9]9Y^H
[.].Y^H                    ——————    returns period (.)
[,],Y^H                    ——————    returns comma (,)
[]                         ——————    for space in the screen menu
[X]XX^H                    ——————    returns X
[Y]YX^H                    ——————    returns Y
[Z]ZZ^H                    ——————    returns Z
```

Note

You can also use a backslash after the character to continue selecting more characters. For example, [2]2\ will return 2 and then pause for the user input, which could be another character. The following file uses a backslash (\) to let the user enter another character.

```
[-KEYPAD- ]
[               ]
[               ]
[               ]
[0]0\
[1]1\
[2]2\
[3]3\
[4]4\
[5]5\
[6]6\
[7]7\
[8]8\
[9]9\
[.].\
[,],\
[]
[X]X\
[Y]Y\
[Z]Z\
```

MENU ITEMS WITH SINGLE OBJECT SELECTION MODE

The Single options for object selection, combined with the menu item repetition, can provide a powerful editing tool.

[ERASE]*^C^CERASE Single

Where ***** ------------------ Menu item repetition

            Single ------------ Single object selection option

In this menu item the asterisk that follows the menu item label repeats the command. ^C^C cancels the existing command. **ERASE** is an AutoCAD command that erases the selected objects. The Single option enables the user to select the objects and then AutoCAD automatically terminates the object selection process. The **ERASE** command then erases the selected objects.

If the point you pick is in a blank area (if there is no object), AutoCAD automatically defaults to crossing or window option. If you pull the window to the left, it is the crossing option and if you pull the box to the right, it is the window option for selecting the objects. The asterisk in front of the menu item repeats the command.

USE OF AUTOLISP IN MENUS

You can combine AutoLISP variables and expressions with the menu items as a command definition. When you select that item from the menu, it will evaluate all the expressions and generate the necessary output. The following examples illustrate the use of AutoLISP variables and expressions. (For more information on AutoLISP, see Chapters 12 and 13.)

Example 10

Write an AutoLISP program that draws a square and then write a screen menu that utilizes the AutoLISP variables and expressions to draw a square.

The following file is the listing of the AutoLISP program that generates a square. The program prompts the user to enter the starting point and the length of the side.

```
(DEFUN C:SQR()
(SETVAR "CMDECHO" 0)
(SETQ P1 (GETPOINT "\n ENTER STARTING POINT: "))
(SETQ S (GETDIST "\n ENTER LENGTH OF SIDE: "))
(SETQ P2 (LIST (+ (CAR P1) S) (CADR P1)))
(SETQ P3 (POLAR P2 (/ PI 2) S))
(SETQ P4 (POLAR P1 (/ PI 2) S))
(COMMAND "PLINE" P1 P2 P3 P4 "C")
(SETVAR "CMDECHO" 1)
(PRINC)
)
```

The following file is the listing of the screen menu file that utilizes the AutoLISP variables and expressions to draw a square:

```
[-SQUARE- ]
[          ]
[          ]
[SQUARE:  ](SETQ P1(GETPOINT "ENTER STARTING POINT:+
 "));\+
(SETQ S (GETDIST "ENTER LENGTH OF SIDE: "));\+
(SETQ P2 (LIST (+ (CAR P1) S) (CADR P1)))+
```

```
(SETQ P3 (POLAR P2 (/ PI 2) S))+
(SETQ P4 (POLAR P1 (/ PI 2) S));+
PLINE !P1 !P2 !P3 !P4 C
```

If you compare the AutoLISP program with the screen menu, you will notice that the statements that prompt the user to enter the information and the statements that do actual calculations are identical. Therefore, you can utilize AutoLISP variables and expressions in the menu file to customize AutoCAD.

DIESEL EXPRESSION IN MENUS

You can also define a DIESEL expression in the screen, tablet, pull-down, or button menu. When you select the menu item, it will automatically assign the value to the **MODEMACRO** system variable and then display the new status line. The following example illustrates the use of DIESEL expression in the screen menu.

Example 11

Write a screen menu that will load the DIESEL macro to display the following information in the status line:

Macro-1	Macro-2	Macro-3
Project name	Pline width	Dimtad
Drawing name	Fillet radius	Dimtix
Current layer	Offset distance	Dimscale

The following file is a listing of the screen menu that contains the definition of three DIESEL macros of Example 11. This menu can be loaded by using AutoCAD's MENU command and then entering the name of the menu file. If you select the first item, DIESEL1, it will automatically display the corresponding status line (Figure 9-10).

```
***screen
[*DIESEL*]
[DIESEL1:]^C^CMODEMACRO;$M=Cust-Acad,N:$(GETVAR,DWGNAME)+
,L:$(GETVAR,CLAYER);
[DIESEL2:]^C^CMODEMACRO;$M=PLWID:$(GETVAR,PLINEWID),+
FRAD:$(GETVAR,FILLETRAD),OFFSET:$(GETVAR,OFFSETDIST),+
LTSCALE:$(GETVAR,LTSCALE);
[DIESEL3:]^C^CMODEMACRO;$M=DTAD:$(GETVAR,DIMTAD),+
DTIX:$(GETVAR,DIMTIX),DSCALE:$(GETVAR,DIMSCALE);
```

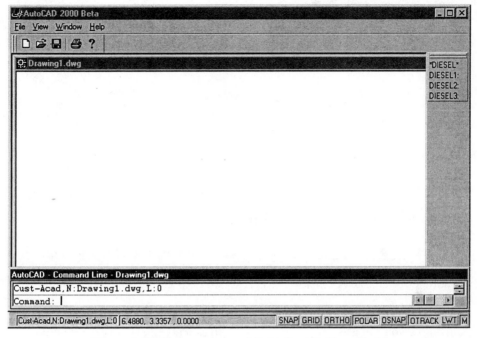

Figure 9-10 *Screen menu and status line for Example 11*

You can also use the **$M=** command to load a DIESEL macro. AutoCAD evaluates the DIESEL string expression that follows the equal sign (=) and the value returned by the DIESEL expression becomes a part of the menu item.

Example
[SCR-MODE]SCREENMODE $M=$(-,1,$(GETVAR,SCREENMODE))

In this example, AutoCAD evaluates the DIESEL string expression $(-,1,$(GETVAR, SCREENMODE)). The **GETVAR** command retrieves the value of the system variable SCREENMODE and subtracts the value from 1. If the value of the system variable **SCREENMODE** is 1, then the DIESEL expression returns 0, and if the value of **SCREENMODE** is 0, then the DIESEL expression returns 1. The value returned by the DIESEL expression is assigned to the system variable **SCREENMODE**. Therefore, this menu item can be used to change the screen display from the text to graphics mode or from the graphics to text mode.

Chapter 9

Review Questions

1. The name of the menu file that comes with the AutoCAD software package is
   _____.

2. The AutoCAD menu file can have up to _____ main sections.

3. The tablet menu can have up to _____ main sections.

4. The section label is designated by _____.

5. A submenu is designated by _____.

6. The part of the menu item that is inside the brackets is for _____ only.

7. Only the first _____ characters can be displayed on the screen menu.

8. If the number of characters inside the bracket exceeds eight, the screen menu _____ work.

9. In a menu file you can use _____ to cancel the existing command.

10. The AutoCAD _____ command is used to load a new menu file.

11. _____ is used to pause for input in a screen menu definition.

12. The menu item label _____ be a combination of uppercase and lowercase characters.

13. Submenu names can be _____ characters long.

14. The maximum number of accessible menu items in a screen menu depends on the _____.

15. Name two text editors that can be used to write menu files. _____
    and , _____.

Exercises

Exercise 5   *General*

Design and write a screen menu for the following AutoCAD commands (file name SME4.MNU).

POLYGON (Center)
POLYGON (Edge)
ELLIPSE (Center)
ELLIPSE (Axis End point)
CHAMFER
EXPLODE
COPY

Exercise 6   *General*

Write a screen menu file for the following AutoCAD commands. Use submenus if required (file name **SME5.MNU**).

ARC	**ROTATE**
-3P	**ARRAY**
-SCE	**DIVIDE**
-CSE	**MEASURE**
BLOCK	
INSERT	**LAYER**
WBLOCK	**SET**
MINSERT	**LIST**

Exercise 7   *General*

Write a screen menu item that will set up the following layers, linetypes, and colors (file name SME6.MNU).

Layer Name	Color	Linetype
0	WHITE	CONTINUOUS
OBJECT	RED	CONTINUOUS
HIDDEN	YELLOW	HIDDEN
CENTER	BLUE	CENTER
DIM	GREEN	CONTINUOUS

Exercise 8   *General*

Write a screen menu for the commands in Figure 9-11. Use the menu item ******** to load Osnaps, and use the menu item MENU-7 to load the root menu (file name SME7.MNU).

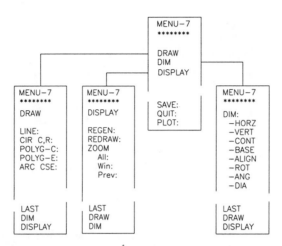

Figure 9-11 *Screen menu displays*

Exercise 9   *General*

Write a coordinated pull-down and screen menu for the following AutoCAD commands. Use a cascading menu for the **LINE** command options in the menu. When the user selects an item from the menu, the corresponding screen menu should be automatically loaded on the screen. The layout of the screen menu and the pull-down menu are shown in Figures 9-12(a) and 9-12(b).

LINE	ZOOM All	TIME
Continue	ZOOM Win	LIST
Close	ZOOM Pre	DISTANCE
Undo	PAN	AREA
.X	DBLIST	
.Y	STATUS	
.Z		
CIRCLE		
ELLIPSE		

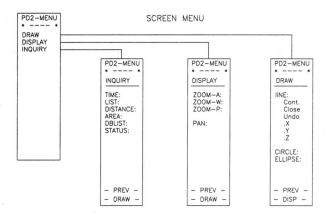

Figure 9-12(a) *Design of screen menu for Exercise 9*

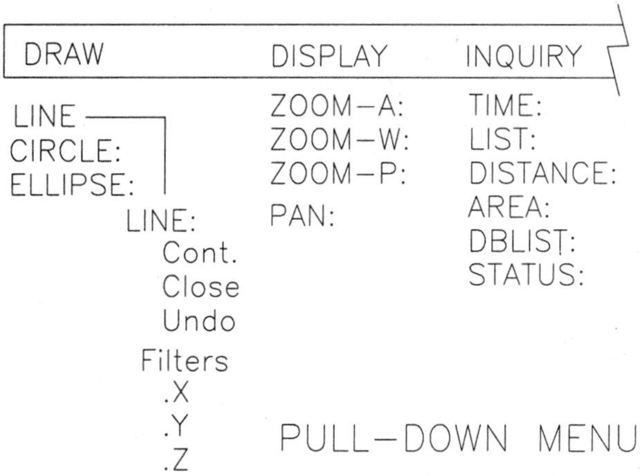

```
DRAW           DISPLAY    INQUIRY

LINE ———       ZOOM-A:    TIME:
CIRCLE:        ZOOM-W:    LIST:
ELLIPSE:       ZOOM-P:    DISTANCE:
       LINE:              AREA:
         Cont.  PAN:      DBLIST:
         Close            STATUS:
         Undo
         Filters
         .X
         .Y     PULL-DOWN  MENU
         .Z
```

Figure 9-12(b) Design of menu for Exercise 9

Exercise 10   *General*

Write a menu, a screen menu, and a tablet menu for the following AutoCAD commands. When you select a command from the template or the menu, the corresponding screen menu should be automatically loaded on the screen. The layout of the screen is shown in Figure 9-13.

LINE	**BLOCK**
PLINE	**WBLOCK**

Chapter 9 (vertical, right margin)

CIRCLE C,R   **INSERT**

CIRCLE C,D   **BLOCK LIST**

ELLIPSE AXIS ENDPOINT   **DDATTDEF**

ELLIPSE CENTER   **DDATTE**

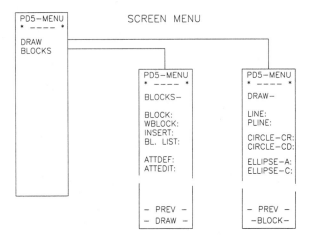

Figure 9-13 *Design of screen menu for Exercise 10*

Exercise 11   *General*

Write a screen, pull-down, and image tile menu for the commands shown in Figure 9-14. When the user selects a command from the menu, it should automatically load the corresponding menu on the screen.

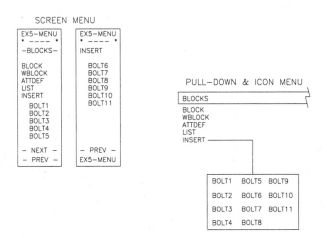

Figure 9-14 *Design of screen, pull-down, and image tile menus for Exercise 11*

Exercise 12
General

Write a screen, tablet, pull-down, and image tile menu for inserting the following commands. B1 to B15 are the block names.

BLOCK
WBLOCK
ATTDEF
LIST
INSERT

BL1	BL6	BL11
BL2	BL7	BL12
BL3	BL8	BL13
BL4	BL9	BL14
BL5	BL10	BL15

Exercise 13
General

Write a pull-down and screen menu for the following AutoCAD commands. Use submenus for the **ARC** and **CIRCLE** commands. When the user selects an item from the menu, the corresponding screen menu should be automatically loaded on the screen.

LINE	**BLOCK**	**QUIT**
PLINE	**INSERT**	**SAVE**
ARC	**WBLOCK**	——
ARC 3P		**PLOT**
ARC SCE		
ARC SCA		
ARC CSE		
ARC CSA		
ARC CSL		
CIRCLE		
CIRCLE C,R		
CIRCLE C,D		
CIRCLE 2P		

The layout shown in Figure 9-15(a) is one of the possible designs for this menu. The **ARC** and **CIRCLE** commands are in separate groups that will be defined as submenus in the menu file. The layout of the screen menu is shown in Figure 9-15(b).

Chapter 9

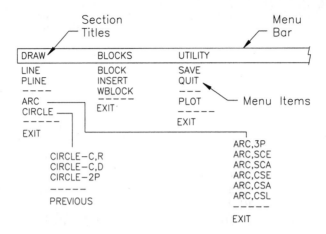

Figure 9-15(a) *Design of pull-down menu for Exercise 13*

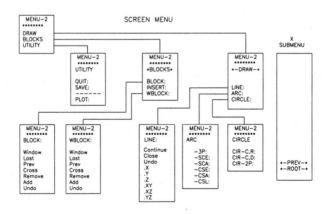

Figure 9-15(b) *Design of screen menu*

Chapter **10**

Customizing the Standard AutoCAD Menu

Learning Objectives

After completing this chapter, you will be able to:
- *Edit the standard AutoCAD menu file, ACAD.MNU.*
- *Load menus and submenus.*
- *Customize tablet areas.*
- *Customize buttons menus.*
- *Customize pull-down and cursor menus.*
- *Customize image tile menus and screen menus.*

THE STANDARD AUTOCAD MENU

The AutoCAD software package comes with a standard menu file: ACAD.MNU. AutoCAD stores the name of the menu that was used last in the system registry. When you start AutoCAD, the last used menu file is automatically loaded (Figure 10-1). The menu sections of a menu file are identified by section labels. The general format of the section label is *****section_name**. For example, if it is the screen menu section, the section label is ***SCREEN. The following is the list of the section labels.

***SCREEN		Screen menu
***TABLET(n)	n is from 1 to 4	Tablet menu
***IMAGE		Image tile menu
***POP(n)	**n** is from 1 to 499	Pull-down menu
	n is 0 and from 500 to 999	For shortcut menus
***BUTTONS(n)	**n** is from 1 to 4	Pointing device menu

***AUX(n)	**n** is from 1 to 4	System pointing device menu
***MENUGROUP		Menu file group name
***TOOLBARS		Toolbar definition
***HELPSTRING		Text displayed in the status bar
***ACCELERATORS		Accelerator key definitions

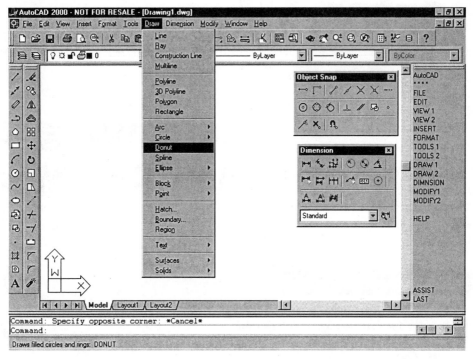

Figure 10-1 Screen display with standard AutoCAD menu loaded

The following file is a partial listing of the ACAD.MNU file that shows some of the section labels:

Note

```
//    AutoCAD 2000 Menu
//    28 April 1997
//    Copyright (C) 1986, 1987, 1988, 1989, 1990, 1991, 1992, 1994, 1996, 1997, 1998
//    by Autodesk, Inc.
```

***MENUGROUP=ACAD
// Begin AutoCAD Digitizer Button Menus

***BUTTONS1
```
// Simple + button
// if a grip is hot bring up the Grips Cursor Menu (POP 500), else send a carriage return
// If the SHORTCUTMENU sysvar is not 0 the first item (for button 1) is NOT USED.
$M=$(if,$(eq,$(substr,$(getvar,cmdnames),1,5),GRIP_),$P0=ACAD.GRIPS $P0=*);
$P0=SNAP $p0=*
^C^C
```

^B
^O
^G
^D
^E
^T

BUTTONS2
// Shift + button
$P0=SNAP $p0=*

BUTTONS3
// Control + button

BUTTONS4
// Control + shift + button
//   Begin System Pointing Device Menus

AUX1
// Simple button
// if a grip is hot bring up the Grips Cursor Menu (POP 500), else send a carriage return
// If the SHORTCUTMENU sysvar is not 0 the first item (for button 1, the "right button")
// is NOT USED.
$M=$(if,$(eq,$(substr,$(getvar,cmdnames),1,5),GRIP_),$P0=ACAD.GRIPS $P0=*);
$P0=SNAP $p0=*
^C^C
^B
^O
^G
^D
^E
^T

AUX2
// Shift + button
$P0=SNAP $p0=*
$P0=SNAP $p0=*

AUX3
// Control + button
$P0=SNAP $p0=*

AUX4
// Control + shift + button
$P0=SNAP $p0=*

***POP0
**SNAP
// Shift-right-click if using the default AUX2 and/or BUTTONS2
// menus. [&Object Snap Cursor Menu]
ID_Tracking [Temporary trac&k point]_tt
ID_From   [&From]_from
ID_MnPointFi [->Poin&t Filters]
ID_PointFilx [.X].X
ID_PointFily [.Y].Y
ID_PointFilz [.Z].Z

***POP1
**FILE
ID_MnFile   [&File]
ID_New   [&New...\tCtrl+N]^C^C_new
ID_Open   [&Open...\tCtrl+O]^C^C_open
ID_FILE_CLOSE [&Close]
ID_PartialOp [$(if,$(eq,$(getvar,fullopen),0),,~)Pa&rtial Load]^C^C_partiaload
ID_Save   [&Save\tCtrl+S]^C^C_qsave
ID_Saveas   [Save &As...]^C^C_saveas
ID_Export   [&Export...]^C^C_export
ID_PlotSetup [Pa&ge Setup...]^C^C_pagesetup
ID_PlotMgr   [Plotter &Manager...]^C^C_plottermanager

***POP2
**EDIT
ID_MnEdit   [&Edit]
ID_U   [&Undo\tCtrl+Z]^C^C_u
ID_Redo   [&Redo\tCtrl+Y]^C^C_redo
             [—]
ID_Cutclip   [Cu&t\tCtrl+X]^C^C_cutclip
ID_Copyclip [&Copy\tCtrl+C]^C^C_copyclip
ID_Copybase [Copy with &Base Point]^C^C_copybase
ID_Copylink [Copy &Link]^C^C_copylink

***POP3
**VIEW
ID_MnView   [&View]
ID_Redrawall [&Redraw]'_redrawall
ID_Regen   [Re&gen]^C^C_regen
ID_Regenall [Regen &All]^C^C_regenall

```
ID_ZoomRealt  [&Realtime]'_zoom ;
ID_ZoomPrevi  [&Previous]'_zoom _p
ID_ZoomWindo  [&Window]'_zoom _w
ID_ZoomDynam  [&Dynamic]'_zoom _d
ID_ZoomScale  [&Scale]'_zoom _s
ID_ZoomCente  [&Center]'_zoom _c




***POP4
**INSERT
ID_MnInsert  [&Insert]
ID_Ddinsert  [&Block...]^C^C_insert
ID_Xattach   [E&xternal Reference...]^C^C_xattach
ID_Imageatta [Raster &Image...]^C^C_imageattach




***POP5
**FORMAT
ID_MnFormat [F&ormat]
ID_Layer    [&Layer...]'_layer
ID_Ddcolor  [&Color...]'_color
ID_Linetype [Li&netype...]'_linetype
ID_Linewt   [Line&weight...]'_lweight




***POP6
**TOOLS
ID_MnTools  [&Tools]
ID_Spell    [Sp&elling]'_spell
ID_Qselect  [&Quick Select...]^C^C_qselect
ID_MnOrder  [->Display &Order]
ID_DrawordeF  [Bring to &Front]^C^C^P(ai_draworder "_f") ^P
ID_DrawordeB  [Send to &Back]^C^C^P(ai_draworder "_b") ^P




***POP7
**DRAW
ID_MnDraw   [&Draw]
ID_Line     [&Line]^C^C_line
ID_Ray      [&Ray]^C^C_ray
ID_Xline    [Cons&truction Line]^C^C_xline
ID_Mline    [&Multiline]^C^C_mline
ID_Pline    [&Polyline]^C^C_plineID_3dpoly  [&3D Polyline]^C^C_3dpoly
```

ID_Polygon   [Pol&ygon]^C^C_polygon
ID_Rectang   [Rectan&gle]^C^C_rectang

***POP8**
**DIMENSION
ID_MnDimensi [Dime&nsion]
ID_QDim   [&QDIM]^C^C_qdim
        [—]
ID_Dimlinear [&Linear]^C^C_dimlinear
ID_Dimaligne [Ali&gned]^C^C_dimaligned
ID_Dimordina [&Ordinate]^C^C_dimordinate
        [—]
ID_Dimradius [&Radius]^C^C_dimradius
ID_Dimdiamet [&Diameter]^C^C_dimdiameter
ID_Dimangula [&Angular]^C^C_dimangular

***POP9**
**MODIFY
ID_MnModify [&Modify]
ID_Ai_propch [&Properties]^C^C_properties
ID_Matchprop [&Match Properties]'_matchprop
ID_MnObject [->&Object]
ID_MnExterna [->&External Reference]
ID_Xbind   [&Bind...]^C^C_xbind
ID_Xclipfram   [<-
$(if,$(eq,$(getvar,xclipframe),1),!.)&Frame]$M=$(if,$(eq,$(getvar,xclipframe),1),^C^C_xclipframe
0,^C^C_xclipframe 1)
ID_MnImage   [->&Image]
ID_Imageadju [&Adjust...]^C^C_imageadjust
ID_Imagequal [&Quality]^C^C_imagequality
ID_Transpare [&Transparency]^C^C_transparency

***POP10**
**WINDOW
ID_MnWindow [&Window]
ID_WINDOW_CASCADE   [&Cascade]
ID_WINDOW_TILE_HORZ [Tile &Horizontally]
ID_WINDOW_TILE_VERT [&Tile Vertically]
ID_WINDOW_ARRANGE   [&Arrange Icons]

***POP11**
**HELP
ID_MnHelp   [&Help]
ID_Help   [AutoCAD &Help\tF1]'_help
ID_PlotHelp   [Fast   Track   to   &Plotting]^C^C^P(help   "acad_plt"
['Fast_Track_to_Plotting_Your_First_Drawing") ^P
ID_Wnew   [&What's New](help "acad_ug" "whatsnew")
ID_L_acla [&Learning Assistance]^C^C^P(if (findfile "alalink.exe") (progn (princ)(startapp
"alalink.exe") (princ) (la_support_assistance_alert)) ^P
ID_Support [&Support Assistance]^C^C^P(if (findfile "asa_main.hlp") (progn (princ)(help
(findfile "asa_main.hlp") "contents")(princ)) (ai_support_assistance_alert)) ^P

***POP500**
**GRIPS
// When a grip is hot, then display the following shortcut menu for grips. See also AUX1
menu.
        [&Grips Cursor Menu]
ID_Enter   [&Enter];
ID_GripMove [&Move]_move
ID_GripMirro [M&irror]_mirror
ID_GripRotat [&Rotate]_rotate
ID_GripScale [Sca&le]_scale

***POP501**
**CMDEFAULT   [Context menu for default mode]
ID_CMNonLast [&Repeat %s]^C^C;
 ID_Cutclip   [Cu&t]^C^C_cutclip
ID_Copyclip [&Copy]^C^C_copyclip
ID_Copybase [Copy with &Base Point]^C^C_copybase
ID_Pasteclip [&Paste]^C^C_pasteclip
ID_Pastebloc [Paste as Bloc&k]^C^C_pasteblock
ID_Pasteorig [Paste to Original Coor&dinates]^C^C_pasteorig

***POP502**
**CMEDIT [Context menu for edit mode]
ID_CMSelLast [&Repeat %s]^C^C;
ID_Cutclip   [Cu&t]^C^C_cutclip
ID_Copyclip [&Copy]^C^C_copyclip
ID_Copybase [Copy with &Base Point]^C^C_copybase
ID_Pasteclip [&Paste]^C^C_pasteclip
ID_Pastebloc [Paste as Bloc&k]^C^C_pasteblock

ID_Pasteorig [Paste to Original Coor&dinates]^C^C_pasteorig

***POP503**
**CMCOMMAND [Context menu for command mode]
ID_Enter   [&Enter];
ID_Cancel   [&Cancel]^C

***POP504**
**OBJECTS_DIMENSION   [Context Menu for Dimension Objects]
ID_DimText   [->Dim Te&xt position]
ID_DimAbove   [&Above dim line]^C^C_ai_dim_textabove
ID_DimTxtCen [&Centered]^C^C_ai_dim_textcenter
ID_DimHome   [&Home text]^C^C_ai_dim_texthome
ID_DimTxtMove2 [&Move text alone]^C^C_aidimtextmove _2

***POP505**
**OBJECT_VIEWPORT   [Context menu for Viewport Object]
ID_VpClip   [&Viewport Clip]^C^C_vpclip
ID_Vport_disp   [->&Display Viewport Objects]
ID_Vport_dispon [&Yes]^C^C_-vports _on _p;;
ID_Vport_dispoff [<-&No]^C^C_-vports _off _p;;
ID_Vport_lock [->Display &Locked]

***POP506**
**OBJECTS_XREF [Context menu for XREF Objects]
ID_Xclip   [Xref Cl&ip]^C^C_xclip
ID_XRef   [Xref Ma&nager...]^C^C_xref

***POP507**
**OBJECT_MTEXT [Context menu for MTEXT Object]
ID_Mtedit   [Mtext Ed&it...]^C^C_mtedit

***POP508**
**OBJECT_TEXT [Context menu for TEXT object]
ID_Ddedit   [Text Ed&it...]^C^C_ddedit

***POP509
**OBJECT_HATCH [Context menu for HATCH object]
ID_Hatchedit   [&Hatch Edit...]^C^C_hatchedit

***POP510
**OBJECT_LWPOLYLINE [Context menu for PLINE object]
ID_Pedit   [Polyline Ed&it]^C^C_pedit

***POP511
**OBJECT_SPLINE [Context menu for SPLINE object]
ID_Splinedit [Spline Ed&it]^C^C_splinedit

***TOOLBARS
**TB_DIMENSION
ID_TbDimensi [_Toolbar("Dimension", _Floating, _Hide, 100, 130, 1)]
ID_Dimlinear [_Button("Linear Dimension", ICON_16_DIMLIN,
ICON_16_DIMLIN)]^C^C_dimlinear
ID_Dimaligne [_Button("Aligned Dimension", ICON_16_DIMALI,
ICON_16_DIMALI)]^C^C_dimaligned
ID_Dimordina [_Button("Ordinate Dimension", ICON_16_DIMORD,
ICON_16_DIMORD)]^C^C_dimordinate
ID_Dimradius [_Button("Radius Dimension", ICON_16_DIMRAD,
ICON_16_DIMRAD)]^C^C_dimradius
ID_Dimdiamet [_Button("Diameter Dimension", ICON_16_DIMDIA,
ICON_16_DIMDIA)]^C^C_dimdiameter
ID_Dimangula [_Button("Angular Dimension", ICON_16_DIMANG,
ICON_16_DIMANG)]^C^C_dimangular

**TB_DRAW
ID_TbDraw [_Toolbar("Draw", _Left, _Show, 0, 0, 1)]
ID_Line   [_Button("Line", ICON_16_LINE, ICON_16_LINE)]^C^C_line
ID_Xline   [_Button("Construction   Line",   ICON_16_XLINE,
ICON_16_XLINE)]^C^C_xline
ID_Mline   [_Button("Multiline", ICON_16_MLINE, ICON_16_MLINE)]^C^C_mline

**TB_INQUIRY
ID_TbInquiry [_Toolbar("Inquiry", _Floating, _Hide, 100, 170, 1)]

Chapter 10

ID_Dist   [_Button("Distance", ICON_16_DIST, ICON_16_DIST)]'_dist
ID_Area   [_Button("Area", ICON_16_AREA, ICON_16_AREA)]^C^C_area
ID_Massprop [_Button("Mass Properties", ICON_16_MASSPR,
ICON_16_MASSPR)]^C^C_massprop
ID_List   [_Button("List", ICON_16_LIST, ICON_16_LIST)]^C^C_list
ID_Id   [_Button("Locate Point", ICON_16_ID, ICON_16_ID)]'_id

**TB_INSERT
ID_TbInsert [_Toolbar("Insert", _Floating, _Hide, 100, 190, 1)]
ID_Ddinsert [_Button("Insert Block", ICON_16_DINSER,
ICON_16_DINSER)]^C^C_insert
ID_Xref   [_Button("External Reference", ICON_16_XREATT,
ICON_16_XREATT)]^C^C_xref

**TB_LAYOUTS
ID_TbLayouts [_Toolbar("Layouts", _Floating, _Hide, 100, 350, 1)]
ID_LayNew   [_Button("New Layout", ICON_16_LAYNEW,
ICON_16_LAYNEW)]^C^C-layout _n
ID_LayTemp   [_Button("Layout from Template", ICON_16_LAYTEM,
ICON_16_LAYTEM)]^C^C-layout _t
ID_PlotSetup [_Button("Page Setup", ICON_16_PLTSET,
ICON_16_PLTSET)]^C^C_pagesetup
ID_VpDialog [_Button("Display Viewports Dialog", ICON_16_VPDLG,
ICON_16_VPDLG)]^C^C_vports

**TB_MODIFY
ID_TbModify [_Toolbar("Modify", _Left, _Show, 1, 0, 1)]
ID_Erase   [_Button("Erase", ICON_16_ERASE, ICON_16_ERASE)]^C^C_erase
ID_Copy   [_Button("Copy Object", ICON_16_COPYOB,
ICON_16_COPYOB)]$M=$(if,$(eq,+
$(substr,$(getvar,cmdnames),1,4),grip),_copy,^C^C_copy)
ID_Mirror   [_Button("Mirror", ICON_16_MIRROR,
ICON_16_MIRROR)]$M=$(if,$(eq,+

**TB_MODIFY_II
ID_TbModifII [_Toolbar("Modify II", _Floating, _Hide, 100, 270, 1)]
ID_Draworder
[_Button("Draworder",ICON_16_DRWORD,ICON_16_DRWORD)]^C^C_draworder
ID_Hatchedit [_Button("Edit Hatch", ICON_16_HATEDI,

ICON_16_HATEDI)]^C^C_hatchedit

**TB_OBJECT_PROPERTIES
ID_TbObjectP [_Toolbar("Object Properties", _Top, _Show, 0, 1, 1)]
ID_Ai_molc [_Button("Make Object's Layer Current", ICON_16_MOLC,
ICON_16_MOLC)]^C^C_ai_molc
ID_Layer   [_Button("Layers", ICON_16_LAYERS, ICON_16_LAYERS)]'_layer
ID_CtrlLayer [_Control(_Layer)]
ID_CtrlColor [_Control(_Color)]

**TB_OBJECT_SNAP
ID_TbOsnap [_Toolbar("Object Snap", _Floating, _Hide, 100, 210, 1)]
ID_Tracking [_Button("Temporary Tracking Point", ICON_16_OSNTMP,
ICON_16_OSNTMP)]_tt
ID_From   [_Button("Snap From", ICON_16_OSNFRO, ICON_16_OSNFRO)]_from
ID_OsnapEndp [_Button("Snap to Endpoint", ICON_16_OSNEND,
ICON_16_OSNEND)]_endp

**TB_ORBIT
ID_TbOrbit [_Toolbar("3D Orbit", _Floating, _Hide, 100, 350, 1)]
ID_3dpan   [_Button("3D Pan", ICON_16_3DPAN, ICON_16_3DPAN)]^C^C_3dpan
ID_3dzoom   [_Button("3D Zoom", ICON_16_3DZOOM,
ICON_16_3DZOOM)]^C^C_3dzoom
ID_3dorbit [_Button("3D Orbit", ICON_16_3DORBIT,
ICON_16_3DORBIT)]^C^C_3dorbit

**TB_SHADING
ID_TbShading [_Toolbar("Shade", _Floating, _Hide, 100, 390, 1)]
ID_2doptim [_Button("2D Wireframe", ICON_16_2DOPTIM,

**TB_REFEDIT
ID_TbRefedit [_Toolbar("Refedit", _Floating, _Hide, 100, 170, 1)]
ID_RefEditor   [_Button("Edit block or Xref", ICON_16_REFED_EDIT,
ICON_16_REFED_EDIT)]^C^C_refedit;

Chapter 10

**TB_REFERENCE
ID_TbReferen [_Toolbar("Reference", _Floating, _Hide, 100, 370, 1)]
ID_Xref   [_Button("External Reference", ICON_16_XREATT,
ICON_16_XREATT)]^C^C_xref
ID_Xattach [_Button("External Reference Attach", ICON_16_XATTAC,
ICON_16_XATTAC)]^C^C_xattach
ID_Xclip   [_Button("External Reference
Clip",ICON_16_XRECLI,ICON_16_XRECLI)]^C^C_xclip

**TB_RENDER
ID_TbRender [_Toolbar("Render", _Floating, _Hide, 100, 230, 1)]
ID_Hide   [_Button("Hide", ICON_16_HIDE, ICON_16_HIDE)]^C^C_hide
ID_Render   [_Button("Render", ICON_16_RENDER,

**TB_SOLIDS
ID_TbSolids [_Toolbar("Solids", _Floating, _Hide, 100, 250, 1)]
ID_Box   [_Button("Box", ICON_16_BOX, ICON_16_BOX)]^C^C_box
ID_Sphere   [_Button("Sphere", ICON_16_SPHERE, ICON_16_SPHERE)]^C^C_sphere
ID_Cylinder [_Button("Cylinder", ICON_16_CYLIND,
ICON_16_CYLIND)]^C^C_cylinder

**TB_SOLIDS2
ID_TbSolids2 [_Toolbar("Solids Editing", _Floating, _Hide, 100, 250, 1)]
ID_Union   [_Button("Union", ICON_16_UNION, ICON_16_UNION)]^C^C_union
ID_Subtract [_Button("Subtract", ICON_16_SUBTRA,
ICON_16_SUBTRA)]^C^C_subtract
ID_Intersect [_Button("Intersect", ICON_16_INTERS,

**TB_STANDARD
ID_TbStandar [_Toolbar("Standard Toolbar", _Top, _Show, 0, 0, 1)]
ID_New   [_Button("New", ICON_16_NEW, ICON_16_NEW)]^C^C_new
ID_Open   [_Button("Open", ICON_16_OPEN, ICON_16_OPEN)]^C^C_open
ID_Save   [_Button("Save", ICON_16_SAVE, ICON_16_SAVE)]^C^C_qsave

**TB_SURFACES
ID_TbSurface [_Toolbar("Surfaces", _Floating, _Hide, 100, 290, 1)]
ID_Solid   [_Button("2D Solid", ICON_16_SOLID, ICON_16_SOLID)]^C^C_solid

ID_3dface [_Button("3D Face", ICON_16_3DFACE, ICON_16_3DFACE)]^C^C_3dface

**TB_UCS
ID_TbUcs [_Toolbar("UCS", _Floating, _Hide, 100, 310, 1)]
ID_Ucs [_Button("UCS", ICON_16_UCS, ICON_16_UCS)]^C^C_ucs
ID_Dducs [_Button("Display UCS Dialog", ICON_16_DDUCS,
ICON_16_DDUCS)]^C^C_+ucsman 0

**TB_UCSII
ID_TbUcsName [_Toolbar("UCS II", _Floating, _Hide, 100, 300, 1)]
ID_Dducs [_Button("Display UCS Dialog", ICON_16_DDUCS,
ICON_16_DDUCS)]^C^C_+ucsman 0
ID_UcsMove [_Button("Move UCS Origin", ICON_16_UCSMOVE,
ICON_16_UCSMOVE)]^C^C_ucs _move
ID_UcsCombo [_Control(_UCSManager)]

**TB_VIEWPOINT
ID_TbViewpoi [_Toolbar("View", _Floating, _Hide, 100, 330, 1)]
ID_Ddview [_Button("Named Views", ICON_16_DDVIEW,
ICON_16_DDVIEW)]^C^C_view
ID_VpointTop [_Button("Top View", ICON_16_VIETOP, ICON_16_VIETOP)]^C^C_-
view
ID_VpointBot [_Button("Bottom View", ICON_16_VIEBOT, ICON_16_VIEBOT)]^C^C_-

ID_TbVpCreat [_Toolbar("Viewports", _Floating, _Hide, 100, 350, 1)]
ID_VpDialog [_Button("Display Viewports Dialog", ICON_16_VPDLG,
ICON_16_VPDLG)]^C^C_vports
ID_VpSingle [_Button("Single Viewport", ICON_16_VPONE,
ICON_16_VPONE)]$M=$(if,$(eq,$(getvar,tilemode),1),^C^C_-vports _si,^C^C_vports)

**TB_WEB
ID_TbWeb [_Toolbar("Web", _Floating, _Hide, 100, 380, 1)]
ID_HlnkBack [_Button("Go Back", ICON_16_HLNK_BACK,
ICON_16_HLNK_BACK)]'_hyperlinkBack
ID_HlnkFwd [_Button("Go Forward", ICON_16_HLNK_FWD,
ICON_16_HLNK_FWD)]'_hyperlinkFwd
ID_HlnkStop [_Button("Stop Navigation", ICON_16_HLNK_STOP,

ICON_16_HLNK_STOP)]'_hyperlinkStop
ID_Browser [_Button("Browse the Web", ICON_16_WEB, ICON_16_WEB)]^C^C_browser

***image
**image_3DObjects
[3D Objects]
[acad(Box3d,Box3d)]^C^C_ai_box
[acad(Pyramid,Pyramid)]^C^C_ai_pyramid
[acad(Wedge,Wedge)]^C^C_ai_wedge

**image_poly
[Set Spline Fit Variables]
[acad(pm-quad,Quadric Fit Mesh)]'_surftype 5
[acad(pm-cubic,Cubic Fit Mesh)]'_surftype 6
[acad(pm-bezr,Bezier Fit Mesh)]'_surftype 8
[acad(pl-quad,Quadric Fit Pline)]'_splinetype 5
[acad(pl-cubic,Cubic Fit Pline)]'_splinetype 6

**image_vporti
[Tiled Viewport Layout]
[acad(vport-1,Single)]^C^C(ai_tiledvp 1 nil)
[acad(vport-3v,Three: Vertical)]^C^C(ai_tiledvp 3 "_V")
[acad(vport-3h,Three: Horizontal)]^C^C(ai_tiledvp 3 "_H")
[acad(vport-4,Four: Equal)]^C^C(ai_tiledvp 4 nil)

***SCREEN
**S
[AutoCAD]^C^C^P(ai_rootmenus) ^P
[* * * *]$S=ACAD.OSNAP
[FILE   ]$S=ACAD.01_FILE
[EDIT   ]$S=ACAD.02_EDIT
[VIEW 1]$S=ACAD.03_VIEW1
[VIEW 2]$S=ACAD.04_VIEW2
[INSERT]$S=ACAD.05_INSERT
[FORMAT]$S=ACAD.06_FORMAT
[TOOLS 1]$S=ACAD.07_TOOLS1
[TOOLS 2]$S=ACAD.08_TOOLS2
[DRAW 1]$S=ACAD.09_DRAW1
[DRAW 2]$S=ACAD.10_DRAW2

[DIMNSION]$S=ACAD.11_DIMENSION
[MODIFY1]$S=ACAD.12_MODIFY1

// The SNAP_TO menu can be called on high resolution displays to paint the snap
// options on the bottom of the screen menu.
// To add lines to the bottom of the Root menu delete the next line.

**SNAP_TO 28
[Endpoint]_endp
[Midpoint]_mid
[Intersec]_int
[App Int]_appint

**ASSIST 3
[Last   ]_l
[Previous]_p
[All   ]_all
[Cpolygon]_cp
[Wpolygon]_wp

**02_EDIT 3
[Undo   ]^C^C_u
[Redo   ]^C^C_redo
[Cut   ]^C^C_cutclip
[Copy   ]^C^C_copyclip
[CopyBase]^C^C_copybase
[CopyLink]^C^C_copylink
[Paste   ]^C^C_pasteclip
[PasteBlk]^C^C_pasteblock
[PasteOri]^C^C_pasteorig
[PasteSpe]^C^C_pastespec
[OLElinks]^C^C_olelinks

**03_VIEW1 3
[Redraw]'_redraw
[Redrawal]'_redrawall
[Regen]^C^C_regen
[Regenall]^C^C_regenall
[Zoom   ]'_zoom

Chapter 10

[Pan   ]'_pan
[Dsviewer]'_dsviewer
[Tilemode]^C^C_tilemode
[Mspace]^C^C_mspace
[Pspace]^C^C_pspace

**04_VIEW2 3
[Hide   ]^C^C_hide
[Shade   ]^C^C_shademode

[Render]^C^C_render
[Scene   ]^C^C_scene
[Light   ]^C^C_light
[Mapping]^C^C_setuv
[Backgrnd]^C^C_background
[Fog   ]^C^C_fog

**06_FORMAT 3
[Layer   ]'_layer
[Color]'_color
[Linetype]'_linetype

[Style   ]'_style
[Ddim   ]'_dimstyle
[Ddptype]'_ddptype
[Mlstyle]^C^C_mlstyle

[Units]'_units
[Thicknes]'_thickness
[Limits]'_limits

[Rename]^C^C_rename
**ENDSCREEN

***TABLET1
**TABLET1STD
[A-1]\
[A-2]\
[A-3]\
[A-4]\
[A-5]\

[A-6]\
[A-7]\
[A-8]\
[A-9]\
[A-10]\

|
|

***TABLET2**
TABLET2STD
// Row J View
^C^C_regen
'_zoom _e
'_zoom _a
'_zoom _w
'_zoom _p
[Draw]\

|
|

***TABLET3**
TABLET3STD
// Row 1
\
\
\
<<135
<<135
<<90
<<90
<<45
<<45

|
|

***TABLET4**
TABLET4STD
// Row S

|
|

***HELPSTRINGS**
ID_2doptim   [Set viewport to 2D wireframe: SHADEMODE 2]
ID_3darray   [Creates a three-dimensional array: 3DARRAY]
ID_3dclip   [Starts 3DORBIT and opens the Adjust Clipping Planes window: 3DCLIP]
ID_3dclipbk [Toggles the back clipping plane on or off in the 3D Orbit Adjust Clipping
Planes window: DVIEW ALL CL B]

ID_3dclipfr [Toggles the front clipping plane on or off in the 3D Orbit Adjust Clipping Planes window: DVIEW ALL CL F]

ID_3dcorbit [Starts the 3DORBIT command with continuous orbit active in the 3D view: 3DCORBIT]

ID_3ddistanc [Starts the 3DORBIT command and makes objects appear closer or farther away: 3DDISTANCE]

ID_3dface [Creates a three-dimensional face: 3DFACE]

***ACCELERATORS
// Bring up hyperlink dialog
ID_Hyperlink [CONTROL+"K"]
// Toggle Orthomode
[CONTROL+"L"]^O
// Next Viewport
[CONTROL+"R"]^V
ID_Copyclip [CONTROL+"C"]
ID_New   [CONTROL+"N"]
ID_Open   [CONTROL+"O"]
ID_Print   [CONTROL+"P"]
ID_Save   [CONTROL+"S"]
ID_Pasteclip [CONTROL+"V"]
ID_Cutclip [CONTROL+"X"]
ID_Redo   [CONTROL+"Y"]
ID_U   [CONTROL+"Z"]
ID_Modify   [CONTROL+"1"]
ID_Content [CONTROL+"2"]
ID_dbConnect [CONTROL+"6"]
ID_VBARun   [ALT+"F8"]
ID_VBAIDE   [ALT+"F11"]

SUBMENUS

The number of items in a submenu section can be so large that they cannot be accommodated on one screen. For example, the maximum number of items that can be displayed on some of the display devices is 21. If the menu or the screen menu has more than 21 items, the menu items in excess of 21 are not displayed on the screen and cannot be accessed. Similarly, the maximum number of assignable blocks in tablet area-1 is 225. If the number of items in the TABLET1 section is more than 225, the menu items in excess of 225 are not assigned to any block on the template and are therefore inaccessible. The user can overcome this problem by using submenus that let the user define smaller groups of items within a menu section. When a submenu is selected, it loads the submenu items and displays them on the screen. In the case of a tablet menu the commands are assigned to different blocks on the template.

Submenu Definition

A submenu definition consists of two asterisk signs (**) followed by the name of the submenu. A menu can have any number of submenus and each submenu should have a unique name. The items that follow a submenu, up to the next section label or submenu label, belong to that submenu. The following is the format of a submenu label:

> **Name
>
> Where   ** ---------------- Two asterisk signs (**) designate a submenu
>   Name ----------- Name of the submenu

Note

The submenu name can be up to 31 characters long.

The submenu name can consist of letters, digits, and the special characters like: $ (dollar), - (hyphen), and _ (underscore).

The submenu name should not have any embedded blanks.

The submenu names should be unique in a menu file.

Submenu Reference

The submenu reference is used to reference or load a submenu. It consists of a "$" sign followed by a letter that specifies the menu section. For example, the letter S specifies a screen menu, B specifies a button menu, I specifies an image tile menu, and Pn specifies a menu where n designates the number of the menu. Similarly, the letter that specifies a tablet menu section is Tn, where n designates the number of the tablet menu section. The menu section is followed by an "=" sign and the name of the submenu that the user wants to activate. The submenu name should be without "**". The following is the format of a submenu reference:

> $Section=Submenu
>
> Where $ ------------------ "$" sign
>   Section ---------- Menu section specifier
>   = ----------------- "=" sign
>   Submenu ------- Name of submenu

> Example
> $S=BLOCKS
>
> Where   S ------------------ S specifies screen menu
>   BLOCKS -------- Name of submenu

Loading Screen Menus

You can load a menu that is defined in the screen menu section from any menu by using the following load command:

> $S=(name1) $S=(name2)
>
> Where "name1" and "name2" are the names of the submenus

Example
$S=X $S=INSERT

> Where   **S** ------------------- S specifies screen menu
> **X** ------------------ Submenu name defined in screen menu section
> **INSERT** --------- Submenu name defined in screen menu section

The first load command ($S=X) loads the submenu X that has been defined in the screen menu section of the menu file. The X submenu, defined in the screen menu section, contains 21 blank lines. When it is loaded, it clears the screen menu. The second load command ($S=INSERT) loads the submenu INSERT that has also been defined in the screen menu section of the menu file.

Loading Pull-down Menus

You can load a pull-down menu from any menu by using the following command:

$P(n)=(name) $P(n)=*

> Where "n" ranges from 1 to 10 (POP1 — POP10) and "name" is the name of the submenu defined in the pull-down menu.

Example
$P1=P1A $P1=*

> Where   **$P1=P1A** ------- Loads the submenu P1A
> **$P1=*** ----------- Forces the display of new menu items

The first load command $P1=P1A loads the submenu P1A that has been defined in the POP1 section of the menu file. The second load command $P1=* is a special AutoCAD command that forces the new menu items to be displayed on the screen.

Loading Image Tile Menus

You can also load an image tile menu from any menu by using the following load command:

$I=(name) $I=*

> Where "name" is the name of the submenu defined in the image submenu

Example
$I=IMAGE1 $I=*

> Where   **$I=IMAGE1** --- Load the submenu IMAGE1
> **$I=*** ------------- To display the dialog box

This menu item consists of two load commands. The first load command $I=IMAGE1 loads the image submenu IMAGE1 that has been defined in the image tile menu section of the file. The second load command $I=* displays the new dialog box on the screen.

CUSTOMIZING TABLET AREA-1

The tablet menu has four sections TABLET1, TABLET2, TABLET3, and TABLET4 (Figure 10-2). Tablet area-1 of the template has 25 columns numbered 1 to 25 and 9 rows designated by letters A through I. The total number of blocks for tablet area-1 is 225 (25 x 9 = 225). You can utilize this area to customize AutoCAD's tablet menu by assigning commands or macros to different blocks. Figure 10-3 shows tablet area-1 of the standard AutoCAD template. Before making any changes or additions to the tablet menu it is very important to figure out the commands that you want to add to the tablet menu and determine their location on the template. A well thought-out tablet design will save a lot of revision time in the long run. Figure 10-3 shows a drawing that has 25 columns and 9 rows. You may draw a similar drawing and make several copies of it to arrange the commands and design the template.

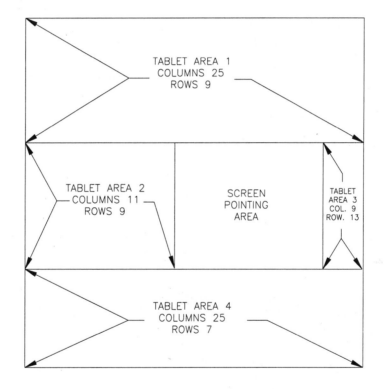

Figure 10-2 *Four tablet areas of the AutoCAD template*

Note

Before making any changes to the menu file make sure that the original file has been properly saved. You may also make a copy of the ACAD.MNU file and then make the changes in the new file. Use the following command to make a copy of the ACAD.MNU file. The name of the new menu file is CUSTOM.MNU. After you make the necessary changes you may load the new menu for testing by using AutoCAD's MENU command.

        COPY ACAD.MNU CUSTOM.MNU

Chapter 10

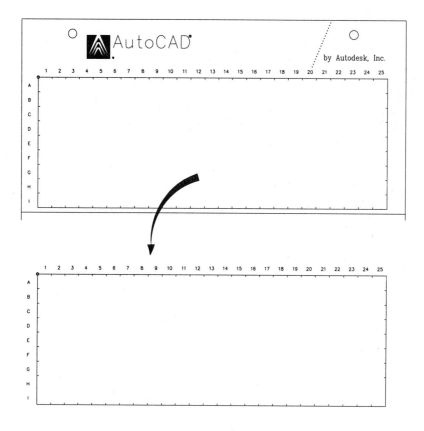

Figure 10-3 *Tablet area-1 with 25 columns and 9 rows*

Example 1

Add the commands shown in Figure 10-4 to the TABLET1 section of CUSTOM.MNU file. B10 through B25 are the names of the blocks. The user should be able to insert the blocks with X-scale and Y-scale factors of 1.0 and a rotation angle of 0.

Before making any changes in the file, you need to determine the location of each block on the template. For example, the WBlock B10 is to be assigned to a block that is in row B and column number 22. Similarly, WBlock B17 is in row C and column 25. The following table shows the location of WBlocks in tablet area-1 of the template:

WBLOCK NAME	ROW	COLUMN	LOCATION IN TABLET1
B10	B	22	B-22
B11	B	23	B-23
B12	B	24	B-24

B13	B	25	B-25
B14	C	22	C-22
B15	C	23	C-23
B16	C	24	C-24
B17	C	25	C-25
B18	D	22	D-22
B19	D	23	D-23
B20	D	24	D-24
B21	D	25	D-25
B22	E	22	E-22
B23	E	23	E-23
B24	E	24	E-24
B25	E	25	E-25

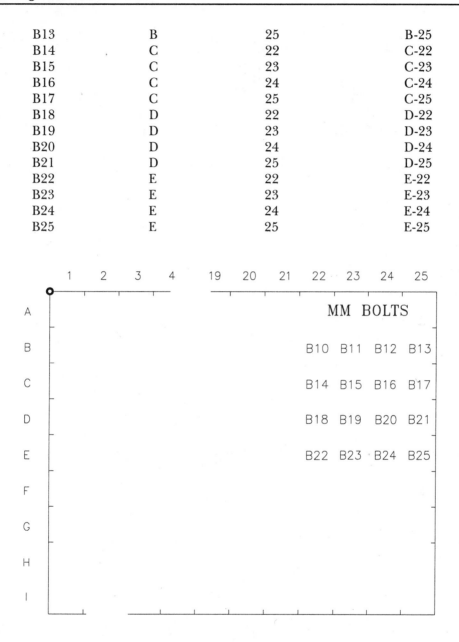

Figure 10-4 *Commands assigned to tablet area-1*

You can use any word processor or text editor to load and edit the file. After the CUSTOM.MNU file is loaded, search for ***TABLET1 section label. The letters and the numbers inside the brackets indicate the row and column. For example, in [A-1], A is for row number and 1 is for column number in the template area-1.

[A-1]

Where **A** ------------------ Row

1 ------------------- Column number

The location of the first WBlock B10 in the menu file is B-22. Locate the menu item B-22 in the menu file and add the **INSERT** command to the menu item command definition. The following file is a partial listing of the TABLET1 section of the menu file after editing:

***TABLET1
**TABLET1STD
[A-1]
[A-2]
[A-3]
[A-4]
[A-5]
[A-6]
[A-7]
[A-8]
[A-9]
[A-10]
[A-11]
[A-12]
[A-13]
[A-14]
[A-15]
[A-16]
[A-17]
[A-18]
[A-19]
[A-20]
[A-21]
[A-22]
[A-23]
[A-24]
[A-25]
[B-1]
[B-2]
[B-3]
[B-4]
[B-5]
[B-6]
[B-7]
[B-8]
[B-9]
[B-10

[B-22]^C^CINSERT;B10;\1.0;1.0;0
[B-23]^C^CINSERT;B11;\1.0;1.0;0
[B-24]^C^CINSERT;B12;\1.0;1.0;0
[B-25]^C^CINSERT;B13;\1.0;1.0;0
[C-1]

|

[C-21]
[C-22]^C^CINSERT;B14;\1.0;1.0;0
[C-23]^C^CINSERT;B15;\1.0;1.0;0
[C-24]^C^CINSERT;B16;\1.0;1.0;0
[C-25]^C^CINSERT;B17;\1.0;1.0;0
[D-1]

|

[D-20]
[D-21]
[D-22]^C^CINSERT;B18;\1.0;1.0;0
[D-23]^C^CINSERT;B19;\1.0;1.0;0
[D-24]^C^CINSERT;B20;\1.0;1.0;0
[D-25]^C^CINSERT;B21;\1.0;1.0;0

|

[E-21]
[E-22]^C^CINSERT;B22;\1.0;1.0;0
[E-23]^C^CINSERT;B23;\1.0;1.0;0
[E-24]^C^CINSERT;B24;\1.0;1.0;0
[E-25]^C^CINSERT;B25;\1.0;1.0;0

Note
To load the new menu CUSTOM.MNU, use AutoCAD's MENU command.

> Command: **MENU**
> Menu file name <ACAD>: **CUSTOM**

When you select the block insert command from the template, AutoCAD will prompt you to enter an insertion point. The X-scale and Y-scale factors and the rotation angle are already defined in the command definition.

^C^CINSERT;B22;\1.0;1.0;0
> Where **B22** --------------- Block or Wblock name
> \-------------------- Pause for user input
> **1.0** ---------------- X-scale factor
> **;** ------------------- Enter

1.0 ---------------- Y-scale factor
; ------------------ Enter
0 ------------------ Rotation angle

SUBMENUS

The number of items in the TABLET1STD section of the menu file can be so large that they may not be accommodated on one template. For example, the maximum number of assignable blocks in tablet area-1 is 225. If the number of items in the TABLET1STD section is more than 225, the menu items in excess of 225 are not assigned to any block on the template and are therefore inaccessible. The user can overcome this problem by using submenus that let the user define any number of menu items in the tablet area-1 of the template. When the user selects the submenu, AutoCAD automatically loads the new submenu and assigns the commands to different blocks in template area-1. The format of the submenu reference is:

$Section=Submenu
Where   **$** ------------------- "**$**" sign
        Section ---------- Menu section specifier
        = ----------------- "**=**" sign
        Submenu ------- Name of submenu

Example
$T1=TAB1
Where   **$T1** -------------- T1 specifies TABLET1 menu section
        Tab1 ------------- Name of submenu

Example 2

Edit the CUSTOM.MNU file to add the commands shown in Figure 10-5. Use the submenus and make a provision for swapping the submenus. When the user selects a block insert command from the tablet menu, AutoCAD should automatically load the corresponding screen menu.

In this example it is assumed that there is no room for adding more commands to tablet area-1; therefore, submenus have to be created to make room for the commands in excess of 225. Two submenus TABA and TABB have been defined in the TABLET1 section of the menu file CUSTOM.MNU. When you load the menu file CUSTOM.MNU, the first submenu TABA is automatically loaded and you can select the block insert commands from the template. You can load the submenu TABB by selecting the **Load TABB** block from the template. AutoCAD loads the submenu TABB and now you can select the commands from the new submenu. If you want to load the submenu TABA back, select the **Load TABA** from the template. The following file is a partial listing of the TABLET1 section of the menu file after inserting the new command definitions:

```
***TABLET1
**TABLET1STD
**TABA
[A-1]
[A-2]
[A-3]
```

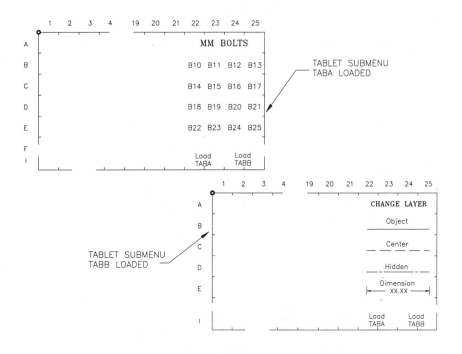

Figure 10-5 *Commands assigned to tablet area-1*

[A-4]
[A-5]
[A-6]
[A-7]
[A-8]
[A-9]
[A-10]

[B-18]
[B-19]
[B-20]
[B-21]
[B-22] ^ C ^ C$S=X $S=INSBLK INSERT;B10;\1.0;1.0;0
[B-23] ^ C ^ C$S=X $S=INSBLK INSERT;B11;\1.0;1.0;0
[B-24] ^ C ^ C$S=X $S=INSBLK INSERT;B12;\1.0;1.0;0
[B-25] ^ C ^ C$S=X $S=INSBLK INSERT;B13;\1.0;1.0;0
[C-1]

[C-21]
[C-22] ^ C ^ C$S=X $S=INSBLK INSERT;B14;\1.0;1.0;0
[C-23] ^ C ^ C$S=X $S=INSBLK INSERT;B15;\1.0;1.0;0
[C-24] ^ C ^ C$S=X $S=INSBLK INSERT;B16;\1.0;1.0;0
[C-25] ^ C ^ C$S=X S=INSBLK INSERT;B17;\1.0;1.0;0
[D-1]

|

[D-20]
[D-21]
[D-22] ^ C ^ C$S=X $S=INSBLK INSERT;B18;\1.0;1.0;0
[D-23] ^ C ^ C$S=X $S=INSBLK INSERT;B19;\1.0;1.0;0
[D-24] ^ C ^ C$S=X $S=INSBLK INSERT;B20;\1.0;1.0;0
[D-25] ^ C ^ C$S=X $S=INSBLK INSERT;B21;\1.0;1.0;0
[E-1]

|

[E-21]
[E-22] ^ C ^ C$S=X $S=INSBLK INSERT;B22;\1.0;1.0;0
[E-23] ^ C ^ C$S=X $S=INSBLK INSERT;B23;\1.0;1.0;0
[E-24] ^ C ^ C$S=X $S=INSBLK INSERT;B24;\1.0;1.0;0
[E-25] ^ C ^ C$S=X $S=INSBLK INSERT;B25;\1.0;1.0;0
[F-1]

|

[H-19]
[H-20]
[H-21]
[H-22] ^ C ^ C$T1=TABA
[H-23] ^ C ^ C$T1=TABA
[H-24] ^ C ^ C$T1=TABB
[H-25] ^ C ^ C$T1=TABB

**TABB
[A-1]
[A-2]
[A-3]
[A-4]
[A-5]
[A-6]
[A-7]
[A-8]

|

[B-18]
[B-19]
[B-20]
[B-21]
[B-22]^C^CLAYER;SET;OBJECT;;
[B-23]^C^CLAYER;SET;OBJECT;;
[B-24]^C^CLAYER;SET;OBJECT;;
[B-25]^C^CLAYER;SET;OBJECT;;
[C-1]

|
|
|

[C-21]
[C-22]^C^CLAYER;SET;CENTER;;
[C-23]^C^CLAYER;SET;CENTER;;
[C-24]^C^CLAYER;SET;CENTER;;
[C-25]^C^CLAYER;SET;CENTER;;
[D-1]

|
|
|

[D-20]
[D-21]
[D-22]^C^CLAYER;SET;HIDDEN;;
[D-23]^C^CLAYER;SET;HIDDEN;;
[D-24]^C^CLAYER;SET;HIDDEN;;
[D-25]^C^CLAYER;SET;HIDDEN;;
[E-1]

|
|
|

[E-21]
[E-22]^C^CLAYER;SET;DIM;;
[E-23]^C^CLAYER;SET;DIM;;
[E-24]^C^CLAYER;SET;DIM;;
[E-25]^C^CLAYER;SET;DIM;;
[F-1]

|
|
|

[H-19]
[H-20]
[H-21]
[H-22]^C^C$T1=TABA
[H-23]^C^C$T1=TABA
[H-24]^C^C$T1=TABB
[H-25]^C^C$T1=TABB

Note

When you load the submenu TABB, the template overlay for tablet area-1 must be changed.

The menu items defined in H-22, H23, H-24, and H-25 load the submenus TABA and TABB respectively.

$$/ \char94 C \char94 C\$T1=TABA$$
Where   ^ **C** ^ **C** ---------- Cancels the existing command twice
       $T1=TABA ---- Loads tablet submenu TABA

CUSTOMIZING TABLET AREA-2

Tablet area-2 has 11 columns and 9 rows. The columns are numbered from 1 to 11 and the rows are designated by the letters J through R as shown in Figure 10-6. The total number of blocks in tablet area-2 is 99 (11 x 9 = 99). There are no empty blocks in this area like tablet area-1 and the commands that have been assigned to the blocks are defined in the TABLET2 menu sections of the standard menu file. You can change or delete the command definition assigned to these blocks. You can even delete the entire TABLET2 menu sections and write your own menu that best fits your needs. However, you must be careful in developing a new menu and you may need quite some time to come up with a reliable menu.

Exercise 1   *General*

Assign the following commands to the blocks in the tenth column of tablet area-2 (Figure 10-6).

       LINE
       PLINE
       ARC,CSE
       ARC,SCE
       ARC,CSA
       CIRCLE-C,R
       CIRCLE-C,D
       CIRCLE-2P

CUSTOMIZING TABLET AREA-3

Tablet area-3 has 9 columns and 13 rows. In Figure 10-7 the columns are numbered from 1 to 9, and the rows are numbered from 1 to 13. The total number of blocks in tablet area-3 are 117 (13 x 9 = 117). The size of the blocks in this area is smaller than the blocks in other sections of the tablet template. Also, the blocks are rectangular in shape, whereas in other tablet areas the blocks are square. You can modify or delete the command definitions assigned to these blocks. You can even delete the entire TABLET3 menu sections and write a menu that best fits your needs. However, you have to be careful when you develop a new menu. The following example illustrates the editing process for adding commands to the TABLET3 section of the CUSTOM.MNU file.

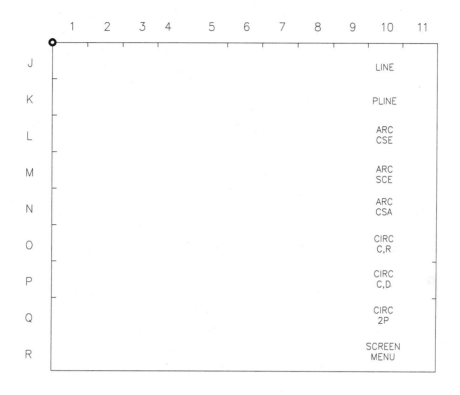

Figure 10-6 Commands assigned to tablet area-2

Example 3

Add the following angles to the TABLET3 section of CUSTOM.MNU file. The layout of the tablet area-3 is shown in Figure 10-7.

ANGLES
30
120
210
330

Use your word processor or text editor to load CUSTOM.MNU file and search for **TABLET3 section of the menu file. Locate the lines that you want to edit and then assign the required command definitions to these lines. The following file is a partial listing of the TABLET3 section of the CUSTOM.MNU file after editing:

**TABLET3

```
<<30
<<30
<<135
<<135
<<90
<<90
<<45
<<45
;
<<120
<<120
<<180
<<180
<\
<\
<<0
<<0
;
<<210
<<210
<<225
<<225
<<270
<<270
<<315
<<315
;
<<330
<<330
^H
^H
^H
^H
^H
^H
;
;
;
m\
m\
cm\
cm\
mm\
mm\
;
;
;
```

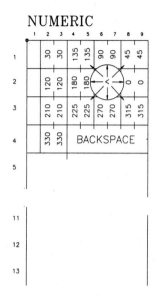

Figure 10-7 *Angles assigned to tablet area-3*

```
.\
.\
+\
+\
%%d\
%%d\
;
;
;
,\
,\
%%p\
%%p\
%%c\
%%c\
```

CUSTOMIZING TABLET AREA-4

Tablet area-4 has 25 columns and 7 rows. As shown in Figure 10-8, the columns are numbered from 1 to 25 and the rows are designated by the letters S through Y. The total number of blocks in tablet area-4 is 175 (7 * 25 = 175). You can modify or delete the command definition assigned to these blocks. You can even delete the entire TABLET4 menu sections and write a menu that best fits your needs. However, you have to be careful when you develop a new menu.

Exercise 2   *General*

Add the following commands to the TABLET4 section of CUSTOM.MNU file. The partial layout of tablet area-4 is shown in Figure 10-8.

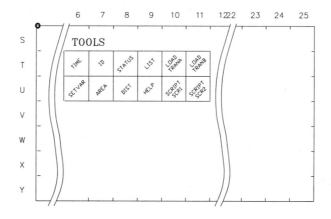

Figure 10-8 *Commands assigned to tablet area-4*

1. Load and run the AutoLISP; routines TRANA.LSP and TRANB.LSP. (It is assumed that the files TRANA.LSP and TRANB.LSP are predefined.)

2. Run the script files SCR1 and SCR2.

Note

^C^C(LOAD "TRANA");TRANA *In this menu, item (LOAD "TRANA") loads the AutoLISP file TRANA. The TRANA that is outside the parenthesis, after the semicolon, executes the function TRANA.*

^C^C(LOAD "TRANA");TRANA
        Where **LOAD**------------ Loads AutoLISP program TRANA
              TRANA --------- Name of the AutoLISP function

^C^CSCRIPT;SCR1 *In this menu item SCR1 is the name of the script file, and **SCRIPT** is an AutoCAD command for running a script file.*

^C^CSCRIPT;SCR1
        Where **SCRIPT** --------- AutoCAD's **SCRIPT** command
              SCR1 ------------ Name of the script file

CUSTOMIZING BUTTONS AND AUXILIARY MENUS

The standard AutoCAD menu file ACAD.MNU contains buttons and auxiliary menu sections. The buttons and auxiliary menu sections are identical and the first section has nine menu items. The following file is a listing of the buttons and auxiliary menu sections of the standard AutoCAD menu file, ACAD.MNU:

```
***BUTTONS1
$M=$(if,$(eq,$(substr,$(getvar,cmdnames),1,5),GRIP_),$P0=ACAD.GRIPS $P0=*);
$P0=SNAP $p0=*
^C^C
^B
^O
^G
^D
^E
^T


***BUTTONS2
$P0=SNAP $p0=*


***AUX1
$M=$(if,$(eq,$(substr,$(getvar,cmdnames),1,5),GRIP_),$P0=ACAD.GRIPS $P0=*);
$P0=SNAP $p0=*
^C^C
^B
```

^O
^G
^D
^E
^T

***AUX2
$p0=SNAP $p0=*

Note
The following table shows the function of the menu items that are defined in the BUTTONS section of the CUSTOM.MNU file.

MENU ITEM	FUNCTION
***BUTTONS	Section label
$p0=*	Displays the cursor menu
^C^C	Cancels the existing command twice
^B	Snap On/off (CTRL B)
^O	Ortho On/off (CTRL O)
^G	Grid On/off (CTRL G)
^D	Coordinate Dial On/off (CTRL D)
^E	Isoplane (CTRL E)
^T	Tablet On/off (CTRL T)

Like any other menu you can make changes in the buttons menu to assign different commands to the buttons of the pointing device. The pointing devices generally come in 4-button or 10-button configuration. The first button is always the pick button and cannot be used for any other purpose. AutoCAD commands can be assigned to the remaining buttons. The following example describes the editing procedure for the BUTTONS section of CUSTOM.MNU file (CUSTOM.MNU file is a copy of ACAD.MNU file).

Example 4

Change the buttons sections of CUSTOM.MNU to assign the following commands to the four buttons of the pointing device (Figure 10-9):

1. PICK	3. ZOOM Win
2. Enter	4. ZOOM Prev

The ZOOM Window command assigned to button number 3 should automatically zoom in an area that is 2 units from the selected point.

Before making any changes it is very important to know the commands and the prompt entries that are associated with the commands. Use your word processor to load the CUSTOM.MNU file and search for the ***BUTTONS1 section. The lines that follow the ***BUTTONS1 section are the lines that need to be edited to assign new commands to the buttons of the pointing device. The following file is a listing of the button menu section of the CUSTOM.MNU file after making the required changes:

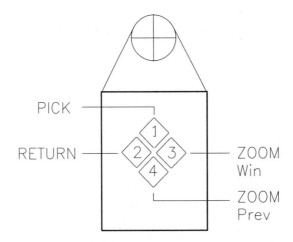

Figure 10-9 *Commands assigned to 4 buttons of a pointing device*

***BUTTONS1**	1
;	2
'ZOOM;WIN;\@2,2	3
'ZOOM;PRE	4
^B	5
^O	6
^G	7
^D	8
^E	9
^T	10

Line 1
***BUTTONS1
This is the section label for the BUTTONS1 menu section.

Line 2
;
The semicolon (;) causes an ENTER. It is like pressing the Enter key from the keyboard or template.

Line 3
'ZOOM;WIN;\@2,2
When you select this menu item, it will zoom as shown in Figure 10-10. The first point you select becomes the first corner of the zoom window and the other corner of the zoom window is 2.0,2.0 units from the selected point.

'ZOOM;WIN;\@2,2
   Where '-------------------- Makes the ZOOM command transparent
       \-------------------- Selected point becomes first corner

2,2 ---------------- Other corner

Before selecting the third button of the pointing device for this command, move the screen cross-hairs to the point where you want to zoom and then click the button. AutoCAD will zoom in the area since the two corners of the window are defined in the menu item. The single quote (') in front of **ZOOM** makes the **ZOOM Window** command transparent.

Line 4
'ZOOM;PRE
This menu item defines the **ZOOM** command with Previous option. The single quote (') in front of the **ZOOM** command makes the **ZOOM Previous** command transparent.

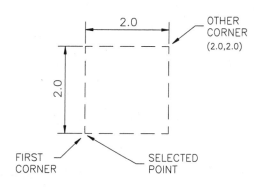

Figure 10-10 Zoom area

Note

*The commands on the first three lines (2 - 4), excluding the ***BUTTONS1 line, will be assigned to the second, third, and fourth button of the four-button pointing device. The remaining items will be ignored and do not affect the commands that are assigned to other buttons. Therefore, you can leave them in the file.*

Example 5

Edit the buttons and the auxiliary menu sections of the menu file CUSTOM.MNU to add the following AutoCAD commands. The pointing device has 10 buttons (Figure 10-11) and button number 1 is used as a pick button. The blocks should be inserted with a scale factor of 1.00 and a rotation angle of 0 degrees. (The CUSTOM.MNU file is a copy of the ACAD.MNU menu file.)

1. PICK BUTTON	2. ENTER	3. CANCEL
4. OSNAPS	5. INSERT B1	6. INSERT B2
7. INSERT B3	8. **ZOOM Window**	9. **ZOOM All**
10. **ZOOM Previous**		

Note

B1, B2, B3 are the names of the blocks or wblocks that have already been created.

It is assumed that the cursor menu POP0 has already been defined in the menu file.

Use the transparent ZOOM command for ZOOM Previous and ZOOM Window.

The following file is a listing of the buttons menu section of the CUSTOM.MNU file after making the required changes:

***BUTTONS

Chapter 10

1

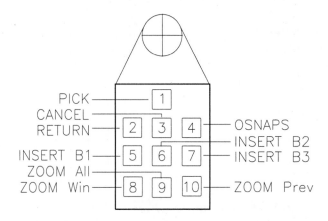

Figure 10-11 Commands assigned to different buttons of a 10-button pointing device

;	2
^C^C	3
$P0=*	4
^C^CINSERT;B1;\1.0;1.0;0	5
^C^CINSERT;B2;\1.0;1.0;0	6
^C^CINSERT;B3;\1.0;1.0;0	7
'ZOOM;Win	8
^C^CZOOM;All	9
'ZOOM;Prev	10
***AUX1	11
;	12
^C^C	13
$P0=*	14
^C^CINSERT;B1;\1.0;1.0;0	15
^C^CINSERT;B2;\1.0;1.0;01	16
^C^CINSERT;B3;\1.0;1.0;0	17
'ZOOM;Win	18
^C^CZOOM;All	19
'ZOOM;Prev	20

Line 3
^C^C
This command definition is assigned to button number 3. It cancels the existing command twice.

Line 4
$P0=*
In this menu item $P0=* is a special command that you can use to access the cursor menu. When you select this item, AutoCAD will display the cursor menu on the screen near the cursor location. The cursor menu that contains the object snap mode commands is defined in the POP0 menu

section of ACAD.MNU file. This command definition is assigned to button number 4 of the pointing device.

   $S=X $S=OSNAPS1

              Where **=** ----------------- Loads submenu X defined in screen menu section

                          OSNAPS1 ------ Loads submenu OSNAPS1 defined in screen
                                        menu section

Line 5

^ C ^ CINSERT;B1;\1.0;1.0;0

In this menu item ^ C ^ C cancels the existing command twice. **INSERT** is an AutoCAD command that can be used to insert a Block or Wblock. B1 is the name of the block and the backslash (\\) pauses for the user input. In this menu item, it is the insertion point of the block. The first 1.0 is the X-scale factor and the second 1.0 is the Y-scale factor of the block. The 0 at the end is for the rotation angle of the block.

   ^ C ^ CINSERT;B1;\1.0;1.0;0

             Where **INSERT** --------- AutoCAD command

                       B1 ----------------- Block name

                       **** -------------------- Pause for block insertion point

                       1.0 ---------------- X-scale factor

                       1.0 ---------------- Y-scale factor

                       ; ------------------ ENTER

                       ; ------------------ ENTER

                       0 ------------------ Rotation angle

Line 9

^ C ^ CZOOM;All

The command definition of this menu item is assigned to button number 9 of the pointing device. If you select this key, it will enter a **ZOOM All** command.

   ^ C ^ CZOOM;All

             Where **ZOOM** ----------- AutoCAD's **ZOOM** command

                       ; ------------------ ENTER

                       All ---------------- All option of **ZOOM** command

Line 10

'ZOOM;Prev

This menu item defines the transparent **ZOOM Previous** command. It is assigned to button number 10 of the pointing device.

   'ZOOM;Prev

             Where ' -------------------- Single quote makes **ZOOM Previous** command
                                   transparent

             Prev   -------------------- **Previous** option of **ZOOM** command

Line 11

***AUX1

In this line AUX1 is the section label for auxiliary menu section. Lines 12 through 20 are defined in this section.

CUSTOMIZING PULL-DOWN AND SHORTCUT MENUS

The pull-down and shortcut menus are a part of the standard AutoCAD menu file (ACAD.MNU) that comes with the AutoCAD software package. The ACAD.MNU file is automatically loaded when you get into the Drawing Editor, provided the standard configuration of AutoCAD has not been changed. In addition to POP0, the menu can have a maximum of 499 sections defined as POP1, POP2, POP3 — — — — POP499. The standard AutoCAD menu uses POP0 and POP500 through POP999 for shortcut menus. The shortcut menu can be loaded and displayed by using the command $P0=*.

CASCADING SUBMENUS IN MENUS

The cascading feature of AutoCAD allows the pull-down and cursor menus to be displayed in a hierarchical order that makes it easier to select submenus. To use the cascading feature in the pull-down and cursor menus, AutoCAD has provided some special characters. For example, the characters -> defines a cascaded submenu and <- designates the last item in the menu. The following table lists some of the frequently used characters that can be used with the pull-down or cursor menus:

Character	Character Description
--	The item label consisting of two hyphens automatically expands to fill the entire width of the pull-down menu. Example: [--]
+	Used to continue the menu item to the next line. This character has to be the last character of the menu item. Example: [Triang:]^C^CLine;1,1;+ 3,1;2,2;
->	This label character defines a cascaded submenu and it must precede the name of the submenu. Example: [->Draw]
<-	This label character designates the last item of the cascaded pull-down or cursor menu. The character must precede the label item. Example: [<-CIRCLE 3P]^C^CCIRCLE;3P
<-<-...	This label character designates the last item of the pull-down or cursor menu and also terminates the parent menu. The character must precede the label item. Example: [<-<-Center Mark]^C^C_dim;_center
$(	This label character can be used with the pull-down and cursor menus to evaluate a DIESEL expression. The character must precede the label item. Example: "$(if,$(getvar,orthomode),Ortho)"
~	This item grays-out the label item. The character must precede the item. Example: [~—]

The length of each menu bar title can be up to 14 characters long. Most display devices provide space for a maximum of 80 characters. Therefore, if there are 16 menus, the length of each menu title should be up to 5 characters long. If the combined length of all menu bar titles exceeds 80 characters, AutoCAD automatically truncates the characters from the longest menu title until it fits all menu titles in the menu bar. The following is a list of some additional features of the menu:

1. The section labels of the menus are ***POP1 through ***POP16. The menu bar titles are displayed in the menu bar.

2. The menus can be accessed by selecting the menu title from the menu bar at the top of the screen.

3. A maximum of 999 menu items can be defined in the menu. This includes the items that are defined in the submenus. The menu items in excess of 999 are ignored.

4. The number of menu items that can be displayed on the screen depends on the display device that you are using. If the cursor or the menu contains more items than what can be accommodated on the screen, the excess items are truncated. For example, if your system can display 21 menu items, then the menu items in excess of 21 are automatically truncated.

5. If AutoCAD is not configured to show the status line, the menus, cursor menus, and the menu bar are automatically disabled.

Example 6

Edit the POP4 section of the menu to add a new insert command with the label NEW-INSERT (Figure 10-12). When the user selects NEW-INSERT from the POP4 menu, it should load and display a cascading submenu that contains the commands for inserting the following blocks:

INSERT-BLOCKS	—————	Cascading submenu title
DOOR1		
DOOR2		
- - - - - - - -		
WINDOW1		
WINDOW2		

Note

The doors and windows are saved as WBLOCKS in the SYMBOLS subdirectory on the D drive.

The user should be able to insert the block at the selected point with X-scale and Y-scale factors of 1.25 and rotation angle of 0.

Do not edit the ACAD.MNU file. Make a copy of the ACAD.MNU file and then edit the new file (CUSTOM.MNU).

Use your word processor to load the menu file CUSTOM.MNU and search for the **INSERT** command in the menu section ***POP4. Now, insert a line that contains the definition of the INSERT-BLOCKS cascading submenu as shown in Figure 10-12.

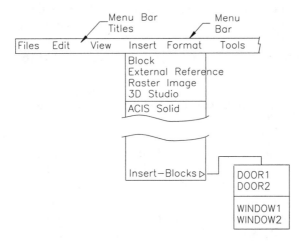

Figure 10-12 POP4 *section of menu*

The following file is a partial listing of CUSTOM.MNU file after editing the POP4 section of the menu and adding a new submenu:

***POP1
**FILE
ID_MnFile   [&File]
ID_New   [&New...\tCtrl+N] ^ C ^ C_new
ID_Open   [&Open...\tCtrl+O] ^ C ^ C_open
ID_FILE_CLOSE [&Close]
ID_PartialOp [$(if,$(eq,$(getvar,fullopen),0),,~)Pa&rtial Load] ^ C ^ C_partialload
ID_Save   [&Save\tCtrl+S] ^ C ^ C_qsave
ID_Saveas   [Save &As...] ^ C ^ C_saveas
ID_Export   [&Export...] ^ C ^ C_export
ID_PlotSetup [Pa&ge Setup...] ^ C ^ C_pagesetup
ID_PlotMgr [Plotter &Manager...] ^ C ^ C_plottermanager

***POP2
**EDIT
ID_MnEdit   [&Edit]
ID_U   [&Undo\tCtrl+Z] ^ C ^ C_u
ID_Redo   [&Redo\tCtrl+Y] ^ C ^ C_redo
ID_Cutclip [Cu&t\tCtrl+X] ^ C ^ C_cutclip
ID_Copyclip [&Copy\tCtrl+C] ^ C ^ C_copyclip
ID_Copybase [Copy with &Base Point] ^ C ^ C_copybase
ID_Copylink [Copy &Link] ^ C ^ C_copylink

***POP3
**VIEW
ID_MnView   [&View]
ID_Redrawall [&Redraw]'_redrawall
ID_Regen   [Re&gen]^C^C_regen
ID_Regenall [Regen &All]^C^C_regenall
ID_MnZoom   [->&Zoom]
ID_ZoomRealt [&Realtime]'_zoom ;
ID_ZoomPrevi [&Previous]'_zoom _p
ID_ZoomWindo [&Window]'_zoom _w
ID_ZoomDynam [&Dynamic]'_zoom _d
ID_ZoomScale [&Scale]'_zoom _s
ID_ZoomCente [&Center]'_zoom _c

***POP4
**INSERT
ID_MnInsert [&Insert]
ID_Ddinsert [&Block...]^C^C_insert
ID_Xattach   [E&xternal Reference...]^C^C_xattach
ID_Imageatta [Raster &Image...]^C^C_imageattach

[->Insert-Blocks]
 [Door1]^C^C_insert;d:/symbols/door1;\1.25;1.25;0
 [Door2]^C^C_insert;d:/symbols/door2;\1.25;1.25;0
 [--]
[Window1]^C^C_inserd;d:/symbols/window1;\1.25;1.25;0
[<-Window2]^C^C_insert;d:/symbols/window2;\1.25;1.25;0

Note
[DOOR1]^C^CINSERT;D:/SYMBOLS/DOOR1;\1.25;1.25;

When you define the search path in the menu file, replace the backslashes (\) by forward slashes (/). To load the file DOOR1 from the SYMBOLS subdirectory in the D drive, the search path normally will be defined as D:\SYMBOLS\DOOR1. But the same statement in a menu file will be defined as D:/SYMBOLS/DOOR1. In the menu file, backslash is used for user input.

[->Insert-Blocks]

In this menu item, the character -> defines a cascading submenu. When you select this item from the POP3 menu, AutoCAD will display the menu on the side of the menu.

[<-Window2]^C^C_insert;d:/symbols/window2;\1.25;1.25;0
In this menu item, the character <- defines the last menu item in the cascading submenu.

SHORTCUT MENUS

The shortcut menus are similar to the pull-down menus, except that the shortcut menu can contain only 499 items compared to 999 items in the pull-down menu. The section label of the cursor menu must be ***POP0 or POP500 through POP999. The shortcut menus are displayed near or at the cursor location. Therefore, it can be used to provide a convenient and quick access to some of the frequently used commands. The following is a list of some of the features of the cursor menu:

1. The section label of the cursor menu is ***POP0 and POP500 through POP999. The menu bar title defined under these section labels are not displayed in the menu bar.

2. On most systems, the menu bar title is not displayed at the top of the cursor menu. However, for compatibility reasons it is recommended to give a dummy menu bar title.

3. The POP0 menu can be accessed through the **$P0=*** menu command. The shortcut menus POP500 through POP999 must be referenced by their alias names. The reserved alias names for AutoCAD use are GRIPS, CMDEFAULT, CMEDIT, and CMCOMMAND. For example, to reference POP500 for grips, use the **GRIPS command line under POP5000. This commands can be issued by a menu item in another menu, such as the button menu, auxiliary menu, or the screen menu. The command can also be issued from an AutoLISP or ADS program.

4. A maximum of 499 menu items can be defined in the cursor menu. This includes the items that are defined in the cursor submenus. The menu items in excess of 499 are ignored.

5. The number of menu items that can be displayed on the screen depends on the system that you are using. If the cursor or the menu contains more items than what can be accommodated on the screen, the excess items are truncated. For example, if your system can display 21 menu items, then the menu items in excess of 21 are automatically truncated.

SUBMENUS

The number of items in a menu file can be so large that the items cannot be accommodated on one screen. For example, the maximum number of items that can be displayed on some of the display devices is 21. If the menu has more than 21 items, the menu items in excess of 21 are not displayed on the screen and cannot be accessed. The user can overcome this problem by using cascading menus or by swapping the submenus that let the user define smaller groups of items within a menu section. When a submenu is selected, it loads the submenu items and displays them on the screen.

Swapping Menus

The menus that use AutoCAD's cascading feature are the most efficient and easy to write. The submenus follow a logical pattern that are easy to load and use without causing any confusion. It is strongly recommended to use the cascading menus whenever you need to write the pull-down or cursor menus. However, AutoCAD provides the option to swap the submenus in the menus. These menus can sometimes cause distraction, because the original menu is completely replaced by the submenu when swapping the menus.

$Section=Submenu

    Where **$** ------------------- "$" sign

                Section ---------- Menu section specifier

                = ----------------- "=" sign

                Submenu ------- Name of submenu

Example

$P1=P1A

    Where **$P1** -------------- P1 specifies menu section 1

                P1A -------------- Name of submenu

CUSTOMIZING IMAGE TILE MENUS

The image tile menus are extremely useful for inserting blocks and selecting a hatch pattern or a text font. You can also use the image tile menus to load an AutoLISP routine or a predefined macro. Therefore, the image tile menu is a powerful tool for customizing AutoCAD.

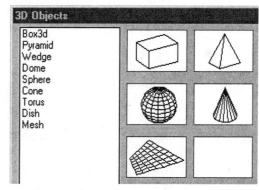

The image tile menus can be accessed from the pull-down, tablet, button, or screen menu. However, the image tile menus cannot be loaded by entering the command from the keyboard. When the user selects an image, a dialog box is displayed on the screen that contains **twenty images**. The names of the slide files associated with images appear on the left side of the dialog box with the scrolling bar that can be used to scroll the filenames. The title of the image tile menu is displayed at the top of the dialog box. When you activate the image tile menu, an arrow appears on the screen that can be moved to select any image. You can select an image by selecting the slide file name from the dialog box and then choosing the **OK** button from the dialog box or double-clicking on the slide file name.

When you select the slide file, AutoCAD highlights the corresponding image by drawing a rectangle around the image. You can also select an image by moving the arrow in the desired image and then pressing the pick button of the pointing device. The corresponding slide file name will be automatically highlighted and if you choose the **OK** button or doubleclick on the image, the command associated with that menu item will be executed. You can cancel an image tile menu by pressing the CTRL and C keys or the ESCAPE key on the keyboard, or selecting an image from the dialog box.

An image tile menu will work only if the system is configured so that the status line is not disabled. Otherwise, the image tile menus or the menus cannot be used. The image tile menu consists of a section label ***IMAGE. There is only one image tile menu section in a file and all the image tile menus are defined in this section.

Chapter 10

***IMAGE

Where ******* ---------------- Three asterisks (***) designate a section label

IMAGE ---------- Section label for an image

SUBMENUS

You can define an unlimited number of menu items in the image tile menu, but only 20 images will be displayed at a time. If the number of items exceeds 20, you can use the **Next** or **Previous** **buttons** of the dialog box to page through different pages of images. You can also define submenus that let the user define smaller groups of items within an image tile menu section. When a submenu is selected, it loads the submenu items and displays them on the screen.

$Section=Submenu

Where **$** ------------------- "$" sign

Section ---------- Menu section specifier

= ----------------- "=" sign

Submenu ------- Name of submenu

Example
$I=IMAGE1

Where **$I** ---------------- I specifies image tile menu

IMAGE1 -------- Name of submenu

IMAGE TILE MENU ITEM LABELS

Like screen and pull-down menus, you can use the menu item labels in the pull-down menus. However, the menu item labels in the image tile menus use different formats and each format performs a particular function in the image tile menu. The menu item labels appear in the slide list box of the dialog box. The maximum number of characters that can be displayed in this box is 17. The characters in excess of 17 are not displayed in the list box. However, it does not affect the command that is defined with the menu item. The following are the different formats of the menu item labels:

[slidename]
In this menu item label format, **slidename** is the name of the slide that is displayed in the image. This name (slidename) is also displayed in the list box of the corresponding dialog box.

[slidename,label]
In this menu item label format, **slidename** is the name of the slide that is displayed in the image. However, unlike the previous format, the **slidename** is not displayed in the list box. It is the label text that is displayed in the list box. For example, if the menu item label is **[BOLT1,1/2-24UNC-3LG]**, **BOLT1** is the name of the slide and **1/2-24UNC-3LG** is the label that will be displayed in the list box.

[slidelib(slidename)]
In this menu item label format, **slidename** is the name of the slide in the slide library file **slidelib**. The slide (slidename) is displayed in the image and the slide filename (slidename) is also displayed in the list box of the corresponding dialog box.

[slidelib(slidename,label)]
In this menu item label format, **slidename** is the name of the slide in the slide library file **slidelib**. The slide (slidename) is displayed in the image and the **label** text is displayed in the list box of the corresponding dialog box.

[blank]
This menu item will draw a line that extends through the width of the list box. It also displays a blank image in the dialog box.

[label]
If the **label** text is preceded by a space, AutoCAD does not look for a slide. The **label** text is displayed in the list box only. For example, if the menu item label is [EXIT]^C, the label text (EXIT) will be displayed in the list box. If you select this item, the cancel command (^C) defined with the item will be executed. The **label** text is not displayed in the image of the dialog box.

Example 7

Write an image tile menu that can be accessed from POP12 for inserting the following blocks (see Figure 10-13):

BL1	BL4
BL2	BL5
BL3	BL6

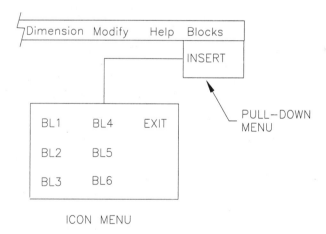

Figure 10-13 *Commands defined in the image tile menu*

Note
It is assumed that the slides have already been created and the names of the slides are the same as blocks.

Do not edit the ACAD.MNU file. Make a copy of the ACAD.MNU file and then edit the new file (CUSTOM.MNU).

Chapter 10

Use your word processor to load the CUSTOM.MNU file and search for the ***IMAGE section label. The standard ACAD.MNU file does not have the POP12 section in the pull-down menu. Therefore, you can define POP12 just before the ***IMAGE section label. Similarly, you can define the INSTBLK image tile menu after the image section label ***IMAGE. The following file is a partial listing of the CUSTOM.MNU file after editing the file.

```
***POP12
[BLOCKS]
[INSERT]^C^C$I=INSTBLK $I=*

***image
**INSTBLK
[INSERT CUSTOMIZED BLOCKS]
[BL1]^C^CINSERT;BL1;\1.0;1.0;0
[BL2]^C^CINSERT;BL2;\1.0;1.0;0
[BL3]^C^CINSERT;BL3;\1.0;1.0;0
[BL4]^C^CINSERT;BL4;\1.0;1.0;0
[BL5]^C^CINSERT;BL5;\1.0;1.0;0
[BL6]^C^CINSERT;BL6;\1.0;1.0;0
[ EXIT]^C^C
```

Note

[INSERT]^C^C$I=INSTBLK $I=*

In this menu item, $I=INSTBLK loads the image submenu INSTBLK that has been defined in the image tile menu section. $I=* forces the display of the dialog box with the images.

[INSERT]^C^C$I=INSTBLK $I=*

Where **$I=INSTBLK** - Loads the submenu INSTBLK that
                            has been defined in the image tile menu section
       $I=* ------------- Displays dialog box with images

***IMAGE
This is the section label of the Image tile menu. All image tile menus have to be defined in this section (***IMAGE) of the menu file.

**INSTBLK
This is the submenu label. The name of the submenu is INSTBLK.

[INSERT CUSTOMIZED BLOCKS]
This menu item describes the contents of the image tile menu and it is displayed at the top of the dialog box.

[BL1]^C^CINSERT;BL1;\1.0;1.0;0
*The BL1 that is inside the brackets is the name of the slide that is displayed in one of the images of the dialog box. ^C^C cancels the existing command twice and the **INSERT** command inserts the block BL1 with the X,Y scale factor of 1.0 and the rotation angle of 0 degrees.*

```
[BL1]^C^CINSERT;BL1;\1.0;1.0;0
```

Where **[BL1]** ------------- Name of the slide

BL1 --------------- Block name

\-------------------- User input (Insertion point)

1.0 ---------------- X-scale factor

1.0 ---------------- Y-scale factor

0 ------------------- Rotation angle

CUSTOMIZING THE SCREEN MENU

Like other menu sections, the screen menu section can be edited to modify or delete the existing commands or add new commands and submenus. Before writing a menu, it is very important to design a menu, know the exact sequence of AutoCAD commands, and know the prompts associated with the commands. You can edit the standard AutoCAD menu and arrange the commands in a way that provides the user an easy and quick access to most frequently used commands. A careful design will save quite some time in the long run. Therefore, it is strongly recommended to consider several possible alternatives and then select the one that is best suited for the job.

SUBMENUS

The number of items in the screen menu file can be so large that the items cannot be accommodated on one screen. For example, the maximum number of items that can be displayed on some of the display devices is 21. If the screen menu has more than 21 items, the menu items in excess of 21 are not displayed on the screen and therefore cannot be accessed. The user can overcome this problem by using submenus that let the user define smaller groups of items within a menu section. When a submenu is selected, it loads the submenu items and displays them on the screen.

$Section=Submenu

Where **$** -------------------- "$" sign

Section ---------- Menu section specifier

= ------------------ "=" sign

Submenu ------- Name of submenu

Example

$S=EDIT

Where **S** -------------------- S specifies screen menu section

EDIT ------------- Name of submenu

Nested Submenus

When a submenu is activated, the current menu is copied to a stack. If you select another submenu, the submenu that was current will be copied or pushed to the top of the stack. The maximum number of menus that can be stacked is eight. If the stack size increases more than eight, the menu that is at the bottom of the stack is removed and forgotten. You can call the previous submenu by using the nested submenu call. The format of the call is:

$S=

Where **$** -------------------- "$" sign

S -------------------- Screen menu specifier

= ------------------ "=" sign

Chapter 10

The maximum number of nested submenu calls that AutoCAD can have is eight. Each time you call a submenu, this pops the last item off the stack and reactivates it.

Note

To load the original menu (ACAD.MNU), load the menu file by using the MENU command.

Command: **MENU**
Enter menu file name or [. (for none)]<SM1>: **ACAD.MNU**

If you need to use input from a keyboard or a pointing device use backslash "\". The system will pause for the user to enter data.

There should be no space after the backslash "\".

The menu items, menu labels, and the command definition can be in uppercase, lowercase, or mixed.

You can introduce spaces between the menu items to improve the readability of the menu file.

If there are more items in the menu than the number of spaces available, the excess items are not displayed on the screen. For example, in the screen menu if the display device limits the number of items to 21, the items in excess of 21 will not be displayed on the screen, and are therefore inaccessible.

If you configure AutoCAD and turn the screen prompt area off, you can increase the number of lines that can be displayed on the screen menu. On some devices it is 24 lines.

Example 8

Edit the standard AutoCAD menu to add the commands as shown in Figure 10-14.

Note

*It is assumed that the image submenu **INSTBLK** has already been defined in the image tile menu section of the menu file. Do not edit the ACAD.MNU file. Make a copy of the ACAD.MNU file and then edit the new file (CUSTOM.MNU).*

You can use your word processor to load the menu file CUSTOM.MNU and search for *****SCREEN** section label. Add the new menu item at the end of the submenu ****S** and then define the submenu ****CUSTOM** and the menu items as shown in Figure 10-14. The following file is a partial listing of the **CUSTOM.MNU** file after editing:

```
***IMAGE
**INSTBLK
[INSERT CUSTOMIZED BLOCKS]
[BL1]^C^CINSERT;BL1;\1.0;1.0;0
[BL2]^C^CINSERT;BL2;\1.0;1.0;0
[BL3]^C^CINSERT;BL3;\1.0;1.0;0
[BL4]^C^CINSERT;BL4;\1.0;1.0;0
```

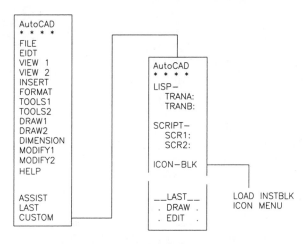

Figure 10-14 *Modified screen menu*

[BL5]^C^CINSERT;BL5;\1.0;1.0;0
[BL6]^C^CINSERT;BL6;\1.0;1.0;0
[EXIT]^C^C

***SCREEN
**S
[AutoCAD]^C^C^P(ai_rootmenus) ^P
[* * * *]$S=ACAD.OSNAP
[FILE   ]$S=ACAD.01_FILE
[EDIT   ]$S=ACAD.02_EDIT
[VIEW 1]$S=ACAD.03_VIEW1
[VIEW 2]$S=ACAD.04_VIEW2
[INSERT]$S=ACAD.05_INSERT

**ASSIST 3
[Last   ]_l
[Previous]_p
[All   ]_all
[Cpolygon]_cp
[Wpolygon]_wp

[CUSTOM] ^ C ^ C$S=X $S=CUSTOM

**CUSTOM 3
[LISP-]
[TRANA:] ^ C ^ C(LOAD "TRANA");TRANA
[TRANB:] ^ C ^ C(LOAD "TRANB");TRANB
[   ]
[SCRIPT-]
[SCR1:] ^ C ^ CSCRIPT;SCR1
[SCR2:] ^ C ^ CSCRIPT;SCR2
[   ]
[IMAGE-BLK] ^ C ^ C$I=INSTBLK $I=*

Note
[CUSTOM] ^ C ^ C$S=X $S=CUSTOM   *In this menu item, $S=X loads the submenu X that has been defined in the screen menu section. $S=CUSTOM loads the submenu CUSTOM that has also been defined in the screen menu section of the menu file CUSTOM.MNU.*

    [CUSTOM] ^ C ^ C$S=X $S=CUSTOM
          Where   **$S=X** ----------- Loads submenu X
                     $S=Custom ---- Loads submenu CUSTOM

**CUSTOM 3   *In this menu item, CUSTOM is the name of the submenu. The 3 that follows the submenu name prints the menu items defined in the submenu CUSTOM from the line number 3. Nothing is printed on the first two lines; therefore, the first two lines AutoCAD and * * * * stay on the screen.*

    **CUSTOM 3
          Where   **Custom** ---------- Submenu name
                  3 ------------------- Start printing from line number 3

[IMAGE-BLK] ^ C ^ C$I=INSTBLK $I=*   *In this menu item, $I=INSTBLK loads the image submenu INSTBLK that has been defined in the image section of the menu file. $I=* forces the display of the new image tile menu on the screen.*

    [IMAGE-BLK] ^ C ^ C$I=INSTBLK $I=*
          Where   **$1=INSTBLK** - Loads the image submenu INSTBLK
                  $SI=* ----------- Forces display of new image tile menu on screen

Review Questions

1. The AutoCAD menu file can have up to _____ sections.

2. The tablet menu can have up to _____ sections.

3. The section label is designated by _____.

4. The submenu label is designated by _____.

5. In a menu file you can use _____ to cancel the existing command.

6. Submenu names can be _____ characters long.

7. You _____ assign the same command to more than one block on the template.

8. AutoCAD's _____ command is used to configure the tablet menu template.

9. AutoCAD's _____ command is used to load a new menu.

10. You need to enter _____ points that are at _____ degrees to configure different tablet areas.

11. The commands are assigned to the buttons of the pointing device in the _____ order in which they appear in the buttons menu.

12. The format of the command used for loading a submenu that has been defined in the screen menu section is _____.

13. The format of the command used for loading a submenu that has been defined in the pull-down menu section is _____.

14. The format of the command used for loading a submenu that has been defined in the image tile menu section is _____.

15. The command that is used to force the display of the current pull-down menu is _____.

Exercises

Exercise 3   *General*

Add the following commands to the TABLET1 section of the standard AutoCAD menu file ACAD.MNU. Figure 10-15 shows the layout of tablet area-1 of the template.

VIEW POINTS

0,0,1	1,0,0	0,1,0
1,-1,1	1,1,1	-1,1,1

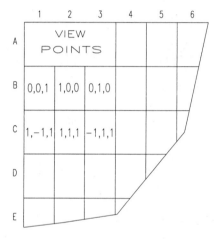

Figure 10-15 *Commands assigned to tablet area-1*

Exercise 4   *General*

Add the following commands to the TABLET1 section of the standard AutoCAD menu file ACAD.MNU. The layout of tablet area-1 is shown in Figure 10-16.

INSERT NO	PLOT 12x18	SETLAYER OBJ
INSERT NC	PLOT 18x24	SETLAYER HID
INSERT COIL	PLOT 24x36	SETLAYER CEN
INSERT RESIS	PRPLOT	SETLAYER DIM

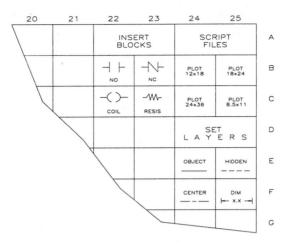

Figure 10-16 Commands assigned to tablet area-1

Exercise 5
General

Write a button menu for the following AutoCAD commands (Figure 10-17). Add the commands to BUTTONS2 section of ACAD.MNU. The pointing device has 10 buttons; button number 1 is used for picking the points. The blocks are to be inserted with a scale factor of 1.00 and a rotation of 0 degrees (Filename BME1.MNU).

1. PICK BUTTON	2. Enter	3. CANCEL
4. OSNAPS	5. END PT	6. CENTER
7. NEAR	8. ZOOM Window	9. ZOOM Prev
10. PAN		

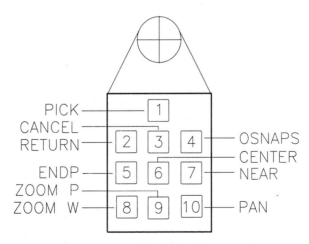

Figure 10-17 Commands assigned to different buttons of a pointing device

Exercise 6   *General*

Add the commands shown in Figure 10-18 to the POP12 section of the standard AutoCAD menu
ACAD.MNU

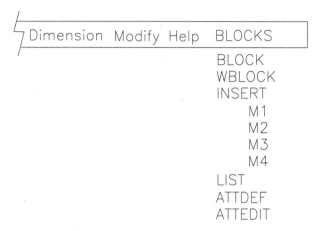

Figure 10-18 POP11 section of menu

Exercise 7   *General*

Write an image tile menu for inserting the following blocks that can be accessed through the
POP13 section. It is assumed that the blocks have already been created.

SYMBOL-X	SYMBOL-Y	SYMBOL-Z
LOGO-1	LOGO-2	LOGO-3
TBLOCK-1	TBLOCK-2	TBLOCK-3

Chapter **11**

Shapes and Text Fonts

Learning Objectives

After completing this chapter, you will be able to:
- *Write shape files.*
- *Use vector length and direction encoding to write shape files.*
- *Compile and load shape/font files.*
- *Use special codes to define a shape.*
- *Write text font files.*

SHAPE FILES

AutoCAD provides a facility to define shapes and text fonts. These files are ASCII files with the extension **.SHP**. You can write these files using any text editor like Notepad.

Shape files contain information about the individual elements that constitute the shape of an object. The basic objects that are used in these files are lines and arcs. You can define any shape using these basic objects, and then insert them anywhere in a drawing. The shapes are easy to insert, and they take less disk space than blocks. However, there are some disadvantages to using shapes. For example, you cannot edit a shape or change it. Blocks, on the other hand, can be edited after exploding them with the AutoCAD **EXPLODE** command.

SHAPE DESCRIPTION

Shape description consists of the following two parts: a header and a shape specification.

Header

The header line has the following format:

*SHAPE NUMBER, DEFBYTES, SHAPE NAME

*201,21,HEXBOLT

        Where ***201** --------------- Shape number

               21 ----------------- Number of data bytes in shape specification

               HEXBOLT ----- Shape name

Every header line starts with an asterisk (*), followed by the **SHAPE NUMBER**. The shape number is any number between 1 and 255 in a particular file, and these numbers cannot be repeated within the same file. However, the numbers can be repeated in another shape file with a different name. **DEFBYTES** is the number of data bytes used by the shape specification and includes the terminating zero. **SHAPE NAME** is the name of a shape, in uppercase letters. The name is ignored if the letters are lowercase. The file must not contain two shapes with the same name.

Shape Specification

The shape specification line contains the complete definition of the shape of an object. The shape is described with special codes, hexadecimal numbers, and decimal numbers. A hexadecimal number is designated by a leading zero (012), and a decimal number is a regular number without a leading zero (12). The data bytes are separated by a comma (,). The maximum number of data bytes is 2,000 bytes per shape, and in a particular shape file there can be more than one shape. The shape specification can have multiple lines. You should define the shape in some logical blocks and enter each block on a separate line. This makes it easier to edit and debug the files. The number of characters on any line must not exceed 80. The shape specification is terminated with a zero.

VECTOR LENGTH AND DIRECTION ENCODING

Figure 11-1 shows the vector direction codes. All vectors in this figure have the same length specification. The diagonal vectors have been extended to match the closest orthogonal vector. Let us assume that the endpoint of vector 0 is two grid units from the intersection point of vectors. The endpoint of vector 1 is one grid directly above the endpoint of vector 0. Therefore, the angle of vector 1 is 26.565 degrees (Tan-1 1/2 = Tan-1 0.5 = 26.565). Similarly, vector 2 is at 45 degrees (Tan-1 2/2 = Tan-1 1 = 45).

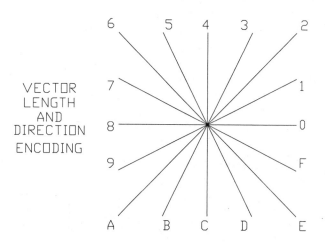

Figure 11-1 *Vector length and direction encoding*

All these vectors have the same magnitude or length specification. In other words, although their actual lengths vary, they are all considered one unit in definition. To define a vector you need its magnitude and direction. That means each shape specification byte contains a vector length and a direction code. The maximum length of the vector is 15 units. Example 1 illustrates the use of vectors.

Example 1

Write a shape file for the resistor shown in Figure 11-2. The name of the file is **SH1.SHP**, and the shape name is RESIS.

The following two lines define the shape file for the given resistor:

    *201,8,RESIS
    020,023,04D,043,04D,023,020,0

The first line is the **header line**; the second line is the **shape specification**.

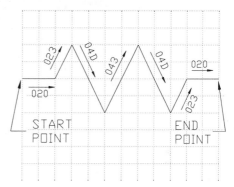

Figure 11-2   *Resistor*

Header Line

    *201,8,RESIS

*201 is the shape number, and 8 is the number of data bytes contained in the shape specification line. RESIS is the name of the shape.

Shape Specification

    <u>**020**</u>,023,04D,043,04D,023,020,0

             Where   **0** ------------------ Hexadecimal notation
                    2 ------------------ Vector length
                    0 ------------------ Direction code

Each data byte in this line, except the terminating zero, has three elements. The first element (0) is the hexadecimal, the second is the length of the vector, and the third element is the direction code. For the first data byte, 020, the length of the vector is 2, and the direction is along the direction vector 0. Similarly, for the second data byte, 023, the first element, 0, is for hexadecimal; the second element, 2, is the length of the vector; and the third element, 3, is the direction code for the vector.

COMPILING AND LOADING SHAPE/FONT FILES

You can compile the shape file or the font file by using the **COMPILE** command.

    Command: **COMPILE**

When you enter this command, AutoCAD will display the **Select Shape** or **Font File** dialog box

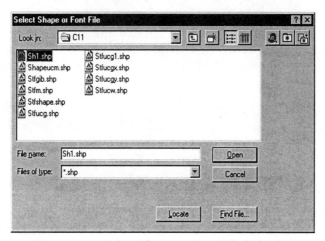

Figure 11-3 Select Shape or Font File dialog box

(Figure 11-3). From this dialog box, select the shape file that you want to compile. If the system variable **FILEDIA** is set to 0, you can enter the shape file name from the command line as follows:

Command: **COMPILE**
Enter NAME of shape file: **SH1**

AutoCAD will compile the file, and if the compilation process is successful, the following prompt will be displayed on the screen:

Compilation successful
Output file name.shx contains nn bytes

For Example 1, the name of the compiled output file is **SH1.SHX** and the number of bytes is 51. This is the file that is loaded when you use the AutoCAD **LOAD** command to load a shape. If AutoCAD encounters an error in compiling a shape file, an error message will be displayed, indicating the type of error and the line number where the error occurred.

To insert a shape in the drawing, you have to be in the drawing editor, and then use the **LOAD** command to load the shape file. You can select the file from the Select Shape File dialog box (Figure 11-4) or enter the name at the Command prompt (FILEDIA=0).

Command: **LOAD**
Name of shape file to load (or ?): *Name of file.* **SH1**

SH1 is the name of the shape file for Example 1. Do not include the extension **.SHX** with the name because AutoCAD automatically assumes the extension. If the shape file is present, AutoCAD will display the shape names that are loaded. To insert the loaded shapes, use the AutoCAD **SHAPE** command:

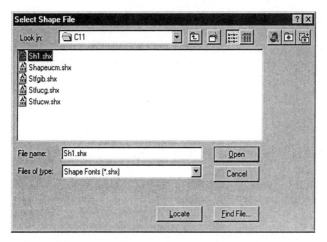

Figure 11-4 Select Shape File dialog box

Command: **SHAPE**
Shape name (or ?)<default>: *Shape name.*
Start point: *Shape origin.*
Height<1.0>: *Number or point.*
Rotation angle<0.0>: *Number or point.*

For Example 1, the shape name is RESIS. After you enter the information about the start point, height, and rotation, the shape will be displayed on the screen (Figure 11-5).

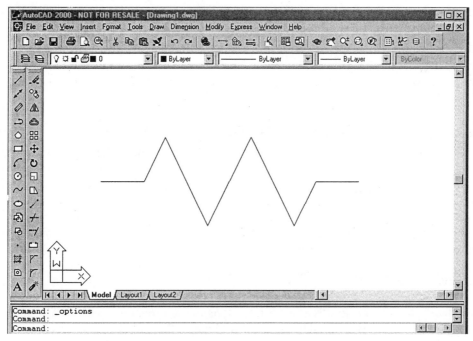

Figure 11-5 The shape (RESIS) inserted in the drawing

SPECIAL CODES

Generating shapes with the direction vectors has certain limitations. For example, you cannot draw an arc or a line that is not along the standard direction vectors. These limitations can be overcome by using special codes that add flexibility and give you better control over the shapes you want to create.

Standard Codes

000	End of shape definition
001	Activate draw mode (pen down)
002	Deactivate draw mode (pen up)
003	Divide vector lengths by next byte
004	Multiply vector lengths by next byte
005	Push current location from stack
006	Pop current location from stack
007	Draw subshape numbers given by next byte
008	X-Y displacement given by the next two bytes
009	Multiply X-Y displacement, terminated by (0,0)
00A or 10	Octant arc defined by next two bytes
00B or 11	Fractional arc defined by next five bytes
00C or 12	Arc defined by X-Y displacement and bulge
00D or 13	Multiple bulge-specified arcs
00E or 14	Process next command only if vertical text style

Code 000: End of Shape Definition

This code marks the end of a shape definition.

Code 001: Activate Draw Mode

This code turns the draw mode on. When you start a shape, the draw mode is on, so you do not need to use this code. However, if the draw mode has been turned off, you can use code 001 to turn it on.

Code 002: Deactivate Draw Mode

This code turns the draw mode off. It is used when you want to move the pen without drawing a line.

  1   2   3   4

Let us say the distance from point 1 to point 2, from point 2 to point 3, and from point 3 to point 4 is 2 units each. The shape specification for this line is:

020,002,020,001,020,0

The first data byte, 020, generates a line 2 units long along direction vector 0. The second data byte, 002, deactivates the draw mode; and the third byte, 020, generates a blank line 2 units long. The fourth data byte, 001, activates the draw mode; and the next byte, 020, generates a line that is

2 units long along direction vector 0. The last byte, 0, terminates the shape description.

Example 2

Write a shape file to generate the character "G" as shown in Figure 11-6.

You can use any text editor to write a shape file. The name of the file is **CHRGEE** and the shape name is GEE. **In the following file, the line numbers are not a part of the file; they are for reference only.**

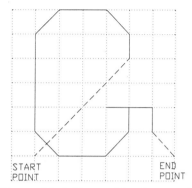

Figure 11-6 Shape of the character "G"

*215,20,GEE	1
002,042,	2
001,014,016,028,01A,	3
04C,01E,020,012,014,	4
002,018,	5
001,020,01C,	6
002,01E,0	7

Explanation
Line 1
*215,20,GEE
The first data byte contains an asterisk (*) and shape number 215. The second data byte is the number of data bytes contained in the shape specification, including the terminating 0. GEE is the name of the shape.

Line 2
002,042,
The data byte 002 deactivates the draw mode (pen up), and the next data byte defines a vector 4 units long along direction vector 2.

Line 3
001,014,016,028,01A,
The data byte 001 activates the draw mode (pen down), and 014 defines a vector that is 1 unit long at 90 degrees (direction vector 4). The data byte 016 defines a vector that is 1 unit long along direction vector 6. The data byte 028 defines a vector that is 2 units long along direction vector 8 (180 degrees). The data byte 01A defines a unit vector along direction vector A.

Line 4
04C,01E,020,012,014,

The data byte 04C defines a vector that is 4 units long along direction vector C. The data byte 01E defines a direction vector that is 1 unit along direction vector E. The data byte 020 defines a direction vector that is 2 units long along direction vector 0 (0 degrees). The data byte 012 defines a direction vector that is 1 unit long along direction vector 2. Similarly, 014 defines a vector that is 1 unit long along direction vector 4.

Line 5
002,018,
The data byte 002 deactivates the pen (pen up), and 018 defines a vector that is 1 unit long along direction vector 8.

Line 6
001,020,01C,
The data byte 001 activates the pen (pen down), and 020 defines a vector that is 2 units long along direction vector 0. The data byte 01C defines a vector that is 1 unit long along direction vector C.

Line 7
002,01E,0
The data byte 002 deactivates the pen, and the next data byte, 01E, defines a vector that is 1 unit long along direction vector E. The data byte 0 terminates the shape specification.

Code 003: Divide Vector Lengths by Next Byte

This code is used if you want to divide a vector by a certain number. In Example 2, if you want to divide the vectors by 2, the shape description can be written as:

> **003,2,020,002,020,001,020,0**

The first byte, 003, is the division code, and the next byte, 2, is the number by which all the remaining vectors are divided. The length of the lines and the gap between the lines will be equal to 1 unit now:

_____   _____

Also, the scale factors are cumulative within a shape. For example, if we insert another code, 003, in the preceding shape description, the length of the last vector, 020, will be divided by 4 (2 x 2):

<u>003,2</u>,020,002,020,001,<u>003,2</u>,020,0
              Where   **003,2** ------------- All the vectors are divided by 2
                      003,2 ------------- All the remaining vectors are divided by 4 (2 x 2)

Here is the output of this shape file:

_____   ___

Code 004: Multiply Vector Lengths by Next Byte

This code is used if you want to multiply the vectors by a certain number. It can also be used to reverse the effect of code 003.

003,2,020,002,020,001,004,2,020,0

Where **003,2** ------------- Divides all the vectors on the right by 2

004,2 ------------- Multiplies all the vectors on the right by 2

In this example, the code 003 divides all the vectors to the right by 2. Therefore, a vector that was 1 unit long will be 0.5 units long now. The second code, 004, multiplies the vectors to the right by 2. We know the scale factors are cumulative; therefore, the vectors that were divided by 2 earlier will be multiplied by 2 now. Because of this cumulative effect, the length of the last vector remains unchanged. This file will produce the following shape:

Codes 005 and 006: Location Save/Restore

Code 005 lets you save the current location of the pen, and code 006 restores the saved location. The following example illustrates the use of codes 005 and 006.

Example 3

Figure 11-7(a) shows three lines that are unit vectors and intersect at one point. After drawing the first line, the pen has to return to the origin to start a second vector. This is done using code 005, which saves the starting point (origin) of the first vector, and code 006, which restores the origin. Now, if you draw another vector, it will start from the origin. Since there are three lines, you need three code 005s and three code 006s. The following file shows the header line and the shape specification for generating three lines as shown in Figure 11-7(a):

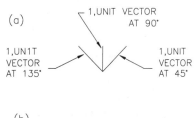
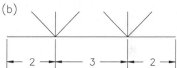

Figure 11-7 *(a) Three unit vectors intersecting at a point; (b) repeating predefined subshapes*

*210,10,POP1
005,005,005,012,006,014,006,016,006,0

Where **005** --------------- Saves origin three times

012 --------------- Generates first vector

006 --------------- Restores origin

014 --------------- Generates second vector

006 --------------- Restores origin

The number of saves (code 005) has to equal the number of restores (code 006). If the number of saves (code 005) is more than the number of restores (code 006), AutoCAD will display the following message when the shape is drawn:

Position stack overflow in shape (shape number)

Similarly, if the number of restores (code 006) is more than the number of saves (code 005), the following message will be displayed:

Position stack underflow in shape (shape number)

The maximum number of saves and restores you can use in a particular shape definition is four.

Code 007: Subshape

You can define a subshape like a subroutine in a program. To reference a subshape, the subshape code, 007, has to be followed by the shape number of the subshape. The subshape has to be defined in the same shape file, and the shape number has to be from 1 to 255.

> ***210,10,POP1**
> 005,005,005,012,006,014,006,016,006,0
> ***211,8,SUB1**
> **020,007,210,030,007,210,020,0**
>   Where   ***210** -------------- Shape number
>   **007** --------------- Subshape reference
>   **210** -------------- Shape number

The shape that this example generates is shown in Figure 11-7(b).

Code 008: X-Y Displacement

In the previous examples, you might have noticed some limitations with the vectors. As mentioned earlier, you can draw vectors only in the 16 predefined directions, and the length of a vector cannot exceed 15 units. These restrictions make the shape files easier and more efficient, but at the same time, they are limiting. Therefore, codes 008 and 009 allow you to generate nonstandard vectors by entering the displacements along the X and Y directions. The general format for Code 008 is:

> **008, XDISPLACEMENT, YDISPLACEMENT**
> or
> **008, (XDISPLACEMENT, YDISPLACEMENT)**

X and Y displacements can range from +127 to -128. Also, a positive displacement is designated by a positive (+) number, and a negative displacement is designated by a negative (-) number. The leading positive sign (+) is optional in a positive number. The parentheses are used to improve readability, but they have no effect on the shape specification.

Code 009: Multiple X-Y Displacements

Whereas code 008 allows you to generate nonstandard vectors by entering a single X and Y displacement, code 009 allows you to enter multiple X and Y displacements. It is terminated by a pair of 0 displacements (0,0). The general format is:

> **009,(XDISPL, YDISPL), (XDISP, YDISPL), . . ., (0,0)**

Code 00A or 10: Octant Arc

If you divide 360 degrees into eight equal parts, each angle will be 45 degrees. Each 45-degree angle segment is called an octant, and the two lines that contain an octant are called an **octant boundary**. The octant boundaries are numbered from 0 to 7, as shown in Figure 11-8. The general format is:

10,(R,+/-0SN)

Where **R** ----------------- Radius of arc
       +/- ---------------- Defines direction, + Counterclockwise, - Clockwise
       0 ----------------- Hexadecimal notation
       S ------------------ Starting octant boundary
       N ----------------- Number of octants

10,(3,-043)

The first number, 10, is the code 00A for the octant arc. The second number, 3, is the radius of the octant arc. The negative sign indicates that the arc is to be generated in a clockwise direction. If it is positive (+) or if there is no sign, the arc will be generated in a counterclockwise direction. Zero is the hexadecimal notation, and the following number, 4, is the number of the octant boundary where the octant arc will start. The next element, 3, is the number of octants that this arc will extend. This example will generate the arc in Figure 11-9. The following is the listing of the shape file that will generate the shape shown in Figure 11-9:

*214,5,FOCT1
001,10,(3,-043),0

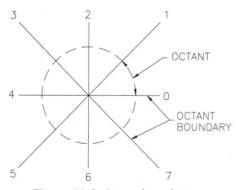

Figure 11-8 *Octant boundaries*

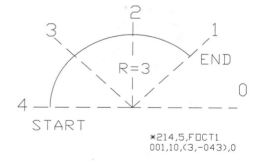

Figure 11-9 *Octant arc*

Code 00B or 11: Fractional Arc

You can generate a nonstandard fractional arc by using code 00B or 11. This code will allow you to start and end an arc at any angle. The definition uses five bytes, and the general format is:

11,(START OFFSET, END OFFSET, HIGHRADIUS, LOWRADIUS, +/-0SN)

The START OFFSET represents how far from an octant boundary the arc starts, and the END OFFSET represents how far from an octant boundary the arc ends. The HIGHRADIUS is zero if the radius is equal to or less than 255 units, and the LOWRADIUS is the radius of the arc. The positive (+) or negative (-) sign indicates, respectively, whether the arc is drawn counterclockwise or clockwise. The next element, S, is the number of the octant where the arc starts, and element N is the number of octants the arc goes through. The following example illustrates the fractional arc concept.

Example 4

Draw a fractional arc of radius 3 units that starts at a 20-degree angle and ends at a 140-degree angle (counterclockwise).

The solution involves the following steps:

1. Find the nearest octant boundary whose angle is less than 140 degrees. The nearest octant boundary is the number 3 octant boundary, whose angle is 135 degrees (3 * 45 = 135).

2. Calculate the end offset to the nearest whole number (integer):
   End offset = (140 - 135) * 256/45 = 28.44 = 28

3. Find the nearest octant boundary whose angle is less than 20 degrees. The nearest octant boundary is 0 and its angle is 0 degrees.

4. Calculate the start offset to the nearest whole number:
   Start offset = (20 - 0) * 256/45 = 113.7 = 114

5. Find the number of octants the arc passes through. In this example, the arc starts in the first octant and ends in the fourth octant. Therefore, the number of octants the arc passes through is four (counterclockwise).

6. Find the octant where the arc starts. In this example, it starts in the 0 octant.

7. Substitute the values in the general format of the fractional arc:

   11,(114,28,0,3,004)

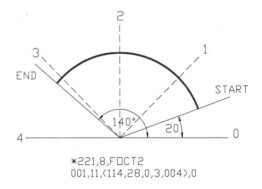

*221,8,FOCT2
001,11,(114,28,0,3,004),0

Figure 11-10 *Fractional arc*

The following shape file will generate the fractional arc shown in Figure 11-10:

*221,8,FOCT2
001,11,(114,28,0,3,004),0

Code 00C or 12: Arc Definition by Displacement and Bulge

Code C can be used to define an arc by specifying the displacement of the endpoint of an arc and the bulge factor. X and Y displacements may range from -127 to +127, and the bulge factor can also range from -127 to +127. A semicircle will have a bulge factor of 127, and a straight line will have a bulge factor of 0. If the bulge factor has a negative sign, the arc is drawn clockwise.

Bulge factor = ((2 * H)/D) * 127
             Where   **H** ------------------ Height of arc
                     D ------------------ Displacement

For a semicircle, 2H   = D
Therefore, bulge   = (D/D) * 127 = 127
For a straight line, H   = 0
Therefore, bulge   = (0/D) * 127 = 0

In Figure 11-11, the distance between the start point and the endpoint of an arc is 4 units, and the height is 1 unit. Therefore, the bulge can be calculated by substituting the values in the previously mentioned relation:

Bulge = (2 * 1/4) * 127 = 63.5 = 63
(integer)

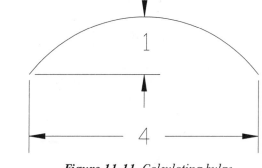

The following shape description will generate the arc shown in Figure 11-11.

*213,5,BULGE1
12,(4,0,-63),0

Figure 11-11 Calculating bulge

             Where   **4** ------------------- X displacement
                     0 ------------------- Y displacement
                     - ------------------- Negative (-), generates clockwise arc
                     63 ----------------- Bulge factor

Code 00D or 13: Multiple Bulge-Specified Arcs

Code 00D or 13 can be used to generate multiple arcs with different bulge factors. It is terminated by a (0,0). The following shape description defines the arc configuration of Figure 11-12:

*214,16,BULGE2
13,(4,0,-111),
(0,4,63),
(-4,0,-111),
(0,-4,63),(0,0),0

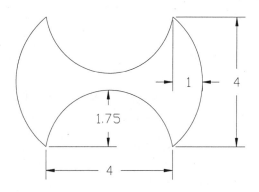

Figure 11-12 Different arc configuration

Code 00E or 14: Flag Vertical Text

Code 00E or 14 is used when the same text font description is to be used in both the horizontal and vertical orientations. If the text is drawn in a horizontal direction, the vector next to code 14 is ignored. If the text is drawn in a vertical position, the vector next to code 14 is not ignored. This lets you generate text in a vertical or horizontal direction with the same shape file.

For horizontal text, the start point is the lower left point, and the endpoint is on the lower right. In vertical text, the start point is at the top center, and the endpoint is at the bottom center of the text, as shown in Figure 11-13. At first, it appears that you need two separate shape files to define the shape of a horizontal and a vertical text. However, with code 14 you can avoid the dual shape definition.

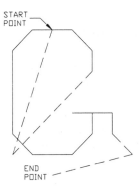

Figure 11-13 shows the pen movements for generating the text character "G". If the text is horizontal, the line that is next to code 14 is automatically ignored. However, if the text

Figure 11-13 *Pen movement for generating the character "G"*

is vertical, the line is not ignored, resulting in resetting the start and endpoints appropriately for vertically aligned text.

```
1*15,28,FLAG
002,14, ─────────────────        If text is horizontal, code 14
008,(-2,-6),                      automatically ignores the next line.
042,001,                          008, (-2,-6),
014,016,028,01A,
04C,01E,020,012,014,
002,018,
001,020,01C,
002,01E,
14, ──────────────────            If text is horizontal, code 14
008,(-4,-1),                      automatically ignores the next line.
0                                 008,(-4,-1),
```

Example 5

Write a shape file for a hammer as shown in Figure 11-14(a). (The name of the shape file is HMR.SHP and the shape name is HAMMER.) The following file defines the shape of a hammer. The line numbers are not a part of the file; they are for reference only.

```
*204,34,HAMMER          1
003,22,                 2
002,8,(2,-1),           3
001,024,                4
8,(-1,4),               5
```

00A,(1,004),	6
8,(-1,-4),06C,	7
00C,(4,0,63),	8
044,8,(17,-1),	9
00C,(0,4,63),	10
8,(-17,-1),0	11

Explanation

Line 1

*204,34,HAMMER

This is the header line that consists of shape number (204), number of data bytes in the shape specification (34), and name of the shape (HAMMER).

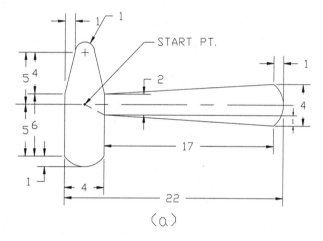

Figure 11-14(a) Dimensions of hammer in units

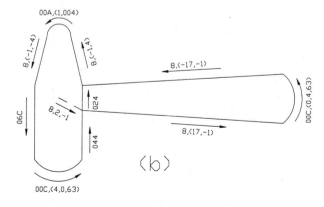

Figure 11-14(b) Pen movements for generating the shape of a hammer

Line 2
003,22,
Data byte 003 has been used to divide the vectors by the next data byte, 22. This reduces the hammer shape to a unit size that facilitates the scaling operation when you insert the shape in a drawing.

Line 3
002,8,(2,-1),
Data byte 002 deactivates the pen (pen up), and the next data byte (code 008) defines a vector that has X-displacement of 2 units and Y-displacement of 1 unit. No line is drawn because the pen is deactivated.

Line 4
001,024,
Data byte 001 activates the pen (pen down) and the next data byte defines a vector that is 2 units long along the direction vector 4.

Line 5
8,(-1,-4),
The first data byte, 8 (code 008), defines a vector whose X-displacement and Y-displacement is given by the next two data bytes. The X-displacement of this vector is -1 and the Y-displacement is -4 units.

Line 6
00A,(1,004),
Data byte 00A defines an octant arc that has a radius of 1 unit as defined by the next data byte. The first element (0) of data byte 004 is a hexadecimal notation. The second element (0) defines the starting octant of the arc, and the third element (4) defines the ending octant of the arc.

Line 7
8,(-1,-4),06C,
Data byte 8 (code 008) defines a vector that has X-displacement of -1 unit, and Y-displacement of -4 units. The next data byte defines a vector that is 6 units long along the direction vector C.

Line 8
00C,(4,0,63),
The first data byte (00C) defines an arc that has X-displacement of 4 units, Y-displacement of 0 units, and a bulge factor of 63.

$$
\begin{aligned}
\text{Bulge factor} \quad &= (2 * H)/D * 127 \\
&= (2 * 1)/4 * 127 \\
&= 63.5 \\
&= 63 \text{ (integer)}
\end{aligned}
$$

Line 9
044,8,(17,-1),
The first data byte defines a vector that is 4 units long along the direction vector 4. The second

data byte (8) defines a vector that has X-displacement of 17 units and Y-displacement of -1 unit.

Line 10
00C,(0,4,63),
The first data byte (00C) defines an arc that has X-displacement of 0 units, Y-displacement of 4 units, and a bulge factor of 63.

$$
\begin{aligned}
\text{Bulge factor} \quad &= (2 * H)/D * 127 \\
&= (2 * 1)/2 * 127 \\
&= 63.5 \\
&= 63 \text{ (integer)}
\end{aligned}
$$

Line 11
8,(-17,-1),0
The first data byte (8) defines a vector that has X-displacement of -17 units and Y-displacement of -1 unit. The data byte 0 terminates the shape definition.

TEXT FONT FILES

In addition to shape files, AutoCAD provides a facility to create new text fonts. After you have created and compiled a text font file, text can be inserted in a drawing like regular text and using the new font. These text files are regular shape files with some additional information describing the text font and the line feed. The following is the general layout of the text font file:

> **Text font description**
> **Line feed**
> **Shape definition**

Text Font Description

The text font description consists of two lines:

> ***0,4,font name**
> **ABOVE, BELOW, MODES, 0**

Where		
*0 -----------------	Special shape number for text font
4 -------------------	Number of data bytes
font name -------	Name of font in lowercase
ABOVE ----------	Upper distance
BELOW ---------	Lower distance
MODES ---------	0 for horizontal text 2 for dual orientation
0 -------------------	Terminating zero

For example, if you are writing a shape definition for an uppercase M, the text font description would be:

> ***0,4,ucm**
> **10,4,2,0**

In the first line, the first data byte (0) is a special shape number for the text font, and every text font file will have this shape number. The next data byte (4) is the number of data bytes in the next line, and ucm is the shape name (name of font). The shape names in all text font files should be lowercase so that the computer does not have to save the names in memory. You can still reference the shape names for editing.

In the second line, the first data byte (10) specifies the height of an uppercase letter above the baseline. For example, in Figure 11-15 (page 11-20), the height of the letter M above the baseline is 10 units. The next data byte (4) specifies the distance of lowercase letters below the baseline. AutoCAD uses this information to scale the text automatically. For example, if you enter the height of text as 1 unit, the text will be 1 unit, although it was drawn 10 units high in the text font definition. The third data byte (2) defines the mode. It can have one of only two values, 0 or 2. If the text is horizontal, the mode is 0; if the text has dual orientation (horizontal and vertical), the mode is 2. The fourth data byte (0) is the zero that terminates the definition.

Line Feed

The line feed is used to space the lines so that characters do not overlap and so that a desired distance is maintained between the lines. AutoCAD has reserved the ASCII number 10 to define the line feed.

```
*10,5,lf
2,8,(0,-14),0
```

Where ***10** ---------------- (10) ASCII number reserved for line feed
       5 ------------------- (5) Number of data bytes in shape specification
       lf ------------------ (lf) Shape name
       2 ------------------- Deactivate pen (pen up)
       (0,-14) ------------ Line feed of 14 units
       0 ------------------- Terminating zero

In the first line, the first data byte (10) is the shape number reserved for line feed, and the next data byte (5) is the number of characters in the shape specification. The data byte lf is the name of the shape.

In the second line, the first data byte (2) deactivates the pen. The next data byte (8) is a special code, 008, that defines a vector by X displacement and Y displacement. The third and fourth data bytes (0,-14) are the X displacement and Y displacement of the displacement vector; they produce a line feed that is 14 units below the baseline. The fifth data byte (0) is the zero that terminates the shape definition.

Shape Definition

The shape number in the shape definition of the text font corresponds to the ASCII code for that character. For example, if you are writing a shape definition for an uppercase M, the shape number is 77.

```
*77,50,ucm
```

Where ***77** ---------------- Shape number - ASCII code of uppercase "M"
       50 ----------------- Number of data bytes

ucm -------------- Shape name

The ASCII codes can be obtained from the ASCII character table, which gives the ASCII codes for all characters, numbers, and punctuation marks.

32	space	56	8	80	P	104	h
33	!	57	9	81	Q	105	i
34	"	58	:	82	R	106	j
35	#	59	;	83	S	107	k
36	$	60	<	84	T	108	l
37	%	61	=	85	U	109	m
38	&	62	>	86	V	110	n
39	,	63	?	87	W	111	o
40	(	64	@	88	X	112	p
41	)	65	A	89	Y	113	q
42	*	66	B	90	Z	114	r
43	+	67	C	91	[	115	s
44	,	68	D	92	\	116	t
45	-	69	E	93	]	117	u
46	.	70	F	94	^	118	v
47	/	71	G	95	_	119	w
48	0	72	H	96	`	120	x
49	1	73	I	97	a	121	y
50	2	74	J	98	b	122	z
51	3	75	K	99	c	123	{
52	4	76	L	100	d	124	I
53	5	77	M	101	e	125	}
54	6	78	N	102	f	126	~
55	7	79	O	103	g		

Example 6

Write a text font shape file (UCM) for an uppercase M as shown in Figure 11-15. The font file should be able to generate horizontal and vertical text. Each grid is 1 unit, and the directions of vectors are designated with leader lines. **In the following file, the line numbers at the right are not a part of the file; they are for reference only.**

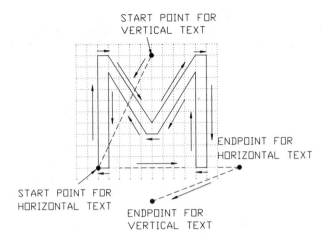

Figure 11-15 *Shape and pen movement of uppercase "M"*

*0,4,uppercase m	1
10,0,2,0	2
*10,13,lf	3
002,8,(0,-14),14,9,(0,14),(14,0),(0,0),0	4
*77,51,ucm	5
2,14,8,(-5,-10),	6
001,009,(0,10),(1,0),(4,-6),(4,6),(1,0),	7
(0,-10),(-1,0),(0,0),	8
003,2,	9
009,(0,17),(-7,-11),(-2,0),(-7,11),	10
(0,-17),(-2,0),(0,0),	11
002,8,(28,0),	12
004,2,	13
14,8,(-9,-4),0	14

Explanation

Line 1

*0,4,uppercase m

The first data byte (0) is the special shape number for the text font file. The next data byte (4) is the number of data bytes, and the third data byte is the name of the shape.

Line 2

10,0,2,0

The first data byte (10) represents the total height of the character M, and the second data byte (0) represents the length of the lowercase letters that extend below the baseline. Data byte 2 is the text mode for dual orientation of the text (horizontal and vertical). If the text was required in the horizontal direction only, the mode would be 0. The fourth data byte (0) is the zero that terminates the definition of this particular shape.

Line 3
*10,13,lf
The first data byte (10) is the code reserved for line feed, and the second data byte (13) is the number of data bytes in the shape specification. The third data byte (lf) is the name of the shape.

Line 4
002,8,(0,-14),14,9,(0,14),(14,0),(0,0),0
The first data byte (002 or 2) is the code to deactivate the pen (pen up). The next three data bytes [8,(0,-14)] define a displacement vector whose X displacement and Y displacement are 0 and -14 units, respectively. This will cause a carriage return that is 14 units below the text insertion point of the first text line. This will work fine if the text is drawn in the horizontal direction only. However, if the text is vertical, the carriage return should produce a displacement to the right of the existing line. This is accomplished by the next seven data bytes. Data byte 14 ignores the next code if the text is horizontal. If the text is vertical, the next code is processed. The next set of data bytes (0,14) defines a displacement vector that is 14 units below the previous point, D1 in Figure 11-16.

Data bytes (14,0) define a displacement vector that is 14 units to the right, D2 in Figure 11-16. These four data bytes combined will result in a carriage return that is 4 units to the right of the existing line. The next set of data bytes (0,0) terminates the code 9, and the last data byte (0) terminates the shape specification.

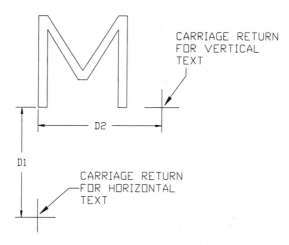

Figure 11-16 Carriage return for vertical and horizontal texts

Line 5
*77,51,ucm
The first data byte (77) is the ASCII code of the uppercase M. The second data byte (51) is the number of data bytes in the shape specification. The next data byte (ucm) is the name of the shape file in lowercase letters.

Line 6
2,14,8,(-5,-10),

The first data byte code (2) deactivates the pen (pen up), and the next data byte code (14) will cause the next code to be ignored if the text is horizontal. In the horizontal text, the insertion point of the text is the starting point of that text line (Figure 11-15). However, if the text is vertical, the starting point of the text is the upper middle point of the character M. This is accomplished by the next three data bytes [8,(-5,-10)], which displace the starting point of the text 5 units to the left (width of character M is 10) and 10 units down (height of character M is 10).

Lines 7,8
001,009,(0,10),(1,0),(4,-6),(4,6),(1,0),
(0,-10),(-1,0),(0,0),
The first byte (001) activates the draw mode (pen down), and the remaining bytes define the next seven vectors.

Lines 9,10,11
003,2,
009,(0,17),(-7,-11),(-2,0),(-7,11),
(0,-17),(-2,0),(0,0),
The inner vertical line of the right leg of the character M is 8.5 units long, and you cannot define a vector that is not an integer. However, you can define a vector that is 2 x 8.5 = 17 units long and then divide that vector by 2 to get a vector 8.5 units long. This is accomplished by code 003 and the next data byte, 2. All the vectors defined in the next two lines will be divided by 2.

Line 12
002,8,(28,0),
The first data byte (002) deactivates the draw mode, and the next three data bytes define a vector that is 28/2 = 14 units to the right. This means that the next character will start 14 - 10 = 4 units to the right of the existing character so that it will produce a horizontal text.

Line 13
004,2,
The code 004 multiplies the vectors that follow it by 2; therefore, it nullifies the effect of code 003,2.

Line 14
14,8,(-9,-4),0
If the text is vertical, the next letter should start below the previous letter. This is accomplished by data bytes 8,(-9,-4), which define a vector that is -9 units along the X axis and -4 units along the Y axis. The data byte 0 terminates the definition of the shape.

To load the shape file, use the **LOAD** command as discussed in Example 1. Use the **STYLE** command to define a text style file that uses the font (UCM) created in this example. Now, you can use the **TEXT** or **DTEXT** command and enter the text as shown in Figure 11-17.

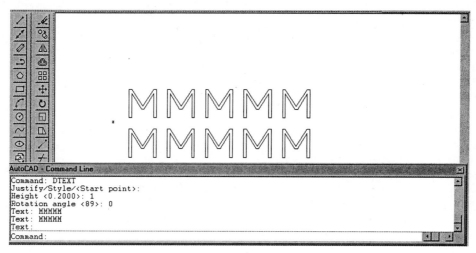

Figure 11-17 Using the defined text font

Example 7

Write a text font shape file for the lowercase m as shown in Figure 11-18. The font file should be able to generate horizontal and vertical text. Each grid is 1 unit, and the direction of the vectors is designated with the leader lines.

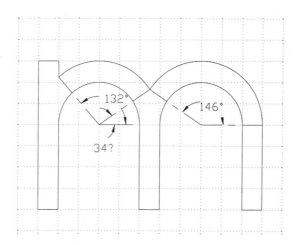

Figure 11-18 Shape of lowercase character, m

The following file is a listing of the text font shape file for Example 7. The line numbers are not a part of the file; they are shown here for reference only.

```
*0,4,lower-case m                                                                    1
14,3,2,0                                                                             2
```

```
*10,13,lf                                                   3
002,8,(0,-18),14,9,(0,18),(27,0),(0,0),0                    4
*109,57,lcm                                                 5
2,14,8,(-11,-14),                                           6
005,005,001,020,084,                                        7
00A,(4,-044),                                               8
08C,020,084,                                                9
00A,(4,-044),                                               10
08C,020,084,                                                11
00B,(0,62,0,6,004),                                         12
00B,(193,239,0,6,003),                                      13
006,9,(0,14),(2,0),(0,0),                                   14
003,5,07C,004,5                                             15
006,2,8,(27,0),                                             16
14,8,(-16,-5),0                                             17
```

The format of most of the lines is the same as in the previous example, except for the following lines, which use save/restore origin, octant, and fractional arcs.

Line 7
005,005,001,020,084,
The first and second data bytes (005) are used to save the location of the point twice. The remaining data bytes activate the draw mode and define the vectors.

Line 8
00A,(4,-044),
The first data byte code (00A) is the code for octant arc and the second data byte (4) defines the radius of the arc. The negative sign (-) in the third data byte generates an arc in a clockwise direction. The first element (0) is a hexadecimal notation. The second element defines the starting octant and the third element (4) defines the number of octants that the arc passes through.

Line 12
00B,(0,62,0,6,004),
The first data byte (00B) is the code for the fractional arc that is defined by the next five data bytes. The second data byte (0) is the starting offset of the first arc, as shown in the following calculations:

1st Arc

Starting angle	= 0	Ending angle = 146
Starting octant	= 0	Ending octant= 4
Starting offset	= (0-0)*256/45	
	= 0	
Ending offset	= (146-135)*256/45	
	= 62.57	
	= 62 (integer)	

The third data byte (62) is the ending offset of the arc and the fourth data byte (0) is the high radius. The fifth data byte (6) defines the radius of the arc. The second element (0) of the next data byte is the starting octant and the third element (4) is the number of octants the arc goes through.

Line 13
00B,(193,239,0,6,003),
The first data byte (00B) is the code for the fractional arc. The remaining data bytes define various parameters of the fractional arc as explained earlier. The offset angles have been obtained from the following calculations:

2nd Arc

Starting angle	$= 34$ Ending angle $= 132$
Starting octant	$= 0$ Ending octant $= 3$
Starting offset	$= (34-0)*256/45$
	$= 193.4$
	$= 193$ (integer)
Ending offset	$= (132-90)*256/45$
	$= 238.9$
	$= 239$ (integer)

Note
Since the offset values have been rounded, it is not possible to describe an arc that is very accurate. Therefore, in this example the origin has been restored after two arcs were drawn. This origin was then used to draw the remaining lines.

Line 14
006,9,(0,14),(2,0),(0,0),
The first data byte (006) restores the previously saved point, and the remaining data bytes define the vectors using code 009.

Review Questions

1. The basic objects used in shape files are _____ and _____.

2. Shapes are easy to insert and take less disk space than _____. However, there are certain disadvantages to using shapes. For example, you cannot _____ a shape.

3. The shape number could be any number between 1 and _____ in a particular file, and these numbers cannot be repeated within the same file.

4. The shape file may not contain two _____ with the same name.

5. A hexadecimal number is designated by a leading _____.

6. The maximum number of data bytes is _____ bytes per shape.

7. To define a vector, you need its magnitude and _____.

8. Do not include the extension with the _____.

9. To load the shape file, use the AutoCAD _____ command.

10. Generating shapes with direction vectors has some limitations. For example, you cannot draw an arc or a line that is not along the _____ vectors. These limitations can be overcome by using the _____, which add a lot of flexibility and give you better control over the shapes you want to create.

11. Code 001 activates the _____ mode, and code _____ deactivates the draw mode.

12. The byte that follows the division code divides the _____ vectors.

13. Code 004 is used if you want to multiply the vectors by a certain number. It can also be used to _____ the effect of code 003.

14. Scale factors are _____.

15. The number of saves (code 005) must be equal to the number of _____ code _____.

16. The maximum number of saves and restores you can use in a particular shape definition is _____.

17. You can define a subshape like a subroutine in a program. To reference the subshape, use code _____.

18. Vectors can be drawn in the 16 predefined directions only, and the length of the vector cannot exceed _____ units.

19. A nonstandard fractional arc can be generated by using code 00B or _____.

20. Code _____ can be used to define an arc by specifying the displacement of the endpoint of an arc and the bulge factor.

21. Bulge factor can range from -127 to _____.

22. Code 00E or _____ is used when the same text font description is to be used in both horizontal and vertical orientation.

23. The text files are regular _____ files with some additional information about the text font description and the line feed.

24. The shape names in all text font files should be lowercase so that the computer does not have to save the names in its _____.

25. The line feed is used to space the lines so that the characters do not _____.

26. The shape number in the shape definition of the text font corresponds to the _____ code for that character.

Exercises

Exercise 1 *General*

Write a shape file for the uppercase M shown in Figure 11-19.

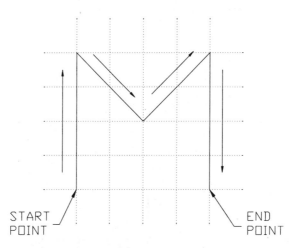

Figure 11-19 *Uppercase letter M*

Exercise 2 *General*

Write a shape file for generating the tapered gib-head key shown in Figure 11-20.

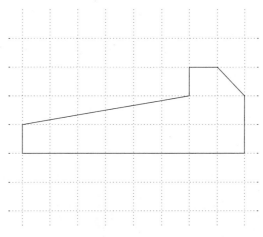

Figure 11-20 *Tapered gib-head key*

Exercise 3   *General*

Write a text font shape file for the uppercase G shown in Figure 11-21.

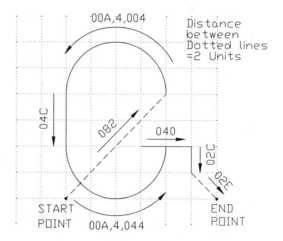

Figure 11-21 *Uppercase letter G*

Exercise 4   *General*

Write a text font shape file for the uppercase W shown in Figure 11-22. The font file should be able to generate horizontal and vertical text.

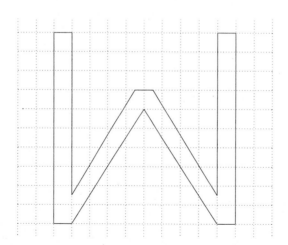

Figure 11-22 *Uppercase letter W*

Chapter 12

AutoLISP

Learning Objectives

After completing this chapter, you will be able to:
- *Perform mathematical operations using AutoLISP.*
- *Use trigonometrical functions in AutoLISP.*
- *Understand the basic AutoLISP functions and their applications.*
- *Load and run AutoLISP programs.*
- *Use **Load/Unload Applications** dialog box.*
- *Use flowcharts to analyze problems.*
- *Test a condition using conditional functions.*

ABOUT AutoLISP

Developed by Autodesk, Inc., **AutoLISP** is an implementation of the **LISP** programming language (LISP is an acronym for **LISt Processor**.) The first reference to LISP was made by John McCarthy in the April, 1960, issue of *The Communications of the ACM*.

Except for **FORTRAN** and **COBOL**, most of the languages developed in the early 1960s have become obsolete. But LISP has survived and has become a leading programming language for artificial intelligence (AI). Some of the dialects of the LISP programming language are Common LISP, BYSCO LISP, ExperLISP, GCLISP, IQLISP, LISP/80, LISP/88, MuLISP, TLCLISP, UO-LISP, Waltz LISP, and XLISP. XLISP is a public-domain LISP interpreter. The LISP dialect that resembles AutoLISP is Common LISP. The AutoLISP interpreter is embedded within the AutoCAD software package. However, AutoCAD LT and AutoCAD Versions 2.17 and lower lack the AutoLISP interpreter; therefore, you can use the AutoLISP programming language only with AutoCAD Release 2.18 and up.

The AutoCAD software package contains most of the commands used to generate a drawing. However, some commands are not provided in AutoCAD. For example, AutoCAD has no command to draw a rectangle or make global changes in the drawing text objects. With AutoLISP you can write a program in the AutoLISP programming language that will draw a rectangle or make global or selective changes in the drawing text objects. As a matter of fact, you can use AutoLISP to write any program or embed it in the menu and thus customize your system to make it more efficient.

The AutoLISP programming language has been used by hundreds of third-party software developers to write software packages for various applications. For example, the author of this text has developed a software package, **SMLayout**, that generates flat layouts of various geometrical shapes like transitions, intersection of pipes and cylinders, elbows, cones, and tank heads. There is demand for AutoLISP programmers as consultants for developing application software and custom menus.

This chapter assumes that you are familiar with AutoCAD commands and AutoCAD system variables. However, you need not be an AutoCAD or programming expert to begin learning AutoLISP. This chapter also assumes that you have no prior programming knowledge. If you are familiar with any other programming language, learning AutoLISP may be easy. A thorough discussion of various functions and a step-by-step explanation of the examples should make it fun to learn. This chapter discusses the most frequently used AutoLISP functions and their application in writing a program. For those functions not discussed in this chapter, refer to the **AutoLISP Programmers Reference Manual** from Autodesk. AutoLISP does not require any special hardware. If your system runs AutoCAD, it will also run AutoLISP. To write AutoLISP programs you can use any text editor.

MATHEMATICAL OPERATIONS

A mathematical function constitutes an important feature of any programming language. Most of the mathematical functions commonly used in programming and mathematical calculations are available in AutoLISP. You can use AutoLISP to add, subtract, multiply, and divide numbers. You can also use it to find the sine, cosine, and arctangent of angles expressed in radians. There is a host of other calculations you can do with AutoLISP. This section discusses the most frequently used mathematical functions supported by the AutoLISP programming language.

Addition

Format **(+ num1 num2 num3 - - -)**

This function (+) calculates the sum of all the numbers to the right of the plus (+) sign (num1 + num2 + num3 + . . .). The numbers can be integers or real. If the numbers are integers, then the sum is an integer. If the numbers are real, the sum is real. However, if some numbers are real and some are integers, the sum is real. In the following display, all numbers in the first two examples are integers, so the result is an integer. In the third example, one number is a real number (50.0), so the sum is a real number.

Examples
Command: (+ 2 5)   returns 7
Command: (+ 2 30 4 50)   returns 86
Command: (+ 2 30 4 50.0)   returns 86.0

Subtraction

Format **(- num1 num2 num3 - - -)**

This function **(-)** subtracts the second number from the first number (num1 - num2). If there are more than two numbers, the second and subsequent numbers are added and the sum is subtracted from the first number [num1 - (num2 + num3 + . . .)]. In the first of the following examples, 14 is subtracted from 28 and returns 14. Since both numbers are integers, the result is an integer. In the third example, 20 and 10.0 are added, and the sum of these two numbers (30.0) is subtracted from 50, returning a real number, 20.0.

Examples
Command: (- 28 14)   returns 14
Command: (- 25 7 11)   returns 7
Command: (- 50 20 10.0)   returns 20.0
Command: (- 20 30)   returns -10
Command: (- 20.0 30.0)   returns -10.0

Multiplication

Format **(* num1 num2 num3 - - -)**

This function **(*)** calculates the product of the numbers to the right of the asterisk (num1 x num2 x num3 x . . .). If the numbers are integers, the product of these numbers is an integer. If one of the numbers is a real number, the product is a real number.

Examples
Command: (* 2 5)   returns 10
Command: (* 2 5 3)   returns 30
Command: (* 2 5 3 2.0)   returns 60.0
Command: (* 2 -5.5)   returns -11.0
Command: (* 2.0 -5.5 -2)   returns 22.0

Division

Format **(/ num1 num2 num3 - - -)**

This function (/) divides the first number by the second number (num1/num2). If there are more than two numbers, the first number is divided by the product of the second and subsequent numbers [num1 / (num2 x num3 x . . .)]. In the fourth of the following examples, 200 is divided by the product of 5.0 and 4 [200 / (5.0 * 4)].

Examples
Command: (/ 30)   returns 30
Command: (/ 3 2)   returns 1
Command: (/ 3.0 2)   returns 1.5
Command: (/ 200.0 5.0 4)   returns 10.0
Command: (/ 200 -5)   returns -40
Command: (/ -200 -5.0)   returns 40.0

INCREMENTED, DECREMENTED, AND ABSOLUTE NUMBERS

Incremented Number
Format **(1+ number)**

This function **(1+)** adds 1 (integer) to **number** and returns a number that is incremented by 1. In the second example below, 1 is added to -10.5 and returns -9.5.

Examples
(1+ 20)	returns 21
(1+ -10.5)	returns -9.5

Decremented Number
Format **(1- number)**

This function **(1-)** subtracts 1 (integer) from the **number** and returns a number that is decremented by 1. In the second example below, 1 is subtracted from -10.5 and returns -11.5.

Examples
(1- 10)	returns 9
(1- -10.5)	returns -11.5

Absolute Number
Format **(abs num)**

The **abs** function returns the absolute value of a number. The number may be an integer number or a real number. In the second example below, the function returns 20 because the absolute value of -20 is 20.

Examples
(abs 20)	returns 20
(abs -20)	returns 20
(abs -20.5)	returns 20.5

TRIGONOMETRIC FUNCTIONS

sin
Format **(sin angle)**

The **sin** function calculates the sine of an angle, where the angle is expressed in radians. In the second of the following examples, the **sin** function calculates the sine of pi (180 degrees) and returns 0.

Examples

Command: (sin 0) returns 0.0
Command: (sin pi) returns 0.0
Command: (sin 1.0472) returns 0.866027

cos

Format **(cos angle)**

The **cos** function calculates the cosine of an angle, where the angle is expressed in radians. In the third of the following examples, the **cos** function calculates the cosine of pi (180 degrees) and returns -1.0.

Examples

Command: (cos 0) returns 1.0
Command: (cos 0.0) returns 1.0
Command: (cos pi) returns -1.0
Command: (cos 1.0) returns 0.540302

atan

Format **(atan num1)**

The **atan** function calculates the arctangent of **num1**, and the calculated angle is expressed in radians. In the second of the following examples, the **atan** function calculates the arctangent of 1.0 and returns 0.785398 (radians).

Examples

Command: (atan 0.5) returns 0.463648
Command: (atan 1.0) returns 0.785398
Command: (atan -1.0) returns -0.785398

You can also specify a second number in the atan function:

Format **(atan num1 num2)**

If the second number is specified, the function returns the arctangent of (num1/num2) in radians. In the first of the following examples the first number (0.5) is divided by the second number (1.0), and the **atan** function calculates the arctangent of the dividend (0.5/1.0 = 0.5).

Examples

Command: (atan 0.5 1.0) returns 0.463648 radians
Command: (atan 2.0 3.0) returns 0.588003 radians
Command: (atan 2.0 -3.0) returns 2.55359 radians
Command: (atan -2.0 3.00) returns -0.588003 radians
Command: (atan -2.0 -3.0) returns -2.55359 radians
Command: (atan 1.0 0.0) returns 1.5708 radians
Command: (atan -0.5 0.0) returns -1.5708 radians

angtos

Format **(angtos angle [mode [precision]])**

The **angtos** function returns the angle expressed in radians in a string format. The format of the string is controlled by the **mode** and **precision** settings.

Examples

(angtos 0.588003 0 4)	returns "33.6901"
(angtos 2.55359 0 4)	returns "146.3099"
(angtos 1.5708 0 4)	returns "90.0000"
(angtos -1.5708 0 2)	returns "270.00"

Note

*In **(angtos angle [mode [precision]])***
angle   is angle in radians
mode   is the angtos mode that corresponds to the AutoCAD system variable AUNITS
The following modes are available in AutoCAD:

ANGTOS MODE	EDITING FORMAT
0	*Decimal degrees*
1	*Degrees/minutes/seconds*
2	*Grads*
3	*Radians*
4	*Surveyor's units*

*Precision is an integer number that controls the number of decimal places. Precision corresponds to the AutoCAD system variable **AUPREC**. The minimum value of **precision** is zero and the maximum is four.*

In the first example above, angle is 0.588003 radians, mode is 0 (angle in degrees), and precision is 4 (four places after decimal). The function will return 33.6901.

RELATIONAL STATEMENTS

Programs generally involve features that test a particular condition. If the condition is true, the program performs certain functions, and if the condition is not true, then the program performs other functions. For example, the relational statement (if (< x 5)) tests true if the value of the variable x is less than 5. This type of test condition is frequently used in programming. The following section discusses various relational statements used in AutoLISP programming.

Equal to

Format **(= atom1 atom2 - - - -)**

This function (=) checks whether the two atoms are equal. If they are equal, the condition is true and the function will return T. Similarly, if the specified atoms are not equal, the condition is false and the function will return nil.

Examples

(= 5 5)	returns T
(= 5 4.9)	returns nil
(= 5.5 5.5 5.5)	returns T
(= "yes" "yes")	returns T
(= "yes" "yes" "no")	returns nil

Not equal to
Format (/= atom1 atom2 - - - -)

This function (/=) checks whether the two atoms are not equal. If they are not equal, the condition is true and the function will return T. Similarly, if the specified atoms are equal, the condition is false and the function will return nil.

Examples

(/= 50 4)	returns T
(/= 50 50)	returns nil
(/= 50 -50)	returns T
(/= "yes" "no")	returns T

Less than
Format (< atom1 atom2 - - - -)

This function (<) checks whether the first atom (atom1) is less than the second atom (atom2). If it is true, then the function will return T. If it is not, the function will return nil.

Examples

(< 3 5)	returns T
(< 5 3 4 2)	returns nil
(< "x" "y")	returns T

Less than or equal to
Format (<= atom1 atom2 - - - -)

This function (<=) checks whether the first atom (atom1) is less than or equal to the second atom (atom2). If it is, the function will return T. If it is not, the function will return nil.

Examples

(<= 10 15)	returns T
(<= "c" "b")	returns nil
(<= -2.0 0)	returns T

Greater than
Format (> atom1 atom2 - - - -)

This function (>) checks whether the first atom (**atom1**) is greater than the second atom (**atom2**). If it is, the function will return **T**. If it is not, then the function will return **nil**. In the first example below, 15 is greater than 10. Therefore, this relational function is true and the function will return T. In the second example, 10 is greater than 9, but this number is not greater than the second 9 therefore, this function will return **nil**.

Examples

(> 15 10)	returns T
(> 10 9 9)	returns nil
(> "c" "b")	returns T

Greater than or equal to

Format (> = atom1 atom2 - - - -)

This function (> =) checks whether the numerical value of the first atom (**atom1**) is greater than or equal to the second atom (atom2). If it is, the function returns **T,** otherwise, it will return **nil**. In the first example below, 78 is greater than 50, but 78 is not equal to 50 therefore, it will return **T**.

Examples

(> = 78 50)	returns T
(> = "x" "y")	returns nil

defun, setq, getpoint, AND Command FUNCTIONS

defun

The **defun** function is used to define a function in an AutoLISP program. The format of the **defun** function is:

    (defun name [argument])
                Where **Name** ------------ Name of the function
                           Argument ------- Argument list

Examples
(defun ADNUM ( )
Defines a function ADNUM with no arguments or local symbols. This means that all variables used in the program are global variables. A global variable does not lose its value after the programs ends.

(defun ADNUM (a b c)
Defines a function ADNUM that has three arguments: **a, b,** and **c**. The variables **a, b,** and **c** receive their value from outside the program.

(defun ADNUM (/ a b)
Defines a function ADNUM that has two local variables: **a** and **b**. A local variable is one that retains its value during program execution and can be used within that program only.

(defun C:ADNUM ( )

With **C:** in front of the function name, the function can be executed by entering the name of the function at the AutoCAD Command: prompt. If **C:** is not used, the function name has to be enclosed in parentheses.

Note

AutoLISP contains some built-in functions. Do not use any of those names for function or variable names. The following is a list of some of the names reserved for AutoLISP built-in functions. (Refer to the AutoLISP Programmer's Reference manual for a complete list of AutoLISP built-in functions.)

abs	ads	alloc
and	angle	angtos
append	apply	atom
ascii	assoc	atan
atof	atoi	distance
equal	fix	float
if	length	list
load	member	nil
not	nth	null
open	or	pi
read	repeat	reverse
set	type	while

setq

The **setq** function is used to assign a value to a variable. The format of the **setq** function is:

(setq Name Value [Name Value].......)

> Where **Name** ------------ Name of variable
>
> **Value** ------------ Value assigned to variable

The value assigned to a variable can be any expression (numeric, string, or alphanumeric). If the value is a string, the string length cannot exceed 100 characters.

Command: **(setq X 12)**
Command: **(setq X 6.5)**
Command: **(setq X 8.5 Y 12)**

In this last expression, the number 8.5 is assigned to the variable **X**, and the number 12 is assigned to the variable **Y**.

Command: **(setq answer "YES")**

In this expression the string value "YES" is assigned to the variable **answer**.

The **setq** function can also be used in conjunction with other expressions to assign a value to a variable. In the following examples the **setq** function has been used to assign values to different variables.

(setq pt1 (getpoint "Enter start point: "))
(setq ang1 (getangle "Enter included angle: "))
(setq answer (getstring "Enter YES or NO: "))

Note

*AutoLISP uses some built-in function names and symbols. Do not assign values to any of those functions. The following functions are valid ones, but must never be used because the **pi** and **angle** functions that are reserved functions will be redefined.*

(setq pi 3.0)
(setq angle (. . .))

getpoint

The **getpoint** function pauses to enable you to enter the X, Y coordinates or X, Y, Z coordinates of a point. The coordinates of the point can be entered from the keyboard or by using the screen cursor. The format of the **getpoint** function is:

(getpoint [point] [prompt])
                Where   **Point** ------------- Enter a point, or select a point
                           Prompt ---------- Prompt to be displayed on screen

Example
(setq pt1 (getpoint))
(setq pt1 (getpoint "Enter starting point"))

Note
You cannot enter the name of another AutoLISP routine in response to the getpoint function.

A 2D or a 3D point is always defined with respect to the current user coordinate system (UCS).

Command

The **Command** function is used to execute standard AutoCAD commands from within an AutoLISP program. The AutoCAD command name and the command options have to be enclosed in double quotation marks. The format of the **Command** function is:

(Command "commandname")
                Where   **Command** ----------------- AutoLISP function
                           commandname ---------- AutoCAD command

Example
(Command "line" pt1 pt2 "")
                Where   **"Line"** ----------- AutoCAD LINE Command
                           Pt1 ---------------- First point
                           Pt2 ---------------- Second point
                           "" ---------------- "" for Return

Note

*Prior to AutoCAD Release 12, the **Command** function **could not be used** to execute the AutoCAD PLOT command. For example: (Command "plot" . . .) was not a valid statement. In AutoCAD2000 Release 14 and Release 13, you can use plot with the Command function (Command "plot" . . .).*

*The **Command** function cannot be used to enter data with the AutoCAD DTEXT or TEXT command. (You can issue the DTEXT and TEXT command with the **Command** function. You can also enter text height and text rotation, but you cannot enter the text when DTEXT or TEXT prompts for text entry.)*

*You cannot use the input functions of AutoLISP with the **Command** function. The input functions are **getpoint, getangle, getstring**, and **getint**. For example, (Command "getpoint" . . .) or (Command "getangle" . . .) are not valid functions. If the program contains such a function, it will display an error message when the program is loaded.*

Example 1

Write a program that will prompt you to select three points of a triangle and then draw lines through those points to generate the triangle shown in Figure 12-1.

Most programs consist of essentially three parts: **input, output,** and **process**. **Process** includes what is involved in generating the desired output from the given input (Figure 12-2). Before writing a program, you must identify these three parts. In this example, the **input** to the program is the coordinates of the three points. The desired **output** is a triangle. The **process** needed to generate a triangle is to draw three lines from P1 to P2, P2 to P3, and P3 to P1. Identifying these three sections makes the programming process less confusing.

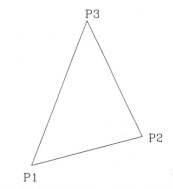

Figure 12-1 Triangle P1, P2, P3

The process section of the program is vital to the success of the program. Sometimes it is simple, but sometimes it involves complicated calculations. If the program involves many calculations, divide them into sections (and perhaps subsections) that are laid out in a logical and systematic order. Also, remember that programs need to be edited from time to time, perhaps by other programmers. Therefore, it is wise to document the programs as clearly and unambiguously as possible so that other programmers can understand what the program is do-

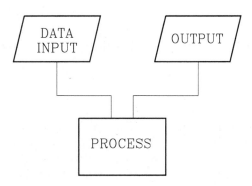

Figure 12-2 Three elements of a Program

Chapter 12

ing at different stages of its execution. Give sketches, and identify points where possible.

Input	**Output**
Location of point P1	
Location of point P2	Triangle P1, P2, P3
Location of point P3	

Process
Line from P1 to P2
Line from P2 to P3
Line from P3 to P1

The following file is a listing of the AutoLISP program for Example 1. **The line numbers at the right are not a part of the program; they are shown here for reference only.**

```
;This program will prompt you to enter three points                    1
;of a triangle from the keyboard, or select three points               2
;by using the screen cursor. P1, P2, P3 are triangle corners.          3
                                                                       4
(defun c:TRIANG1()                                                     5
    (setq P1 (getpoint "\n Enter first point of Triangle: "))          6
    (setq P2 (getpoint "\n Enter second point of Triangle: "))         7
    (setq P3 (getpoint "\n Enter third point of Triangle: "))          8
    (Command "LINE" P1 P2 P3 "C")                                      9
)                                                                      10
```

Explanation

Lines 1-3
The first three lines are comment lines describing the function of the program. These lines are important because they make it easier to edit a program. Comments should be used when needed. All comment lines must start with a semicolon (;). These lines are ignored when the program is loaded.

Line 4
This is a blank line that separates the comment section from the program. Blank lines can be used to separate different modules of a program. This makes it easier to identify different sections that constitute a program. The blank lines have no effect on the program.

Line 5
(defun c:TRIANG1( )
In this line, **defun** is an AutoLISP function that defines the function **TRIANG1**. **TRIANG1** is the name of the function. The c: in front of the function name, **TRIANG1**, enables it to be executed like an AutoCAD command. If the c: is missing, the **TRIANG1** command can be executed only by enclosing it in parentheses (TRIANG1). The TRIANG1 function has three global variables (P1, P2, and P3). When you first write an AutoLISP program, it is a good practice to keep the variables global, because after you load and run the program, you can check the values of these variables by entering exclamation point (!) followed by the variable name at AutoCAD Command prompt

(Command: !P1). Once the program is tested and it works, you should make the variable local, (defun c:TRIANG1(/ P1 P2 P3)).

Line 6
(setq P1 (getpoint "\n Enter first point of Triangle: "))
In this line, the **getpoint** function pauses for you to enter the first point of the triangle. The prompt, **Enter first point of Triangle,** is displayed in the prompt area of the screen. You can enter the coordinates of this point at the keyboard or select a point by using the screen cursor. The setq function then assigns these coordinates to the variable **P1. \n** is used for the carriage return so that the statement that follows \n is printed on the next line ("n" stands for "newline").

Lines 7 and 8
(setq P2 (getpoint "\n Enter second point of Triangle: "))
(setq P3 (getpoint "\n Enter third point of Triangle: "))
These two lines prompt you to enter the second and third corners of the triangle. These coordinates are then assigned to the variables **P2** and **P3. \n** causes a carriage return so that the input prompts are displayed on the next line.

Line 9
(Command "LINE" P1 P2 P3 "C")
In this line, the **Command** function is used to enter the AutoCAD LINE command and then draw a line from P1 to P2 and from P2 to P3. "C" (for "close" option) joins the last point, P3, with the first point, P1. All AutoCAD commands and options, when used in an AutoLISP program, have to be enclosed in double quotation marks. The variables P1, P2, P3 are separated by a blank space.

Line 10
This line consists of a closing parenthesis that completes the definition of the function, TRIANG1. This parenthesis could have been combined with the previous line. It is good practice to keep it on a separate line so that any programmer can easily identify the end of a definition. In this program there is only one function defined, so it is easy to locate the end of a definition. But in some programs a number of definitions or modules within the same program might need to be clearly identified. The parentheses and blank lines help to identify the start and end of a definition or a section in the program.

LOADING AN AUTOLISP PROGRAM

There are generally two names associated with an AutoLISP program: the program file name and the function name. For example, **TRIANG.LSP** is the name of the file, not a function name. All AutoLISP file names have the extension **.LSP**. An AutoLISP file can have one or several functions defined within the same file. For example, TRIANG1 in Example 1 is the name of a function. To execute a function, the AutoLISP program file that defines that function must be loaded. Use the following command to load an AutoLISP file when you are in the drawing editor:

Command: **(load "[path]file name")**
              Where **Command** ------ AutoCAD command prompt
                       load ------------- Loads an AutoLISP program file
                       [path]file name Path and name of AutoLISP program file

The AutoLISP file name and the optional path name must be enclosed in double quotes. The **load** and **file name** must be enclosed in parentheses. If the parentheses are missing, AutoCAD will try to load a shape or a text font file, not an AutoLISP file. The space between **load** and **file name** is not required. If AutoCAD is successful in loading the file, it will display the name of the function in the Command: prompt area of the screen.

To run the program, type the name of the function at the AutoCAD Command: prompt, and press ENTER (Command: **TRIANG1**). If the function name does not contain **C:** in the program, you can run the program by enclosing the function name in parentheses:

    Command: **TRIANG1** or Command: **(TRIANG1)**

Note

Use a forward slash when defining the path for loading an AutoLISP program. For example, if the AutoLISP file TRIANG is in the LISP subdirectory on the C drive, use the following command to load the file. You can also use a double backslash (\\) in place of the forward slash.
Command: **(load "c:/lisp/triang")** or Command: **(load "c:\\lisp\\triang")**

You can also load an application by using the standard Windows drag and drop technique. To load a LISP program, select the file in the Windows Explorer and then drag and drop it in the graphics window of AutoCAD. The selected program will be automatically loaded. Another way to load an AutoLISP program using the **Load/Unload Application** dialog box (Figure 12-3). This dialog box can be invoked by selecting Load Applications in the **Tools** menu or by entering **APPLOAD** at the AutoCAD Command prompt.

Load/Unload Application Dialog Box

The **Load/Unload Application** dialog box (Figure 12-3) can be used to load LSP, VLX, FAS, VBA, DBX, and ObjectARX applications. VBA, DBX, and ObjectARX files are loaded immediately when you select a file. LSP, VLX, and FAS files are qued and loaded when you close the **Load/Unload Application** dialog box. The top portion of the dialog box lists the files in the selected directory. The file type can be changed by entering the file type (*.lsp) in the **Files of type:** edit box or by selecting the file type in the pop-down list. A file can be loaded by selecting the file and then choosing the Load button. The following is the description of other features of the **Load/Unload Application** dialog box:

Load

The **Load** button can be used to load and reload the selected files. The files can be selected from the file list box, Load Application tab, or History List tab. The ObjecrARX files cannot be re-loaded. You must first unload the ObjectARX file and then reload it.

Load Application Tab

When you choose the **Load Application** tab, AutoCAD displays the applications that are currently loaded. You can add files to this list by dragging the files names from the file list box and then dropping them in the Load Application list.

History List Tab

When you choose the **History List** tab, AutoCAD displays the list of files that have been previously

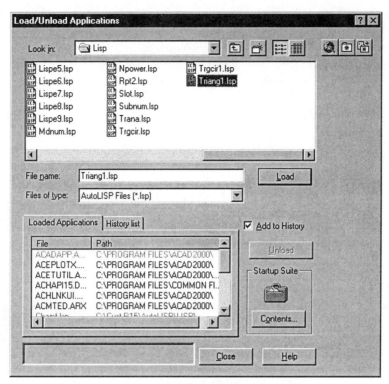

Figure 12-3 *Loading AutoLISP files using the* **Load/**
Unload Applications *dialog box*

loaded with Add to History selected. If the Add to History is not selected and you drag and drop
the files in the History list, the files are loaded but not added to the History List.

Add to History
When Add to History is selected, it automatically adds the files to the History List when you drag
and drop the files in the History List.

Unload
The **Unload** button appears when you choose the **Loaded Application** tab. To unload an applica-
tion, select the file name in the Loaded Application file list and then choose the **Unload** button.
LISP files and the ObjectARX files that are not registered for unloading cannot be unloaded.

Remove
The **Remove** button appears when you choose the **History List** tab. To remove a file from the
History List, select the file in the History List and then choose the **Remove** button.

Startup Suit
The files in the Startup Suit are automatically loaded each time you start AutoCAD. When you
choose the Startup Suit, AutoCAD displays the **Startup Suit** dialog box that contains a list of files.
You can add files to the list by choosing the **Add** button. You can also drag the files from the file list
box and drop them in the Startup Suit. To add files from the History List, right click on the file.

Exercise 1

Write an AutoLISP program that will draw a line between two points (Figure 12-4). The program must
prompt the user to enter the X and Y coordinates of
the points.

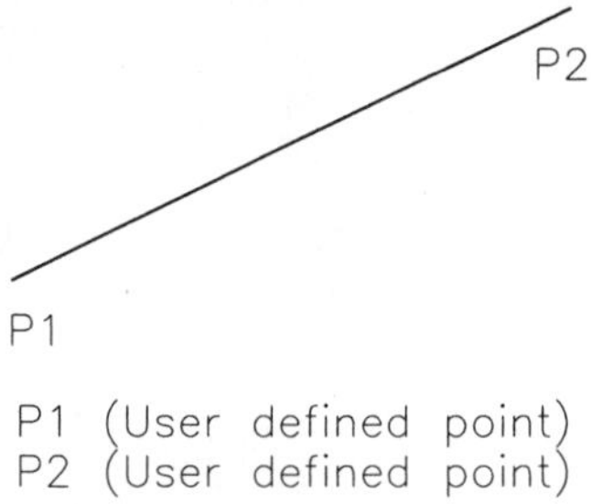

P2

P1

P1 (User defined point)
P2 (User defined point)

Figure 12-4 Draw line from point P1 to P2

getcorner, getdist, AND setvar FUNCTIONS

getcorner

The **getcorner** function pauses for you to enter the coordinates of a point. The coordinates of the
point can be entered at the keyboard or by using the screen crosshairs. This function requires a
base point, and it displays a rectangle with respect to the base point as you move the screen
crosshairs around the screen. The format of the **getcorner** function is:

(getcorner point [prompt])

          Where **point** ------------------------ Base point
                   prompt --------------------- Prompt displayed on screen

Examples
(getcorner pt1)
(setq pt2 (getcorner pt1))
(setq pt2 (getcorner pt1 "Enter second point: "))

Note
*The base point and the point that you select in response to the getcorner function are located with
respect to the current UCS.*
*If the point you select is a 3D point with X, Y, and Z coordinates, the Z coordinate is ignored. The
point assumes current elevation as its Z coordinate.*

getdist

The **getdist** function pauses for you to enter distance, and it then returns the distance as a real
number. The format of the **getdist** function is:

(getdist [point] [prompt])

          Where **Point** ------------- First point for distance
                   Prompt ---------- Any prompt that needs to be displayed on screen

Examples

(getdist)
(setq dist (getdist))
(setq dist (getdist pt1))
(setq dist (getdist "Enter distance"))
(setq dist (getdist pt1 "Enter second point for distance"))

The distance can be entered by selecting two points on the screen. For example, if the assignment is **(setq dist (getdist))**, you can enter a number or select two points. If the assignment is **(setq dist (getdist pt1))**, where the first point (pt1) is already defined, you need to select the second point only. The **getdist** function will always return the distance as a real number. For example, if the current setting is architectural and the distance is entered in architectural units, the **getdist** function will return the distance as a real number.

setvar

The **setvar** function assigns a value to an AutoCAD system variable. The name of the system variable must be enclosed in double quotes. The format of the **setvar** function is:

(setvar "variable-name" value)
Where **Variable-name** - AutoCAD system variable
          Value ------------- Value to be assigned to the system variable

Examples

(setvar "cmdecho" 0)
(setvar "dimscale" 1.5)
(setvar "ltscale" 0.5)
(setvar "dimcen" -0.25)

Example 2

Write an AutoLISP program that will generate a chamfer between two given lines by entering the chamfer angle and the chamfer distance. To generate a chamfer, AutoCAD uses the values assigned to system variables **CHAMFERA** and **CHAMFERB**. When you select the AutoCAD CHAMFER command, the first and second chamfer distances are automatically assigned to the system variables **CHAMFERA** and **CHAMFERB**. The CHAMFER command then uses these assigned values to generate a chamfer. However, in most engineering drawings, the preferred way to generate the chamfer is to enter the chamfer length and the chamfer angle, as shown in Figure 12-5.

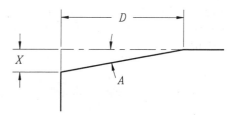

Figure 12-5 *Chamfer with angle A and distance D*

Input	Output
First chamfer distance (D)	Chamfer between any two
Chamfer angle (A)	selected lines

Process
1. Calculate second chamfer distance.
2. Assign these values to the system
   variables CHAMFERA and CHAMFERB.
3. Use AutoCAD CHAMFER command.
   to generate chamfer

Calculations

x/d = tan a

x=   d * (tan a)

   = d * [(sin a) / (cos a)]

The following file is a listing of the program for Example 2. **The line numbers on the right are not a part of the file; they are for reference only.**

```
;This program generates a chamfer by entering                    1
;the chamfer angle and the chamfer distance                      2
;                                                                3
(defun c:chamf (/ d a)                                           4
(setvar "cmdecho" 0)                                             5
   (graphscr)                                                    6
   (setq d (getdist "\n Enter chamfer distance: "))             7
   (setq a (getangle "\n Enter chamfer angle: "))              8
   (setvar "chamfera" d)                                        9
   (setvar "chamferb" (* d (/ (sin a) (cos a))))              10
   (command "chamfer")                                         11
   (setvar "cmdecho" 1)                                        12
   (princ)                                                     13
)                                                              14
```

Explanation

Line 7
(setq d (getdist "\n Enter chamfer distance: "))
The **getdist** function pauses for you to enter the chamfer distance, then the **setq** function assigns that value to variable d.

Line 8
(setq a (getangle "\n Enter chamfer angle: "))
The **getangle** pauses for you to enter the chamfer angle, then the **setq** function assigns that value to variable a.

Line 9
(setvar "chamfera" d)
The **setvar** function assigns the value of variable d to AutoCAD system variable **chamfera**.

Line 10
(setvar "chamferb" (* d (/ (sin a) (cos a))))

The **setvar** function assigns the value obtained from the expression **(* d (/ (sin a) (cos a)))** to the AutoCAD system variable **chamferb**.

Line 11
(command "chamfer")
The **Command** function uses the AutoCAD **CHAMFER** command to generate a chamfer.

Exercise 2   *General*

Write an AutoLISP program that will generate the drawing shown in Figure 12-6. The program should prompt the user to enter points P1 and P2 and diameters D1 and D2.

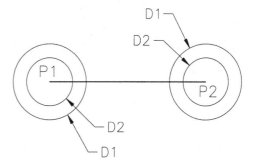

Figure 12-6 Concentric circles with connecting line

list FUNCTION

In AutoLISP the **list** function is used to define a 2D or 3D point. The list function can also be designated by using the single quote character ('), if the expression does not contain any variables or undefined items.

Examples
(Setq x (list 2.5 3 56))   returns 2.5, 3.56
(Setq x '(2.5 3.56))   returns 2.5, 3.56

car, cdr, AND cadr FUNCTIONS
car

The **car** function returns the first element of a list. If the list does not contain any elements, the function will return **nil**. The format of the **car** function is:

(car list)
          Where   **car** ---------------- Returns the first element
                  list ---------------- List of elements

Examples
(car '(2.5 3 56))   returns 2.5
(car '(x y z))   returns X

(car '((15 20) 56))   returns (15 20)
(car '( ))   returns nil

The single quotation mark signifies a list.

cdr

The **cdr** function returns a list with the first element removed from the list. The format of the **cdr** function is:

(cdr list)

> Where   **cdr** ---------------- Returns a list with the first element removed
>   **list** ---------------- List of elements

Examples

(cdr '(2.5 3 56)   returns (3 56)
(cdr '(x y z))   returns (Y Z)
(cdr '((15 20) 56))   returns (56)
(cdr '( ))   returns nil

cadr

The **cadr** function performs two operations, **cdr** and **car**, to return the second element of the list. The **cdr** function removes the first element, and the **car** function returns the first element of the new list. The format of the **cadr** function is:

(cadr list)

> Where   **cadr** -------------- Performs two operations (car (cdr '(x y z))
>   **list** ---------------- List of elements

Examples

(cadr '(2 3))   returns 3
(cadr '(2 3 56))   returns 3
(cadr '(x y z))   returns y
(cadr '((15 20) 56 24))   returns 56

In these examples, **cadr** performs two functions:

(cadr '(x y z))   = (car (cdr '(x y z)))
                                       = (car (y z)) returns y

Note

In addition to the functions car, cdr, and cadr, several other functions can be used to extract different elements of a list. Following is a list of these functions, where the function f consists of a list '((x y) z w)).

(setq f '((x y) z w))
(caar f) = (car (car f))   *returns x*

$(cdar\,f)= (cdr\ (car\,f))$ *returns y*
$(cadar)= (car\ (cdr\ (car\,f)))$ *returns y*
$(cddr\,f)= (cdr\ (cdr\,f))$ *returns w*
$(caddr\ f)= (car\ (cdr\ (cdr\ f)))$ returns w

graphscr, textscr, princ, AND terpri FUNCTIONS

graphscr

The **graphscr** function switches from the text window to the graphics window, provided the system has only one screen. If the system has two screens, this function is ignored.

textscr

The **textscr** function switches from the graphics window to the text window, provided the system has only one screen. If the system has two screens, this function is ignored.

princ

The **princ** function prints (or displays) the value of the variable. If the variable is enclosed in double quotes, the function prints (or displays) the expression that is enclosed in the quotes. The format of the **princ** function is:

    (princ [variable or expression])

Examples

(princ)   prints a blank on screen
(princ a)   prints the value of variable a on screen
(princ "Welcome")   prints Welcome on screen

terpri

The **terpri** function prints a new line on the screen, just as **\n** does. This function is used to print the line that follows the **terpri** function.

Examples

(setq p1 (getpoint "Enter first point: "))(terpri)
(setq p2 (getpoint "Enter second point: "))

The first line (Enter first point:) will be displayed in the screen's command prompt area. The **terpri** function causes a carriage return therefore, the second line (Enter second point:) will be displayed on a new line, just below the first line. If the terpri function is missing, the two lines will be displayed on the same line (Enter first point: Enter second point:).

Example 3

Write a program that will prompt you to enter two opposite corners of a rectangle and then draw the rectangle on the screen as shown in Figure 12-7.

Input
Coordinates of point P1
Coordinates of point P3

Output
Rectangle

Process
1. Calculate the coordinates
   of the points P2 and P4.
2. Draw the following lines.
   Line from P1 to P2
   Line from P2 to P3
   Line from P3 to P4
   Line from P4 to P1

The X and Y coordinates of points P2 and P4 can be calculated using the **car** and **cadr** functions. The **car** function **extracts** the X coordinate of a given list, and the **cadr** function extracts the Y coordinate.

X coordinate of point p2
x2 = x3
x2 = car (x3 y3)
x2 = car p3

Y coordinate of point p2
y2 = y1
y2 = CADR (x1 y1)
y2 = CADR p1

X coordinate of point p4
x4 = x1
x4 = car(x1 y1)
x4 = car p1

Y coordinate of point p4
y4 = y3
y4 = cadr (x3 y3)
y4 = cadr p3

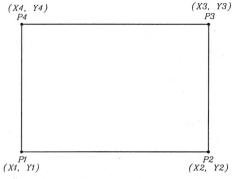

Figure 12-7 *Rectangle P1 P2 P3 P4*

Therefore, points p2 and p4 are:
p2 = (list (car p3) (cadr p1))
p4 = (list (car p1) (cadr p3))

The following file is a listing of the program for Example 3. **The line numbers at the right are for reference only; and are not a part of the program.**

```
;This program will draw a rectangle. User will                        1
;be prompted to enter the two opposite corners                        2
;                                                                      3
(defun c:RECT1(/ p1 p2 p3 p4)                                         4
```

```
     (graphscr)                                              5
     (setvar "cmdecho" 0)                                    6
     (prompt "RECT1 command draws a rectangle")(terpri)      7
     (setq p1 (getpoint "Enter first corner"))(terpri)       8
     (setq p3 (getpoint "Enter opposite corner"))(terpri)    9
     (setq p2 (list (car p3) (cadr p1)))                    10
     (setq p4 (list (car p1) (cadr p3)))                    11
  (command "line" p1 p2 p3 p4 "c")                          12
  (setvar "cmdecho" 1)                                      13
  (princ)                                                   14
  )                                                         15
```

Explanation

Lines 1-3

The first three lines are comment lines that describe the function of the program. All comment lines that start with a semicolon are ignored when the program is loaded.

Line 4

(defun c:RECT1(/ p1 p2 p3 p4)
The **defun** function defines the function **RECT1**.

Line 5

(graphscr)
This function switches the text screen to the graphics screen, if the current screen happens to be a text screen. Otherwise, this function has no effect on the display screen.

Line 6

(setvar "cmdecho" 0)
The **setvar** function assigns the value 0 to the AutoCAD system variable **cmdecho**, which turns the echo off. When **cmdecho** is off, AutoCAD command prompts are not displayed in the command prompt area of the screen.

Line 7

(prompt "RECT1 command draws a rectangle")(terpri)
The **prompt** function will display the information in double quotes ("RECT1 command draws a rectangle"). The function **terpri** causes a carriage return so that the next text is printed on a separate line.

Line 8

(setq p1 (getpoint "Enter first corner"))(terpri)
The **getpoint** function pauses for you to enter a point (the first corner of the rectangle), and the **setq** function assigns that value to variable p1.

Line 9

(setq p3 (getpoint "Enter opposite corner"))(terpri)
The **getpoint** function pauses for you to enter a point (the opposite corner of the rectangle), and the **setq** function assigns that value to variable p3.

Line 10

(setq p2 (list (car p3) (cadr p1)))

The **cadr** function extracts the Y coordinate of point p1, and the car function extracts the X coordinate of point p3. These two values form a list, and the setq function assigns that value to variable p2.

Line 11

(setq p4 (list (car p1) (cadr p3)))

The **cadr** function extracts the Y coordinate of point p3, and the **car** function extracts the X coordinate of point p1. These two values form a list, and the **setq** function assigns that value to variable p4.

Line 12

(command "line" p1 p2 p3 p4 "c")

The command function uses the AutoCAD **LINE** command to draw lines between points p1, p2, p3, and p4. The c (close) joins the last point, p4, with the first point, p1.

Line 13

(setvar "cmdecho" 1)

The **setvar** function assigns a value of 1 to the AutoCAD system variable **cmdecho**, which turns the echo on.

Line 14

(princ)

The princ function prints a blank on the screen. If this line is missing, AutoCAD will print the value of the last expression. This value does not affect the program in any way. However, it might be confusing at times. The **princ** function is used to prevent display of the last expression in the command prompt area.

Line 15

The closing parenthesis completes the definition of the function **RECT1** and ends the program.

 Note

In this program the rectangle is generated after you define the two corners of the rectangle. The rectangle is not dragged as you move the screen crosshairs to enter the second corner. However, the rectangle can be dragged by using the getcorner function, as shown in the following program listing:

```
;This program will draw a rectangle with the
;drag mode on and using getcorner function
;
(defun c:RECT2(/ p1 p2 p3 p4)
    (graphscr)
    (setvar "cmdecho" 0)
    (prompt "RECT2 command draws a rectangle")(terpri)
    (setq p1 (getpoint "enter first corner"))(terpri)
    (setq p3 (getcorner p1 "Enter opposite corner" ))(terpri)
    (setq p2 (list (car p3) (cadr p1)))
```

```
        (setq p4 (list (car p1) (cadr p3)))
(command "line" p1 p2 p3 p4 "c")
(setvar "cmdecho" 1)
(princ)
)
```

getangle AND getorient FUNCTIONS

getangle

The **getangle** function pauses for you to enter the angle and; then it returns the value of that angle in radians. The format of the **getangle** function is:

(getangle [point] [prompt])
> Where **point** ------------- First point of the angle
> **prompt** ---------- Any prompt that needs to be displayed on screen

Examples
(getangle)
(setq ang (getangle))
(setq ang (getangle pt1))———————————— pt1 is a predefined point
(setq ang (getangle "Enter taper angle"))
(setq ang (getangle pt1 "Enter second point of angle"))

The angle you enter is affected by the angle setting. The angle settings can be changed using the AutoCAD **UNITS** command or by changing the value of the AutoCAD system variables, **ANGBASE** and **ANGDIR**. Following are the default settings for measuring an angle:

> The angle is measured with respect to the positive X-axis (3 o'clock position). The value of this setting is saved in the AutoCAD system variable **ANGBASE**.

> The angle is positive if it is measured in the counterclockwise direction and is negative if it is measured in the clockwise direction. The value of this setting is saved in the AutoCAD system variable **ANGDIR**.

If the angle has a default setting [Figure 12-8(a)], the getangle function will return 2.35619 radians for an angle of 135.

Examples
(setq ang (getangle "Enter angle"))   returns 2.35619 for an angle of 135 degrees

Figure 12-8(b) shows the new settings of the angle, where the Y-axis is 0 degrees and the angles measured clockwise are positive. The **getangle** function will return 3.92699 for an angle of 135 degrees. The getangle function calculates the angle in the counterclockwise direction, **ignoring the direction set in the system variable ANGDIR**, with respect to the angle base as set in the system variable ANGBASE [Figure 12-9(b)].

Chapter 12

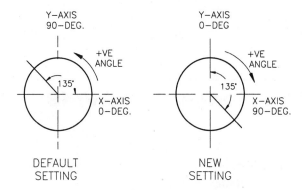

Figure 12-8(a)   Figure 12-8(b)

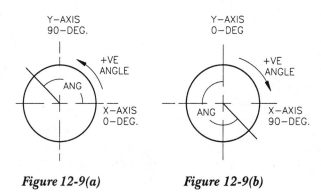

Figure 12-9(a)   Figure 12-9(b)

Examples
(setq ang (getangle "Enter angle"))   returns 3.92699

getorient

The **getorient** function pauses for you to enter the **angle**, then it returns the value of that angle in radians. The format of the **getorient** function is:

(getorient [point] [prompt])
                Where **point** ------------ First point of the angle
                            prompt ---------- Any prompt that needs to be displayed on the screen

Examples
(getorient)
(setq ang (getorient))
(setq ang (getorient pt1))
(setq ang (getorient "Enter taper angle"))
(setq ang (getorient pt1 "Enter second point of angle"))

The **getorient** function is just like the **getangle** function. Both return the value of the angle in radians. However, the **getorient** function always measures the angle with a positive X-axis (3 o'clock position) and in a counterclockwise direction. **It ignores the ANGBASE and ANGDIR settings**. If the settings have not been changed, as shown in Figure 12-10(a) (default settings for ANGDIR and ANGBASE), for an angle of 135 degrees the **getorient** function will return 2.35619 radians. If the settings are changed, as shown in Figure 12-10(b), for an angle of 135 degrees the **getorient** function will return 5.49778 radians. Although the settings have **been** changed where the angle is measured with the positive Y-axis and in a clockwise direction, the **getorient** function ignores the new settings and measures the angle from positive X-axis and in a counterclockwise direction.

Note

For the getangle and getorient functions you can enter the angle by typing the angle at the keyboard or by selecting two points on the screen. If the assignment is (setq ang (getorient pt1)), where the first point pt1 is already defined, you will be prompted to enter the second point. You can enter this point by selecting a point on the screen or by entering the coordinates of the second point.

180 degrees is equal to pi (3.14159) radians. To calculate an angle in radians, use the following relation:

Angle in radians = (pi x angle)/180

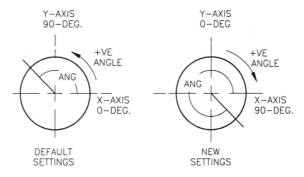

Figure 12-10(a)   *Figure 12-10(b)*

getint, getreal, getstring, AND getvar FUNCTIONS

getint

The **getint** function pauses for you to enter an integer. The function always returns an integer, even if the number you enter is a real number. The format of the **getint** function is:

(getint [prompt])
            Where **prompt** ---------- Optional prompt that you want to display on screen

Examples
(getint)
(setq numx (getint))

(setq numx (getint "Enter number of rows: "))
(setq numx (getint "\n Enter number of rows: "))

getreal

The **getreal** function pauses for you to enter a real number and it always returns a real number, even if the number you enter is an integer. The format of the **getreal** function is:

(getreal [prompt])
                    Where **prompt** ---------- Optional prompt that is displayed on screen

Examples
(getreal)
(setq realnumx (getreal))
(setq realnumx (getreal "Enter num1: "))
(setq realnumx (getreal "\n Enter num2: "))

getstring

The **getstring** function pauses for you to enter a string value, and it always returns a string, even if the string you enter contains numbers only. The format of the **getstring** function is:

(getstring [prompt])
                    Where **prompt** ---------- Optional prompt that is displayed on screen

Examples
(getstring)
(setq answer (getstring))
(setq answer (getstring "Enter Y for yes, N for no: "))
(setq answer (getstring "\n Enter Y for yes, N for no: "))

Note
The maximum length of the string is 132 characters. If the string exceeds 132 characters, the excess is ignored.

getvar

The **getvar** function lets you retrieve the value of an AutoCAD system variable. The format of the **getvar** function is:

(getvar "variable")
                    Where **variable** ---------- AutoCAD system variable name

Examples
(getvar)
(getvar "dimcen")   returns 0.09
(getvar "ltscale")   returns 1.0
(getvar "limmax")   returns 12.00,9.00
(getvar "limmin")   returns 0.00,0.00

Note

The system variable name should always be enclosed in double quotes.

You can retrieve only one variable value in one assignment. To retrieve the values of several system variables, use a separate assignment for each variable.

polar AND sqrt FUNCTIONS

polar

The **polar** function defines a point at a given angle and distance from the given point (Figure 12-11). The angle is expressed in radians, measured positive in the counterclockwise direction (assuming default settings for **ANGBASE** and **ANGDIR**). The format of the **polar** function is:

(polar point angle distance)
Where **point** ------------ Reference point
angle ------------ Angle the point makes with the referenced point
distance --------- Distance of the point from the referenced point

Examples
(polar pt1 ang dis)
(setq pt2 (polar pt1 ang dis))
(setq pt2 (polar '(2.0 3.25) ang dis))

sqrt

The **sqrt** function calculates the square root of a number, and the value this function returns is always a real number. The format of the **sqrt** function is:

(sqrt number)
Where **number** ---------- Number you want to find the square root of
(real or integer)

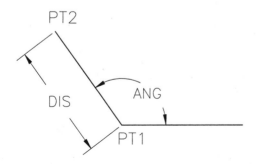

*Figure 12-11 Using the **polar** function to define a point*

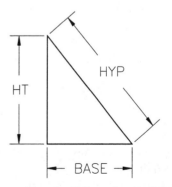

(setq hyp (sqrt (+ (* base base) (* ht ht))))
*Figure 12-12 Application of the **sqrt** function*

Examples

(sqrt 144)	returns 12.0
(sqrt 144.0)	returns 12.0
(setq x (sqrt 57.25))	returns 7.566373
(setq x (sqrt (* 25 36.5)))	returns 30.207615
(setq x (sqrt (/ 7.5 (cos 0.75))))	returns 3.2016035
(setq hyp (sqrt (+ (* base base) (* ht ht))))	

Example 4

Write an AutoLISP program that will draw an equilateral triangle outside a circle (Figure 12-13). The sides of the triangle are tangent to the circle. The program should prompt you to enter the radius and the center point of the circle.

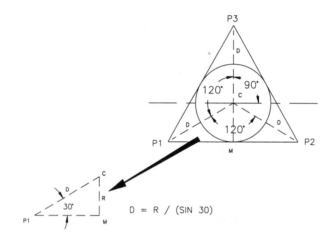

Figure 12-13 *Equilateral triangle outside a circle*

The following file is the listing of the AutoLISP program for Example 4.

```
;This program will draw a triangle outside
;the circle with the lines tangent to circle
:
(defun dtr (a)
  (* a (/ pi 180.0))
)
(defun c:trgcir(/ r c d p1 p2 p3)
(setvar "cmdecho" 0)
(graphscr)
  (setq r(getdist "\n Enter circle radius: "))
  (setq c(getpoint "\n Enter center of circle: "))
  (setq d(/ r (sin(dtr 30))))
  (setq p1(polar c (dtr 210) d))
  (setq p2(polar c (dtr 330) d))
  (setq p3(polar c (dtr 90) d))
```

```
(command "circle" c r)
(command "line" p1 p2 p3 "c")
(setvar "cmdecho" 1)
(princ)
)
```

Exercise 3   *General*

Write an AutoLISP program that will draw an isosceles triangle P1,P2,P3. The base of the triangle (P1,P2) makes an angle B with the positive X-axis (Figure 12-14). The program should prompt you to enter the starting point, P1, length L1, and angles A and B.

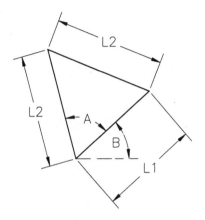

Figure 12-14 Isosceles triangle at an angle

Exercise 4   *General*

Write a program that will draw a slot with centerlines. The program should prompt you to enter slot length, slot width, and the layer name for the centerlines (Figure 12-15).

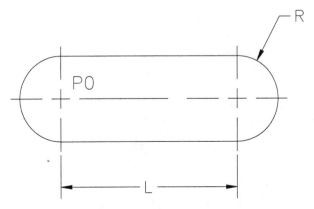

Figure 12-15 Slot of length L and radius R

itoa, rtos, strcase, AND prompt FUNCTIONS

itoa

The **itoa** function changes an integer into a string and returns the integer as a string. The format of the **itoa** function is:

(itoa number)

           Where **Number** --------- The integer number that you want to convert into a string

Examples
(itoa 89)   returns "89"
(itoa -356)   returns "-356"

(setq intnum 7)
(itoa intnum)   returns "7"

(setq intnum 345)
(setq intstrg (itoa intnum))   returns "345"

rtos

The **rtos** function changes a real number into a string and the function returns the real number as a string. The format of the **rtos** function is:

(rtos realnum)

           Where **Realnum** -------- The real number that you want to convert into a string

Examples
(rtos 50.6)   returns "50.6"
(rtos -30.0) returns "-30.0"
(setq realstrg (rtos 5.25))   returns "5.25"

(setq realnum 75.25)
(setq realstrg (rtos realnum))   returns "75.25"

The **rtos** function can also include mode and precision. The format of the **rtos** function with mode and precision is:

(rtos realnum [mode] [precision])

           Where **realnum** --------- Real number
                          mode ------------- Unit mode, like decimal, scientific
                          precision -------- Number of decimal places or denominator of fractional units

strcase

The **strcase** function converts the characters of a string into uppercase or lowercase. The format of the **strcase** function is:

(strcase string [true])
Where **String** ------------ String that needs to be converted to uppercase or
lowercase
True ------------- If it is not nil, all characters are converted to lowercase

The **true** is optional. If it is missing or if the value of **true** is nil, the string is converted to uppercase. If the value of true is not nil, the string is converted to lowercase.

Examples
(strcase "Welcome Home")   returns "WELCOME HOME"

(setq t 0)
(strcase "Welcome Home" t)   returns "welcome home"

(setq answer (strcase (getstring "Enter Yes or No: ")))

prompt
The **prompt** function is used to display a message on the screen in the command prompt area.
The contents of the message must be enclosed in double quotes. The format of the **prompt** func-
tion is:

(prompt message)
Where **message** --------- Message that you want to display on the screen

Examples
(prompt "Enter circle diameter: ")
(setq d (getdist (prompt "Enter circle diameter: ")))

Note
On a two-screen system, the prompt function displays the message on both screens.

Example 5

Write a program that will draw two circles of radii r1 and r2, representing two pulleys that are
separated by a distance d. The line joining the centers of the two circles makes an angle a with the
X-axis, as shown in Figure 12-16.

Input	Output
Radius of small circle - r1	Small circle of radius - r1
Radius of large circle - r2	Large circle of radius - r2
Distance between circles - d	Lines tangent to circles
Angle of center line - a	
Center of small circle - c1	

Process
1. Calculate distance x1, x2.
2. Calculate angle ang.

3. Locate point c2 with respect to point c1.
4. Locate points p1, p2, p3, p4.
5. Draw small circle with radius r1 and center c1.
6. Draw large circle with radius r2 and center c2.
7. Draw lines p1 to p2 and p3 to p4.

Calculations

$x1 = r2 - r1$
$x2 = SQRT\ [\ d \char`^ 2 - (r2 - r1) \char`^ 2]$
$tan\ ang = x1 / x2$
$ang = atan\ (x1 / x2)$
$a1a = 90 + a + ang$
$a1b = 270 + a - ang$
$a2a = 90 + a + ang$
$a2b = 270 + a - ang$

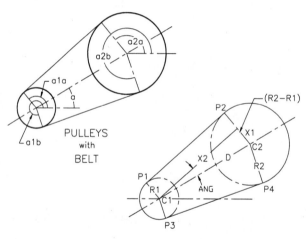

Figure 12-16 *Two circles with tangent lines*

The following file is a listing of the AutoLISP program for Example 5. **The line numbers on the right are not a part of the file. These numbers are for reference only.**

```
;This program draws a tangent (belt) over two        1
;pulleys that are separated by a given distance.      2
                                                      3
;This function changes degrees into radians           4
(defun dtr (a)                                        5
  (* a (/ pi 180.0))                                  6
  )                                                   7
;End of dtr function                                  8
```

```
;The belt function draws lines that are tangent to circles              9
(defun c:belt(/ r1 r2 d a c1 x1 x2 c2 p1 p2 p3 p4)                      10
  (setvar "cmdecho" 0)                                                  11
  (graphscr)                                                            12
  (setq r1(getdist "\n Enter radius of small pulley: "))               13
  (setq r2(getdist "\n Enter radius of larger pulley: "))              14
  (setq d(getdist "\n Enter distance between pulleys: "))              15
  (setq a(getangle "\n Enter angle of pulleys: "))                     16
  (setq c1(getpoint "\n Enter center of small pulley: "))              17
  (setq x1 (- r2 r1))                                                   18
  (setq x2 (sqrt (- (* d d) (* (- r2 r1) (- r2 r1)))))                 19
  (setq ang (atan (/ x1 x2)))                                          20
  (setq c2 (polar c1 a d))                                             21
  (setq p1 (polar c1 (+ ang a (dtr 90)) r1))                          22
  (setq p3 (polar c1 (- (+ a (dtr 270)) ang) r1))                     23
  (setq p2 (polar c2 (+ ang a (dtr 90)) r2))                          24
  (setq p4 (polar c2 (- (+ a (dtr 270)) ang) r2))                     25
                                                                        26
  ;The following line draw circles and lines                           27
  (command "circle" c1 p3)                                             28
  (command "circle" c2 p2)                                             29
  (command "line" p1 p2 "")                                            30
  (command "line" p3 p4 "")                                            31
  (setvar "cmdecho" 1)                                                 32
  (princ))                                                             33
```

Explanation

Line 5
(defun dtr (a)
In this line, the **defun** function defines a function, **dtr (a)**, that converts degrees into radians.

Line 6
(* a (/ pi 180.0))
(/ pi 180) divides the value of **pi** by 180, and the product is then multiplied by angle a (180 degrees is equal to **pi** radians).

Line 10
(defun c:belt(/ r1 r2 d a c1 x1 x2 c2 p1 p2 p3 p4)
In this line, the function **defun** defines a function, c:belt, that generates two circles with tangent lines.

Line 18
(setq x1 (- r2 r1))
In this line, the function **setq** assigns a value of r2 - r1 to variable x1.

Line 19
(setq x2 (sqrt (- (* d d) (* (- r2 r1) (- r2 r1)))))

In this line, **(- r2 r1)** subtracts the value of r1 from r2 and **(* (- r2 r1) (- r2 r1))** calculates the square of (- r2 r1). **(sqrt (- (* d d) (* (- r2 r1) (- r2 r1))))** calculates the square root of the difference, and **setq x2** assigns the product of this expression to variable x2.

Line 20
(setq ang (atan (/ x1 x2)))
In this line, **(atan (/ x1 x2))** calculates the arctangent of the product of **(/ x1 x2)**. The function **setq ang** assigns the value of the angle in radians to variable **ang**.

Line 21
(setq c2 (polar c1 a d))
In this line, **(polar c1 a d)** uses the **polar** function to locate point c2 with respect to c1 at a distance d and making an angle a with the positive X-axis.

Line 22
(setq p1 (polar c1 (+ ang a (dtr 90)) r1))
In this line, **(polar c1 (+ ang a (dtr 90)) r1))** locates point p1 with respect to c1 at a distance r1 and making an angle **(+ ang a (dtr 90))** with the positive X-axis.

Line 28
(command "circle" c1 p3)
In this line, the **Command** function uses the AutoCAD **CIRCLE** command to draw a circle with center c1 and a radius defined by the point p3.

Line 30
(command "line" p1 p2 "")
In this line, the **Command** function uses the AutoCAD **LINE** command to draw a line from p1 to p2. The pair of double quotes ("") at the end introduces a Return, which terminates the LINE command.

Exercise 5   *General*

Write an AutoLISP program that will draw two lines tangent to two circles, as shown in Figure 12-17. The program should prompt you to enter the circle diameters and the center distance between the circles.

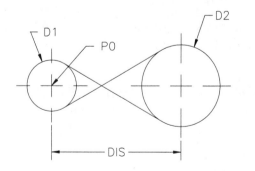

Figure 12-17 Circles with two tangent lines

FLOWCHARTS

A **flowchart** is a graphical representation of an algorithm. It can be used to analyze a problem systematically. It gives a better understanding of the problem, especially if the problem involves some conditional statements. It consists of standard symbols that represent a certain function in the program. For example, a rectangle is used to represent a process that takes place when the program is executed. The blocks are connected by lines indicating the sequence of operations. Figure 12-18 gives the standard symbols that can be used in a flowchart.

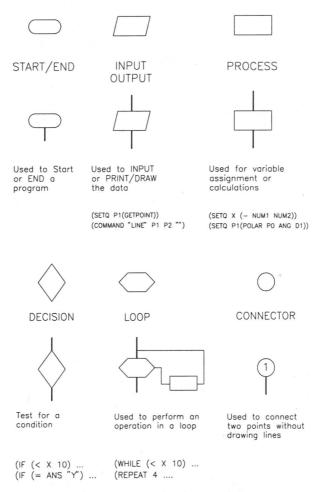

Figure 12-18 *Flowchart symbols*

CONDITIONAL FUNCTIONS

The relational functions discussed earlier in the chapter establish a relationship between two atoms. For example, (< x y) describes a test condition for an operation. To use such functions in a meaningful way, a conditional function is required. For example, (if (< x y) (setq z (- y x)) (setq z (-

x y))) describes the action to be taken when the condition is true (T) and when it is false (nil). If the condition is true, then z = y - x. If the condition is not true, then z = x - y. Therefore, conditional functions are very important for any programming language, including AutoLISP.

if

The **if** function (Figure 12-19) evaluates the first expression (then), if the specified condition returns "true". It evaluates the second expression (else), if the specified condition returns "nil". The format of the **if** function is:

(if condition then [else])

Where   **condition** ------- Specified conditional statement
            then ------------- Expression evaluated if the condition returns T
            else ------------- Expression evaluated if the condition returns nil

Examples

(if (= 7 7) ("true"))   returns "true"
(if (= 5 7) ("true") ("false"))   returns "false"

(setq ans "yes")
(if (= ans "yes") ("Yes") ("No"))   returns "Yes"

(setq num1 8)
(setq num2 10)
(if (> num1 num2)
  (setq x (- num1 num2))
  (setq x (- num2 num1))
)   returns 2

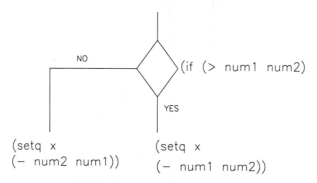

Figure 12-19 if function

Example 6

Write an AutoLISP program that will subtract a smaller number from a larger number. The program should also prompt you to enter two numbers.

Input
Number (num1)
Number (num2)

Output
x = num1 - num2
or
x = num2 - num1

Process
If num1 > num2 then x = num1 - num2
If num1 < num2 then x = num2 - num1

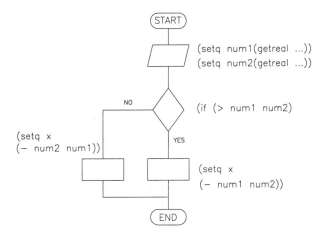

Figure 12-20 Flowchart for Example 6

The flowchart in Figure 12-20 describes the process involved in writing the program using standard flowchart symbols.

The following file is a listing of the program for Example 6. **The line numbers are not a part of the file; they are for reference only.**

```
;This program subtracts smaller number                    1
;from larger number                                       2
;                                                         3
(defun c:subnum( )                                        4
   (setvar "cmdecho" 0)                                   5
   (setq num1 (getreal "\n Enter first number: "))        6
   (setq num2 (getreal "\n Enter second number: "))       7
   (if (> num1 num2)                                       8
      (setq x (- num1 num2))                               9
      (setq x (- num2 num1))                              10
      )                                                   11
```

```
(setvar "cmdecho" 1)                                          12
(princ)                                                       13
)                                                             14
```

Explanation

Line 8
(if (> num1 num2)
In this line, the **if** function evaluates the test expression **(> num1 num2)**. If the condition is true, it returns **T**; if the condition is not true, it returns nil.

Line 9
(setq x (- num1 num2))
This expression is evaluated if the test expression **(if (> num1 num2)** returns T. The value of variable num2 is subtracted from num1, and the resulting value is assigned to variable x.

Line 10
(setq x (- num2 num1))
This expression is evaluated if the test expression **(if (> num1 num2)** returns nil. The value of variable num1 is subtracted from num2, and the resulting value is assigned to variable x.

Line 11
)
The closing parenthesis completes the definition of the if function.

Example 7

Write an AutoLISP program that will enable you to multiply or divide two numbers (Figure 12-21). The program should prompt you to enter the choice of multiplication or division. The program should also display an appropriate message if you do not enter the right choice. The following file is a listing of the AutoLISP program for Example 7:

```
;This program multiplies or divides two given numbers
(defun c:mdnum()
    (setvar "cmdecho" 0)
    (setq num1 (getreal "\n Enter first number: "))
    (setq num2 (getreal "\n Enter second number: "))
    (prompt "Do you want to multiply or divide. Enter M or D: ")
    (setq ans (strcase (getstring)))
    (if (= ans "M")
       (setq x (* num1 num2))
    )
    (if (= ans "D")
       (setq x (/ num1 num2))
    )
    (if (and (/= ans "D")(/= ans "M"))
       (prompt "Sorry! Wrong entry, Try again")
    )
```

```
(setvar "cmdecho" 1)
(princ))
```

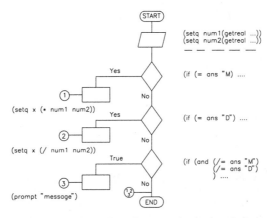

Figure 12-21 *Flow diagram for Example 7*

progn

The **progn** function can be used with the **if** function to evaluate several expressions. The format of **progn** function is:

(progn expression expression . . .)

The **if** function evaluates only one expression if the test condition returns "true". The **progn** function can be used in conjunction with the **if** function to evaluate several expressions.

Example
```
(if (= ans "yes")
   (progn
     (setq x (sin ang))
     (setq y (cos ang))
     (setq tanang (/ x y))
   ))
```

while

The **while** function (Figure 12-22) evaluates a test condition. If the condition is true (expression does not return nil), the operations that follow the while statement are repeated until the test expression returns nil. The format of the **while** function is:

(while testexpression operations)
Where **Testexpression** ----------- Expression that tests a condition
Operations --------------- Operations to be performed until the test expression returns nil

Example

```
(while (= ans "yes")
  (setq x (+ x 1))
  (setq ans (getstring "Enter yes or no: "))
)
(while (< n 3)
    (setq x (+ x 10))
    (setq n (1+ n))
)
```

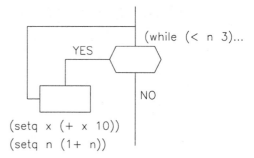

Figure 12-22 While function

Example 8

Write an AutoLISP program that will find the nth power of a given number. The power is an integer. The program should prompt you to enter the number and the nth power (Figure 12-23).

Input
Number x
nth power n

Output
product x^n

Process
1. Set the value of t = 1 and c = 1.
2. Multiply t * x and assign that value to the variable t.
3. Repeat the process until the counter c is less than or equal to n.

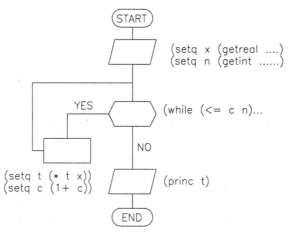

Figure 12-23 Flowchart for Example 8

The following file is a listing of the AutoLISP program for Example 8.

```
;This program calculates the nth
;power of a given number
(defun c:npower()
    (setvar "cmdecho" 0)
    (setq x(getreal "\n Enter a number: "))
    (setq n(getint "\n Enter Nth power-integer number: "))
    (setq t 1) (setq c 1)
    (while (<= c n)
      (setq t (* t x))
      (setq c (1+ c))
    )
    (setvar "cmdecho" 1)
    (princ t)
)
```

Example 9

Write an AutoLISP program that will generate the holes of a bolt circle (Figure 12-24). The program should prompt you to enter the center point of the bolt circle, the bolt circle diameter, the bolt circle hole diameter, the number of holes, and the start angle of the bolt circles.

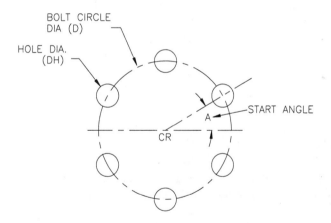

Figure 12-24 *Bolt circle with six holes*

```
;This program generates the bolt circles
;
(defun c:bcl( )
(graphscr)
(setvar "cmdecho" 0)
    (setq cr(getpoint "\n Enter center of Bolt-Circle: "))
    (setq d(getdist "\n Dia of Bolt-Circle: "))
    (setq n(getint "\n Number of holes in Bolt-Circle: "))
```

```
        (setq a(getangle "\n Enter start angle: "))
        (setq dh(getdist "\n Enter diameter of hole: "))
        (setq inc(/ (* 2 pi) n))
        (setq ang 0)
        (setq r (/ dh 2))
    (while (< ang (* 2 pi))
        (setq p1 (polar cr (+ a inc) (/ d 2)))
        (command "circle" p1 r)
        (setq a (+ a inc))
        (setq ang (+ ang inc))
        )
    (setvar "cmdecho" 1)
    (princ)
    )
```

repeat

The **repeat** function evaluates the expressions **n** number of times as specified in the **repeat** function (Figure 12-25). The variable **n** must be an integer. The format of the **repeat** function is:

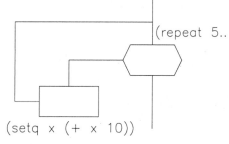

Figure 12-25 Repeat function

repeat n

Where **n** ------------------- n is an integer that defines the number of times the expressions are to be evaluated

Example
```
(repeat 5
  (setq x (+ x 10))
  )
```

Example 10

Write an AutoLISP program that will generate a given number of concentric circles. The program should prompt you to enter the center point of the circles, the start radius, and the radius increment (Figure 12-26).

The following file is a listing of the AutoLISP program for Example 10.
```
    ;This program uses the repeat function to draw
    ;a given number of concentric circles.
    (defun c:concir( )
    (graphscr)
    (setvar "cmdecho" 0)
    (setq c (getpoint "\n Enter center point of circles: "))
    (setq n (getint "\n Enter number of circles: "))
    (setq r (getdist "\n Enter radius of first circle: "))
    (setq d (getdist "\n Enter radius increment: "))
```

```
(repeat n
  (command "circle" c r)
  (setq r (+ r d))
  )
(setvar "cmdecho" 1)
(princ)
)
```

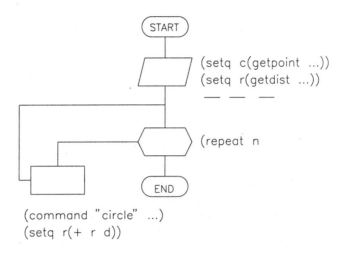

Note

*AutoCAD allows you to load the specified AutoLISP programs automatically, each time you start AutoCAD. For example, if you are working on a project and you have loaded an AutoLISP program, the program will autoload if you start another drawing. You can enable this feature by adding the name of the file to the Startup Suite of the **Load/Unload Application** dialog box. For details, see the Load/Unload Application discussed earlier in this chapter.*

Example 11

Write an AutoLISP program that will generate a flat layout drawing of a transition and then dimension the layout. The transition and the layout without dimensions are shown in Figure 12-27.

The following file is a listing of the AutoLISP program for Example 11. Programs do not need to be in lowercase letters. They can be uppercase or a combination of uppercase and lowercase.

```
;This program generates flat layout of
;a rectangle to rectangle transition
;
(defun c:TRANA(/)
(graphscr)
(setvar "cmdecho" 0)
```

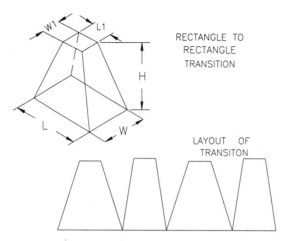

RECTANGLE TO
RECTANGLE
TRANSITION

LAYOUT OF
TRANSITON

Figure 12-27 *Flat layout of a transition*

```
(setq L (getdist "\n Enter length of bottom rectangle: "))
(setq W (getdist "\n Enter width of bottom rectangle: "))
(setq H (getdist "\n Enter height of transition: "))
(setq L1 (getdist "\n Enter length of top rectangle: "))
(setq W1 (getdist "\n Enter width of top rectangle: "))

(setq x1 (/ (- w w1) 2))
(setq y1 (/ (- l l1) 2))
(setq d1 (sqrt (+ (* h h) (* x1 x1))))
(setq d2 (sqrt (+ (* d1 d1) (* y1 y1))))
(setq s1 (/ (- l l1) 2))
(setq p1 (sqrt (- (* d2 d2) (* s1 s1))))
(setq s2 (/ (- w w1) 2))
(setq p2 (sqrt (- (* d2 d2) (* s2 s2))))

(setq t1 (+ l1 s1))
(setq t2 (+ l w))
(setq t3 (+ l s2 w1))
(setq t4 (+ l s2))
(setq pt1 (list 0 0))
(setq pt2 (list s1 p1))
(setq pt3 (list t1 p1))
(setq pt4 (list l 0))
(setq pt5 (list t4 p2))
(setq pt6 (list t3 p2))
(setq pt7 (list t2 0))
(command "layer" "make" "ccto" "c" "1" "ccto" "")
(command "line" pt1 pt2 pt3 pt4 pt5 pt6 pt7 "c")
```

```
(setq sf (/ (+ l w) 12))
(setvar "dimscale" sf)
(setq c1 (list 0 (- 0 (* 0.75 sf))))
(setq c7 (list (- 0 (* 0.75 sf)) 0))
(setq c8 (list (- l (* 0.75 sf)) 0))

(command "layer" "make" "cctd" "c" "2" "cctd" "")
(command "dim" "hor" pt1 pt2 c1 "" "base" pt3 "" "base" pt4 "" "exit")
(command "dim" "hor" pt4 pt5 c1 "" "base" pt6 "" "base" pt7 "" "exit")
(command "dim" "vert" pt1 pt2 pt2 "" "exit")
(command "dim" "vert" pt4 pt5 pt5 "" "exit")
(command "dim" "aligned" pt1 pt2 c7 "" "exit")
(command "dim" "aligned" pt4 pt5 c8 "" "exit")
(setvar "cmdecho" 1)
(princ))
```

Example 12

Write an AutoLISP program that can generate a flat layout of a cone as shown in Figure 12-28. The program should also dimension the layout.

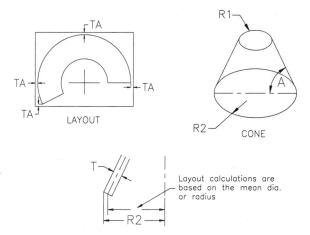

Figure 12-28 Flat layout of a cone

The following file is a listing of the AutoLISP program for Example 12:

```
;This program generates layout of a cone
;
;DTR function changes degrees to radians
(defun DTR (a)
 (* PI (/ A 180.0))
 )
;RTD Function changes radians to degrees
```

```
(defun rtd (a)
(* a (/ 180.0 pi))
)

(defun tan (a)
(/ (sin a) (cos a))
)
(defun c:cone-1p(/)
(graphscr)
(setvar "cmdecho" 0)
(setq r2 (getdist "\n enter outer radius at larger end: "))
(setq r1 (getdist "\n enter inner radius at smaller end: "))
(setq t (getdist "\n enter sheet thickness:-"))
(setq a (getangle "\n enter cone angle:-"))
;this part of the program calculates various parameters
;needed in calculating the strip layout
(setq x0 0)
(setq y0 0)
(setq sf (/ r2 3))
(setvar "dimscale" sf)
(setq ar a)
(setq tx (/ (* t (sin ar)) 2))
(setq rx2 (- r2 tx))
(setq rx1 (+ r1 tx))
(setq w (* (* 2 pi) (cos ar)))
(setq rl1 (/ rx1 (cos ar)))
(setq rl2 (/ rx2 (cos ar)))

;this part of the program calculates the x-coordinate
;of the points
  (setq x1 (+ x0 rl1)
      x3 (+ x0 rl2)
      x2 (- x0 (* rl1 (cos (- pi w))))
      x4 (- x0 (* rl2 (cos (- pi w))))
  )

;this part of the program calculates the y-coordinate
;of the points
  (setq y1 y0
      y3 y0
      y2 (+ y0 (* rl1 (sin (- pi w))))
      y4 (+ y0 (* rl2 (sin (- pi w))))
      )

  (setq p0 (list x0 y0)
      p1 (list x1 y1)
      p2 (list x2 y2)
```

```
        p3 (list x3 y3)
        p4 (list x4 y4)
        )
  (command "layer" "make" "ccto" "c" "1" "ccto" "")
  (command "arc" p1 "c" p0 p2)
  (command "arc" p3 "c" p0 p4)
  (command "line" p1 p3 "")
  (command "line" p2 p4 "")

  (setq f1 (/ r2 24))
  (setq f2 (/ r2 2))
  (setq d1 (list (+ x3 f2) y3))
  (setq d2 (list x0 (- y0 f2)))

(command "layer" "make" "cctd" "c" "2" "cctd" "")
(setvar "dimtih" 0)
(command "dim" "hor" p0 p1 d2 "" "baseline" p3 "" "baseline" p2 "" "baseline" p4 "" "exit")
(command "dim" "vert" p0 p2 d1 "" "baseline" p4 "" "exit")
(setvar "dimscale" 1)
(setvar "cmdecho" 1)
(princ)
)
```

Chapter 12

Review Questions

1. Evaluate the following AutoLISP functions:

    Command: (+ 2 30 5 50)   returns _____
    Command: (+ 2 30 4 55.0)   returns _____

    (- 20 40)   returns _____
    (- 30.0 40.0)   returns _____

    (* 72 5 3 2.0)   returns _____
    (* 7 -5.5)   returns _____

    (/ 299 -5)   returns _____
    (/ -200 -9.0)   returns _____

    (1- 99)   returns _____
    (1- -18.5)   returns _____

(abs -90)   returns _____
(abs -27.5)   returns _____

(sin pi)   returns _____
(sin 1.5)   returns _____
(cos pi)   returns _____
(cos 1.2)   returns _____
(atan 1.1 0.0)   returns _____ radians
(atan -0.4 0.0)   returns _____ radians
(angtos 1.5708 0 5)   returns _____
(angtos -1.5708 0 3)   returns _____

(< "x" "y")   returns _____
(>= 80 90 79)   returns _____

2. The **setq** function is used to assign a value to _____.

3. The _____ function pauses to enable you to enter the X, Y coordinates or X, Y, Z coordinates of a point.

4. The _____ function is used to execute standard AutoCAD commands from within an AutoLISP program.

5. In an AutoLISP expression, the AutoCAD command name and the command options have to be enclosed in double quotation marks. (T/F).

6. The getdist function pauses for you to enter a _____ and it then returns the distance as a real number.

7. The _____ function assigns a value to an AutoCAD system variable. The name of the system variable must be enclosed in _____.

8. The **cadr** function performs two operations, _____ and _____, to return the second element of the list.

9. The _____ function prints a new line on the screen just as \n.

10. The _____ function pauses for you to enter the angle, then it returns the value of that angle in radians.

11. The _____ function always measures the angle with a positive X-axis and in a counter-clockwise direction.

12. The _____ function pauses for you to enter an integer. The function always returns an integer, even if the number that you enter is a real number.

13. The _____ function lets you retrieve the value of an AutoCAD system variable.

14. The _____ function defines a point at a given angle and distance from the given point (Figure 12-10).

15. The _____ function calculates the square root of a number and the value this function returns is always a real number.

16. The _____ function changes a real number into a string and the function returns the real number as a string.

17. The **if** function evaluates the test expression (**> num1 num2**). If the condition is true it returns _____, if the condition is not true, it returns_____.

18. The _____ function can be used with the **if** function to evaluate several expressions.

19. The **while** function evaluates the test condition. If the condition is true (expression does not return nil) the operations that follow the while statement are _____ until the test expression returns _____.

20. The **repeat** function evaluates the expressions n number of times as specified in the **repeat** function. The variable n must be a real number. (T/F).

Exercises

Exercise 6   *General*

Write an AutoLISP program that will draw three concentric circles with center C1 and diameters D1, D2, D3 (Figure 12-29). The program should prompt you to enter the coordinates of center point C1 and the circle diameters D1, D2, D3.

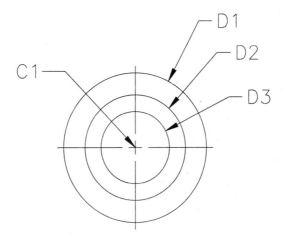

Figure 12-29 *Three concentric circles with diameters D1, D2, D3*

Exercise 7   *General*

Write an AutoLISP program that will draw a line from point P1 to point P2 (Figure 12-30). Line P1, P2 makes an angle A with the positive X-axis. Distance between the points P1 and P2 is L. The diameter of the circles is D1 (D1 = L/4).

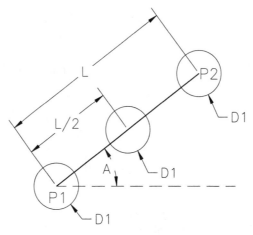

Figure 12-30 *Circles and line making an angle A with X-axis*

Exercise 8   *General*

Write an AutoLISP program that will draw an isosceles triangle P1, P2, P3 (Figure 12-31). The program should prompt you to enter the starting point P1, length L1, and the included angle A.

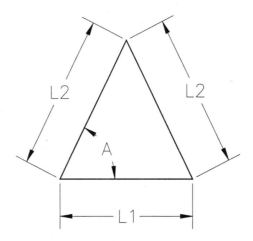

Figure 12-31 *Isosceles triangle*

Exercise 9   *General*

Write an AutoLISP program that will draw a parallelogram with sides S1, S2 and angle W as shown in Figure 12-32. The program should prompt you to enter the starting point PT1, lengths S1, S2, and the included angle W.

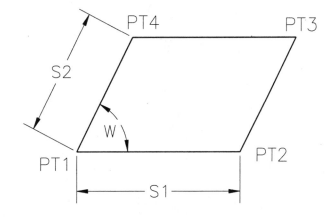

Figure 12-32 Parallelogram with sides S1, S2, and angle W

Exercise 10   *General*

Write an AutoLISP program that will draw a square of sides S and a circle tangent to the four sides of the square as shown in Figure 12-33. The base of the square makes an angle, ANG, with the positive X-axis. The program should prompt you to enter the starting point P1, length S, and angle ANG.

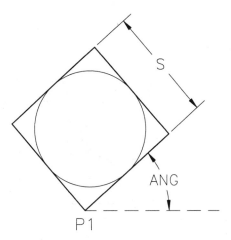

Figure 12-33 Square of side S at an angle ANG.

Chapter 12

Exercise 11   *General*

Write an AutoLISP program that will draw an equilateral triangle inside the circle (Figure 12-34). The program should prompt you to enter the radius and the center point of the circle.

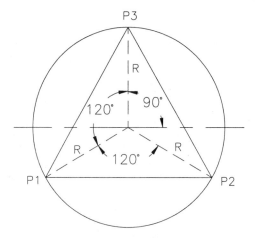

Figure 12-34 Equilateral triangle inside a circle

Exercise 12   *General*

Write an AutoLISP program that will delete all objects contained within the upper (limmax) and lower (limmin) limits. Use AutoCAD's SETVAR and ERASE commands to delete the objects.

Exercise 13   *General*

Write an AutoLISP program that will draw two lines tangent to two circles as shown in Figure 12-35. The program should prompt you to enter the circle diameters and the center distance between the circles.

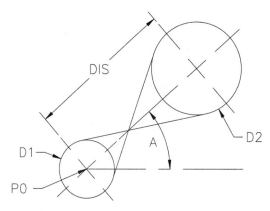

Figure 12-35 Circle with tangent lines at angle A

Exercise 14

Write a program that will draw a slot with center lines. The program should prompt you to enter slot length, slot width, and the layer name for center lines (Figure 12-36).

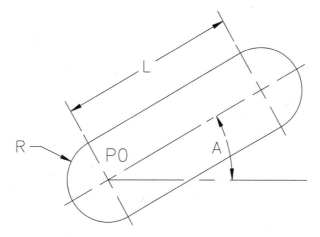

Figure 12-36 Slot of length L and radius R

Exercise 15

Write an AutoLISP program that will draw a line and then generate a given number of lines (N), parallel to the first line (Figure 12-37).

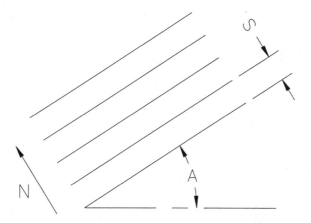

Figure 12-37 N number of lines offset at a distance S

Exercise 16   *General*

Write an AutoLISP program to draw a circle with center lines. The program should prompt for the diameter of circle, center of circle, and the angle of center lines as shown in Figure 12-38.

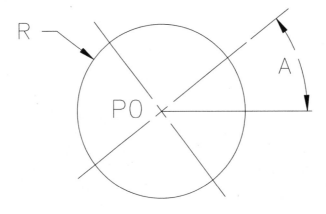

Figure 12-38 Circle with center lines at an angle A

Exercise 17   *General*

Write a program to draw a keyway slot. The program should prompt you to enter the width of slot, depth of slot, angle of slot, and starting point (Figure 12-39).

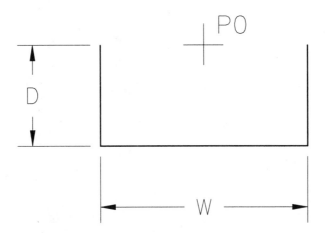

Figure 12-39 Keyway slot of width W and depth D

Exercise 18   *General*

Write an AutoLISP program that will draw the figure as shown in Figure 12-40 with center lines and dimensions. Assume L5=D1, L3=1.5*D1, L6=10*D1, L1= L6-D1, L4=L3+D1.

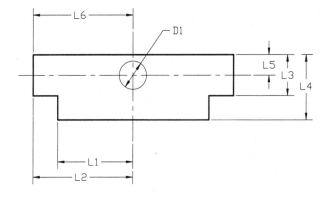

Figure 12-40 *Drawing for Exercise 18*

Exercise 19   *General*

Write an AutoLISP program that will draw a hub with the key slot as shown in Figure 12-41. The program should prompt the user to enter the values for P0 (Center of hub/shaft, D1 (Diameter of shaft), D2 (Outer diameter of hub), W (Width of key), and H (Height of key). The program should also draw the center lines in the Center layer (Green color) and draw dimension T and W in the Dim layer (Magenta color).

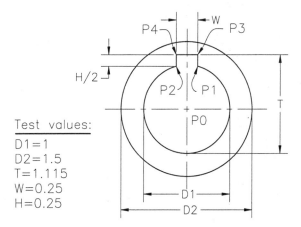

Figure 12-41 *Drawing for Exercise 19*

Exercise 20   *General*

Write an AutoLISP program that will draw the two views of a bushing as shown in Figure 12-42. The program should prompt you to enter the starting point P0, lengths L1, L2, and the bushing diameters ID, OD, HD. The distance between the front view and the side view of bushing is DIS (DIS = 1.25 * HD). The program should also draw the hidden lines in the HID layer and center lines in the CEN layer. The center lines should extend 0.75 units beyond the object line.

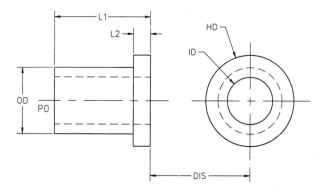

Figure 12-42 *Two views of bushing*

Chapter **13**

AutoLISP: Editing
the Drawing Database

Learning Objectives

After completing this chapter, you will be able to:
- *Edit the drawing database using AutoLISP.*
- *Use the* **ssget, sslength, ssname, entget, assoc, cons, subst,** *and* **entmod** *functions.*
- *Retrieve information from the drawing database.*
- *Edit the database and substitute the values back into the drawing database.*

EDITING THE DRAWING DATABASE

In addition to writing programs to create new commands, you can use the AutoLISP programming language to edit the drawing database. This is a powerful tool to make changes in a drawing.
For example, you can write a program that will delete all text objects in the drawing or change the
layer and color of all circles just by entering one command. Once you understand how AutoCAD
stores the information of the drawing objects and how it can be retrieved and edited, you can
manipulate the database any way you want to, limited only by your imagination.

This chapter discusses some of the commands that are frequently used to edit the drawing database. For other commands, not discussed in this section, please refer to the online reference manual,
AutoLISP Programmers Reference, published by Autodesk.

ssget

The **ssget** function enables you to select any number of objects in a drawing. The object selection
modes (window, crossing, previous, last, etc.) and the points that define the corners of the window
can be included in the **ssget** assignment. The format of the **ssget** function is:

(ssget [<u>selection-mode</u>] [point1 point2])

Where **selection-mode** ----------- Object selection mode (w,c,l,p,etc.)
       point 1 --------------------- First point of window (optional)
       point 2 --------------------- Second point of window (optional)

Examples

(ssget)   For general object selection
(ssget "L")   For selecting last object
(ssget "p")   For selecting previous selection set
(ssget "w" (list 0 0) (list 12.0 9.0))   Object selection using window object selection mode,
                                           where the window is defined by points 0,0 and 12.0,9.0
(ssget "c" pt1 pt2)   Object selection using crossing object selection mode,
                                           where the window is defined by predefined points pt1
                                           and pt2

Example 1

Write an AutoLISP program that will erase all objects within the drawing limits, limmax, and limmin. Use the **ssget** function to select the objects.

The following file is a listing of the AutoLISP program for Example 1. The line numbers are not a part of the program; they are shown here for reference only.

```
;This program will delete all objects                    1
;that are within the drawing limits                      2
;                                                         3
(defun c:delall()                                        4
   (setvar "cmdecho" 0)                                  5
   (setq pt1 (getvar "limmin"))                          6
   (setq pt2 (getvar "limmax"))                          7
   (setq ss1 (ssget "c" pt1 pt2))                        8
   (command "erase" ss1 "")                              9
   (command "redraw")                                    10
   (setvar "cmdecho" 1)                                  11
   (princ)                                               12
)                                                        13
```

Lines 1-3
The first three lines are comment lines that describe the function of the program. Notice that all comment lines start with a semicolon (;).

Line 4
(defun c:delall()
In this line the **defun** function defines function delall.

Line 6
(setq pt1 (getvar "limmin"))

The **getvar** function secures the value of the lower left corner of the drawing limits (.i.limmin;) and the **setq** function assigns that value to variable pt1.

Line 7
(setq pt2 (getvar "limmax"))
The **getvar** function secures the value of the upper right corner of the drawing limits (limmax) and the **setq** function assigns that value to variable pt2.

Line 8
(setq ss1 (ssget "c" pt1 pt2))
The **ssget** function uses the "crossing" objection selection mode to select the objects that are within or touching the window defined by points pt1 and pt2. The **setq** function then assigns this object selection set to variable ss1.

Line 9
(command "erase" ss1 "")
The **command** function uses AutoCAD's **ERASE** command to erase the predefined object selection set ss1.

Line 10
(command "redraw")
In this line the **command** function uses AutoCAD's **REDRAW** command to redraw the screen and get rid of the blip marks left after erasing the objects.

ssget "X"

The **ssget "X"** function enables you to select specified types of objects in the entire drawing database, even if the layers are frozen or turned off. The format of the **ssget "X"** function is:

(ssget "X" specified-criteria)
        Where **X** ----------------------------- Filter mode of the ssget function
                specified-criteria -------- List of the specified criteria for selecting the objects

Examples

(ssget "X" (list (cons 0 "TEXT")))	returns a selection set that consists of all TEXT objects in the drawing
(ssget "X" (list (cons 7 "ROMANC")))	returns a selection set that consists of all TEXT objects in the drawing with the text style name ROMANC
(ssget "X" (list (cons 0 "LINE")))	returns a selection set that consists of all LINE objects in the drawing
(ssget "X" (list (cons 8 "OBJECT")))	returns a selection set that consists of all objects in the OBJECT layer

The **ssget "X"** function can contain more than one selection criteria. This option can be used to select a specific set of objects in a drawing. For example, if you want to select the LINE objects in the OBJECT layer, there are two selection criteria. The first is that the object has to be a LINE; the second is that the LINE object has to be in the OBJECT layer. As shown in Example 1, these two selection criteria can be combined to filter out the objects that satisfy these two conditions.

> (ssget "X" (list (cons 0 "LINE")(cons 8 "OBJECT")))

Group Codes for ssget "X"

The following table is a list of AutoCAD group codes that can be used with the function ssget **"X"**:

Group Code	Code Function
0	Object type
2	Block name for block reference
3	Dimension object DIMSTYLE name
6	Linetype name
7	Text style name
8	Layer name
38	Elevation
39	Thickness
62	Color number
66	Attributes
210	3D extrusion direction

Example 2

Write an AutoLISP program that will erase all text objects in a drawing on a specified layer. Use the filter option of the **ssget** function (ssget "X") to select text objects in the specified layer.

The following file is a listing of the AutoLISP program for Example 2:

```
;This program will delete all text
;in the user-specified layer
;
(defun c:deltext()
   (setvar "cmdecho" 0)
   (setq layer (getstring "\n Enter layer name: "))
   (setq ss1 (ssget "x" (list (cons 8 layer) (cons 0 "text"))))
   (command "erase" ss1 "")
   (command "redraw")
   (setvar "cmdecho" 1)
   (princ))
```

sslength

The **sslength** function determines the number of objects in a selection set and returns an integer corresponding to the number of objects found. The format of the **sslength** function is:

(sslength selection-set)
> Where **selection-set** --- Name of the selection set

Examples

(setq ss1 (ssget))
(setq num (sslength ss1))   returns the number of objects in the predefined selection set ss1

(setq ss2 (ssget "l"))
(setq num (sslength ss2))   returns the number of objects (1) in selection set ss2, where selection set ss2 has been defined as the last object in the drawing

ssname

The **ssname** function returns the name of the object, from a predefined selection set, as referenced by the index that designates the object number. The name of the object returned by this function is in the hexadecimal format (such as 60000014). The format of the **ssname** function is:

(ssname selection-set index)
> Where **selection-set** --- A predefined selection set
> **index** ------------- Index designates the object number in a selection set

Examples

(setq ss1 (ssget))
(setq index 0)
(setq entname (ssname ss1 index))   returns the name of the first object contained in the pre defined selection set ss1

Note

If the index is 0, the ssname function returns the name of the first object in the selection set. Similarly, if the index is 1, it returns the name of the second object.

entget

The **entget** function retrieves the object list from the object name. The name of the object can be obtained by using the function ssname. The format of the **entget** function is:

(entget object-name)
> Where **object-name** ---- Name of the object obtained by ssname

Examples

(setq ss1 (ssget))
(setq index 0)
(setq entname (ssname ss1 index))
(setq entlist (entget entname))   returns the list of the first object from the variable, entname, and assigns the list to the variable, entlist

assoc

The **assoc** function searches for a specified code in the object list and returns the element that contains that code. The format of the **assoc** function is:

(assoc code <u>object-list</u>)

      Where **code** -------------- AutoCAD's object code
           object-list ------- List of an object

Examples

(setq ss1 (ssget))
(setq index 0)
(setq entname (ssname ss1 index))
(setq entlist (entget entname))
(setq entasso (assoc 0 entlist))   returns the element associated with AutoCAD's object code
                   0, from the list defined by the variable entlist

cons

The **cons** function constructs a new list from the given elements or lists. The format of the **cons** function is:

(cons <u>first-element</u> <u>second-element</u>)

      Where **first-element** -------------- First element or list
           second-element ---------- Second element or list

Examples

(cons 'x 'y)   returns (X . Y)
(cons '(x y) 'z)   returns ((X Y) . Z)
(cons '(x y z) '(0.5 5.0))   returns ((X Y Z) 0.5 5.0)

subst

The **subst** function substitutes the new item in place of old items. The old items can be a single item or multiple items, provided they are in the same list. The format of the **subst** function is:

(subst <u>new-item</u> <u>old-item</u> <u>object-list</u>)

      Where **new-item** -------- New item that will replace old items
          old-item --------- Old items that are to be replaced
          object-list ------- Object list name

Examples

(setq entlist '(x y x))
(setq newlist (subst '(z) '(x) entlist))   returns (z y z); the subst function replaces x in
                       the object list (entlist) by z

entmod

The **entmod** function updates the drawing by writing the modified list back to the drawing database. The format of the **entmod** function is:

(entmod <u>object-list</u>)
                Where **object-list** ------- Name of the modified object list

Example 3

Write an AutoLISP program that will enable you to change the height of a text object. The program should prompt you to enter the new height of the text.

<u>Input</u>
New text height
Text object

<u>Output</u>
Text with new text height

Process

1. Select the text object; obtain the name of the object using the function **ssname**.
2. Extract the list of the object using the function **entget**.
3. Separate the element associated with AutoCAD object code 0 from the list using the function **assoc**.
4. Construct a new element where the height of text is changed to a new height using the function **cons**.
5. Substitute the new element back into the original list using the **subst** function.
6. Update the drawing database using the **entmod** function.

The following file is a listing of the AutoLISP program for Example 3. The line numbers are not a part of the program; they are shown here for reference only.

```
;This program changes the height of the                    1
;selected text, only one text at a time.                   2
;                                                          3
(defun c:chgtext1()                                        4
(setvar "cmdecho" 0)                                       5
(setq newht (getreal "\n Enter new text height: "))        6
(setq ss1 (ssget))                                         7
(setq name (ssname ss1 0))                                 8
(setq ent (entget name))                                   9
(setq oldlist (assoc 40 ent))                             10
(setq conlist (cons (car oldlist) newht))                 11
(setq newlist (subst conlist oldlist ent))                12
(entmod newlist)                                          13
(setvar "cmdecho" 1)                                      14
(princ)                                                   15
)                                                         16
```

Chapter 13

HOW THE DATABASE IS RETRIEVED AND EDITED

To change the objects in a drawing you need to understand the structure of the drawing database and how it can be manipulated. Once you understand this concept, it is easy and sometimes fun to edit the drawing database and the drawing. The following step-by-step explanation describes the process involved in changing the height of a selected text object in a drawing. Assume that the text that needs to be edited is "CHANGE TEXT" and that this text is already drawn on the screen. The height of the text is 0.3 units. Before going through the following steps, load the AutoLISP program from Example 3, and run it so that the variables are assigned a value.

Step 1
Select the text using the function **ssget** or **ssget "X"** and assign it to variable ss1. AutoCAD creates a selection set that could have one or more objects. In line 7 **(setq ss1 (ssget))** of the program for Example 3, the selection set is assigned to variable ss1. Use the following command to check the variable ss1:

> Command: **!ss1**
> <Selection set: 2>

Step 2
There could be several objects in a selection set and these objects need to be separated, one at a time, before any change is made to an object. This is made possible using the function **ssname** which extracts the name of an object. The index number used in the function **ssname** determines the object whose name is being extracted. For example, if the index is 0 the **ssname** function will extract the name of the first object, if the index is 1 the **ssname** function will extract the name of second object, and so on. In line 8 **(setq name (ssname ss1 0))** of the program, the **ssname** extracts the name of the first object and assigns it to the variable name. Use the following command to check the variable, name:

> Command: **!name**
> <Object name: 60000018>

Step 3
Extract the object list using the function **entget**. In line 9 **(setq ent (entget name))** of the program, the value of the list has been assigned to the variable ent. Use the following command to check the value of the variable ent.

> Command: **!ent**
> ((-1.<Object name: 600000018> (0 . "TEXT") (8 . "0") (10 4.91227 5.36301 0.0) **(40 . 0.3)**
> (1 . "CHANGE TEXT") (50 .0.0) (41 . 1.0) (51 .0.0) (7 . "standard") (71 .0)) (72 . 1) (11
> 6.51227 5.36302 0.0) (210 0.0 0.0 1.0))

This list contains all the information about the selected text object (CHANGE TEXT), but you are only interested in changing the height of the text. Therefore, you need to identify the element that contains the information about the text height (40 . 0.3) and separate that from the list.

Step 4
Use the function **assoc** to separate the element that is associated with code 40 (text height). The

statement in line 10 **(setq oldlist (assoc 40 ent))** of the program uses the assoc function to separate the value and assign it to the variable oldlist. Use the following command to check the value of this variable:

Command: **!oldlist**
(40 . 0.3)

Step 5

The **(40 . 0.3)** element consists of the code for text (40), and the text height (0.3). To change the height of the old text, the text height value (0.3) needs to be replaced by the new value. This is accomplished by constructing a new list as described in line 11 **(setq conlist (cons (car oldlist) newht))** of the program. This line also assigns the new element to variable conlist. For example, if the value assigned to variable newht is 0.5, the new element will be (40 . 0.5). Use the following command to check the value of conlist.

Command: **!conlist**
(40 . 0.5)

Step 6

After constructing the new element, use the subst function to substitute the new element back into the original list, ent. This is accomplished by line 12 **(setq newlist (subst conlist oldlist ent))** of the program. Use the following command to check the value of the variable newlist:

Command: **!newlist**
((-1.<Object name: 600000018> (0 . "TEXT") (8 . "0") (10 4.91227 5.36301 0.0) **(40 . 0.5)**
(1 . "CHANGE TEXT") (50 .0.0) (41 . 1.0) (51 .0.0) (7 . "standard") (71 .0)) (72 . 1) (11
6.51227 5.36302 0.0) (210 0.0 0.0 1.0))

Step 7

The last step is to update the drawing database. Do this by using the function **entmod** as shown in line 13 **(entmod newlist)** of the program.

Example 4

Write an AutoLISP program that will enable you to change the height of all text objects in a drawing (Figure 13-1). The program should prompt you to enter the new text height.

The following file is a listing of the AutoLISP program for this example. The line numbers are not a part of the program they are for reference only.

```
;This program changes the height of                              1
;all text objects in a drawing.                                  2
;                                                                 3
(defun c:chgtext2()                                              4
    (setvar "cmdecho" 0)                                        5
    (setq newht (getreal "\n Enter new text height: "))         6
    (setq ss1 (ssget "x" (list (cons 0 "text"))))               7
```

Chapter 13

```
      (setq index 0)                                         8
      (setq num (sslength ss1))                              9
      (repeat num                                            10
         (setq name (ssname ss1 index))                      11
         (setq ent (entget name))                            12
         (setq oldlist (assoc 40 ent))                       13
         (setq conlist (cons (car oldlist) newht))           14
         (setq newlist (subst conlist oldlist ent))          15
         (entmod newlist)                                    16
         (setq index (1+ index))                             17
      )                                                      18
      (setvar "cmdecho" 1)                                   19
      (princ)                                                20
   )                                                         21
```

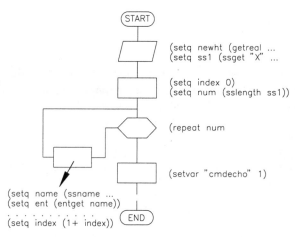

Figure 13-1 *Flowchart for Example 4*

Line 7
(setq ss1 (ssget "x" (list (cons 0 "text"))))
The **ssget "X"** function filters the text objects from the drawing database. The **setq** function assigns that selected set of text objects to variable **ss1**.

Line 8
(setq index 0)
The **setq** function sets the value of the **index** variable to 0. This variable is used later to select different objects.

Line 9
(setq num (sslength ss1))
The function **sslength** determines the number of objects in the selection set ss1 and the setq function assigns that number to the **num** variable.

Line 10
(repeat num
The **repeat** function will repeat the processes defined within the repeat function **num** number of times.

Example 5

Write an AutoLISP program that will enable you to change the height of the selected text objects in a drawing (Figure 13-2). The program should prompt you to enter the new text height.

```
;This program changes the height of the
;selected text objects.
(defun c:chgtext3()
(setvar "cmdecho" 0)
(setq newht (getreal "\n Enter new text height: "))
(setq ss1 (ssget))
(setq index 0)
(setq num (sslength ss1))
(repeat num
  (setq name (ssname ss1 index))
  (setq ent (entget name))
  (setq ass (assoc 0 ent))
  (setq index (1+ index))
  (If (= "TEXT" (cdr ass))
      (progn
      (setq oldlist (assoc 40 ent))
      (setq conlist (cons (car oldlist) newht))
      (setq newlist (subst conlist oldlist ent))
      (entmod newlist)
      )
    )
  )
(setvar "cmdecho" 1)
(princ)
)
```

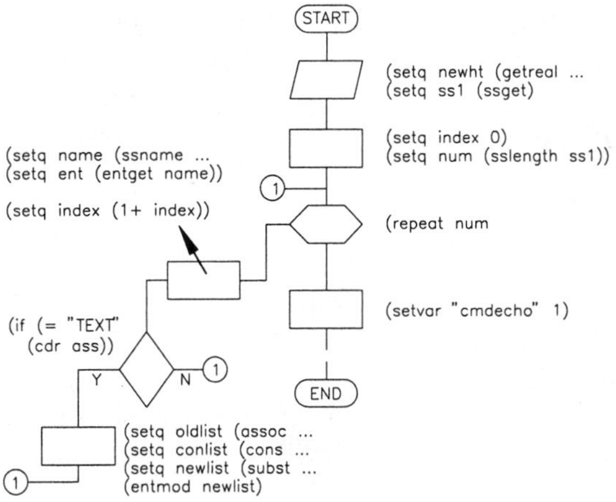

Figure 13-2 Flowchart of Example 5

Review Questions

1. In addition to writing programs to create new commands, you can use the AutoLISP programming language to edit the drawing database. (T/F) (T)

2. The _____ ssget function enables you to select any number of objects in a drawing.

3. The _____ ssget "X" function enables you to select specified types of objects in the entire drawing database, even if the layers are frozen or turned off.

4. The _____ sslength function determines the number of objects in a selection set and returns an integer corresponding to the number of objects found.

5. The _____ ssname function returns the name of the object, from a predefined selection set, as referenced by the index that designates the object number.

6. The _____ entget function retrieves the object list from the object name.

7. The _____ assoc function searches for a specified code in the object list and returns the element that contains that code.

8. The _____ cons function constructs a new list from the given elements or lists.

9. The _____ subst function substitutes the new item in place of old items.

Exercises

Exercise 1   *General*

Write an AutoLISP program that will enable you to change the layer of the selected objects in a drawing. The program should prompt you to enter the new layer name.

Exercise 2   *General*

Write an AutoLISP program that will change the text style name of the selected text objects in a drawing. The program should prompt you to enter the new text style.

Exercise 3   *General*

Write an AutoLISP program that will change the layer of the selected objects in a drawing to a new layer. You should be able to enter the new layer by selecting an object in that layer.

Chapter **14**

Visual LISP

Learning Objectives

After completing this chapter, you will be able to:
- *Start Visual LISP in AutoCAD.*
- *Use the Visual LISP Text Editor.*
- *Load and run Visual LISP programs.*
- *Load exisiting AutoLISP files.*
- *Use the Visual LISP Console.*
- *Use the Visual LISP formatter and change formatting options.*
- *Debug a Visual LISP program and trace the variables.*

VISUAL LISP

Visual LISP is another step forward to extend the customizing power of AutoCAD. Since AutoCAD Release 2.0 in the mid 1980s, AutoCAD users have used AutoLISP to write programs for their applications, architecture, mechanical, electrical, air conditioning, civil, sheet metal, and hundreds of other applications. AutoLISP has some limitations. For example, when you use a text editor to write a program, it is difficult to locate and check parentheses, AutoLISP functions, and variables. Debugging is equally problematic, because it is hard to find out what the program is doing and what is causing an error in the program. Generally, programmers will add some statements in the program to check the values of variables at different stages of the program. Once the program is successful, the statements are removed or changed into comments. Balancing parentheses, and formatting the code are other problems with traditional AutoLISP programming.

Visual LISP has been designed to make programming easier and efficient. It has a powerful text editor and a formatter. The text editor allows color coding of parentheses, function names, variables and other components of the program. The formatter formats the code in a easily readable format. It also has a watch facility that allows you to watch the values of variables and expressions.

Visual LISP has an interactive and intelligent console that has made programming easier. It also provides context sensitive help for AutoLISP functions and apropos features for symbol name search. The debugging tools have made it easier to debug the program and check the source code. These and other features have made Visual LISP the preferred tool for writing LISP programs. Visual LISP works with AutoCAD and it contains its own set of windows. AutoCAD must be running when you are working with Visual LISP.

Overview of Visual LISP

In this overview you will learn how to start Visual LISP, use the Visual LISP Text Editor to write a LISP Program, and then how to load and run the programs. As you go through the following steps, you will notice that Visual LISP has some new capabilities and features that are not available in AutoLISP.

Starting Visual LISP

Pull-down:	Tools > AutoLISP > Visual LISP Editor
Command:	VLIDE

1. Start AutoCAD.

2. In the **Tools** menu select **AutoLISP** and then select **Visual LISP Editor**. You can also start Visual LISP by entering the **VLIDE** command at the Command prompt. AutoCAD displays the Visual LISP window, Figure 14-1. If it is the first time, in the current drawing session, you are loading Visual LISP, the Trace window might appear displaying information about the current release of Visual LISP and errors that might be encountered when loading Visual LISP. The Visual LISP window has four areas; Menu, Toolbars, Console window, and Status bar, Figure 14-1. The following is a short description of these areas:

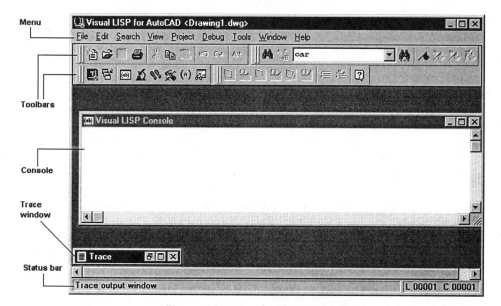

Figure 14-1 *Visual LISP window*

Menu. The menu bar is at the top of the Visual LISP window and lists different menu items. If you select a menu, AutoCAD displays the items in that menu. Also, the function of the menu item is displayed in the status bar that is located at the bottom of the Visual LISP window.

Toolbars: The toolbars provide a quick and easy way to invoke Visual LISP commands. Visual LISP has 5 toolbars; Debug, Edit, Find, Inspect, and Run. Depending on the window that is active, the display of the toolbars changes. If you move the mouse pointer over a tool and position it there for few seconds, a tool tip is displayed that indicates the function of that tool. A more detailed description of the tool appears in the status bar.

Console window. The Visual LISP Console window is contained within the Visual LISP window. It can be scrolled and positioned anywhere in the Visual LISP window. The Console window can be used to enter AutoLISP or Visual LISP commands. For example, if you want to add two numbers, enter (+ 2 9.5) next to $ sign and then press the **Enter** key. Visual LISP will return the result and display it in the same window.

Status bar. When you select a menu item or a tool in the toolbar, Visual LISP displays the related information in the status bar that is located near the bottom of the Visual LISP window.

Using the Visual LISP Text Editor
1.   Select **New File** from the **File** menu. The Visual LISP Text Editor window is displayed, Figure 14-2. The default file name is Untitled-0 as displayed at the top of the editor window.

2.   To activate the editor, click anywhere in the Visual LISP Text Editor.

3.   Type the following program and notice the difference between the Visual LISP Text Editor and the text editors you have been using to write AutoLISP programs (like Notepad).

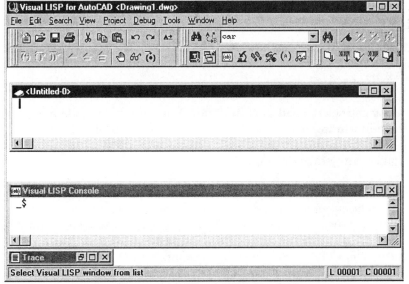

Figure 14-2 Visual LISP Text Editor

```
;;;This program will draw a triangle
(defun tr1 ()
(setq p1(list 2 2))
(setq p2(list 6.0 3.0))
(setq p3(list 4.0 7.0))
(command "line" p1 p2 p3 "c")
)
```

5. Choose the **File** menu and select **Save** or **Save As**. In the **Save As** dialog box enter the name of the file, **Triang1.lsp**. After you save the file, the file name is displayed at the top of the editor window, Figure 14-3.

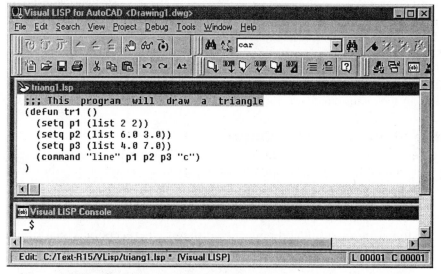

Figure 14-3 Writing program code in the Visual LISP Text Editor

Loading and Running Programs

1. Make sure the Visual Lisp Text Editor window is active. If not, click a point anywhere in the window to make it active.

2. In the **Tools** menu select **Load Text in Editor**. You can also load the program by choosing **Load active edit window** in the **Tools** toolbar. Visual LISP displays a message in the Console window indicating that the program has been loaded. If there was a problem in loading the program, an error message will appear.

3. To run the program, enter the function name **(tr1)** at the console prompt (-$ sign). The function name must be enclosed in parentheses. The program will draw a triangle in AutoCAD. To view the program output, switch to AutoCAD by choosing **Activate AutoCAD** in the **View** toolbar. You can also run the program from AutoCAD. Switch to AutoCAD and enter the name of the function **TR1** at the Command prompt. AutoCAD will run the program and draw a triangle on the screen, Figure 14-4.

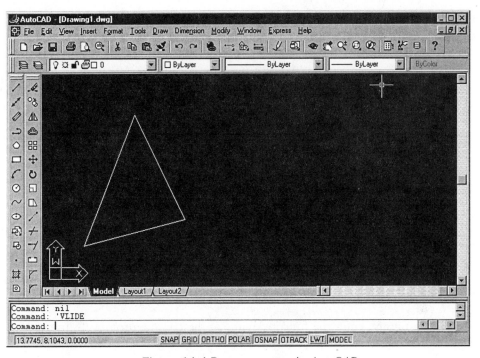

Figure 14-4 Program output in AutoCAD

Exercise 1   *General*

Write a LISP program that will draw an equilateral triangle. The program should prompt the user to select the starting point (bottom left corner) and the length of one of the sides of the triangle, Figure 14-5.

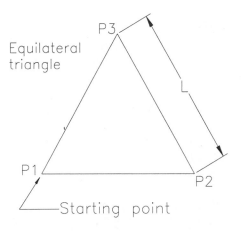

Figure 14-5 Drawing for Exercise 1

Chapter 14

Loading Existing AutoLISP Files

You can also load the AutoLISP files in Visual LISP and then edit, debug, load, and run the programs from Visual LISP.

1. Start AutoCAD. In the **Tools** menu select **AutoLISP** and then select the **Visual LISP Editor**. You can also start Visual AutoLISP by entering the **VLIDE** command at the Command prompt.

2. In the Visual LISP Text Editor, choose **Open File** in the **File** menu. The **Open file to edit/view** dialog box is displayed on the screen.

3. Select the AutoLISP program that you want to load and then choose **Open**. The program is loaded in the Visual LISP Text Editor.

4. To format the program in the edit window, select the **Format edit window** tool from the **Tools** toolbar. The program will automatically get formatted.

5   To load the program, choose **Load active edit window** in the **Tools** toolbar or choose **Load Text in Editor** from the **Tools** menu.

6. To run the program, enter the function name at the console prompt ($) in the Visual LISP Console window.

VISUAL LISP CONSOLE

Most of the programming in Visual LISP is done in the Visual LISP Text Editor. However, you can also enter the code in the Visual LISP console and see the results immediately. For example, if you enter **(sqrt 37.2)** at the console prompt (after the -$ sign) and press the **Enter** key, Visual LISP will evaluate the statement and return the value 6.09918. Similarly, enter the code **(setq x 99.3)**, **(+ 38 23.44)**, **(- 23.786 35)**, **(abs -37.5)**, **(sin 0.333)** and see the results. Each statement must be entered after the $ sign as shown in Figure 14-6.

Features of Visual LISP Console

1. In the Visual LISP console you can enter more than one statement on a line and then press the Enter key. Visual LISP will evaluate all statements and then return the value. Enter the following statement after the $ sign:
    -$(setq x 37.5) (setq y (/ x 2))
    Visual LISP will display the value of X and Y in the console, Figure 14-7

2. If you want to find the value of a variable in the code being developed, enter the variable name after the $ sign and Visual LISP will return the value and display it in the console. For example, if you want to find the value of X, enter X after the $ sign.

3. Visual LISP allows you to write an AutoLISP expression on more than one line by pressing **Ctrl+Enter** at the end of the line. For example, if you want to write the following two expressions on two lines, enter the first line and then press the **Ctrl+Enter** keys. You will notice that Visual LISP does not display the $ sign in the next line, indicating continuation of the statement on the previous line. Now enter the next line and press the **Enter** key, Figure 14-7.

$ (setq n 38)   Press the **Ctrl+Enter** keys
(setq counter (- n 1))

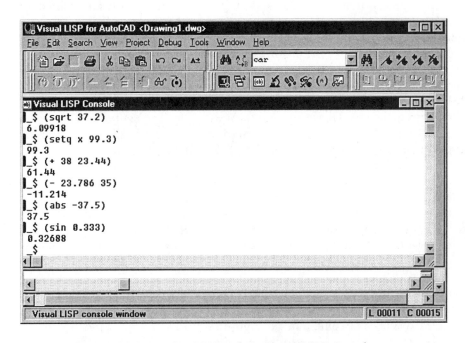

Figure 14-6 *Entering LISP code in Visual LISP Console*

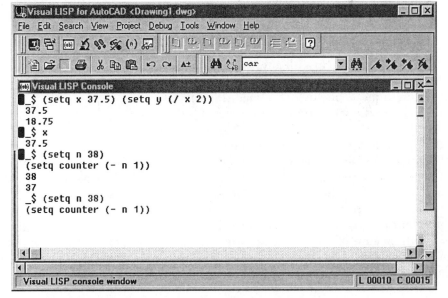

Figure 14-7 *Entering LISP code in Visual LISP Console*

4. To retrieve the text that has been entered in the previous statement, press the **Tab** key while at the Console prompt (-$). Each time you press the **Tab** key, the previously entered text is displayed. You can continue pressing the **Tab** key to cycle through the text that has been entered in the Visual LISP Console. If the same statement has been entered several times, pressing the Tab key will display the text only once and then jump to the statement before that. Pressing the **Shift+Tab** key displays the text in the opposite direction.

5. Pressing the **Ese** key clears the text following the Console prompt (-$). However, if you press the **Shift+Esc** keys, the text following the Console prompt is not evaluated and stays as is. The cursor jumps to the next Console prompt. For example, if you enter the expression **(setq x 15)** and then press the **Esc** key, the text is cleared. However, if you press the **Shift+Esc** keys, the expression (setq x 15) is not evaluated and the cursor jumps to the next Console prompt without clearing the text.

6. The **Tab** key also lets you do an associative search for the expression that starts with the specified text string. For example, if you want to search the expression that started with (sin, enter it at the Console prompt (-$) and then press the **Enter** key. Visual LISP will search the text in the Console window and return the statement that starts with (sin. If it cannot find the text, nothing happens, except maybe a beep.

7. If you click the **right mouse** button (right-click) anywhere in the Visual LISP Console or press the **Shift+F10** keys, Visual LISP displays the context menu, Figure 14-8. You can use the context menu to perform several tasks like Cut, Copy Paste, Clear Console Window, Find, Inspect, Add Watch, Apropros Window, Symbol Service, Undo, Redo, AutoCAD Mode, and Toggle Console Log.

8. One of the important features of Visual LISP is the facility of context-sensitive help available for the Visual LISP functions. If you need help with any Visual LISP function, enter or select the function and then choose the Help button in the **Tools** toolbar. Visual LISP will display the help for the selected function, Figure 14-9.

Figure 14-8 Context menu displayed *when you click the right mouse button*

9. Visual LISP also allows you to keep a record of the Visual LISP Console activities by saving them on the disk as a log files (.log). To create a log file, select **Toggle Console Log** in the **File** menu or select **Toggle Console Log** from the context menu. The Visual LISP Console window must be active to create a log file.

10. As you enter the text, Visual LISP automatically assigns a color to it. The colors are assigned based on the nature of the text string. For example, if the text is a built-in function or a protected symbol, the color assigned is Blue. Similarly, text strings are Magenta, integers are green, parentheses are red, real numbers are teal, and comments are magenta.

11. Although native AutoLISP and Visual LISP are separate programming environments, you can transfer any text entered in the Visual LISP Console to AutoCAD Command line. To accom-

plish this, enter the code at the Console prompt and then select **AutoCAD Mode** in the context menu or from the **Tools** menu. The Console prompt is replaced with Command prompt. Press the **Tab** key to display the text that was entered at the Console prompt. Press the ENTER key to switch to AutoCAD screen. The text is displayed at AutoCAD's Command prompt.

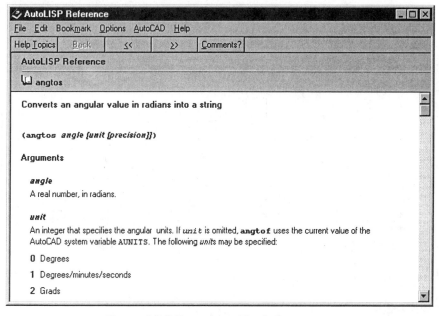

Figure 14-9 Context-sensitive help screen

Exercise 2   *General*

Enter the following statements in the Visual LISP Console at the console prompt (-$):
```
(+ 2 30 4 38.50)
(- 20 39 32)
(* 2.0 -7.6 31.25)
(/ -230 -7.63 2.15)
(sin 1.0472)
(atan -1.0)
(< 3 10)
(<= -2.0 0)
(setq variable_a 27.5)
(setq variable_b (getreal "Enter a value:"))
```

USING VISUAL LISP TEXT EDITOR

You can use the **Visual LISP Text Editor** to enter programming code. It has several features that are not available in other text editors. For example, when you enter text, it is automatically assigned a color based on the nature of the text string. If the text is a parentheses, it is assigned red color. If the text is a Visual LIPS function, it is assigned blue color. The Visual LISP Text Editor also allows you to execute an AutoLISP function without leaving the text editor, and also checks for

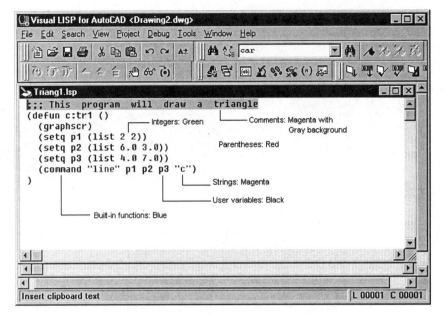

Figure 14-10 Color coding for LISP code

balanced parentheses. These features make it an ideal tool for writing Visual LISP programs.

Color Coding

Visual LISP Text Editor provides color coding for files that are recognized by Visual LISP. Some of the files that it recognizes are LISP, DCL, SQL, and C language source files. When you enter text in the text editor, Visual LISP automatically determines the color and assigns it to the text, Figure 14-10. The following table shows the default color scheme:

Visual LISP Text Element	Color
Parentheses	Red
Built-in functions	Blue
Protected symbols	Blue
Comments	Magenta with gray background
Strings	Magenta
Real numbers	Teal
Integers	Green
Unrecognized items	Black (like variables)

Words in the Visual LIPS Text Editor

In most text editors the words are separated by space. In the Visual LISP Text Editor, the word has a different meaning. A word is a set of characters that are separated by one or more of the following special characters:

Character Description	Visual LISP Special Characters
Space	
Tab	

Single quote	'
Left parentheses	(
Right parentheses	)
Double quotes	"
Semicolon	;
Newline	\n

For example, if you enter the text (Command"Line"P1 P2 P3"c), Visual LISP treats Command and Line as separate words because they are separated by double quotes. In a regular text editor (Command"Line" would have been treated as a single word. Similarly, in **(setq p1,** setq and p1 are separate words because they are separated by a space. If there was no space, Visual LISP will treat **(setqp1** as one word.

Context Sensitive Help

One of the important features of Visual LISP is the facility of context-sensitive help available for the Visual LISP functions. If you need help with any Visual LISP function, enter or select the function in the text editor and then choose the **Help** button in the **Tools** toolbar. Visual LISP will display the help for the selected function.

Context Menu

If you click the **right mouse** button (right-click) anywhere in the Visual LISP Text Editor or press the **Shift+F10** keys, Visual LISP displays the context menu, Figure 14-11. You can use the context menu to perform several tasks like Cut, Copy Paste, Clear Console Window, Find, Inspect, Add Watch, Apropos Window, Symbol Service, Undo, Redo, AutoCAD Mode, and Toggle Console Log. The following table contains a short description of these functions:

Figure 14-11 *Context menu displayed when you click the right*

Functions in the context Menu	Description of the function
Cut	Moves the text to the clipboard
Copy	Copies the text to the clipboard
Paste	Pastes the contents of the clipboard at the cursor position
Find	Finds the specified text
Go to Last Edited	Moves the cursor to the last edited position
Toggle Breakpoint	Sets or removes a breakpoint at the cursor position
Inspect	Opens the **Inspect** dialog box
Add Watch	Opens the **Add Watch** dialog box
Apropos Window	Opens **Apropos options** dialog box
Symbol Service	Opens **Symbol Service** dialog box
Undo	Reverses the last operation
Redo	Reverses the previous Undo

VISUAL LISP FORMATTER

You can use the Visual LISP Formatter to format the text entered in the text editor. When you format the text, the formatter automatically arranges the AutoLISP expressions in way that makes it easy to read and understand the code. It also makes it easier to debug the program when the program does not perform the intended function.

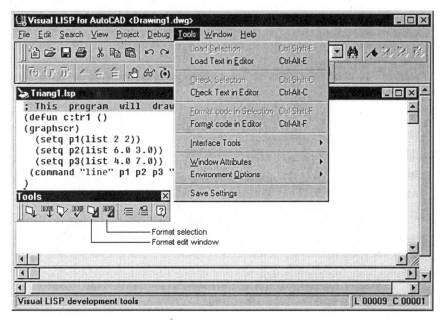

Figure 14-12 Using Format selection or Format window tools

Running the Formatter

With the formatter you can format all the text in the Visual LISP text editor window or the selected text. To format the text, the text editor window must be active. If it is not, click a point anywhere in the window to make it active. To format all the text, select the **Format edit window** tool in the **Tools** toolbar or select **Format code in Editor** from the **Tools** menu. To format part of the text in the text editor, select the text and then select the **Format selection** tool in the **Tools** toolbar or select **Format code in Selection** from the **Tools** menu. Figure 14-12 shows the **Tools** menu and **Tools** toolbar.

Formatting Comments

In Visual LISP you can have five types of comments and depending on the comment in the program code, Visual LISP will format and position the comment text according to its type. Following is the list of comment types available in Visual LISP:

Comment formatting	Description
; Single semicolon	The single semicolon is indented as specified in the Single-semicolon comment indentation of the AutoLISP Format option.

;; Current column comment   The comment is indented at the same position as the previous line of the program code.

;;; Heading comment   The comment appears without any indentation.

;| Inline comment|;   The Inline comments appears as any other expression.

;_ Function closing comment   The function closing comment appears just after the previous function, provided the Insert form-closing comment is on in the AutoLISP Format options.

Two text editor windows are shown in Figure 14-13. The window at the top has the code without formatting. The text window at the bottom has been formatted. If you compare the two, you will see the difference formatting makes in the text display.

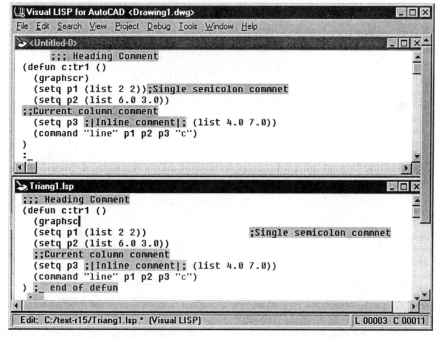

Figure 14-13 *The top window shows code without formatting and the bottom window shows the same code after formatting the Programming code*

Chapter 14

Changing Formatting Options

You can change the formatting options through the **Format options** dialog box (Figure 14-14) that can be invoked by choosing the **Tools** menu, **Environment Options**, and then **Visual LISP Format Options**. In this dialog box if you select the **More** options button, Visual LISP will display another dialog box (Figure 14-15) listing more formatting options.

DEBUGGING THE PROGRAM

When you write a program, it is rare that the program will run the first time. Even if runs, it may not perform the desired function. The error could be a syntax error, missing parentheses, misspelled function name, improper use of a function, or simply a typographical error. It is time

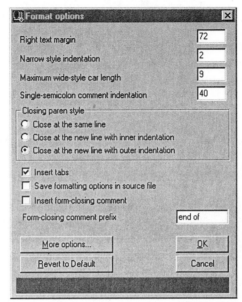

Figure 14-14 Format options dialog box

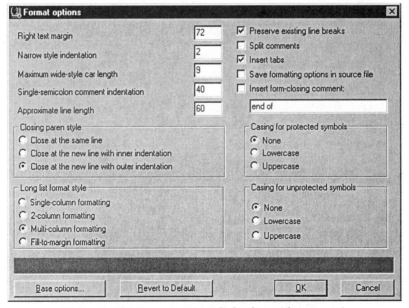

Figure 14-15 Format options dialog box with more options

consuming and sometimes difficult to locate the source of the problem. Visual LISP comes with several debugging tools that makes it easier to locate the problem and debug the code. Make sure the Debug toolbar is displayed on the screen. If it is not, in the Visual LISP window click on the **View** menu and then select **Toolbars** to display the **Toolbars** dialog box. Select the **Debug** check box, choose the **Apply** button, and then choose the **Close** button to exit the dialog box. The **Debug** toolbar will appear on the screen. Figure 14-16 shows the toolbar with tool tips.

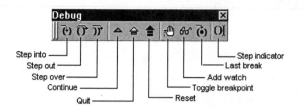

Figure 14-16 Tools in Debug toolbar

1. To try some of the debugging tools, enter the following program in the Visual LISP Text Editor and then save the file as **triang2.lsp**:

    ```
    ;;; This program will draw a triangle and an arc
    (defun tr2 ()
      (setq p1 (getpoint "\nEnter first point p1:"))
      (setq p2 (getpoint "\nEnter second point p2:"))
      (setq p3 (getpoint "\nEnter third point p3:"))
      (command "arc" p1 p2 p3)
      (command "line" p1 p2 p3 "c")
    )
    ```

2. To format the code, choose the **Format edit window** tool in the **Tools** toolbar. You can also format the code by selecting **Format code in Editor** in the Tools menu.

3. Load the program by choosing the **Load active edit window** tool in the **Tools** toolbar. You can also load the program from the **Tools** menu by selecting **Load Text in Editor**.

4. In the Visual LISP Text Editor, position the cursor in front of the following line and then choose the **Toggle breakpoint** tool in the **Debug** toolbar. You can also select it from the **Debug** menu. Visual LISP inserts a breakpoint at the cursor location.

    ```
    |(setq p2 (getpoint "\Enter second point p2:"))
    ```

5. In the Visual LISP Console, at the Console prompt, enter the name of the function and then press the **Enter** key to run the program.

    ```
    -$ (tr2)
    ```

    AutoCAD window will appear and the first prompt in the program (Enter first point p1:) will be displayed in the Command prompt area. Pick a point and the Visual LISP window will appear. Notice, the line that follows the breakpoint is highlighted, as shown in Figure 14-17.

6. Choose the **Step Into** button in the **Debug** toolbar or select **Step Into** from the **Debug** menu. You can also press the function key **F8** to invoke the **Step Into** command. The program starts

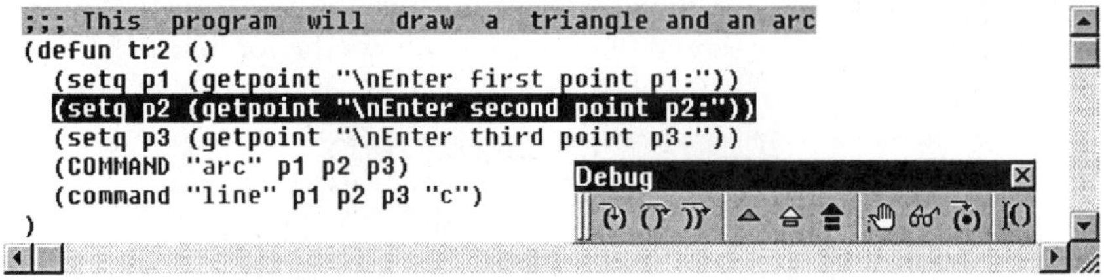

```
;;; This program will draw a triangle and an arc
(defun tr2 ()
   (setq p1 (getpoint "\nEnter first point p1:"))
   (setq p2 (getpoint "\nEnter second point p2:"))
   (setq p3 (getpoint "\nEnter third point p3:"))
   (COMMAND "arc" p1 p2 p3)
   (command "line" p1 p2 p3 "c")
)
```

Figure 14-17 Using the Step Into command to start execution until it encounters the next expression

execution until it encounters the next expression. The next expression that is contained within the inside parentheses is highlighted, Figure 14-18. Notice the position of the vertical bar in the Step indicator tool in the Debug toolbar. It is on the left of the Step indicator tool. Also, the cursor is just before the expression ("getpoint "\nEnter second point:).

```
;;; This program will draw a triangle and an arc
(defun tr2 ()
   (setq p1 (getpoint "\nEnter first point p1:"))
   (setq p2 (getpoint "\nEnter second point p2:"))
   (setq p3 (getpoint "\nEnter third point p3:"))
   (COMMAND "arc" p1 p2 p3)
   (command "line" p1 p2 p3 "c")
)
```

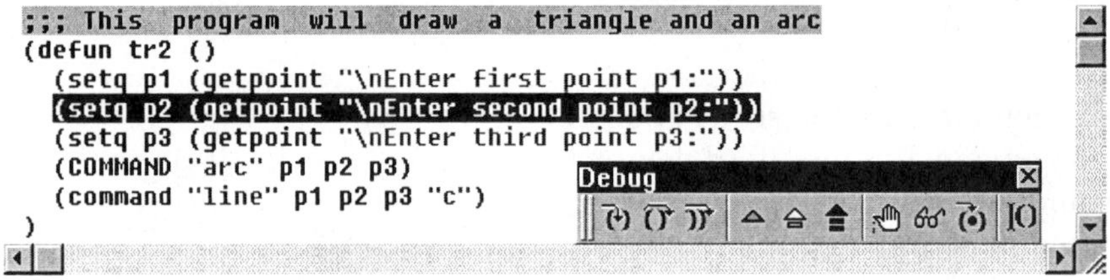

Figure 14-18 Using the Step Into command again, the cursor jumps to the end of the expression

7. Choose the **Step Into** button again. The AutoCAD window appears and the second prompt in the program (Enter second point p2:) is displayed in the Command prompt area. Pick a point, and the Visual LISP window returns, Figure 14-19. Notice the position of the vertical bar in the **Step indicator** tool in the **Debug** toolbar. It is just to the right of the Step Indicator. Also, the cursor is just after the expression ("getpoint "\nEnter second point:).

8. Choose the **Step Into** button again to step through the program. The entire line is highlighted and the cursor moves to the end of the expression (getpoint "Enter second point:). Notice the position of the vertical bar in the **Step indicator** tool in the **Debug** toolbar.

```
Triang2.LSP
;;; This program will draw a triangle and an arc
(defun tr2 ()
   (setq p1 (getpoint "\nEnter first point p1:"))
   (setq p2 (getpoint "\nEnter second point p2:"))
   (setq p3 (getpoint "\nEnter third point p3:"))
   (COMMAND "arc" p1 p2 p3)
   (command "line" p1 p2 p3 "c")
)
```

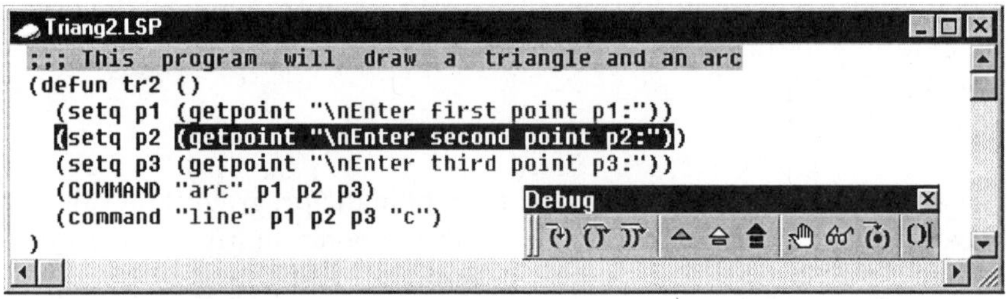

Figure 14-19 Using the Step Into command again, the vertical bar is displayed to the right of Step Indicator symbol

9. Choose the **Step Into** button again; the next line of the program is highlighted and the cursor is at the start of the statement.

10 To step over the highlighted statement, choose the **Step over** tool in the **Debug** toolbar, or select **Step Over** in the Debug menu. You can also press the **Shift+F8** keys to invoke this command. The AutoCAD window is displayed and the prompt (Enter third point:) is displayed in the Command prompt area. Select a point, and the Visual LISP window appears.

11. Choose the **Step over** button again; Visual LISP highlights the next line. If you choose the Step over button again, the entire expression is executed.

12. Choose the **Step over** button until you go through all the statements and the entire program is highlighted.

Tracing Variables

Sometimes it becomes necessary to trace the values of the variables used in the program. For example, let us assume the program crashes or does not perform as desired. Now you want to locate where the problem occurred. One of the ways you can do it is by tracing the value of the variables used in the program. The following steps illustrate this procedure:

1. Open the file containing the program code in the Visual LISP Text Editor. You can open the file by selecting **Open File** from the **File** menu.

2. Load the program by selecting **Load Text in Editor** from the **Tools** menu or by selecting the **Load active edit window** tool in the **Tools** toolbar.

3. Run the program by entering the name of the function (tr2) at the Console prompt (-$) in the Visual LISP Console. The program will perform the operations as defined in the program. In this program, it will draw a triangle and an arc between the user-specified points. Now we want to find out the coordinates of the points p1, p2, and p3. To accomplish this we can use the **Watch window** tool in the **View** toolbar.

4. Highlight the variable **p1** and then choose the **Watch window** tool in the **View** toolbar. You can also select **Watch Window** in the **View** menu. Visual LISP displays the **Watch** window that lists the X, Y, and Z coordinates of point p1.

5. Position the cursor near (in front, middle, or end) the variable p2 and then choose the **Add Watch** tool in the **Watch** window. Select the **OK** button in the **Add Watch** window. The value of the selected variable is listed in the **Watch** window, Figure 14-20. Similarly you can trace the value of other variables in the program.

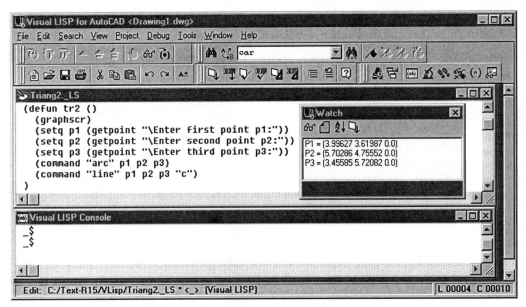

Figure 14-20 Using Add Watch to trace variables in the program

Review Questions

1. What are the different ways to start Visual LISP.
    1. _____   2. _____

2. Name the four areas of the Visual LISP window.
    1. _____   2. _____
    3. _____   4. _____

3. Name the toolbars available in Visual LISP.
    1. _____   2. _____
    3. _____   4. _____
    5. _____

4. What are the different ways of loading a Visual LISP program.
    1. _____   2. _____

5. How can you run a Visual LISP program.

    _____

6. Can you open an AutoLISP file in the Visual LISP Text Editor. _____

7. Explain the function of the Visual LISP Console.

    _____

    _____

8. Can you enter several LISP statements in the Visual LISP Console. If yes, give an example.

    _____

9. In the Visual LISP Console, how can you enter a LISP statements in more than one line.

    _____

10. What is a context menu and how can you display it.

    _____

    _____

11. In the Visual LISP Text Editor, what are the default colors assigned to the following
    Parantheses _____   LISP functions _____
    Comments _____   Integers _____

12. What is the function of the Visual LISP Formatter and how can you format the text.

    _____

    _____

13. How can you get the heading comment and function closing comment.

    _____

14. How can you change the formatting options.

    _____

15. How can you debug a Visual LISP program.

    _____

    _____

16. How can you trace the value of variables in the Visual LISP program.

    _____

    _____

Exercises

Exercise 3   *Architectural*

Write a LISP program that will draw the staircase shown in Figure 14-21. When you run the program, the user should be prompted to enter the rise and run of the staircase and the number of steps. The user should also be prompted to enter the length of the landing area.

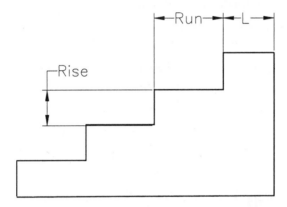

Figure 14-21 Drawing for Exercise 3

Exercise 4   *Architectural*

Write a LISP program that will draw the picture frame of Figure 14-22. The program should prompt the user to enter the length, width, and thickness of the frame.

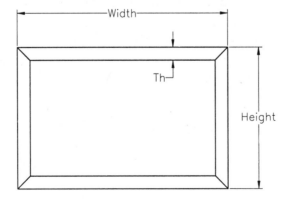

Figure 14-22 Drawing for Exercise 4

Write a LISP program that will draw a three-dimensional table as shown in Figure 14-23. The program should prompt the user to enter the length, width, height, thickness of tabletop, and the size of the legs. The program should also use the UCS command to change the view direction to display the 3D view of the table.

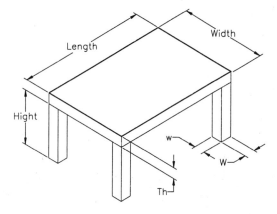

Figure 14-23 *Drawing for Exercise 5*

Chapter 15

Programmable Dialog Boxes Using Dialog Control Language

Learning Objectives

After completing this chapter, you will be able to:
• *Write programs using dialog control language.*
• *Use predefined attributes.*
• *Load a dialog control language (DCL) file.*
• *Display new dialog boxes.*
• *Use standard button subassemblies.*
• *Use AutoLISP functions to control dialog boxes.*
• *Manage dialog boxes with AutoLISP.*
• *Use tiles, buttons, and attributes in DCL programs.*

DIALOG CONTROL LANGUAGE

Dialog control language (**DCL**) files are ASCII files that contain the descriptions of dialog boxes. A DCL file can contain the description of a single or multiple dialog boxes. There is no limit to the number of dialog box descriptions that can be defined in a DCL file. The suffix of a DCL file is **.DCL** (such as **DDOSNAP.DCL.**)

This chapter assumes that you are familiar with AutoCAD commands, AutoCAD system variables, and AutoLISP programming. You need not be a programming expert to learn writing programs for dialog boxes in DCL or to control the dialog boxes through AutoLISP programing. However, knowledge of any programming language should help you to understand and learn DCL. This chapter introduces you to the basic concepts of developing a dialog box, frequently used attributes, and tiles. A thorough discussion of DCL functions and a step-by-step explanation of examples should make it easy for you to learn DCL. For those functions not discussed in this chapter, you can refer to the *AutoCAD Customization Guide from Autodesk*. To write programs in DCL, you need no special software or hardware. If AutoCAD is installed on your computer, you can write DCL files. To write DCL files, you can use any text editor.

DIALOG BOX

A dialog control language (DCL) file contains the description of how the dialog boxes will appear on the screen. These boxes can contain buttons, text, lists, edit boxes, rows, columns, sliders, and images. A sample dialog box is shown in Figure 15-1.

You do not need to specify the size and layout of a dialog box or its component parts. Sizing is done automatically when the dialog box is loaded on the screen. By itself, the dialog box cannot perform the functions it is designed for. The functions of a dialog box are controlled by a program written in the AutoLISP programming language or with the AutoCAD development system (ADS), or ARX. For example, if you load a dialog box and select the Cancel button, it will not perform the cancel operation. The instructions associated with a button or any part of the dialog box are handled through the

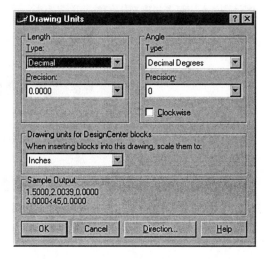

Figure 15-1 Drawing Units dialog box

functions provided in AutoLISP, ADS, or ARX. Therefore, AutoLISP and ADS are needed to control the dialog boxes, and you should have, in addition to an understanding of DCL, a good knowledge of AutoLISP or ADS to develop new dialog boxes or edit existing ones.

Dialog boxes are not dependent on the platform; therefore, they can run on any system that supports AutoCAD. However, depending on the graphical user interface (GUI) of the platform, the appearance of the dialog boxes might change from one system to another. The functions defined in the dialog box will still work without making any changes in the dialog box or the application program (AutoLISP or ADS) that uses these dialog boxes.

DIALOG BOX COMPONENTS

The two major components of a dialog box are the tiles and the box itself. The tiles can be arranged in rows and columns in any desired configuration. They can also be enclosed in boxes or borders to form subassemblies, giving them a tree structure, Figure 15-2(b). The basic tiles, such as buttons, lists, edit boxes, and images, are predefined by the programmable dialog box (PDB) facility of AutoCAD. These buttons are described in the file **base.dcl**. The layout and function of a tile is determined by the attribute assigned to it. For example, the height attribute controls the height of the tile. Similarly, the label attribute specifies the text that is associated with the tile. Some of the components of a dialog box are shown in Figure 15-2(a). Following this figure is a list of the predefined tiles and their format in DCL.

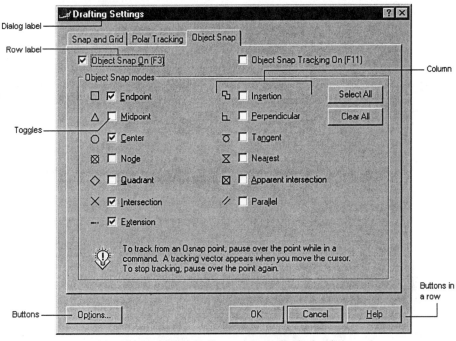

Figure 15-2(a) *Components of a dialog box*

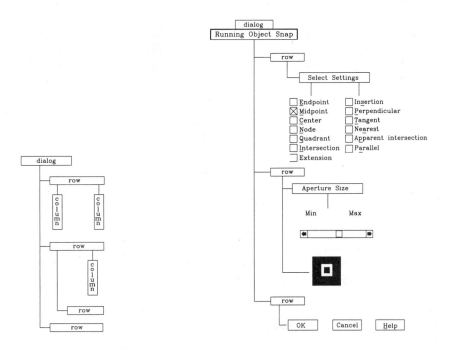

Figure 15-2(b) *Tree structure of a dialog box*

Predefined Tile	DCL Format
Button	button
Edit box	edit_box
Image button	image_button
List box	list_box
Pop-up list	popup_list
Radio button	radio_button
Slider	slider
Toggle	toggle
Column	column
Boxed column	boxed_column
Row	row
Boxed row	boxed_row
Radio column	radio_column
Boxed radio column	boxed_radio_column
Radio row	radio_row
Boxed radio row	boxed_radio_row
Image	image
Text	text
Spacer	spacer

BUTTON AND TEXT TILES

Button Tile

Format in DCL: **button**

The button tile consists of a rectangular box that resembles a push button. The button's label appears inside the button. For example, in the OK button of a dialog box, the label OK appears inside the button. If you select the OK button in a dialog box, it performs the functions defined in the dialog box and clears the dialog box from the screen. Similarly, if you select the cancel button, it cancels the dialog box without taking any action.

Note

A dialog box should contain at least one OK button or a button that is equivalent to it. This allows you to exit the dialog box when you are done using it.

Text Tile

Format in DCL: **text**

The text tile is used to display information or a title in a dialog box. It has limited application in the dialog boxes because most of the tiles have their own label attributes for tiling. However, if you need to display any text string in the dialog box, you can do so with the text tile. The text tile is used extensively in AutoCAD alert boxes to display warnings or error messages.

Note

An alert box must contain an OK button or a Cancel button to end the dialog box.

TILE ATTRIBUTES

The appearance, size, and function performed by a dialog box tile depend on the attributes that have been assigned to the tile. For example, if a button has been assigned the **fixed_width** attribute, the width of the box surrounding the button will not stretch through the entire length of the dialog box. Similarly, the **height** attribute determines the height of the tile, and the key attribute assigns a name to the tile that is then used by the application program. The tile attribute consists of two parts: the name of the attribute and the value assigned to the attribute. For example, in the expression **fixed_width = true, fixed_width** is the name of the attribute and **true** is the value assigned to the attribute. Attribute names are like variable names in programming, and the values assigned to these variables must be of a specific type.

Types of Attribute Values

Integer. Unlike integer values in programming (1, 15, 22), the numeric values assigned to attributes can be both integers and real numbers.

> **Examples**
> width = 15
> height = 10

Real Number. The real values assigned to attributes should always be fractional real numbers with a leading digit.

> **Examples**
> aspect_ratio = 0.75

Quoted String. The string values assigned to an attribute consist of a text string that is enclosed in double quotes.

> **Examples**
> key = "accept"
> label = "OK"

Reserved Word

Dialog control language uses some reserved words as identifiers. These identifiers are alphanumeric characters that start with a letter.

> **Examples**
> is_default = true
> fixed_width = true

Note

Reserved names are case-sensitive. For example, "is_default = true" is not the same as "is_default = True". The reserved word is "true", not "True" or "TRUE".

Like reserved words, attribute names are also case-sensitive. For example, the attribute is "key", not "Key" or "KEY".

Chapter 15

PREDEFINED ATTRIBUTES

To facilitate writing programs in dialog control language, AutoCAD has provided some predefined attributes that are defined in the programmable dialog box (PDB) package that comes with the AutoCAD software. Some of these attributes can be used with any tile and some only with a particular type of tile. The values assigned to various attributes in a DCL file are used by the application program to handle the tiles or the dialog box. Therefore you must use the correct attributes and assign an appropriate value to these attributes. The following is a list of some of the frequently used predefined attributes defined in the PDB facility of AutoCAD:

action	key
alignment	label
allow_accept	layout
aspect_ratio	list
color	max_value
edit_limit	min_value
edit_width	mnemonic
fixed_height	multiple_select
fixed_width	small_increment
height	tabs
is_cancel	value
is_default	width

key, label, AND is_default ATTRIBUTES

key Attribute

Format in DCL
key

Examples
key = "accept"
key = "XLimit"

The **key** attribute assigns a name to a tile. The name must be enclosed in double quotes. This name can then be used by the application program to handle the tile. A dialog box can have any number of key values, but within a particular dialog box the values used for the key attributes must be unique. In the example key = "accept", a string value "accept" is assigned to the **key** attribute. If there is another key attribute in the dialog box, you must assign it a different value.

label Attribute

Format in DCL
label

Examples
label = "OK"
label = "Hello DCL Users"

Sometimes it is necessary to display a label in a dialog box. The label attribute can be used in a boxed column, boxed radio column, boxed radio row, boxed row, button tile, dialog box, or edit box. Some of the frequently used label attributes are described next.

Use of the label Attribute in a Dialog Box. When the label attribute is used in a dialog box, it is displayed in the top border or the title bar of the dialog box. The label must be a string enclosed in double quotes. Use of the label in a dialog box is optional; in case of default, no title is displayed in the dialog box.

Example
welcome : dialog {
     label = "Sample Dialog Box";

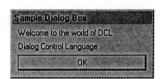

In this example, the label "Sample Dialog Box" is displayed at the top of the dialog box.

Use of the label Attribute in a Boxed Column. When the label attribute is used in a boxed column, the label is displayed within a box in the upper left corner of the column; the box consists of a single line at the top of the column. The label must be a quoted string, and the default is a set of a quoted string (" "). If the default is used, only the box is displayed, without any label.

Example
: text {
   label = "Welcome to the world of DCL";
}

In this example, the label "Welcome to the world of DCL" is displayed within a box in the upper left corner of the column.

Use of the label Attribute in a Button. When the label attribute is used in a button, the label is displayed inside the button. The label must be a quoted string and has no default.

Example
: button {
 key = "accept";
 label = "OK";
}

In this example, the label "OK" is displayed inside and in the center of the button tile.

is_default Attribute

Format in DCL   **Example**
is_default   is_default = true

The **is_default** attribute is used for the button of a dialog box. In the example, the value assigned to the **is_default** attribute is true. Therefore, this button will be automatically selected when you press ENTER. For example, if you load a dialog box on the screen, one way to exit and accept the values of the dialog box is to select the OK button. You can accomplish the same thing by pressing ENTER (accept key). This action is made possible by assigning the value, **true**, to the **is_default** attribute. In a dialog box the default button can be recognized by the thick border drawn around the text string Note.

Chapter 15

Note
In a dialog box, only one button can be assigned the true value for the is_default attribute.

fixed_width AND alignment ATTRIBUTES

fixed_width Attribute

Format in DCL   **Example**
fixed_width   fixed_width = true

This attribute controls the width of the tile. If the value of this attribute is set to true, the width of the tile does not extend across the complete width of the dialog box. The width of the tile is automatically adjusted to the length of the text string that is displayed in the tile.

alignment Attribute

Format in DCL   **Example**
alignment   alignment = centered
                                       alignment = right

The value assigned to the **alignment** attribute determines the horizontal or vertical position of the tile in a row or column. For a row, the values that can be assigned to this attribute are left, right, and centered. The default value of the alignment attribute is left; this forces the tile to be displayed left-justified. For a column, the possible values of the **alignment** attribute are top, bottom, and centered. The default value is centered.

Example 1

Using dialog control language (DCL), write a program for the following dialog box (Figure 15-3). The dialog box has two text labels and an OK button to end the dialog box.

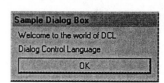

The following file is a listing of the DCL file for the dialog box of Example 1. The name of this DCL file is **dclwel1.dcl. The line numbers are not a part of the file; they are for reference only.**

Figure 15-3 Dialog box for Example 1

```
welcome1 : dialog {                                    1
        label = "Sample Dialog Box";                   2
        : text {                                       3
        label = "Welcome to the world of DCL";         4
        }                                              5
        : text {                                       6
            label = "Dialog Control Language";         7
        }                                              8
        : button {                                     9
            key = "accept";                           10
```

```
            label = "OK";                                            11
        is_default = true;                                           12
    }                                                                13
}                                                                    14
```

Explanation

Line 1
welcome1 : dialog {
In this line, **welcome1** is the name of the dialog and the definition of the dialog box is contained within the braces. The open brace in this line starts the definition of this dialog box.

Line 2
label = "Sample Dialog Box";
In this line, **label** is the label attribute, and **"Sample Dialog Box"** is the string value assigned to the label attribute. This string will be displayed in the title bar of the dialog box. The label description must be enclosed in quotes. If this line is missing, no title is displayed in the title bar of the dialog box.

Lines 3-5
: text {
label = "Welcome to the world of DCL";
}
These three lines define a text tile with the label description. In the first line, **text** refers to the text tile. The line that follows it, **label = "Welcome to the world of DCL";** defines a label for this tile that will be displayed left-justified in the dialog box. The closing brace completes the definition of this tile.

Lines 9-13
: button {
key = "accept";
label = "OK";
is_default = true;
}
These five lines define the attributes of the button tile. In the first line, **button** refers to the button tile. In the second line, **key = "accept";** specifies an ASCII name, **"accept"**, that will be used by the application program to refer to this tile. The next line, **is_default = true;**, specifies that this button is the default button. It is automatically selected if you press ENTER at the keyboard.

The closing brace in line 14 completes the definition of the dialog box.

LOADING A DCL FILE

As with an AutoLISP file, you can load a DCL file from the AutoCAD drawing editor. A DCL file can contain the definition of one or several dialog boxes. There is no limit to the number of dialog boxes you can define in a DCL file. The format of the command for loading a dialog box is:

Note
When you use the load_dialog and new_dialog functions to load and display a DCL program, the AutoCAD screen will freeze. To avoid this, it is recommended to use the AutoLISP program to load and display a DCL program. See "Using AutoLISP Function to Load a DCL File" at the end of this section.)

(load_dialog filename)
        Where **load_dialog** ---- Load command for loading a dialog file
                 filename -------- Name of the DCL file, with or without the file extension
                            (.dcl)

Example
(load_dialog "dclwel1.dcl") or (load_dialog "dclwel1")

In this example, **dclwel1** is the name of the DCL file and **.dcl** is the file extension for DCL files.

The file name can be with or without the DCL file extension (**welcome1** or **welcome1.dcl**). The load function returns an integer value that is used as a handle in the **new_dialog** function and the **unload_dialog** function. This integer will be referred to as **dcl_id** in subsequent sections. You do not need to use the name dcl_id for this integer; it could be any name (dclid, or just id).

DISPLAYING A NEW DIALOG BOX

The load_dialog function loads the DCL file, but it does not display it on the screen. The format of the command used to display a new dialog box is:

(new_dialog dlgname dcl_id)
        Where **new_dialog** ----- Command to display a dialog box
                 dlgname --------- Name of the dialog box
                 dcl_id ----------- Integer returned in the load_dialog function

Example
(new_dialog "welcome1" 1)

In this example, assume that dcl_id is 1, an integer returned by the load_dialog function. You can use the (load_dialog) and (new_dialog) commands to load the DCL file in Example 1. In the following command sequence, assume that the integer (dcl_id) returned by the (load_dialog) function is 3. Figure 15-4 shows the dialog box after entering the following two command lines:

    Command: (load_dialog "dclwel1.dcl")
    3
    Command: (new_dialog "welcome1" 3)

In this example, notice that the OK button stretches across the complete width of the dialog box. This button would look better if its width were limited to the width of the string "OK". This can be accomplished by using the **fixed_width = true;** attribute in the definition of this button. You can

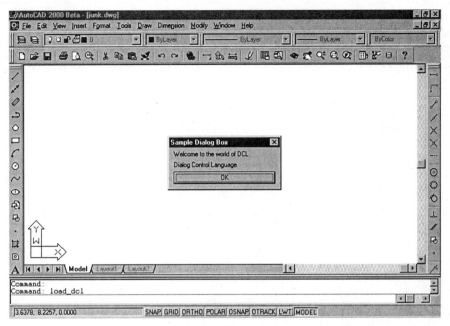

Figure 15-4 Dialog box as displayed on the screen

also use the **alignment = centered** attribute to display the OK button label center-justified. The following file is a listing of the DCL file where the OK label for the OK button is center-justified and the width of the box does not stretch across the width of the dialog box (Figure 15-5).

```
welcome2 : dialog {
    label = "Sample Dialog Box";
    : text {
       label = "Welcome to the world of DCL";
    }
    : text {
    label = "Hello - DCL";
    }
    : button {
    key = "accept";
    label = "OK";
       is_default = true;
    fixed_width = true;                    (Controls width)
    alignment = centered;                  (Controls
       }                              Justification)
}
```

Using the AutoLISP Function to Load a DCL File

You can use the following AutoLISP program to load and display the dialog box without freezing the screen.

Chapter 15

```
(defun c:load_dcl( / dcl_id )
   (setq dcl_id (load_dialog "dclwel1.dcl"))          (Loads the DCL file)
   (new_dialog "welcome1" dcl_id)                     (Initializes the dialog box)
   (start_dialog)                                     (Displays the dialog box)
   (princ)
)
```

Where   -------------------- **load_dcl** is the name of the AutoLISP function.

-------------------- **dclwel1** is the name of the DCL file that you want to load.

-------------------- **welcome1** is the name of the dialog as defined in the DCL file.

Figure 15-5 *Dialog box with fixed width for OK button*

USE OF STANDARD BUTTON SUBASSEMBLIES

Some standard button subassemblies are predefined in the **base.dcl** file. You can use these standard buttons in your DCL file to maintain consistency among various dialog boxes. One such predefined button is **ok_cancel**, which displays the OK and Cancel buttons in the dialog box (Figure 15-6). The following file is the listing of the DCL file of Example 1, using the **ok_cancel** predefined standard button subassembly:

```
welcome3   : dialog {
    label = "Sample Dialog Box";
    : text {
       label = "Welcome to the world of DCL";
    }
    : text {
          label = "Dialog Control Language";
          alignment = right;
    }
    ok_cancel;
}
```

The following is a list of the standard button subassemblies that are predefined in the **base.dcl** file. These buttons are also referred to as **dialog exit buttons** because they are used to exit a dialog box.

OK Button
    Format in DCL: **ok_only**

OK and Cancel
  Format in DCL: **ok_cancel**

OK, Cancel, and Help Buttons
  Format in DCL: **ok_cancel_help**

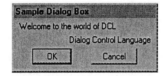

Figure 15-6 *Dialog box with OK and Cancel*

AUTOLISP FUNCTIONS

load_dialog

The AutoLISP function **load_dialog** is used to **load a DCL file** that is specified in the load_dialog function. In the following examples, the name of the file that AutoCAD loads is "dclwel1". The extension (**.dcl**) of the file is optional. When the DCL file is loaded successfully, AutoCAD returns an integer that identifies the dialog box.

Format in DCL	**Examples**
(load_dialog filename)	(load_dialog "dclwel1.dcl")
	(load_dialog "dclwel1")

unload_dialog

The AutoLISP function **unload_dialog** is used to **unload a DCL file** that is specified in the unload_dialog function. The file is specified by the variable (dcl_id) that identifies a DCL file.

Format in DCL	**Example**
(unload_dialog dcl_id)	(unload_dialog dcl_id)

new_dialog

The AutoLISP function **new_dialog** is used to initialize a dialog box and then display it on the screen. In the following file, **"welcome1"** is the name of the **dialog box**. (Note: welcome1 is not the name of the DCL file.) The variable **dcl_id** contains an integer value that is returned when the DCL file is loaded.

Format in DCL	**Example**
(new_dialog "dialogname" dcl_id)	(new_dialog "welcome1" dcl_id)

Chapter 15

start_dialog

The AutoLISP function **start_dialog** is used in an AutoLISP program to accept user input from the dialog box. For example, if you select OK from the dialog box, the start_dialog function retrieves the value of that tile and uses it to take action and to end the dialog box.

> Format in DCL
> (**start_dialog**)

done_dialog

The AutoLISP function **done_dialog** is used to terminate the display of the dialog box from the screen. This function must be defined within the action expression, as shown in the following example.

> Format in DCL   **Example**
> (**done_dialog**)   (action_tile "accept" "(done_dialog)")

action_tile

The AutoLISP function **action_tile** is used to associate an action expression with a tile in the dialog box. In the following example, the **action_tile** function associates the tile "accept" with the action expression (done_dialog) that terminates the dialog box. The "accept" is the name of the tile assigned to the OK button in the DCL file.

> Format in DCL   **Example**
> (**action_tile tile-name action-expression**)   (action_tile "accept" "(done_dialog)")

MANAGING DIALOG BOXES WITH AUTOLISP

When you load the DCL file of Example 1 and select the OK button, it does not perform the desired function (exit from the dialog box). This is because a dialog box cannot, by itself, execute the AutoCAD commands or the functions assigned to a tile. An application program is required to handle a dialog box. These application programs can be written in AutoLISP or ADS. The functions defined in AutoLISP and ADS can be used to load a DCL file, display the dialog box on the screen, prompt user input, set values in tiles, perform action associated with user input, and execute AutoCAD commands. Example 2 describes the use of an AutoLISP program to handle a dialog box.

Example 2

Write an AutoLISP program that will handle the dialog box and perform the functions shown in the dialog box of Example 1.

The following file is a listing of the DCL file of Example 1. This DCL file defines the dialog box **welcome1**, which contains only one action tile: **"OK"**. If you select the **OK** button, you must be able to exit the dialog box. As just mentioned, the dialog box will not perform by itself the function assigned to the **OK** button unless you write an application that will execute the functions defined in the dialog box.

```
welcome1    : dialog {
    label = "Sample Dialog Box";
    : text {
        label = "Welcome to the world of DCL";
    }
    : text {
    label = "Dialog Control Language";
    }
     : button {
        key = "accept";
        label = "OK";
        is_default = true;
    }
}
```

The following file is a listing of the AutoLISP program that loads the DCL file (**dclwel1**) of Example 1, displays the dialog box **welcome1** on, and defines the action for the **OK** button. **The line numbers are not a part of the program; they are for reference only.**

```
(defun C:welcome ( / dcl_id)                                         1
(setq dcl_id (load_dialog "dclwel1.dcl"))                            2
(new_dialog "welcome1" dcl_id)                                       3
(action_tile                                                         4
   "accept"                                                          5
   "(done_dialog)")                                                  6
(start_dialog)                                                       7
(unload_dialog dcl_id)                                               8
(princ)                                                              9
)                                                                   10
```

Explanation

Line 1
(defun C:welcome (/ dcl_id)
In this line, **defun** is an AutoLISP function that defines the function welcome. With the C: in front of the function name, the **welcome** function can be executed like an AutoCAD command. The **welcome** function has one local variable, **dcl_id.**

Line 2
(setq dcl_id (load_dialog "dclwel1.dcl"))
In this line, **(load_dialog "dclwel1.dcl")** loads the DCL file **dclwel1.dcl** and returns a positive integer. The **setq** function assigns this integer value to the local variable **dcl_id.**

Line 3
(new_dialog "welcome1" dcl_id)
In this line, the AutoLISP function **new_dialog** loads the dialog box **welcome1** that is defined in the DCL file (line 1 of DCL file). The variable **dcl_id** is an integer that identifies the DCL file.

Note
The dialog name (welcome1) in the AutoLISP program must be the same as the dialog name in the DCL file (welcome1).

Lines 4-6
(action_tile
  "accept"
  "(done_dialog)")
In the DCL file of Example 1, the **OK** button has been assigned an ASCII name, **accept** (key = "accept"). The first two lines associate the key (OK button) with the action expression. The action_tile initializes the association between the OK button and the action expression (done_dialog). If you select the **OK** button from the dialog box, the AutoLISP program reads that value and performs the function defined in the statement (done_dialog). The done_dialog function ends the dialog box.

Line 7
(start_dialog)
The **start_dialog** function enables the AutoLISP program to accept your input from the dialog box.

Line 8
(unload_dialog dcl_id)
This statement unloads the DCL file identified by the integer value of dcl_id.

Lines 9 and 10
(princ)
)
The **princ** function displays a blank on the screen. You use this to prevent display of the last expression in the command prompt area of the screen. If the **princ** function is not used, AutoCAD will display the value of the last expression. The closing parenthesis in the last line completes the definition of the welcome function.

ROW AND BOXED ROW TILES

Row Tile
Format in DCL: **row**

In a DCL file, several tiles can be grouped together to form a composite row or a composite column that is treated as a single tile. A row tile consists of several tiles grouped together in a horizontal row.

Boxed Row Tile
Format in DCL: **boxed_row**

In a boxed row, the tiles are grouped together in a row and a border is drawn around them, forming a box shape. If the boxed row has a label, it will be displayed left-justified at the top of the box, above the border line. If no label attribute is defined, only the box is displayed around the tile. On some systems, depending on the graphical user interface (GUI), the label may be displayed inside the border.

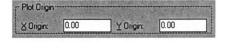

COLUMN, BOXED COLUMN, AND TOGGLE TILES

Column Tile
  Format in DCL: **column**

In a column tile, the tiles are grouped together in a vertical column to form a composite tile.

Boxed Column Tile
  Format in DCL: **boxed_column**

In a boxed column, the tiles are grouped together in a column and a border is drawn around the tiles. If the boxed column has a label, it will be displayed left-justified at the top of the box, above the border line. If there is no label attribute defined, only the box is displayed around the tile. On some systems, depending on the graphical user interface (GUI), the label may be displayed inside the border.

Toggle Tile
  Format in DCL: **toggle**

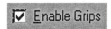

A toggle tile in a dialog box displays a small box on the screen with an optional label on the right of the box. Although the label is optional, you should label the toggle box so that users know what function is assigned to the toggle box. A toggle box has two states: on and off. When the function is turned on, a check mark is displayed in the box. Similarly, when a function is turned off, no check mark is displayed. The on or off state of the toggle box represents a Boolean value. When the toggle box is on, the Boolean value is 1; when the toggle box is off, the Boolean value is 0.

MNEMONIC ATTRIBUTE

Format in DCL	Example
mnemonic	mnemonic = "U"

A dialog box can have several tiles with labels. One of the ways you can select a tile is by using the arrow to highlight the tile and then pressing the accept key. This is possible if you have a pointing device like a digitizer or a mouse. If you do not have a pointing device, it may not be possible to select a tile. However, you can select a tile by using the mnemonic key assigned to the tile. For example, the mnemonic character for the Color tile is C. If you press the C key at the keyboard, it will highlight the Color tile. The mnemonic character is designated by underlining one of the characters in the label. The underlining is done automatically once you define the mnemonic attribute for the tile. You can select only one character **in the label** as a mnemonic character and you must use different mnemonic characters for different tiles. If a dialog box has two tiles with the same mnemonic character, only one tile will be selected when you press the mnemonic key. The mnemonic characters are not case-sensitive; therefore, they can be uppercase (C) or lowercase (c). However, the character in the label you select as a mnemonic character should be capitalized for easy identification.

Note
Some graphical user interfaces (GUIs) do not support mnemonic attributes. On such systems, you cannot select a tile by pressing the mnemonic key. However, you can still use pointing devices to select the tile.

Example 3   *General*

Write a DCL program for the object snap dialog box shown in Figure 15-7(a). The object snap tiles are arranged in two columns in a boxed row.

Before writing a program, especially a DCL program for a dialog box, you should determine the organization of the dialog box. It is given in Example 3. But when you develop a dialog box yourself, you must be careful when organizing the tiles in the dialog box. The structure of the DCL program depends on the desired output. In Example 3, the desired output is shown in Figure 15-7(a). This dialog box has two rows and two columns. The first row has two columns and the second row has no columns, as shown in Figure 15-7(b).

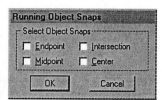

Figure 15-7(a) Dialog box for object snaps

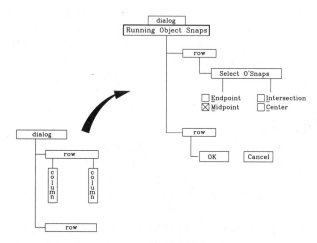

Figure 15-7(b) Dialog box for object snaps

The following file is a listing of the DCL file for Example 3. **The line numbers are not a part of the file; they are shown for reference only.**

```
osnapsh : dialog {                                                      1
    label = "Running Object Snaps";                                     2
  : boxed_row {                                                         3
   label = "Select Osnaps";                                            4
   : column {                                                          5
       : toggle {                                                      6
       label = "Endpoint";                                            7
       key = "Endpoint";                                              8
       mnemonic = "E";                                                9
       fixed_width = true;                                            10
            }                                                          11
       : toggle {                                                      12
       label = "Midpoint";                                            13
       key = "Midpoint";                                              14
       mnemonic = "M";                                                15
       fixed_width = true;                                            16
            }                                                          17
     }                                                                18
   : column {                                                          19
       : toggle {                                                      20
       label = "Intersection";                                        21
       key = "Intersection";                                          22
       mnemonic = "I";                                                23
       fixed_width = true;                                            24
            }                                                          25
       : toggle {                                                      26
       label = "Center";                                              27
       key = "Center";                                                28
       mnemonic = "C";                                                29
       fixed_width = true;                                            30
          }                                                            31
      }                                                                32
    }                                                                  33
     ok_cancel;                                                        34
   }                                                                   35
```

Explanation

Lines 3 and 4
: boxed_row {
 label = "Select Osnaps";
The **boxed_row** is a predefined cluster tile that draws a border around the tiles. The second line, **label = "Select Osnaps";**, will display the label (**Select Osnaps**) left-justified at the top of the box. The **label** is a predefined DCL attribute.

Lines 5-7
: column {
  : toggle {

label = "Endpoint";

The **column** is a predefined tile that will arrange the tiles within it into a vertical column. The **toggle** is another predefined tile that displays a small box in the dialog box with an optional text on the right side of the box. The **label** attribute will display the label (**Endpoint**) to the right of the toggle box.

Lines 8-11
key = "Endpoint";
mnemonic = "E";
fixed_width = true;
}
The key attribute assigns a name (**Endpoint**) to the tile. This name is then used by the application program to handle the tile. The second line, **mnemonic = "E";**, defines the keyboard mnemonic for **Endpoint**. This causes the character **E** of **Endpoint** to be displayed underlined in the dialog box. The attribute **fixed_width** controls the width of the tile. If the value of this attribute is true, the tile does not stretch across the width of the dialog box. The closing brace (}) on the next line completes the definition of the toggle tile.

Lines 31-35
```
            }
        }
      }
   ok_cancel;
}
```
The closing brace on line 32 completes the definition of the toggle tile, and the closing brace on line 33 completes the definition of the column of line 20. The closing brace on line 34 completes the definition of the boxed_row of line 3. The predefined tile, **ok_cancel**, displays the **ok** and **cancel** tiles in the dialog box. The closing brace on the last line completes the definition of the dialog box.

After writing the program, use the following commands to load the DCL file and display the dialog box on the screen. The name of the file is assumed to be **osnapsh.dcl**. If AutoCAD is successful in loading the file, it will return an integer. In the following example, the integer AutoCAD returns is assumed to be 1. The screen display after loading the dialog box is shown in Figure 15-8.

Command: (load_dialog "osnapsh.dcl")
1
Command: (new_dialog "osnapsh" 1)

AUTOLISP FUNCTIONS
logand and logior
These AutoLISP functions, **logand** and **logior**, are used to obtain the result of **logical bitwise AND and logical bitwise inclusive OR** of a list of numbers.

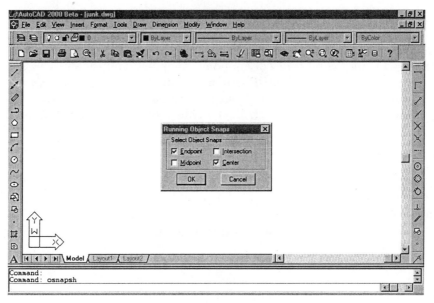

Figure 15-8 *Screen display with dialog box for Example 3*

Examples of logand function

(logand 2 7)	will return	2
(logand 8 15)	will return	8
(logand 6 15 7)	will return	6
(logand 6 15 1)	will return	0
(logand 1 4)	will return	0
(logand 1 5)	will return	1

Examples of logior function

(logior 2 7)	will return	7
(logior 8 15)	will return	15
(logior 6 15 7)	will return	15
(logior 6 15 1)	will return	15
(logior 1 4)	will return	5
(logior 1 5)	will return	5

One application of bit-codes is found in the object snap modes. The following is a list of bit-codes assigned to different object snap modes:

Snap Mode	Bit Code	Snap Mode	Bit Code
None	0	Intersection	32
Endpoint	1	Insertion	64
Midpoint	2	Perpendicular	128
Center	4	Tangent	256
Node	8	Nearest	512
Quadrant	16	Quick	1024

The system variable **OSMODE** can be used to specify a snap mode. For example, if the value assigned to **OSMODE** is 4, the Object snap mode is center. The bit-codes can be combined to produce multiple Object snap modes. For example, you can get endpoint, midpoint, and center object snaps by adding the bit-codes of endpoint, midpoint, and center (1 + 2 + 4 = 7) and assigning that value (7) to the **OSMODE** system variable. In AutoLISP, the existing snap modes information can be extracted by using the logand function. For example, if **OSMODE** is set to 7, the logand function can be used to check different bit-codes that represent object snaps.

(logand 1 7)	will return	1	(Endpoint)
(logand 2 7)	will return	2	(Midpoint)
(logand 4 7)	will return	4	(Center)

atof and rtos Functions

The **atof** function converts a string into a real number and the rtos function converts a number into a string.

Examples
(atof "3.5") will return   3.5
(rtos 8 15)   will return   "8.1500"

get_tile and set_tile Functions

The **get_tile** function retrieves the current value of a dialog box tile and the **set_tile** function sets the value of a dialog box tile.

Examples
(get_tile "midpoint")
(set_tile "xsnap" value)

Example 4

Write an AutoLISP program that will handle the dialog box and perform the functions as described in the dialog box of Example 3.

The following file is a listing of the AutoLISP program for Example 4. **The line numbers are not a part of the program; they are for reference only.**

```
;;Lisp program for setting Osnaps                                   1
;;Dialog file name is osnapsh.dcl                                   2
(defun c:osnapsh ( / dcl_id)                                        3
(setq dcl_id (load_dialog "osnapsh.dcl"))                           4
(new_dialog "osnapsh" dcl_id)                                       5
;;Get the existing value of object snaps and                       6
;;then write the values to the dialog box                          7
(setq osmode (getvar "osmode"))                                     8
(if (= 1 (logand 1 osmode))                                         9
   (set_tile "Endpoint" "1")                                       10
   )                                                               11
```

```
(if (= 2 (logand 2 osmode))                                              12
   (set_tile "Midpoint" "1")                                            13
   )                                                                     14
(if (= 32 (logand 32 osmode))                                          15
   (set_tile "Intersection" "1")                                       16
   )                                                                     17
(if (= 4 (logand 4 osmode))                                            18
   (set_tile "Center" "1")                                             19
   )                                                                     20
                                                                         21
;;Read the values as set in the dialog box and                         22
;;assign those values to AutoCAD variable osmode                        23
(defun setvars ()                                                       24
(setq osmode 0)                                                         25
(if (= "1" (get_tile "Endpoint"))                                      26
   (setq osmode (logior osmode 1))                                     27
   )                                                                     28
(if (= "1" (get_tile "Midpoint"))                                     29
   (setq osmode (logior osmode 2))                                     30
   )                                                                     31
(if (= "1" (get_tile "Intersection"))                                 32
   (setq osmode (logior osmode 32))                                    33
   )                                                                     34
(if (= "1" (get_tile "Center"))                                        35
   (setq osmode (logior osmode 4))                                     36
   )                                                                     37
   (setvar "osmode" osmode)                                            38
   )                                                                     39
                                                                         40
(action_tile "accept" "(setvars) (done_dialog)")                       41
(start_dialog)                                                          42
(princ)                                                                 43
)                                                                        44
```

Explanation

Lines 1 and 2
;;Lisp program for setting Osnaps
;;Dialog file name is osnapsh.dcl
These lines are comment lines, and all comment lines start with a semicolon. AutoCAD ignores the lines that start with a semicolon.

Lines 3-5
(defun c:osnapsh (/ dcl_id)
(setq dcl_id (load_dialog "osnapsh.dcl"))
(new_dialog "osnapsh" dcl_id)
In line 3, **defun** is an AutoLISP function that defines the **osnapsh** function. Because of the c: in front of the function name, the **osnapsh** function can be executed like an AutoCAD command.

The **osnapsh** function has one local variable, **dcl_id**. In line 4, the **(load_dialog "osnapsh.dcl")** loads the DCL file osnapsh.dcl and returns a positive integer. The **setq** function assigns this integer to the local variable, dcl_id. In line 5, the AutoLISP **new_dialog** function loads the dialog **osnapsh that is defined in the DCL file** (line 1 of the DCL file). The variable **dcl_id** has an integer value that identifies the DCL file.

Line 8
(setq osmode (getvar "osmode"))
In this line, **getvar "osmode"** has the value of the AutoCAD system variable **osmode**, and the **setq** function sets the **osmode** variable equal to that value. Note that the first **osmode** is just a variable, whereas the second osmode, in quotes (**"osmode"**), is the system variable.

Lines 9-11
(if (= 1 (logand 1 osmode))
  (set_tile "Endpoint" "1")
)

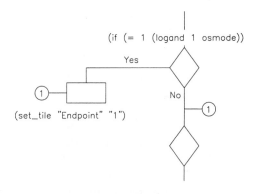

In line 9, the **(logand 1 osmode)** will return 1 if the bit-code of endpoint (1) is a part of the **OSMODE** value. For example, if the value assigned to **OSMODE** is 7 (1 + 2 + 4 = 7), (logand 1 7) will return 1. If **OSMODE** is 6 (2 + 4 = 6), then (logand 1 6) will return 0. The AutoLISP **if** function checks whether the value returned by **(logand 1 osmode)** is 1. If the function returns **T** (true), the instructions described in the second line are carried out. Line 10 sets the value of the **Endpoint** tile to 1; that displays a check mark in the toggle box. The closing parenthesis in line 11 completes the definition of the **if** function. If the expression **(if (= 1 (logand 1 osmode))** returns **nil**, the program skips to line 12 of the program.

Lines 24 and 25
(defun setvars ( )
(setq osmode 0)
Line 24 defines a **setvars** function, and line 25 sets the value of the **osmode** variable to zero.

Lines 26-28
(if (= "1" (get_tile "Endpoint"))
  (setq osmode (logior osmode 1))
)
In line 26, the **(get_tile "Endpoint")** obtains the value of the toggle tile named **Endpoint**. If the Endpoint toggle tile is on, the value it returns is 1; if the toggle tile is off, the value it returns is 0. In line 27, the **setq** function sets the value of **osmode** to the value returned by **(logior osmode 1)**. For example, if the initial value of osmode is 0, then **(logior osmode 1)** will return 1. Similarly, if the initial value of **osmode** is 1, then **(logior osmode 2)** will return 3.

Lines 41 and 42
(action_tile "accept" "(setvars) (done_dialog)")
(start_dialog)
Line 42, **(start_dialog)**, starts the dialog box. In the dialog file the name assigned to the **OK** button is **"accept"**. When you select the **OK** button in the dialog box, the program executes the **setvars** function that updates the value of the **osmode** system variable and sets the selected object snaps.

PREDEFINED RADIO BUTTON, RADIO COLUMN, BOXED RADIO COLUMN, AND RADIO ROW TILES

Predefined Radio Button Tile
Format in DCL: **radio_button**

The **radio button** is a predefined active tile. Radio buttons can be arranged in a row or in a column. The unique characteristic of radio buttons is that only one button can be selected at a time. For example, if there are buttons for scientific, decimal, and engineering units, only one can be selected. If you select the decimal button, the other two buttons will be turned off automatically. Because of this special characteristic, radio buttons must be used only in a radio row or in a radio column. The label for the radio button is optional; in most systems the label appears to the right of the button.

Predefined Radio Column Tile
Format in DCL: **radio_column**

The radio column is an active predefined tile where the radio button tiles are arranged in a column and only one button can be selected at a time. When the radio buttons are arranged in a column, the buttons are next to each other vertically and are easy to select. Therefore, you should arrange radio buttons in a column to make it easy to select a radio button.

Predefined Boxed Radio Column Tile
Format in DCL: **boxed_radio_column**

The **boxed radio column** is an active predefined tile where a border is drawn around the radio column.

Predefined Radio Row Tile
Format in DCL: **radio_row**

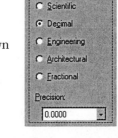

The **radio row** consists of radio button tiles arranged in a row. Only one button can be selected at a time. The radio row can become quite long if there are several radio button tiles and labels. In selecting a radio button, the cursor travel will in-

crease because the buttons are not immediately next to each other. Therefore, you should avoid using radio button tiles in a row, especially if there are more than two.

Example 5

Write a DCL program for a dialog box that will enable you to select different units and unit precision as shown in Figure 15-9. Also, write an AutoLISP program that will handle the dialog box.

The following file is a listing of the DCL program for the dialog box shown in Figure 15-9. The name of the dialog box is **dwgunits. The line numbers are not a part of the file; they are shown here for reference only.**

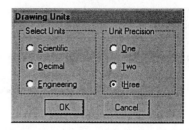

Figure 15-9 *Dialog box for Example 5*

```
dwgunits : dialog {                                             1
  label = "Drawing Units";                                     2
  : row {                                                       3
   : boxed_column {                                             4
    label = "Select Units";                                    5
    : radio_column {                                            6
     : radio_button {                                           7
      key = "scientific";                                       8
      label = "Scientific";                                     9
      mnemonic = "S";                                          10
     }                                                         11
     : radio_button {                                          12
      key = "decimal";                                         13
      label = "Decimal";                                       14
      mnemonic = "D";                                          15
     }                                                         16
     : radio_button {                                          17
      key = "engineering";                                     18
      label = "Engineering";                                   19
      mnemonic = "E";                                          20
     }                                                         21
    }                                                          22
   }                                                           23
   : boxed_column {                                            24
    label = "Unit Precision";                                  25
    : radio_column {                                           26
     : radio_button {                                          27
      key = "one";                                             28
      label = "One";                                           29
      mnemonic = "O";                                          30
     }                                                         31
     : radio_button {                                          32
```

```
              key = "two";                                                    33
              label = "Two";                                                  34
              mnemonic = "T";                                                 35
              }                                                               36
            : radio_button {                                                  37
              key = "three";                                                  38
              label = "Three";                                               39
              mnemonic = "h";                                                 40
              }                                                               41
            }                                                                 42
          }                                                                   43
        }                                                                     44
      ok_cancel;                                                              45
      }                                                                       46
```

Explanation

Lines 4 and 5
: boxed_column {
  label = "Select Units";
The **boxed_column** is an active predefined tile that draws a border around the column. Line 5, **label = "Select Units";**, displays the label **(Select Units)** at the top of the column. The **label** is a predefined DCL attribute.

Lines 6-8
: radio_column {
  : radio_button {
    key = "scientific";
The **radio_column** is a predefined active tile that will arrange the tiles within it in a vertical column and draw a border around the column. The **radio_button** is another active predefined tile that displays a radio button in the dialog box, with an optional text to the right of the button. The key attribute assigns a name (scientific) to the tile. This name is then used by the application program to handle the tile.

Lines 9-11
    label = "Scientific";
    mnemonic = "S";
    }
The label attribute will display the label (Scientific) on the right of the toggle box. The second line, **mnemonic = "S";**, defines the keyboard mnemonic for **Scientific**. This causes the letter S of **Scientific** to be displayed underlined in the dialog box. The closing brace (}) on the next line completes the definition of the toggle tile.

Lines 41-46
```
              }
            }
          }
        }
```

Chapter 15

ok_cancel;
}
The closing brace on line 41 completes the definition of the radio button and the closing brace on the second line completes the definition of the radio column. The closing brace on the next line completes the definition of the boxed_column, and the closing brace on the next line completes the definition of the row on line 3 of the DCL file. The predefined tile, **ok_cancel**, displays the **OK** and **Cancel** tiles in the dialog box. The last closing brace completes the definition of the dialog box.

Use the following commands to load and display the dialog file on the screen. The name of the file is assumed to be **dwgunits.dcl**, and the name of the dialog is **dwgunits**. If AutoCAD is successful in loading the file, it will return an integer. In the following examples, the integer that AutoCAD returns is assumed to be 5. The screen display after loading the dialog box is shown in Figure 15-10.

> **Command:** (load_dialog "osnapsh.dcl")
> **5**
> **Command:** (new_dialog "osnapsh" 5)

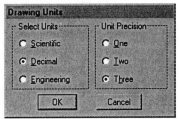

Figure 15-10 Dialog box for Example 5

The following file is a listing of the AutoLISP program that loads, displays, and handles the dialog box for Example 5. **The line numbers are not a part of the file; they are shown here for reference only.**

```
;;Lisp program dwgunits.lsp for setting units                     1
;;and precision. Dialog file name dwgunits.dcl                    2
;                                                                 3
(defun c:dwgunits ( / dcl_id)                                    4
(setq dcl_id (load_dialog "dwgunits.dcl"))                       5
(new_dialog "dwgunits" dcl_id)                                   6
;                                                                 7
;;Get the existing values of lunits and luprec                   8
;;and turn the corresponding radio_button on                     9
;                                                                10
(setq lunits (getvar "lunits"))                                 11
(if (= 1 lunits)                                                12
   (set_tile "scientific" "1")                                  13
   )                                                            14
(if (= 2 lunits)                                                15
```

```
          (set_tile "decimal" "1")                              16
          )                                                     17
     (if (= 3 lunits)                                           18
          (set_tile "engineering" "1")                          19
          )                                                     20
     ;                                                          21
     (setq luprec (getvar "luprec"))                            22
     (if (= 1 luprec)                                           23
          (set_tile "one" "1")                                  24
          )                                                     25
     (if (= 2 luprec)                                           26
          (set_tile "two" "1")                                  27
          )                                                     28
     (if (= 3 luprec)                                           29
          (set_tile "three" "1")                                30
          )                                                     31
                                                                32
     ;;Read the value of the radio_buttons and                 33
     ;;assign it to AutoCAD lunit and luprec variables          34
     ;                                                          35
     (action_tile "scientific" "(setq lunits 1)")              36
     (action_tile "decimal" "(setq lunits 2)")                 37
     (action_tile "engineering" "(setq lunits 3)")             38
     ;                                                          39
     (action_tile "one" "(setq luprec 1)")                     40
     (action_tile "two" "(setq luprec 2)")                     41
     (action_tile "three" "(setq luprec 3)")                   42
     (action_tile "accept" "(done_dialog)")                    43
     ;                                                          44
     (start_dialog)                                            45
     (setvar "lunits" lunits)                                  46
     (setvar "luprec" luprec)                                  47
     (princ)                                                   48
     )                                                         49
```

Explanation

Lines 1-3
;;Lisp program dwgunits.lsp for setting units
;;and precision. Dialog file name dwgunits.dcl
;
The first three lines of this program are comment lines, and all comment lines start with a semicolon. AutoCAD ignores them.

Lines 4-6
(defun c:dwgunits (/ dcl_id)
(setq dcl_id (load_dialog "dwgunits.dcl"))
(new_dialog "dwgunits" dcl_id)

In line 4, **defun** is an AutoLISP function that defines the **dwgunits** function, which has one local variable, **dcl_id.** The **c:** in front of the function name, **dwgunits**, makes the **dwgunits** function act like an AutoCAD command. In the next line, **(load_dialog "dwgunits.dcl")** loads the DCL file **dwgunits.dcl** and returns a positive integer. The **setq** function assigns this integer to the local variable, dcl_id. In the next line, the AutoLISP **new_dialog** function displays the **dwgunits** dialog box that is defined in the DCL file (line 1 of DCL file). The **dcl_id** variable is an integer that identifies the DCL file.

Line 11
(setq lunits (getvar "lunits"))
In this line, **getvar "lunits"** has the value of the AutoCAD system variable, **lunits**, and the **setq** function sets the **lunits** variable equal to that value. The first lunits is a variable, whereas the second **lunits**, in quotes (**"lunits"**), is a system variable.

(if (= 1 luprec)

Yes

No

(set_tile "one" "1")

Lines 12-14
(if (= 1 lunits)
    (set_tile "scientific" "1")
)
The **if** function (AutoLISP function) checks whether the value of the variable **lunits** is 1. If the function returns **T** (true), the instructions described in the next line are carried out. Line 13 sets the value of the tile named **"scientific"** equal to 1; that turns the corresponding radio button on. The closing parenthesis in line 14 completes the **if** function. If the **if** function, **(if (= 1 lunits)**, returns **nil**, the program skips to line number 14.

Line 22
(setq luprec (getvar "luprec"))
In this line, **getvar "luprec"** has the value of the AutoCAD system variable **luprec**, and the **setq** function sets the luprec variable equal to that value. The first **luprec** is a variable, whereas the second luprec, in quotes (**"luprec"**), is a system variable.

Lines 23-25
(if (= 1 luprec)
    (set_tile "one" "1")
)
In line 23, the **if** function checks whether the value of the **luprec** variable is 1. If the function returns **T** (true), the instructions described in the next line are carried out. This line sets the value of the tile named **"one"** to 1; that turns the corresponding radio button on. The closing parenthesis in the next line completes the definition of the **if** function. If the **if** function returns **nil**, the program skips to line 25.

(if (= 1 luprec)

Yes

No

(set_tile "one" "1")

Line 36

(action_tile "scientific" "(setq lunits 1)")

If the radio button tile named **"scientific"** is turned on, the **setq** function sets the value of the AutoCAD system variable **lunits** to 1. The **lunits** system variable controls the drawing units. The following is a list of the integer values that can be assigned to the **lunits** system variable:

1	Scientific	4	Architectural
2	Decimal	5	Fractional
3	Engineering		

Line 40

(action_tile "one" "(setq luprec 1)")

If the radio button tile named **"one"** is turned on, the **setq** function sets the value of the AutoCAD system variable **luprec** to 1. The **lunits** system variable controls the number of decimal places in a decimal number or the denominator of a fractional or architectural unit.

Lines 43-47

(action_tile "accept" "(done_dialog)")

;

(start_dialog)

(setvar "lunits" lunits)

(setvar "luprec" luprec)

In the dialog file, the name assigned to the **OK** button is **"accept"**. When you select the **OK** button in the dialog box, the program executes the function defined in the **dwgunits** dialog box. The **(start_dialog)** function starts the dialog box. The **(setvar "lunits" lunits)** and **(setvar "luprec" luprec)** set the values of the **lunits** and **luprec** system variables to lunits and luprec, respectively.

EDIT BOX TILE

Format in DCL: **edit_box**

X Spacing	0.500
Y Spacing	0.500
Snap Angle	0
X Base	0.000

The **edit box** is a predefined active tile that enables you to enter or edit a single line of text. If the text is longer than the length of the edit box, the text will automatically scroll to the right or left horizontally. The label for the edit box is optional, and it is displayed to the left of the edit box.

width AND edit_width ATTRIBUTES

width Attribute

Format in DCL: **width**   **Example:** width = 22

The **width** attribute is used to keep the width of the tile to a desired size. The value assigned to the **width** attribute can be a real number or an integer that represents the distance in character width. The value assigned to the **width** attribute defines the minimum width of the tile. For example, in this example the minimum width of the tile is 22. However, the tile will automatically stretch if more space is available. It will retain the size of 22 only if the **fixed_width** attribute is assigned to the tile. The **width** attribute can be used with any tile.

edit_width Attribute

Format in DCL: **edit_width**   **Example:** edit_width = 10

The **edit_width** attribute is used with the predefined edit box tiles, and it determines the size of the edit box in character width units. If the width of the edit box is 0, or if it is not specified and the fixed_width attribute is not assigned to the tile, the tile will automatically stretch to fill the available space. When the edit box is stretched, the PDB facility inserts spaces between the edit box and the label so that the box is right justified and the label is left justified.

Example 6   *General*

Write a DCL program for a dialog box (Figure 15-11) that will enable you to turn the snap and grid on and off. You should also be able to edit the X and Y values of snap and grid. Also, write an AutoLISP program that will load, display, and handle the dialog box.

The following file is a listing of the DCL file for Example 6.

```
dwgaids : dialog {
  label = "Drawing Aids";
  : row {
   : boxed_column {
    label = "SNAP";
    fixed_width = true;
    width = 22;
    : toggle {
     label = "On";
     mnemonic = "O";
     key = "snapon";
     }
    : edit_box {
     label = "X-Spacing";
     mnemonic = "X";
     key = "xsnap";
     edit_width = 10;
     }
    : edit_box {
     label = "Y-Spacing";
     mnemonic = "Y";
     key = "ysnap";
     edit_width = 10;
     }
    }
   : boxed_column {
    label = "GRID";
    fixed_width = true;
```

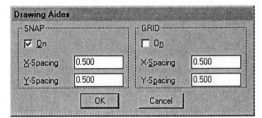

Figure 15-11 *Dialog box for Example 6*

```
        width = 22;
      : toggle {
        label = "On";
        mnemonic = "n";
        key = "gridon";
        }
      : edit_box {
        label = "X-Spacing";
        mnemonic = "S";
        key = "xgrid";
        edit_width = 10;
        }
      : edit_box {
        label = "Y-Spacing";
        mnemonic = "p";
        key = "ygrid";
        edit_width = 10;
        }
      }
    }
  ok_cancel;
  }
```

The following file is a listing of the AutoLISP program for Example 6. When the program is loaded and run, it will load, display, and control the dialog box (Figure 15-12).

```
;;Lisp program for Drawing Aids dialog box
;;Dialog file name is dwgaids.dcl
;
(defun c:dwgaids( / dcl_id snapmode xsnap ysnap
 orgsnapunit gridmode gridsnap xgrid ygrid orggridunit)
(setq dcl_id (load_dialog "dwgaids.dcl"))
(new_dialog "dwgaids" dcl_id)

;;Get the existing value of snapmode and snapunit
;;and write those values to the dialog box
(setq snapmode (getvar "snapmode"))
(if (= 1 snapmode)
   (set_tile "snapon" "1")
   (set_tile "snapon" "0")
   )
(setq orgsnapunit (getvar "snapunit"))
(setq xsnap (car orgsnapunit))
(setq ysnap (cadr orgsnapunit))
(set_tile "xsnap" (rtos xsnap))
(set_tile "ysnap" (rtos ysnap))
;
```

```
;;Get the existing value of gridmode and gridunit
;;and write those values to the dialog box
(setq gridmode (getvar "gridmode"))
(if (= 1 gridmode)
   (set_tile "gridon" "1")
   (set_tile "gridon" "0"))
(setq orggridunit (getvar "gridunit"))
(setq xgrid (car orggridunit))
(setq ygrid (cadr orggridunit))
(set_tile "xgrid" (rtos xgrid))
(set_tile "ygrid" (rtos ygrid))
;;Read the values set in the dialog box and
;;then change the associated AutoCAD variables
(defun setvars ()
(setq xsnap (atof (get_tile "xsnap")))
(setq ysnap (atof (get_tile "ysnap")))
(setvar "snapunit" (list xsnap ysnap))
(if (= "1" (get_tile "snapon"))
   (setvar "snapmode" 1)
   (setvar "snapmode" 0))
(setq xgrid (atof (get_tile "xgrid")))
(setq ygrid (atof (get_tile "ygrid")))
(setvar "gridunit" (list xgrid ygrid))
(if (= "1" (get_tile "gridon"))
   (progn
     (setvar "gridmode" 0)
     (setvar "gridmode" 1))
   (setvar "gridmode" 0)
   ))
(action_tile "accept" "(setvars) (done_dialog)")
(start_dialog)
(princ)
)
```

SLIDER AND IMAGE TILES

Slider Tile

Format in DCL
slider

A slider tile is an active predefined tile that consists of a slider bar (rectangular strip), a small indicator box, and the direction arrows at the ends of the slider bar. The slider tile can be used to obtain a string value. The value is determined by the position of the indicator box in the slider bar. The string value returned by the slider tile can then be used by the application program. For example, you can use the value returned by the aperture slider tile to set AutoCAD's **aperture**

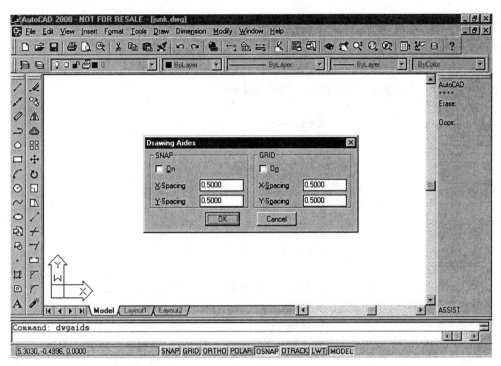

Figure 15-12 Screen display with the dialog box for Example 6

system variable. The indicator box can be dragged to the left or right by positioning the arrow on the indicator box and then moving the arrow while holding the pick button down. It can also be moved in increments by positioning the arrow to the left or right of the indicator box or in the arrow box of the slider, and clicking it. The slider tiles can be horizontal or vertical. If the slider tile is horizontal, the values increase from left to right; if the slider is vertical the values increase from bottom to top. The values returned by the slider tile are always integers.

Image Tile

Format in DCL
image

The image tile is a predefined tile that is used to display graphical information. For example, it can be used to display the aperture box, linetypes, icons, and text fonts in dialog boxes. It consists of a rectangular box and the vector graphics that are displayed within the box.

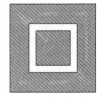

min_value, max_value, small_increment, and big_increment Attributes

min_value and max_value Attributes

Format in DCL	**Examples**
min_value	min_value = 2
max_value	max_value = 15

The **min_value** attribute and the **max_value** attribute are predefined attributes that specify the minimum and maximum value that the **slider** tile will return. In the previous example, the minimum value is 2 and the maximum value is 15. If you do not assign these attributes to a slider tile, the slider automatically assumes the default values. For the **min_value** attribute the default value is 0, for the **max_value** attribute the default value is 10,000.

small_increment and big_increment Attributes

Format in DCL	**Examples**
small_increment	small_increment = 1
big_increment	big_increment = 1

The value assigned to the **small_increment** attribute and the **big_increment** attribute determines the increment value of the slider incremental control. For example, if the increment is 1, the values returned by the slider will be in the increment of 1. If these attributes are not assigned to a slider tile, the slider automatically assumes the default values. The default value of **small_increment** is one one-hundredth (1/100) of the slider range and the default value of **big_increment** is one-tenth (1/10) of the slider range.

aspect_ratio and color Attributes

aspect_ratio Attribute

Format in DCL
aspect_ratio

Example
aspect_ratio = 1

The aspect_ratio is a predefined attribute that can be used with an image. The aspect ratio is the ratio of the width of the image box to the height of the image box (width/height). You can control the size of the image box by assigning an integer value to the width attribute and the height attribute. You can also control the size

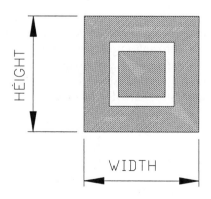

of the image box by assigning an integer value to one of the attributes (width or height) and assigning a real or integer value to the aspect_ratio attribute. For example, if the value assigned to the height attribute is five and the value assigned to the aspect_ratio is 0.5, the width of the image box will be 2.5 (width = height x aspect_ratio). In this case, you do not need to specify the width attribute.

color Attribute

Format in DCL
color

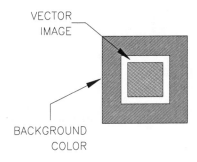

Example
color = 2

The color is a predefined attribute that can be used with the image box to control the background color. The integer number assigned to the color attribute specifies an AutoCAD color index. In the above example, the background color is yellow (color number 2). The color of the vector image that is drawn inside the image box is determined by the color specified in the vector_image attribute.

Example 7   *General*

Write a DCL program for the dialog box in Figure 15-13(a) that will enable you to change the size of the aperture box.

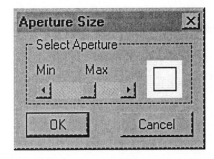

Figure 15-13(a) Dialog box for Example 7

The dialog box shown in Figure 15-13(a) has two rows and two columns. The first column has two items, label (**Min .... Max**) and the **slider bar**. The second column has only one item (the image box). The second row has the **OK** and **Cancel** buttons. As shown in Figure 15-13(b), the dialog box also has a dialog label (Aperture Size) and a row label (Select Aperture). The following file is a listing of the DCL file for Example 7. The line numbers are not a part of the file; they are shown here for reference only.

```
aprtsize : dialog {                                              1
  label = "Aperture Size";                                       2
```

Chapter 15

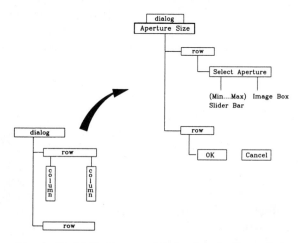

Figure 15-13(b) *Layout of dialog box for Example 7*

```
: boxed_row {                                              3
 label = "Select Aperture";                                4
: column {                                                 5
   fixed_width = true;                                      6
  : text {                                                 7
   label = "Min       Max";                                8
   alignment = centered;                                    9
   }                                                       10
  : slider {                                               11
   key = "aperture_slider";                                12
   min_value = 2;                                          13
   max_value = 17;                                         14
   width = 15;                                             15
   height = 1;                                             16
   big_increment = 1;                                      17
   fixed_width = true;                                     18
   fixed_height = true;                                    19
   }                                                       20
 }                                                         21
 : image {                                                 22
   key = "aperture_image";                                 23
   aspect_ratio = 1;                                       24
   width = 5;                                              25
   color = 2;                                              26
   }                                                       27
 }                                                         28
  ok_cancel;                                               29
}                                                          30
```

Lines 11, 12
: slider {
key = "aperture_slider";
In the first line, **: slider** starts the definition of the slider tile. The **slider** is an active predefined tile that consists of a slider bar, a small indicator box, and the direction arrows at the ends of the slider bar. The second line, **key = "aperture_slider"**, assigns the name, **aperture_slider**, to this slider tile.

Lines 13, 14
min_value = 2;
max_value = 17;
The **min_value** is a predefined attribute that specifies the minimum value that the **slider** tile will return. Similarly, the **max_value** attribute specifies the maximum value that the slider tile will return. In the above two lines, the minimum value is 2 and the maximum value is 17.

Lines 15-17
width = 15;
height = 1;
big_increment = 1;
The first line specifies the width of the slider tile and the second line specifies the height of the slider tile. The third line, **big_increment = 1;**, defines the increment of the indicator box in the slider bar.

Lines 22, 23
: image {
 key = "aperture_image";
The first line, **: image {**, starts the definition of the image box and the second line assigns a name, **aperture_image**, to the image tile. The image is a predefined tile that is used to display graphical information.

Lines 24-26
aspect_ratio = 1;
width = 5;
color = 2;
The first line, **aspect_ratio = 1;**, defines the ratio of the width of the image box to the height of the image box. The second line, **width = 5;**, specifies the width of the image box. Since the aspect ratio is 1, the height of the image box is 5 (width/height = aspect_ratio). The third line, **color = 2;**, assigns the AutoCAD color number 2 (yellow) to the background of the image box.

AUTOLISP FUNCTIONS

dimx_tile and dimy_tile

The AutoLISP function, **dimx_tile**, obtains the width dimension (x_aperture) of the specified image box along the X-axis and the **dimy_tile** function obtains the height dimension (y_aperture) along the Y-axis. In the following examples, "aperture_image" is the name of the image tile that is assigned by using the **key** attribute in the DCL file (key = "aperture_image").

Chapter 15

Format   **Examples**
(**dimx_tile tilename**)   (dimx_tile "aperture_image")
(**dimy_tile tilename**)   (dimy_tile "aperture_image")

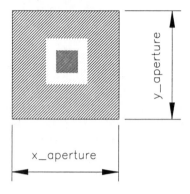

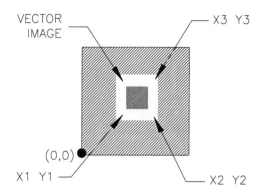

vector_image

The AutoLISP function **vector_image** draws a vector (line) in the active image box, between the points that are defined in the **vector_image** function. In the following example, AutoCAD will draw a line from point 1,1 to point 3,3 of the image box and the color of this vector will be AutoCAD's color number 1 (red).

Format
(**vector_image x1 y1 x2 y2 color**)

Example
(vector_image 1 1 3 3 1)
        Where   **1 1** --------------- X and Y coordinates of first point
                       3 ------------------ X and Y coordinates of second point
                       3 ------------------ AutoCAD color index

fill_image

The AutoLISP function, **fill_image**, fills the rectangle in the active image box with the color specified in the **fill_image** function. For example, in the following example the active image box will be filled with yellow color (AutoCAD's color number 2). The filled rectangle is determined by the coordinates of the first point (x1 y1) and the second point (x2 y2), the two opposite corners of the rectangle.

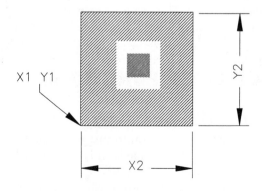

Format
(**fill_image x1 y1 x2 y2 color**)

Example
(fill_image 0 0 x_aperture y_aperture 2)

start_image

The AutoLISP function, **start_image**, starts the image whose name is specified in the **start_image** function. In the following example, the name of the image tile is **"aperture_image"** which is assigned by using **key** attribute in the DCL file (key = "aperture_image").

Format
(start_image)

Example
(start_image "aperture_image")

end_image

The AutoLISP function **end_image** ends the active image whose name is specified in the **end_image** function. In the following example, the name of the active image tile is **"aperture_image"**.

Format
(end_image)

Example
(end_image "aperture_image")

$value

The **$value** function is an AutoLISP action expression that retrieves the string value from the tile of a dialog box. The tile could be an edit box or a toggle box. In Example 8, the **$value** function retrieves the current value of the active tile and the setq function assigns that value to the variable, aprt_size.

Format
$value

Example
(setq aprt_size $value)

Example 8

Write an AutoLISP program that will load, display, and control the dialog box of Example 7 (Figure 15-13(a)). The flowchart and screen display for this example are shown in Figures 15-14 and 15-15.

The following file is a listing of the AutoLISP file for Example 8. The line numbers are not a part of the file; they are shown here for reference only.

```
;;APRTSIZE.LSP, AutoLISP program for Aperture        1
;; Dialog Box. DCL file name — APRTSIZE.DCL          2
;                                                     3
(defun c:aprtsize ( )                                4
  (setq dcl_id (load_dialog "aprtsize"))             5
  (new_dialog "aprtsize" dcl_id)                     6
;                                                     7
;Obtain value of the system variable "aperture",     8
;calculate X and Y values of vector image, and draw  9
;vector image in the image box.                       10
```

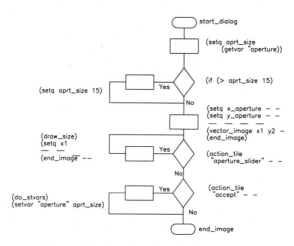

Figure 15-14 *Flowchart for Example 8*

```
(setq aprt_size (getvar "aperture"))                              11
(if (> aprt_size 15)                                              12
  (setq aprt_size 15)                                             13
  )                                                               14
(setq x_aperture (dimx_tile "aperture_image"))                   15
(setq y_aperture (dimy_tile "aperture_image"))                   16
(set_tile "aperture_slider" (itoa aprt_size))                    17
(setq x1 (- (/ x_aperture 2) aprt_size))                         18
(setq x2 (+ (/ x_aperture 2) aprt_size))                         19
(setq y1 (- (/ y_aperture 2) aprt_size))                         20
(setq y2 (+ (/ y_aperture 2) aprt_size))                         21
(start_image "aperture_image")                                   22
(fill_image 0 0 x_aperture y_aperture 2)                         23
(vector_image x1 y1 x2 y1 1)                                      24
(vector_image x2 y1 x2 y2 1)                                      25
(vector_image x2 y2 x1 y2 1)                                      26
(vector_image x1 y2 x1 y1 1)                                      27
(end_image)                                                      28
(action_tile "aperture_slider"                                   29
   "(draw_size (setq aprt_size (atoi $value)))")                 30
(action_tile "accept" "(do_setvars)(done_dialog)")              31
 (start_dialog)                                                  32
(princ)                                                          33
)                                                               34
;Set aperture variable "aperture" equal to aprt_size            35
(defun do_setvars ( )                                            36
 (setvar "aperture" aprt_size)                                   37
 )                                                              38
;                                                               39
```

```
;Calculate the X and Y coordinates of vector image        40
;and draw the aperture image in the image box.            41
(defun draw_size (aprt_size)                              42
  (setq x1 (- (/ x_aperture 2) aprt_size))                43
  (setq x2 (+ (/ x_aperture 2) aprt_size))                44
  (setq y1 (- (/ y_aperture 2) aprt_size))                45
  (setq y2 (+ (/ y_aperture 2) aprt_size))                46
  (start_image "aperture_image")                          47
  (fill_image 0 0 x_aperture y_aperture -2)               48
  (vector_image x1 y1 x2 y1 1)                             49
  (vector_image x2 y1 x2 y2 1)                             50
  (vector_image x2 y2 x1 y2 1)                             51
  (vector_image x1 y2 x1 y1 1)                             52
  (end_image)                                             53
)                                                         54
```

Lines 11-13
(setq aprt_size (getvar "aperture"))
(if (> aprt_size 15)
  (setq aprt_size 15)

In the first line, the AutoLISP function **getvar** obtains the value of the **"aperture"** system variable and the **setq** function assigns that value to the **aprt_size** variable. The second line,**(if (> aprt_size 15)**, checks whether the value of the variable aprt_size is greater than 15. If it is, the third line, **(setq aprt_size 15)**, sets the value of the **aprt_size** variable equal to 15. The value of the "aperture" system variable must be equal to or less than 15 to display the vector image in the image box because the maximum size of the image box in this example is 15.

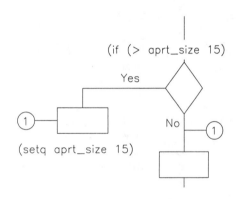

Lines 15, 16
(setq x_aperture (dimx_tile "aperture_image"))
(setq y_aperture (dimy_tile "aperture_image"))
In the first line, the AutoLISP function, **dimx_tile**, retrieves the X-dimension of the image tile and the setq function assigns that value to the **x_aperture** variable. Similarly, the **dimy_tile** function retrieves the Y-dimension of the image tile.

Line 17
(set_tile "aperture_slider" (itoa aprt_size))
The AutoLISP function **itoa** changes the integer value of the **aprt_size** variable into a string and the **set_tile** function assigns that value to **"aperture_slider"**. The **aperture_slider** is the name of the slider tile in the DCL file (DCL file of Example 7).

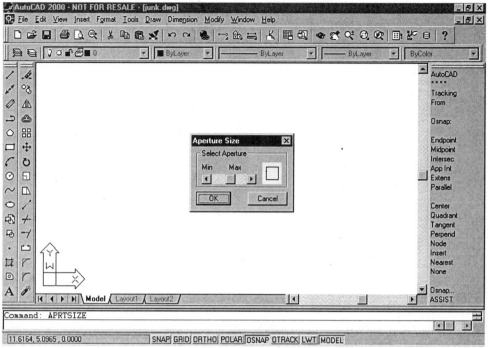

Figure 15-15 Screen display with the dialog box for Example 8

Lines 18-21
(setq x1 (- (/ x_aperture 2) aprt_size))
(setq x2 (+ (/ x_aperture 2) aprt_size))
(setq y1 (- (/ y_aperture 2) aprt_size))
(setq y2 (+ (/ y_aperture 2) aprt_size))
The first line calculates the X-coordinate of the
lower left corner of the aperture box. The value
is calculated by dividing the **x_aperture** distance by 2 and then subtracting the distance
aprt_size from it. The aprt_size has the same
value as that of the system variable, "aperture".
Aperture is the distance of the sides of the aperture box from the crosshair lines, measured
in pixels. Similarly, the second line calculates
the X-coordinate of the lower-right corner. The
next two lines calculate the Y-coordinates of
these two points.

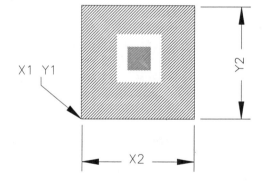

Lines 22, 23
(start_image "aperture_image")
(fill_image 0 0 x_aperture y_aperture 2)
The AutoLISP function **start_image** starts the image box and **aperture_image** is the name of the
image box as defined in the DCL file. The fill_image function fills the image within the specified

coordinates with AutoCAD color index 2 (yellow).

Lines 27, 28
(vector_image x1 y2 x1 y1 1)
(end_image)
The AutoLISP function **vector_image** draws a vector (line) from the point with coordinates x1 y2 to the point with coordinates x1 y1. The **end_image** function ends the image.

Lines 29-31
(action_tile "aperture_slider"
  "(draw_size (setq aprt_size (atoi $value)))")
(action_tile "accept" "(do_setvars)(done_dialog)")
The AutoLISP action expression **$value** retrieves the string value of the active slider tile and the atoi function returns the integer value of the string. This value is then used by the **draw_size** function to draw the vector image of the aperture box. The **aperture_slider** is the name of the slider tile defined in the DCL file. The third line executes the **do_setvars** function and ends the dialog box when you select OK from the dialog box. The **draw_size** and **do_setvars** functions are defined in the program.

Lines 36, 37
(defun do_setvars ( )
  (setvar "aperture" aprt_size)
The defun function defines the **do_setvars** function. The **setvar** function sets the value of the AutoCAD system variable, **"aperture"**, equal to **aprt_size**.

Lines 42, 43
(defun draw_size (aprt_size)
  (setq x1 (- (/ x_aperture 2) aprt_size))
The first line defines the **draw_size** function with one argument, **aprt_size**, and the second line calculates the value of X-coordinate of the first vector.

Review Questions

1. A dialog control language (DCL) file can contain descriptions of multiple files. (T/F)

2. In a DCL file, you do not need to specify the size of a dialog box. (T/F)

3. Dialog boxes are not dependent on the platform. (T/F)

4. Dialog boxes can contain several OK buttons. (T/F)

5. The numeric values assigned to the attributes in a DCL file can be both integers and real numbers. (T/F)

Chapter 15

6. The reserved names in DCL are case-sensitive. (T/F)

7. The label attribute used in a button tile has no default value. (T/F)

8. In a dialog box, only one button can be assigned the true value for the is_default attribute. (T/F)

9. The load_dialog function displays the dialog box on the screen. (T/F)

10. You cannot select a tile by using the mnemonic key assigned to the tile. (T/F)

11. Mnemonic characters are case-sensitive. (T/F)

12. Bit-codes cannot be combined to produce multiple object snaps. (T/F)

13. The radio button is a predefined attribute. (T/F)

14. You should arrange the radio buttons in a column setting. (T/F)

15. The edit box active tile allows you to enter or edit multiple lines of text. (T/F)

16. The width attribute will make the tile stretch automatically to fill the entire width of the dialog box. (T/F)

17. The slider tile returns a string value. (T/F)

18. The aspect ratio is the ratio of the width of the image box to the height of the image box. (T/F)

19. The color attribute can be used with an image box to control the background color. (T/F)

20. The vector_image function can be used to draw vectors in the Drawing Editor. (T/F)

21. The AutoLISP function, $value, retrieves the real value from the tile of a dialog box. (T/F)

22. The basic tiles, such as buttons, edit boxes, and images, are predefined by the _____ facility of AutoCAD.

23. The label of a button appears _____ the button.

24. The _____ attribute assigns a name to the tile.

25. The label attribute in a boxed column must be a _____ string.

26. The _____ attribute controls the width of a tile.

27. The format of the command for loading a dialog box is _____.

28. The AutoLISP function _____ is used to initialize a dialog box and then display it on the screen.

29. The AutoLISP function _____ is used to accept user input from the dialog box.

30. The AutoLISP function _____ is used to associate an action expression with a tile in the dialog box.

31. The AutoLISP function _____ can be used to obtain the result of logical bitwise AND.

32. The edit_width attribute determines the size of the edit box in _____ units.

33. The image tile is used to display _____ information.

34. The default value of the max_value attribute is _____.

35. The default value of the big_increment attribute is _____ of the slider range.

36. The AutoLISP function _____ can be used to fill the image box with a color.

Exercises

Exercise 1   *General*

Using dialog control language (DCL), write a program for the dialog box in Figure 15-16. Also, write an AutoLISP program that will load, display, and control the dialog box and perform the functions shown in the dialog box.

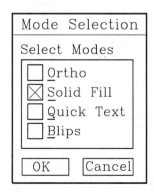

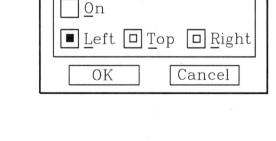

Figure 14-16 Dialog box for mode selection   *Figure 14-17 Dialog box for isometric snap/grid*

Exercise 2   *General*

Write a DCL program for the isometric snap/grid dialog box shown in Figure 15-17. Also, write an AutoLISP program that will load, display, and control the dialog box and perform the functions shown in the dialog box.

Exercise 3   *General*

Write a DCL program for the dialog box in Figure 15-18 that will enable you to insert a block. Also, write an AutoLISP program that will load, display, and control the dialog box. The values shown in the edit boxes for insertion point, scale, and rotation are the default values.

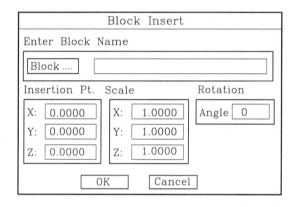

Figure 15-18 Dialog box for inserting blocks

Exercise 4   *General*

Write a DCL program for the dialog box in Figure 15-19 that will enable you to select the angle and angle precision. Also, write an AutoLISP program that will load, display, and control the dialog box.

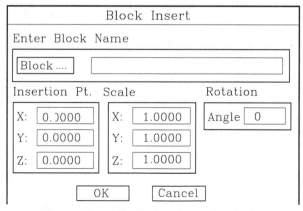

Figure 15-19 Dialog box for angle selection

Chapter 16

DIESEL:
A String Expression Language

Learning Objectives

After completing this chapter, you will be able to:
- *Use DIESEL to customize a status line.*
- *Use the* **MODEMACRO** *system variable.*
- *Write macro expressions using DIESEL.*
- *Use AutoLISP with* **MODEMACRO**.

DIESEL

DIESEL (Direct Interpretively Evaluated String Expression Language) is a string expression language. It can be used to display a user-defined text string (macro expression) in the status line by altering the value of the AutoCAD system variable **MODEMACRO**. The value assigned to **MODEMACRO** must be a string, and the output it generates will be a string. It is fairly easy to write a macro expression in DIESEL, and it is an important tool for customizing AutoCAD. However, DIESEL is slow, and it is not intended to function like AutoLISP or DCL. You can use AutoLISP to write and assign a value to the **MODEMACRO** variable, or you can write the definition of the **MODEMACRO** expression in the menu files. A detailed explanation of the DIESEL functions and the use of DIESEL in writing a macro expression is given later in this chapter.

STATUS LINE

When you are in AutoCAD, a status line is displayed at the bottom of the graphics screen (Figure 16-1). This line contains some useful information and tools that will make it easy to change the status of some AutoCAD functions. To change the status, you must click on the buttons. For example, if you want to display grid lines on the screen, click on the **GRID** button. Similarly, if you

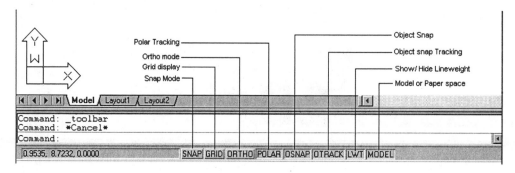

Figure 16-1 *Default status line display*

want to switch to paper space, click on **MODEL**. The status line contains the following information:

Coordinate Display. The coordinate information displayed in the status line can be static or dynamic. If the coordinate display is static, the coordinate values displayed in the status line change only when you specify a point. However, if the coordinate display is dynamic (default setting), AutoCAD constantly displays the absolute coordinates of the graphics cursor with respect to the UCS origin. AutoCAD can also display the polar coordinates (length<angle) if you are in an AutoCAD command.

SNAP. If SNAP is on, the cursor snaps to the snap point.

GRID. If GRID is on, grid lines are displayed on the screen.

ORTHO. If ORTHO is on, a line can be drawn in vertical or horizontal direction.

POLAR. When the POLAR snap is on, the cursor snaps to polar angles as set in the Polar Tracking tab of the **Drafting Settings** dialog box.

OSNAP. If OSNAP is on, you can use the running object snaps. If OSNAP is off, the running object snaps are temporarily disabled. The status of OSNAP (Off or On) does not prevent you from using regular object snaps.

OTRACK. When Object Tracking is on, you can track from a point. The object snap must be set before using Object Tracking.

MODEL and PAPER Space. AutoCAD displays MODEL in the status line when you are working in the model space. If you are working in the paper space, AutoCAD will display PAPER in place of MODEL.

LWT. When Lineweight button is on, the objects are displayed with the assigned width.

TILE. AutoCAD displays TILE, if TILEMODE is on (1). When TILEMODE is on, you are in the model space. If TILE is off, you are in the paper space (TILEMODE is 0).

MODEMACRO SYSTEM VARIABLE

The AutoCAD system variable **MODEMACRO** can be used to display a new text string in the status line. You can also display the value returned by a macro expression using the DIESEL language, which is discussed in a later section of this chapter. MODEMACRO is a system variable and you can assign a value to this variable by entering MODEMACRO at the Command prompt or by using the SETVAR command. For example, if you want to display **Customizing AutoCAD** in the status line, enter **SETVAR** at the Command: prompt and then press ENTER. AutoCAD will prompt you to enter the name of the system variable. Enter **MODEMACRO** and then press ENTER again. Now you can enter the text you want to display in the status line. After you enter **Customizing AutoCAD** and press ENTER, the status line will display the new text.

Command: **MODEMACRO**
or
Command: **SETVAR**
Variable name or ?: **MODEMACRO**
New value for MODEMACRO, or . for none<"">: **Customizing AutoCAD**

You can also enter MODEMACRO at the Command prompt and then enter the text that you want to display in the status line.

Command: **MODEMACRO**
New value for MODEMACRO, or . for none<"">: **Customizing AutoCAD**

Once the value of the **MODEMACRO** variable is changed, it retains that value until you enter a new value, start a new drawing, or open an existing drawing file. If you want to display the standard text in the status line, enter a period (.) at the prompt New value for **MODEMACRO**, or . **for none <"">:**. The value assigned to the **MODEMACRO** system variable is not saved with the drawing, in any configuration file, or anywhere in the system.

Command: **MODEMACRO**
New value for MODEMACRO, or . for none<"">:

CUSTOMIZING THE STATUS LINE

The information contained in the status line can be divided into two parts: toggle functions and coordinate display. The toggle functions part consists of the status of Snap, Grid, Ortho, Polar Tracking, Object Snap, Object Tracking, Lineweight, Model Space, and Tile (Figure 16-1). The coordinate display displays the X, Y, and Z coordinates of the cursor. The status line can be customized to your requirements by assigning a value to the AutoCAD system variable **MODEMACRO**. The value assigned to this variable is displayed leftjustified in the status bar at the bottom of the AutoCAD window. The number of characters that can be displayed in the status line depends on the system display and the size of the AutoCAD window. The coordinate display field cannot be changed or edited.

The information displayed in the status line is a valuable resource. Therefore, you must be careful when selecting the information to be displayed in the status line. For example, when working on a project, you may like to display the name of the project in the status line. If you are using several dimensioning styles, you could display the name of the current dimensioning style (DIMSTYLE) in the status line. Similarly, if you have several text files with different fonts, the name of the current text file (TEXTSTYLE) and the text height (TEXTSIZE) can be displayed in the status line. Sometimes, in 3D drawings, if you need to monitor the viewing direction (VIEWDIR), the camera coordinate information can be displayed in the status line. Therefore, the information that should be displayed in the status line depends on you and the drawing requirements. AutoCAD lets you customize this line and have any information displayed in the status line that you think is appropriate for your application.

MACRO EXPRESSIONS USING DIESEL

You can also write a macro expression using DIESEL to assign a value to the **MODEMACRO** system variable. The macro expressions are similar to AutoLISP functions, with some differences. For example, the drawing name can be obtained by using the AutoLISP statement **(getvar dwgname)**. In DIESEL, the same information can be obtained by using the macro expression **$(getvar,dwgname)**. However, unlike the case with AutoLISP, the DIESEL macro expressions return only string values. The format of a macro expression is:

> **$(function-name,argument1,argument2, ...)**

> **Example**
> $(getvar,dwgname)

Here, **getvar** is the name of the DIESEL string function and **dwgname** is the argument of the function. There must not be any spaces between different elements of a macro expression. For example, spaces between the $ sign and the open parentheses are not permitted. Similarly, there must not be any spaces between the comma and the argument, **dwgname**. All macro expressions must start with a **$** sign.

The following example illustrates the use of a macro expression using DIESEL to define and then assign a value to the **MODEMACRO** system variable.

Example 1

Using the AutoCAD MODEMACRO command, redefine the status line to display the following information in the status line:

> Project name (Cust-Acad)
> Name of the drawing (DEMO)
> Name of the current layer (OBJ)

Note that in this example the project name is Cust-Acad, the drawing name is DEMO, and the current layer name is **OBJ**.

Before entering the **MODEMACRO** command, you need to determine how to retrieve the required information from the drawing database. For example, here the project name **(Cust-Acad)** is a user-defined name that lets you know the name of the current project. This project name is not saved in the drawing database. The name of the drawing can be obtained using the DIESEL string function **GETVAR $(getvar,dwgname)**. Similarly, the **GETVAR** function can also be used to obtain the name of the current layer, **$(getvar,clayer)**. Once you determine how to retrieve the information from the system, you can use the **MODEMACRO** system variable to obtain the new status line. For Example 1, the following DIESEL expression will define the required status line.

> Command: **MODEMACRO**
> New value for MODEMACRO, or . for none<"">: **Cust-Acad N:$(GETVAR,dwgname)**
> **L:$(GETVAR,clayer)**

Explanation
Cust-Acad
Cust-Acad is assumed to be the project name you want to display in the status line.

N:$(GETVAR,dwgname)
Here, N: is used as an abbreviation for the drawing name. The GETVAR function retrieves the name of the drawing from the system variable **dwgname** and displays it in the status line, next to N:.

L:$(GETVAR,clayer)
Here L: is used as an abbreviation for the layer name. The GETVAR function retrieves the name of the current layer from the system variable **clayer** and displays it in the status line.

The new status line is shown in Figure 16-2(a).

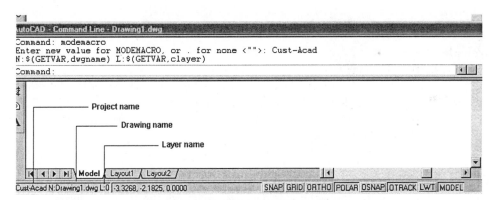

Figure 16-2(a) Status line for Example 1

Example 2

Using the AutoCAD **MODEMACRO** command, redefine the status line to display the following information in the status line, Figure 16-2(b):

Name of the current textstyle
Size of text
User-elapsed time in minutes

In this example the abbreviations for text style, text size, and user-elapsed time in minutes are TSTYLE:, TSIZE:, and ETM:, respectively.

> Command: **MODEMACRO**
> New value for MODEMACRO, or . for none<"">:
>> TSTYLE:$(GETVAR,TEXTSTYLE)TSIZE:$(GETVAR,TEXTSIZE)
>> ETM:$(FIX,$(*,60,$(*,24,$(GETVAR,TDUSRTIMER))))

Explanation

TSTYLE:$(GETVAR,TEXTSTYLE)

The **GETVAR** function obtains the name of the current textstyle from the system variable **TEXTSTYLE** and displays it next to TSTYLE: in the status line.

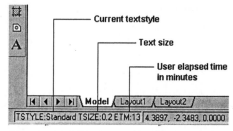

Figure 16-2(b) *Status line for Example 2*

TSIZE:$(GETVAR,TEXTSIZE)

The **GETVAR** function obtains the current size of the text from the system variable **TEXTSIZE** and then displays it next to TSIZE: in the status line.

ETM:$(FIX,$(*,60,$(*,24,$(GETVAR,TDUSRTIMER))))

The **GETVAR** function obtains the user-elapsed time from the system variable **TDUSRTIMER** in the following format:

> **<Number of days>.<Fraction>**

Example

0.03206400 (time in days)

To change this time into minutes, multiply the value obtained from the system variable **TDUSRTIMER** by 24 to change it into hours, and then multiply the product by 60 to change the time into minutes. To express the minutes value without a decimal, determine the integer value using the DIESEL string function FIX.

Example

Assume that the value returned by the system variable TDUSRTIMER is 0.03206400. This time is in days. Use the following calculations to change the time into minutes, and then express the time as an integer:

> 0.03206400 days x 24 = 0.769536 hr
> 0.769536 hr x 60 = 46.17216 min
> integer of 46.17216 min = 46 min

USING AUTOLISP WITH MODEMACRO

Sometimes the DIESEL expressions can be as long as those shown in Example 1 and Example 2. It takes time to type the DIESEL expression, and if you make a mistake in entering the expression, you have to retype it. Also, if you need several different status line displays, you need to type them every time you want a new status line display. This can be time-consuming and sometimes confusing.

To make it convenient to change the status line display, you can use AutoLISP to write a DIESEL expression. It is easier to load an AutoLISP program, and it also eliminates any errors that might be caused by typing a DIESEL expression. The following example illustrates the use of AutoLISP to write a DIESEL expression to assign a new value to the **MODEMACRO** system variable.

Example 3

Using AutoLISP, redefine the value assigned to the **MODEMACRO** system variable to display the following information in the status line:

Name of the current text style
Size of text
User-elapsed time in minutes

In this example the abbreviations for text style, text size, and user-elapsed time in minutes are TSTYLE:, TSIZE:, and ETM:, respectively.

The following file is a listing of the AutoLISP program for Example 3. The name of the file is **ETM.LSP.** The line numbers are not a part of the file; they are shown here for reference only.

```
(defun c:etm ( )                              1
(setvar "MODEMACRO"                           2
(strcat                                       3
  "TSTYLE:$(getvar,textstyle)"                4
  " TSIZE:$(getvar,textsize)"                 5
  " ETM:$(fix,$(*,60,$(*,24,                  6
  $(getvar,tdusrtimer))))"                    7
  )                                           8
  )                                           9
)                                            10
```

Explanation
Line 3
(strcat
The AutoLISP function **strcat** links the string value of lines 4 through 7 and returns a single string that becomes a DIESEL expression for the **MODEMACRO** command.

Line 4
"TSTYLE:$(getvar,textstyle)"
This line is a DIESEL expression in which getvar, a DIESEL string function, retrieves the value of the system variable **textstyle** and **$(getvar,textstyle)** is replaced by the name of the textstyle. For example, if the textstyle is STANDARD, the line will return "TSTYLE:STANDARD". This is a string because it is enclosed in quotes.

Lines 6 and 7
" ETM:$(fix,$(*,60,$(*,24,
$(getvar,tdusrtimer))))"
These two lines return **ETM:** and the time in minutes as a string. The **fix** is a DIESEL string
function that changes a real number to an integer.

To load this AutoLISP file **(ETM.LSP)**, use the following commands. In this example, the file
name and the function name are the same (ETM).

    Command: **(load "ETM")**
    ETM
    Command: **ETM**

DIESEL EXPRESSIONS IN MENUS

You can also define a DIESEL expression in the screen, tablet, pull-down, or button menu. When
you select the menu item, it will automatically assign the value to the MODEMACRO system
variable and then display the new status line. The following example illustrates the use of the
DIESEL expression in the screen menu.

Example 4

Write a DIESEL macro for the screen menu that displays the following information in the status
line (Figure 16-3):

Macro-1	**Macro-2**	**Macro-3**
Project name	Pline width	Dimtad
Drawing name	Fillet radius	Dimtix
Current layer	Offset distance	Dimscale

The following file is a listing of the screen menu that contains the definition of three DIESEL
macros for Example 4. This menu can be loaded using AutoCAD's MENU command and then
entering the name of the file. If you select the first item, DIESEL1, it will automatically display the
new status line.

```
***screen
[*DIESEL*]
[DIESEL1:]^C^CMODEMACRO;$M=Cust-Acad,N:$(GETVAR,DWGNAME)+
,L:$(GETVAR,CLAYER);
[DIESEL2:]^C^CMODEMACRO;$M=PLWID:$(GETVAR,PLINEWID),+
FRAD:$(GETVAR,FILLETRAD),OFFSET:$(GETVAR,OFFSETDIST),+
LTSCALE:$(GETVAR,LTSCALE);
[DIESEL3:]^C^CMODEMACRO;$M=DTAD:$(GETVAR,DIMTAD),+
DTIX:$(GETVAR,DIMTIX),DSCALE:$(GETVAR,DIMSCALE);
```

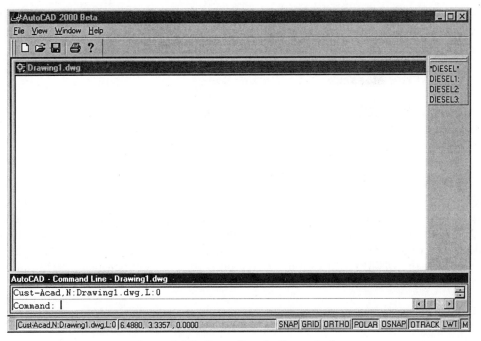

Figure 16-3 *Status line for Example 4*

MACROTRACE SYSTEM VARIABLE

The MACROTRACE variable is an AutoCAD system variable that can be used to debug a DIESEL expression. The default value of this variable is 0 (off). It is turned on by assigning a value of 1 (on). When on, the MACROTRACE system variable will evaluate all DIESEL expressions and display the results in the command prompt area. For example, if you have defined several DIESEL expressions in a drawing session, all of them will be evaluated at the same time and the messages, if any, will be displayed in the command prompt area.

Example
Command: **MODEMACRO**
New value for MODEMACRO, or . for none<"">: **$(getvar,dwgname),$(getvar clayer)**

Note that in this DIESEL expression a comma is missing between getvar and clayer. If the MACROTRACE system variable is on, the following message will be displayed in the Command prompt area:

Eval: $(GETVAR,DWGNAME)
====>UNNAMED
Eval: $(GETVAR CLAYER)
Err: $(GETVAR CLAYER)??

This error message gives you an idea about the location of the error in a DIESEL expression. In the previous example, the first part of the expression successfully returns the name of the drawing (unnamed) and the second part of the expression results in the error message. This confirms that there is an error in the second part of the DIESEL expression. You can further determine the cause of the error by comparing it with the error messages in the following table:

Error Message	Description
$?	Syntax error
$?(func,??)	Incorrect argument to function
$(func)??	Unknown function
$(++)	Output string too long

DIESEL STRING FUNCTIONS

Like AutoLISP, you can use DIESEL functions to do some mathematical operations, retrieve the values from the drawing database, and display the values in the status line in a predetermined order. For example, you can add, subtract, multiply, and divide the numbers. You can also obtain the values of some system variables and display them in the status line. The maximum number of parameters that the DIESEL expression can contain is 10. This number includes the name of the function. The following section discusses some frequently used DIESEL string functions.

Addition

Format **$(+,num1,num2,num3 - - -)**

This function (+) calculates the sum of the numbers that are to the right of the plus (+) sign. The numbers can be integers or real.

Examples

$(+,2,5)   returns 7
$(+,2,5,50.75)   returns 57.75

 Note
You can test the calculations by assigning the DIESEL expression to the MODEMACRO system variable.

Command: **MODEMACRO**
New value for MODEMACRO, or . for none<"">: **$(+,2,5)**

AutoCAD will return a value of 7, which will be displayed in the status line. It is important to note that the values returned by the DIESEL string expressions are string values.

Subtraction

Format **$(-,num1,num2,num3,- - -)**

This function (-) subtracts the second number from the first number. If there are more than two numbers, the second and the subsequent numbers are added and their sum is subtracted from the first number.

Examples

$(-,28,14)	returns 14
$(-,25,7,11.5)	returns 6.5

Multiplication

Format $(*,num1,num2,num3,- - -)

This function (*) calculates the product of the numbers that are to the right of the asterisk.

Examples

$(*,2,5)	returns 10
$(*,2,5,3.0)	returns 30.0
$(*,2,5,3.25)	returns 32.5

Division

Format $(/,num1,num2,num3 - - -)

This function (/) divides the first number by the second number. If there are more than two numbers, the first number is divided by the product of the second and subsequent numbers.

Examples

(/ 3 2)	returns 1.5
(/ 3.0 2)	returns 1.5
(/ 200 5 4)	returns 10
(/ 200.0 5.5)	returns 36.363636
(/ 200 -5)	returns -40
(/ -200 -5.0)	returns 40.0

Relational Statements

Some DIESEL expressions involve features that test a particular condition. If the condition is true the expression performs a certain function; if the condition is not true, the expression performs another function. The following section discusses various relational statements used in DIESEL expressions.

Equal to

Format $(=,num1,num2)

This function (=) checks whether the two numbers are equal. If they are, the condition is true and the function will return **1**. Similarly, if the specified numbers are not equal, the condition is false and the function will return **0**.

Examples

$(=,5,5)	returns 1
$(=,5,4.9)	returns 0
$(=,5,-5)	returns 0

Not equal to

Format **$(!=,num1,num2)**

This function (!=) checks whether the two numbers are **not equal**. If they are not equal, the condition is true and the function will return **1**; otherwise the function will return **0**.

Examples

$(!=,50,4)	returns 1
$(!=,50,50)	returns 0
$(!=,50,-50)	returns 1

Less than

Format **$(<,num1,num2)**

This function (<) checks whether the first number **(num1)** is less than the second number **(num2)**. If it is true, the function will return **1**; otherwise the function will return **0**.

Examples

$(<,3,5)	returns 1
$(<,5,3)	returns 0
$(<,3.0,5)	returns 1

Less than or equal to

Format **$(<=,num1,num2)**

This function (<=) checks whether the first number **(num1)** is less than or equal to the second number **(num2)**. If it is true, the function will return **T**; otherwise the function will return **nil**.

Examples

$(<=,10,15)	returns 1
$(<=,19,10)	returns 0
$(<=,-2.0,0)	returns 1

Greater than

Format **$(>,num1,num2)**

This function (**>**) checks whether the first number (**num1**) is greater than the second number (**num2**). If it is true, the function will return **1**; otherwise the function will return **0**.

Examples

$(>,15,10)	returns 1
$(>,20,30)	returns 0

Greater than or equal to

Format **$(>=,num1,num2)**

This function (**>=**) checks whether the first number (**num1**) is greater than or equal to the second number (**num2**). If it is true, the function will return **1**; otherwise the function will return **0**.

Examples

$(>=,78,50)	returns 1
$(>=,78,88)	returns 0

eq function

Format **$(eq,value1,value2)**

This function (**eq**) checks whether the two string values are equal. If they are, the condition is true and the function will return **1**; otherwise the function will return **0**.

Examples

$(eq,5,5)	returns 1
$(eq,yes,yes)	returns 1
$(eq,yes,no)	returns 0

angtos

Format **$(angtos,angle[,mode,precision])**

The angtos function returns the angle expressed in radians in a string format. The format of the string is controlled by the mode and precision settings.

Examples

$(angtos,0.588003,0,4)	returns 33.6901
$(angtos,-1.5708,1,2)	returns 270d0'

Note

The following modes are available in AutoCAD.

ANGTOS MODE	EDITING FORMAT
0	*Decimal degrees*
1	*Degrees/minutes/seconds*
2	*Grads*
3	*Radian*
4	*Surveyor's units*

Precision *is an integer number that controls the number of decimal places. Precision corresponds to the AutoCAD system variable,* **AUPREC**. *The minimum value of* **precision** *is 0, the maximum is 4.*

eval function

Format **$(eval,string)**

The **eval** function passes the string to the DIESEL evaluator and the result obtained after evaluating the string is returned and displayed in the status line.

Examples

$(eval,welcome)	returns welcome
$(eval,$(getvar,dimscale))	returns value of dimscale

fix function

Format **$(fix,num)**

The **fix** function converts a real number into an integer by truncating the digits after the decimal.

Examples

$(fix,42.573)	returns 42
$(fix,-23.50)	returns -23

getvar function

Format **$(getvar,varname)**

The **getvar** function retrieves the value of an AutoCAD system variable.

Examples

$(getvar,dimtad)	returns value of dimtad
$(getvar,clayer)	returns current layer name

rtos function

Format   **$(rtos,number)**
or   **$(rtos,number,mode,precision)**

The **rtos** function changes a given number into a real number and the format of the real number is determined by the mode and precision values.

Examples

$(rtos,50)   returns 50
$(rtos,1.5,5,4)   returns 1 1/2

Note

The following linear unit modes are available in AutoCAD:

MODE VALUE	STRING FORMAT
1	*Scientific*
2	*Decimal*
3	*Engineering*
4	*Architectural*
5	*Fractional*

Precision is an integer number that controls the number of decimal places. Precision corresponds to the AutoCAD system variable, LUPREC, and the mode corresponds to LUNITS. If the mode and precision values are omitted, AutoCAD uses the current values of LUNITS and LUPREC as set with the units command.

if function

Format **$(if,condition,then,else)**

The **if** function evaluates the first expression (then), if the value returned by the specified condition is non zero. The second expression (else) is evaluated if the specified condition returns 0.

(if condition then [else])
    Where **condition** ------------------ Specified conditional statement
              then ------------------ Expression evaluated if the condition returns non zero
              else ------------------ Expression evaluated if the condition returns 0

Examples

$(if,$(=,7,7),true)   returns true
$(if,$(=,5,7),true,false)   returns false
$(if,1.5,true,false)   returns true

strlen function

Format **$(strlen,string)**

The **strlen** function returns an integer number that designates the number of characters contained in the specified string.

Example
$(strlen,Customizing AutoCAD)   returns 19

linelen function

Format **$(linelen)**

The **linelen** function returns an integer number that lets you know the number of characters that can be accommodated in the status line. This number varies and depends on the system. However, on most systems the number of characters that can be displayed in the layer-mode section of the status line is 240.

Example
$(linelen)   returns 240 (depends on the system)

upper function

Format **$(upper,string)**

The **upper** function returns the specified string in uppercase.

Example
$(upper,Customizing)   returns CUSTOMIZING

edtime function

Format **$(edtime,date,display-format)**

The **edtime** function can be used to edit the date and display it in a format specified by display-format. For example, the Julian date obtained from AutoCAD's system variable, DATE, is obtained in the form 2449013.85156759. The **edtime** function can be used to display this date in an understandable form. The following table gives the format of the phrases that can be used with the **edtime** function:

FORMAT	OUTPUT	FORMAT	OUTPUT
D	5	H	2
DD	05	HH	02
DDD	Tue	MM	23
DDDD	Tuesday	SS	12
M	1	MSEC	325
MO	11	AM/PM	PM
MON	Nov	am/pm	am
MONTH	November	A/P	P
YY	92	a/p	p
YYYY	1992		

Chapter 16

Example

$(edtime,$(getvar,date),DDDD",",DD MONTH YY - HH:MMAM/PM)

returns Monday,25 January 93 - 08:52PM

In the above example, the getvar function retrieves the date in Julian format. The edtime function displays the date as specified by the display-format. The following illustration shows the corresponding fields of the display-format and the value returned by the edtime function:

DDDD",",DD MONTH YY - HH:MMAM/PM

Where **DDDD** ------------------- Monday

, ------------------- ,

DD ------------------- 25

MONTH ------------------ January

YY ------------------- 93

- -------------------- -

HH ------------------- 08

MMAM ------------------- 52

PM ------------------- PM

Review Questions

1. DIESEL (direct interpretively evaluated string expression language) is a string expression language. (T/F)

2. The value assigned to the **MODEMACRO** variable is a string, and the output it generates is not a string. (T/F)

3. You cannot define a DIESEL expression in the screen menu. (T/F)

4. The coordinate information displayed in the status line can be dynamic only. (T/F)

5. Once the value of the **MODEMACRO** variable is changed, it retains that value until you enter a new value, start a new drawing, or open an existing drawing file. (T/F)

6. The number of characters that can be displayed in the layer mode field on any system is 38. (T/F)

7. The coordinate display field cannot be changed or edited. (T/F)

8. You can write a macro expression using DIESEL to assign a value to the **MODEMACRO** system variable. (T/F)

9. In DIESEL, the drawing name can be obtained by using the macro expression **$(getvar,dwgname)**. (T/F)

10. You cannot use AutoLISP to write a DIESEL expression. (T/F)

Fill in the blanks.

11. All macro expressions must start with a _____ sign.

12. The display-format DDD will return Monday as _____.

13. The DIESEL expression, $(upper,AutoCAD), will return _____.

14. The DIESEL expression, $(strlen,AutoCAD), will return _____.

15. The DIESEL expression, $(if,$(=,3,2),yes,no), will return _____.

16. The DIESEL expression, $(fix,-17.75), will return _____.

17. The DIESEL expression, $(eq,Customizing,customizing), will return _____.

18. The DIESEL expression, $(/,81,9,9), will return _____.

19. The MACROTRACE system variable can be used to _____ a DIESEL expression.

20. The status line can be turned off by _____ AutoCAD.

Exercises

Exercise 1 *General*

Using the AutoCAD **MODEMACRO** command, redefine the status line to display the following information in the status line:

Your name
Name of drawing

Exercise 2 *General*

Using the AutoCAD **MODEMACRO** command, redefine the status line to display the following information in the status line:

Name of the current dimension style
Dimension scale factor (dimscale)
User-elapsed time in hours

The abbreviations for dimstyle, dimscale, and user-elapsed time in hours are DIMS:, DIMFAC:, and ETH:, respectively.

Exercise 3 *General*

Using **AutoLISP**, redefine the status line to display the following information in the status line:

Name of the current dimension style
Dimension scale factor (dimscale)
User-elapsed time in hours

Chapter 17

Visual Basic

Learning Objectives

After completing this chapter, you will be able to:
- *Load and run sample VBA projects.*
- *Utilize the Visual Basic Editor.*
- *Understand and use AutoCAD objects.*
- *Use object properties.*
- *Apply and use AutoCAD methods.*

ABOUT Visual Basic

The original BASIC was developed in 1963. BASIC (<u>B</u>eginners <u>A</u>ll-purpose <u>S</u>ymbolic <u>I</u>nstruction <u>C</u>ode) was intended to be an accessible, user-friendly programming language. BASIC was the language supplied with many of the early microcomputers starting in the late 1970s. It continued development until the advent of Windows. In 1991 the first version of **Visual Basic (VB)** appeared followed by new versions spaced about a year apart. The current version 6, like its predecessors, lets even a beginning programmer create a user-friendly graphical user interface (GUI) with a small fraction of the effort required previously. It used to take a team of programmers working with a language like C that came with a stack of reference material nearly a foot thick to do the same thing.

Another advantage inherent to Visual Basic is that engineers may have old BASIC programs used in sizing components. This code, minus old input/output, could be used as the core of new Visual Basic modules to parametrically design and draft the component

Autodesk first licensed **Visual Basic® for Applications (VBA)** from Microsoft for use in R14. Dozens of other large software companies have similarly adopted VBA. The VBA language tools are either identical or very similar to those in the stand-alone Visual Basic. Programmers who know how to write macros for Microsoft Office only have to learn AutoCAD-specific functionality to program VBA in AutoCAD.

A capability going along with VBA is the ability to communicate with other applications such as Microsoft Excel, Microsoft Word, Microsoft Access, and Visual Basic using **ActiveX Automation**. Automation lets you access and manipulate the objects and functionality of AutoCAD from Excel or another application supporting ActiveX. Alternately the objects and functionality of Excel, for example, can be used inside AutoCAD. This cross-application macro programming capability does not exist in AutoLISP. Of course, before you can use AutoCAD's objects in some other software supporting ActiveX you must make the application aware that the **AutoCAD Object Library** is available on the computer.

There are over three million programmers using Visual Basic. A great many more have programmed in BASIC but not VB. This chapter assumes you have some familiarity with BASIC or programming concepts such as those covered in the AutoLISP chapter of this book and are comfortable looking up commands and syntax in the help system.

This is the second iteration of VBA for Autodesk. In this version you can now reference a macro from another project and have more than one macro loaded. You can also create libraries of common functions and macros.

OBJECTS

Visual Basic uses **objects** to facilitate what it can accomplish. Examples of objects include drawings or documents, geometric elements such as lines or circles, and user interface controls to handle input and output to programs or macros. The "visual" part of Visual Basic refers to the familiar user interface controls such as check boxes, scroll bars, and command buttons as used to open or save files seen in Windows applications. These controls are screen objects that may be dragged from a "toolbox" onto the background object called a form. An example of a simple form is the background window of a dialog box. In the course of dragging a control from the Toolbox onto a form, the coding that provides the functionality of the control is simultaneously added to the form object. Several aspects of true object-oriented programming (OOP) languages such as C++ are missing from Visual Basic. So VB is considered an object-based, rather than an object-oriented, programming language. The missing aspects are more than made up for by the development tools and environments available to the user.

Functions known as **methods** have been defined in the AutoCAD object library to perform an action on an object, for example, draw a line in a drawing. The method AddLine adds a line object to a drawing. **Properties** are functions that set or return information about the state of an object. In the example of a line drawn in a drawing object the Color, Layer, StartPoint, and EndPoint would be some of the properties of the Line object.

ADD METHOD

The way to draw in paper space, model space, or in a block is to use the **Add method** such as **AddCircle**, **AddLine**, **AddArc**, and **AddText**. Instead of thinking of drawing in the model space of the current drawing, think in terms of adding a geometric object. **Dot notation** is required to clarify on which object the method is acting. The dot notation reference starts with the most global object first narrowing to the right.

AddCircle

The **AddCircle** method requires a predefined center point and radius. Both of these arguments are required. Other input options used in the AutoCAD circle command such as three-points, two-points, or tangent-tangent-radius require additional programming in VBA to calculate the center point and radius for the AddCircle method. The format of the **AddCircle** method is:

> **ThisDrawing.ModelSpace.AddCircle centerpoint, radius**
> or
> **Circle1=ThisDrawing.ModelSpace.AddCircle(centerpoint, radius)**
> > Where **centerpoint**--------------------------Center point, double precision vector
> > **radius** ---------------------------------Radius, double precision number

Note

The arguments of the first form of the AddCircle method follow a required space. The second form of the add method, where the arguments are inside parentheses, is used where the circle object is assigned to a name for use later in the program. For the examples below, point1 and point2 are predefined points consisting of a vector of variant/double precision or double precision coordinate values. These may be defined with assignment statements as shown in Example 1 on page 17-8. The pound sign (#) signifies a double precision number in Basic.

Examples

ThisDrawing.ModelSpace.AddCircle point1, 3#
Circle2= ThisDrawing.ModelSpace.AddCircle(point2, 4#)

AddLine

The AddLine method requires two predefined endpoints. The AddLine method has a syntax similar to AddCircle:

> **ThisDrawing.ModelSpace.AddLine firstpoint, secondpoint**
> > where **firstpoint**------------------- First point, double precision vector
> > **secondpoint**-------------- Second point, double precision vector

Examples

ThisDrawing.ModelSpace.AddLine point1, point2
Line1=ThisDrawing.ModelSpace.AddLine(point1, point2)

Note

The second form of the add method, where the arguments are inside parentheses, may be used for the AddLine method or any of the other Add methods below where the object is assigned to a name for use later in the program.

AddArc

The **AddArc** method format is:

> **ThisDrawing.ModelSpace.AddArc ctrpt, radius, StartAng, EndAng**
>> Where   **ctrpt** ------------------------- Center point, double precision vector
>>   **radius** ----------------------- Radius, double precision number
>>   **StartAng** ------------------- Arc starting angle in radians, double precision
>>   **EndAng** -------------------- Arc ending angle in radians, double precision

Example

ThisDrawing.ModelSpace.AddArc point1, 4#, 0#, 1.570796327

AddText

The **AddText** method requires a predefined string, insertion point and text height. The **AddText** method syntax is:

> **ThisDrawing.ModelSpace.AddText textString$, point1, textHeight**
>> Where   **textString$** ----------------- Actual text to be displayed
>>   **point1** ----------------------- Position point, double precision vector
>>   **textHeight** ----------------- Text Height, positive double precision number

Example

ThisDrawing.ModelSpace.AddText ".063 TYP, 4 PLACES", point1, 0.25#

Finding Help on Methods and Properties

You will find excellent help in the VBA Integrated Development Environment with good examples showing the exact syntax necessary for the AddLine, AddCircle, or any of the particular methods or properties needed to write a parametric program. You can also find general information such as on Methods, help on Visual Basic key words, and many non-AutoCAD VBA programming examples taken from Excel, Word, or PowerPoint in AutoCAD 2000. The **topics under "AutoCAD Help" from the Help menu including** "ActiveX and VBA Developers Guide," shown in Figure 17-1. It is available through a shortcut from the AutoCAD drawing editor's help system or from the Visual Basic Editor Integrated Design Environment (IDE). Look under VBA and ActiveX Automation. The Index and Find tabs of this help menu are very useful. Another way to get help in the VBA IDE, you can use the **Object Browser**

Figure 17-1 VBA and Active X Developers Guide

which can be accessed from the IDE **View** menu. Select **AcadProject** in the drop-down list box. Select object **ThisDrawing** in the left window and **ModelSpace** in the right one, as shown in Figure 17-2. The question mark help button on the toolbar or **F1** takes you to a screen where you can access the ModelSpace Collection of help screens on methods, properties, and examples. Alternately, the function key **F1** will get help on any word typed in the code window.

Loading and Saving VBA Projects

To start or load a VBA project in the AutoCad Drawing Editor, shown in Figure 17-3, you select under the **Tools** menu, the item **Macro**, and the **flyout labeled Visual Basic Editor,** you will enter the **Integrated Development Environment (IDE).** The IDE consists of a number of useful windows that can be sized, shown, hidden, or otherwise customized to suit your needs. Figure 17-4 shows the IDE with the VBA project for Example 1 loaded. There

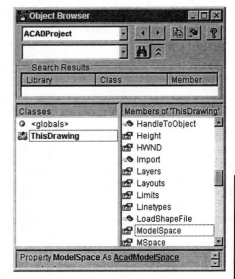

Figure 17-2 Object Browser

are six windows shown. The top left window is the Project Explorer, which allows navigation among windows. Note the View Code, View Object, and Toggle Folders Icons at the top of the window. The bottom left menu is the Properties Window, which shows object properties and is a convenient way to change them.

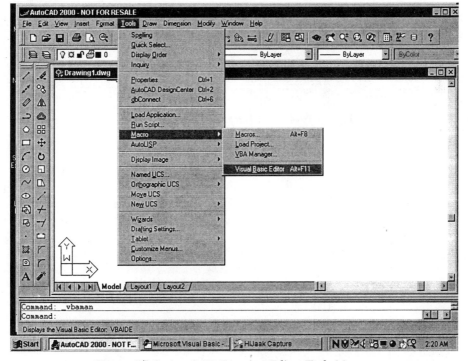

Figure 17-3 AutoCAD Drawing Editor Tools Menu

The top right window is the UserForm Window where the "Visual" user interface is constructed using the Toolbox. The next window down on the right is the UserForm code window where Basic language program code resides, which runs when triggered by events such as a mouse click on a control button. The next window down is the Module code window, where mathematical functions and procedures with the .bas extension are generally kept. The bottom right Immediate and Watch Windows are used for debugging.

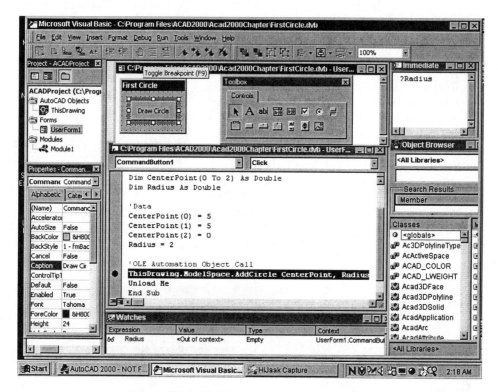

Figure 17-4 *VBA Integrated Development Environment*

It is a good idea to save your VBA program before testing it. To save a project from the VBA Integrated Development Environment use the **File** menu item **Save.** Menu items may be more quickly accessed with Alt and the underlined letter. AutoCAD VBA projects have the file extension DVB.

Example 1

Write a program that will draw a circle centered at 5,5,0 with radius 2, as shown in Figure 17-6.

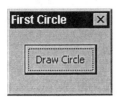

Figure 17-5 *Screen Interface Form for Example 1*

Figure 17-6 *Circle , Center at 5,5,0 Radius 2*

Once you are in the IDE you should insert a Module and a UserForm using the **Insert** menu item **Module** and **UserForm** respectively. The default names will be Module1 and UserForm1. Click the form after insertion to make it the active window. To insert a control such as a command button onto the form, you click the corresponding button on the toolbox. The second button from the left in the sec-ond row of the toolbox inserts the com-mand button control, as shown in Fig-ure 17-7.

Next click the form at the desired position. Resize the UserForm and CommandButton to your taste and edit the caption on the button or in the properties window at the bottom right of Figure 17-6 to say something like "Draw Circle." Click the form background to get out of editing the caption when you are finished.

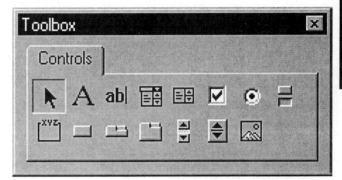

Figure 17-7 *Toolbox*

A **project** is the name given to the forms, controls, modules, and programming making up a Visual Basic program. To finish a project it is necessary to write the code underlying the visual control(s). One way to get to the code window for the CommandButton control is to double click on the button and enter the code for CommandButton1_Click(). Visual Basic is **event driven**. Lines 1 through 17 from Example 1 correspond to the code that runs when the command button is clicked with the mouse.

The Sub DrawCircle behind Module1 can be defined for the purpose of allowing the module to be run from the **VBA** menu item **Run Macro,** as shown in Figure 17-8.

The following file is a listing of the Visual Basic program for Example 1. **The line numbers at the right are for reference only and are not a part of the programming.**

Command Button Code

```
'UserForm1 Code to Draw Circle, Radius 2                                  1
'Centered at 5,5,0. Trigger is mouse click on                            2
'command button marked Draw Circle                                       3
Private Sub CommandButton1_Click()                                       4
```

```
        Dim CenterPoint(0 To 2) As Double                                          5
        Dim Radius As Double                                                       6

        'Data                                                                      7
        CenterPoint(0) = 5                                                         8
        CenterPoint(1) = 5                                                         9
        CenterPoint(2) = 0                                                        10
        Radius = 2                                                               11
                                                                                 12
        'OLE Automation Object Call                                              13
        ThisDrawing.ModelSpace.AddCircle CenterPoint, Radius                     14
        Unload Me                                                                15
        End Sub                                                                  16
```

Explanation

Lines 1 to 3

The first lines are comments or remarks describing the function of the program. Comments make understanding and modifying a program easier and should be used liberally. Comments start with a Rem or use an apostrophe ('). These lines are ignored when the program is run.

Line 4

Private Sub CommandButton1_Click()

This line defines where Sub CommandButton1_Click() starts. The subroutine is executed when CommandButton1 is clicked with the mouse. It is the code that draws the circle when the command button is clicked. There is no need to type this line as VBA generates it automatically as soon as the command button control is added to the form.

Lines 5 and 6

Dim CenterPoint(0 To 2) As Double
Dim Radius As Double

These two lines are necessary to establish the double precision variable type for the arguments needed by the AddCircle method. The default type is Variant, which will not work with AddCircle.

Lines 8 through 11

CenterPoint(0) = 5
CenterPoint(1) = 5
CenterPoint(2) = 0
Radius = 2

Here the circle center and radius are assigned values.

Line 14

ThisDrawing.ModelSpace.AddCircle CenterPoint, Radius

This line applies the AddCircle method to the ModelSpace object, which is part of the ThisDrawing object. Note the required space between the key word AddCircle and the first argument, CenterPoint.

Line 15
Unload Me

This is a method that removes UserForm1 from memory and returns the focus back to AutoCAD.
Line 16
End Sub
Like line 4, this line is generated automatically.

Module1 Code

The following code is entered in the Module1 code window.

```
'Module 1 General Declarations                                    1
Sub DrawCircle()                                                  2
UserForm1.Show                                                    3
End Sub                                                           4
```

Explanation

Line 1 through 4
Sub DrawCircle()
UserForm1.Show
End Sub
This subroutine's function is to create a Macro name, DrawCircle, which shows up in the lower list
box of the Run Macro dialog box run from AutoCAD's **Tools** menu item **Macro** shown in Figure

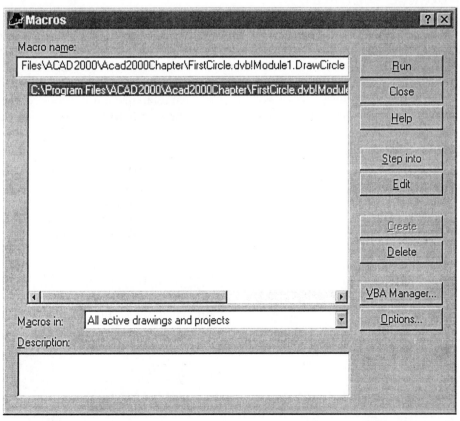

Figure 17-8 *AutoCAD Drawing Editor Tool Menu Macro Items and Macros Flyout*

17-8. These four lines of code belong to the Module1 object. Another way to run a project is to select the menu **R**un item **Run Sub/UserForm** in the VBA IDE.

GetPoint, GetDistance, and GetAngle Methods
GetPoint Method

The **GetPoint method** allows you to enter the X, Y coordinates or X, Y, Z coordinates of a point. The coordinates of the point can be entered from the keyboard or by using the screen cursor. The format of the **GetPoint** method is:

> **P = ThisDrawing.Utility.GetPoint([Point], [Prompt])**
> Enter a point from the keyboard or select a point in the AutoCAD graphics editor
> Where **[Point]** ---------------------- Optional reference point, rubber-band origin
> **[Prompt])** ------------------ Optional prompt to be displayed on screen

Example
pnt1 =ThisDrawing.Utility.GetPoint("Enter 1st Point")
Pt2=ThisDrawing.Utility.GetPoint(Pnt1,"Enter 2nd Point")

GetDistance Method

The **GetDistance method** lets you enter a distance on the command line, a distance from a given point, or two points, and it then returns the distance as a double precision number. The format of the **GetDistance** method is:

> **d = ThisDrawing.Utility.GetDistance([point], [prompt])**
> Where **[point]** ---------------------- Optional reference point, rubber band origin
> **[prompt]** ------------------ Optional prompt to be displayed on screen

GetAngle Method

The **GetAngle method** allows you to enter an angle, either from the keyboard in degrees or by selecting two points. In the case of selecting points, the positive horizontal direction is taken as one leg of the angle, the first point selected as the vertex and the second point defines the second leg. If the point argument is specified, AutoCAD uses this point as the first point or angle vertex. The GetAngle method returns the value of the angle in radians as a double precision value. The format of the **GetAngle** Method is:

> **ang = ThisDrawing.Utility.GetAngle([point], [prompt])**
> Where **ang** --------------------------- Angle in radians
> **[point]** ---------------------- Optional vertex point
> **[prompt]** ------------------ Optional screen prompt to clarify angle selec-
>   tion

Examples
a1 = ThisDrawing.Utility.GetAngle(,"Enter taper angle in degrees")
ang = ThisDrawing.Utility.GetAngle(pt1) 'pt1 is a predefined point
ThisDrawing.Utility.GetAngle pt1, "Enter second point of angle"

Note
*The angle you enter is affected by the angle setting. The angle settings can be changed by changing the value of the AutoCAD system variables **ANGBASE** and **ANGDIR**. The default settings for measuring an angle are as follows:*

*The angle is measured with respect to the positive X-Axis (3 o'clock position). The value for 3 o'clock corresponds to the current value of **ANGBASE**, the AutoCAD system variable being 0. **ANGBASE** could be set in any of 4, 90 degree quadrant directions*

*The angle is positive if it is measured in the counterclockwise direction and is negative if it is measured in the clockwise direction. The value of this setting is saved in the AutoCAD system variable **ANGDIR**. The **GetOrientation** method has the same syntax as the GetAngle method but ignores **ANGBASE** and **ANGDIR** system variables. The 0 angle is always at 3 o'clock, and angles are always positive counterclockwise.*

Example 2

Write a program that will draw a triangle with user supplied vertices P1, P2, and P3 as in Figure 17-9. This program is to use the GetPoint and AddLine methods.

Figure 17-10 Screen Interface Form for Example 2

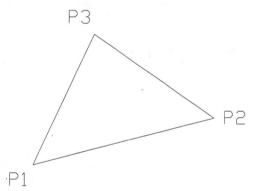

Figure 17-9 Tirangle with User Defined Points

Once again from the IDE a Module and UserForm are inserted using the **Insert** menu item **Module** and **UserForm** respectively. The command button is inserted or dragged over from the toolbar to the UserForm with the mouse to produce an interface form similar to Figure 17-10. The following file is a listing of the project for Example 2. **The line numbers at the right are for reference only and are not part of the program.**

Command Button Code

```
'The function of this routine is to draw                           1
'a triangle form 3 user specified points                          2
'The trigger is a mouse click on the command                      3
'button labeled Start.                                            4
```

'pnt1, pnt2, pnt3 are variant by default	5
'Returns a point in WCS	6
Private Sub CommandButton1_Click()	7
UserForm1.Hide	8
pnt1 = ThisDrawing.Utility.GetPoint(,"Provide the First Point: ")	9
'Returns a variant vector since GetPoint returns a point in WCS	10
'And draws a rubber-band line from the optional first point	11
pnt2 = ThisDrawing.Utility.GetPoint(pnt1, "Second Point? ")	12
'Draw first side of triangle	13
ThisDrawing.ModelSpace.AddLine pnt1, pnt2	14
pnt3 = ThisDrawing.Utility.GetPoint(pnt2, "3rd Point? ")	15
ThisDrawing.ModelSpace.AddLine pnt2, pnt3	16
ThisDrawing.ModelSpace.AddLine pnt3, pnt1	17
Unload Me	18
End Sub	19

Explanation

Lines 1-6, 10, 11, and 13
These lines are comments or remarks describing the function of the following line or lines. Comments make understanding and modifying a program easier and should be used liberally. Comments start with a Rem or use an apostrophe ('). These lines are ignored when the program is run.

Line 7
Private Sub CommandButton1_Click()
This line defines where Sub CommandButton1_Click() starts. The subroutine code is executed when CommandButton1 is clicked with the mouse.

Line 8
UserForm1.Hide
This statement hides the screen interface form and returns focus to AutoCAD. If UsefForm1 from the previous example is still loaded, the UserForm for this example would be Named UserForm2 by default. In this event and you should change the code in line 8 to **Userform2.Hide**.

Line 9, 12 and 15
pnt1 = ThisDrawing.Utility.GetPoint(,"Provide the First Point: ")
pnt2 = ThisDrawing.Utility.GetPoint(pnt1, "Second Point? ")
pnt3 = ThisDrawing.Utility.GetPoint(pnt2, "3rd Point? ")
The GetPoint method is used without a reference point in line 15. The point may be input either with the keyboard or mouse. When used with a reference point we see a rubber-band line from the reference point to the mouse cursor position.

Lines 14, 16, and 17
ThisDrawing.ModelSpace.AddLine pnt1, pnt2
ThisDrawing.ModelSpace.AddLine pnt2, pnt3
ThisDrawing.ModelSpace.AddLine pnt3, pnt1
The **AddLine method** requires the two point arguments to be double precision vectors with 3 components. Note the required space before the first point.

Line 18 and 19
Unload Me
End Sub
These lines remove UserForm1 from memory, return the focus back to AutoCAD, and end the subroutine.

Module1 Code

The following code is entered in the Module1 code window.

```
Option Explicit                                  1
Sub Triangle() 'Module1                          2
UserForm1.Show                                   3
End Sub                                           4
```

Explanation

Line 1
Option Explicit
The inclusion of this line forces explicit declaration of all variable types. This minimizes common inconsistent variable usage errors and typographic errors as they are quickly caught at run time. Otherwise, undeclared variable types would be Variant by default.

Lines 2-4
Sub Triangle() 'Module1
UserForm1.Show
End Sub
This subroutine's function is to create a Macro name, Triangle, that can be run from AutoCAD's **VBA** menu item **Run Macro**. Lines 1-4 are a part of Module1. The remaining lines of code below are a part of the UserForm1 object. Again, if UserForm1 from the previous example is still loaded, the UserForm for this example would be named UserForm2 and Module1 similarly would be named Module2 by default. In this event and you should change the names in lines 2-4 accordingly.

PolarPoint and AngleFromXAxis Methods

PolarPoint Method

The **PolarPoint method** defines a point at a given angle and distance from a given point. It has the syntax:

P = ThisDrawing.Utility.PolarPoint(Point, Angle, Distance)
where   **Point** ------------------------ Reference point, rubber-band origin
              Angle ----------------------- Angle in radians, double precision
              Distance -------------------- Distance from point, double precision

AngleFromXAxis Method

The **AngleFromXAxis method** calculates the angle of a line defined by two points from the horizontal axis in radians. The format of the **AngleFromXAxis method** is:

$$ang = ThisDrawing.Utility.AngleFromXAxis(point1, point2)$$

where   **point1** ----------------------- Start point of the line

           point2 ----------------------- End point of the line

Exercise 1

Write a Visual Basic program that will draw a line between two points, as in Figure 17-12. The user may either enter coordinates or choose points with the mouse. The graphical screen should draw a rubber-band on the screen to the current position of the mouse cursor if the second point is entered with the mouse. Use a graphical user interface similar to the one shown in Figure 17-11.

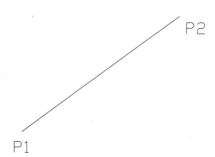

Figure 17-11 Screen Interface Form for Exercise 1

Figure 17-12 Line woth user defined end points

Example 3

Write a program in the AutoCAD VBA IDE that will draw a triangle based on a given line produced from two points P1 and P2, on an included angle and on a length of the second side shown in Figure 17-14.

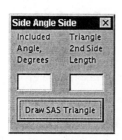

Figure 17-13 Screen Interface Form for Example 3

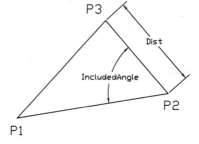

Figure 17-14 Side Angle Side Triangle

A UserForm1 is created by selecting the **Insert** menu item **UserForm**. To create TextBoxes from the toolbox, click on the TextBox button, then click on the form. The labels and control button on the form are created the same way to give the results in Figure 17-13. Insert a Module1 also. The following file is a listing of the Visual Basic program for Example 3. **The line numbers at the right are for reference only and are not a part of the program**.

Command Button Code

```
'(Declarations) (General)                                                        1
Const PI = 3.141592654                                                           2
Public IncludedAngle As Double 'pi-converted input angle                         3
Public Angle As Double  'included angle                                          4
Public Dist As Double   'length of 2nd side                                      5
                                                                                 6
'This procedure draws a triangle from 2 sides and an included angle SAS          7
'Program trigger is the command button labeled "Draw SAS Triangle"
                                                                                 8
'Additional feature is use of text boxes for included angle, 2nd side length input  9
'Included angle is angle between base and 2nd side                               10
                                                                                 11
Private Sub CommandButton1_Click()                                               12
UserForm1.Hide                                                                   13
'p1 is 1st point of base, variant type by default                               14
'p2 is 2nd point of base, variant type by default                               15
p1 = ThisDrawing.Utility.GetPoint(, "Enter or select 1st base point:")           16
p2 = ThisDrawing.Utility.GetPoint(p1, "Enter  select 2nd base point:")           17
ThisDrawing.ModelSpace.AddLine p1, p2 'draw base line                            18
If TextBox1.Text = "" Then                                                       19
Angle = ThisDrawing.Utility.GetAngle(p2,"Key ang. from +horiz. or select incl. angle:")
                                                                                 20
Else                                                                             21
Angle = ThisDrawing.Utility.AngleFromXAxis(p1, p2) + IncludedAngle               22
End If                                                                           23
If TextBox2.Text = "" Then                                                       24
Dist = ThisDrawing.Utility.GetDistance(p2, "Enter  select dist. from base point:")  25
End If                                                                           26
p3 = ThisDrawing.Utility.PolarPoint(p2, Angle, Dist)                             27
ThisDrawing.ModelSpace.AddLine p2, p3                                            28
ThisDrawing.ModelSpace.AddLine p1, p3                                            29
Unload Me                                                                        30
End Sub                                                                          31
                                                                                 32
Private Sub textBox1_Change()                                                    33
IncludedAngle = PI - Val(TextBox1.Text) * PI / 180                               34
End Sub                                                                          35
```

36
37
38
39

```
Private Sub textBox2_Change()
Dist = Val(TextBox2.Text)
End Sub
```

Explanation

Line 2
Const PI = 3.141592654
The constant pi is used in this routine so is declared to demonstrate this type of statement.

Lines 3-5
Public IncludedAngle As Double
Public Angle As Double
Public Dist As Double
These variables are declared to be public, which means any procedure can access them from any module in the project (file) without passing the argument in a parameter list. Also three variables are declared to be double precision as required by the AutoCAD methods in which they will be employed.

Lines 7-12
'This module draws a triangle from 2 sides and an included angle SAS
'Program trigger is the command button labeled "Draw SAS Triangle"
'Additional feature is use of text box for included angle, 2nd side length input
'Included angle is angle between base and 2nd side
Private Sub CommandButton1_Click()
These lines give the purpose and trigger of the procedure CommandButton1_Click, which starts with line 8. Purpose and trigger remarks are useful for all event driven subroutines.

Line 13
UserForm1.Hide
This statement hides the screen interface form and returns focus to AutoCAD. Otherwise the user would have to close the form with the button at the top right corner of the window to proceed with entering input in AutoCAD.

Lines 14-17
'p1 is 1st point of base, variant type by default
'p2 is 2nd point of base, variant type by default
p1 = ThisDrawing.Utility.GetPoint(, "Enter or select 1st base point:")
p2 = ThisDrawing.Utility.GetPoint(p1, "Enter select 2nd base point:")
These lines get input from the AutoCAD graphic screen or command line for the two points defining the base side of the triangle.

Lines 19-23 and 33-35
If textBox1.Text = "" Then
Angle=ThisDrawing.Utility.GetAngle(p2,"Enter angle from Horiz. or select included angle:")
Else
Angle=ThisDrawing.Utility.AngleFromXAxis(FirstPoint,SecondPoint)+IncludedAngle

End If
Private Sub textBox1_Change()
IncludedAngle = PI - Val(textBox1.Text) * PI / 180 'in radians
End Sub

The If ... Then ... Else statement checks TextBox1 for an entry. If no angle has been entered on the UserForm the GetAngle method is used. If a numeric angle was entered on the form it is converted by the subroutine textBox1_Change() to a radian included angle. Adding the radian angle returned by the AngleFromXAxis method gives the AutoCAD polar angle for the vertex (third) point.

Lines 24-26
If textBox2.Text = "" Then
Dist = ThisDrawing.Utility.GetDistance(p2, "Enter or select dist. from base point:")
End If

If no numeric entry for the second side length was entered on the user form these lines use the GetDistance method.

Lines 37-39
Private Sub textBox2_Change()
Dist = Val(textBox2.Text)
End Sub

If a numeric entry for the second side length was entered on the user form these lines convert from text to numeric form.

Lines 18, 28 and 29
ThisDrawing.ModelSpace.AddLine p1, p2
ThisDrawing.ModelSpace.AddLine p2, p3
ThisDrawing.ModelSpace.AddLine p1, p3

These methods draw the base, second, and third sides of the triangle.

Module1 Code

The following code is entered in the Module1 code window.

```
Option Explicit                    1
Sub SAS() 'Module1                 2
UserForm1.Show                     3
End Sub                            4
```

Exercise 2

Write a program in the AutoCAD VBA IDE that will draw a triangle based on a given line produced from two points P1 and P2, on an adjacent angle at one end, and another angle at the other end as shown in Figure 17-16. Employ a UserForm similar to the one shown in Figure 17-15

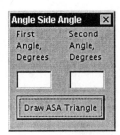

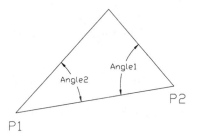

Figure 17-15 *Screen Interface Form for Exercise 2*

Figure 17-16 *Angle Side Angle Triangle with user defined points defining the base*

Exercise 3

Write a program in the AutoCAD VBA IDE that will draw a triangle based on a given line produced from two points P1 and P2, and two distances, which are the lengths of the sides adjoining each end as in Figure 17-18. Employ a UserForm similar to the one shown in Figure 17-17.

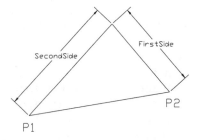

Figure 17-17 *Screen Interface Form for Exercise 3*

Figure 17-18 *Side Side Side Triangle with user defined points defining the base*

Additional VBA examples

More VBA examples are included with AutoCAD 2000 beyond the large number in the help files. These are considered sample files and are located in the **Sample** subdirectory of the ACAD2000 directory and the AutoCAD 2000 CD ROM. Another resource for learning VBA is the web. For example, the Autodesk web site has an active VBA newsgroup that you can access from the adesknews.autodesk.com news server without needing a usenet news server. The VBA news group is autodesk.autocad.customization.vba.

Review Questions

1. _____ lets you access and manipulate the objects and functionality of AutoCAD from Excel or another application supporting ActiveX.

2. Before you can use AutoCAD's objects in some other software supporting ActiveX you must make the application aware that the AutoCAD _____ Library is available on the computer.

3. There are over _____ programmers using Visual Basic.

4. Visual Basic is considered an object-_____, rather than an object-oriented, programming language.

5. Functions known as _____ have been defined in the AutoCAD object library to perform an action on an object, for example, draw a line in a drawing.

6. _____ are functions that set or return information about the state of an object.

7. The term **IDE**, which is the **Visual Basic Editor**, stands for _____ .

8. A _____ (extension .dvb) is the name given to the forms, controls, modules, and programming making up a saved AutoCAD Visual Basic file.

9. _____ forces explicit declaration of all variable types, which minimizes common inconsistent variable usage errors.

10. _____ is the variable type returned by the GetPoint method.

11. The _____ method always measures the angle with a positive X-axis (3 o'clock position) and in a counterclockwise direction.

12. The _____ method allows you to enter the X, Y coordinates or X, Y, Z coordinates of a point.

13. The _____ method lets you retrieve the value of an AutoCAD system variable.

14. The _____ method defines a point at a given angle and distance from the given point.

15. The _____ method lets you enter a distance on the command line, a distance from a given point, or two points.

16. The _____ method allows you to enter an angle, either from the keyboard in degrees or by selecting two points.

17. The _____ method calculates the angle of a line defined by two points from the horizontal axis in radians.

Exercises

Exercise 4

Write a Visual Basic program that will draw an equilateral triangle inside the circle (Figure 17-19).

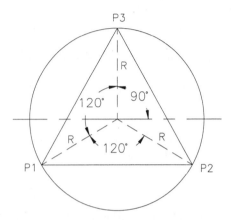

Figure 17-19 Equilateral Triangle inside a circle

Exercise 5

Write a Visual Basic program that will draw a square of sides S and a circle tangent to the four sides of the square as shown in Figure 17-20. The base of the square makes an angle, ANG, with the positive X-axis. The program should allow you to enter the starting point P1, length S, and angle ANG on a user interface form or in the AutoCAD graphical interface.

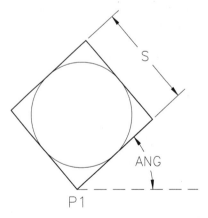

Figure 17-20 Square of side S at an angle ANG

Exercise 6

Write a Visual Basic program that will draw a line at an angle A and then generate the given number of lines (N), parallel to the first line with an offset distance S as entered on a user form or from the keyboard.

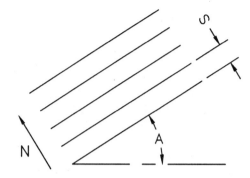

Figure 17-21 Number of lines at a distance of S

Exercise 7

Write a program that will draw a slot (Figure 17-22) with center lines. The program should allow you to enter slot length, slot width, and the layer name for center lines on a user interface form or in the AutoCAD graphical interface.

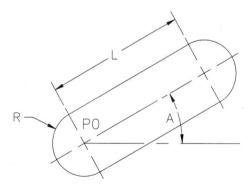

Figure 17-22 Slot of length L and radius R

Exercise 8

Write a Visual Basic program that will draw two lines tangent to two circles, as shown in Figure 17-23. The program should allow you to enter the circle diameters and the center distance between the circles on a user interface form or in the AutoCAD graphical interface.

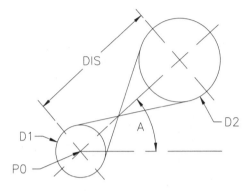

Figure 17-23 *Circles with tangent at an angle A*

Exercise 9

Write a VBA program that will draw a hub with the key slot, as shown in Figure 17-24. The user should enter the four parameters in the note below on a user form or in the AutoCAD graphical interface. Use the program inside a circle of Diameter D1 to produce a keyed bushing with the test dimensions.

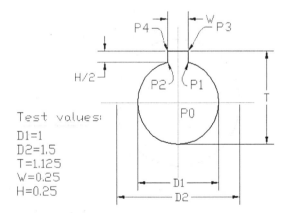

Figure 17-24 *Hub with a Keyway*

Exercise 10

Write a VBA program to draw the tangent arc cam shown in Figure 17-25.

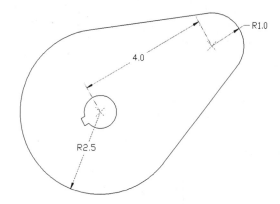

Figure 17-25 *Tangent Arc Flat Plate Cam*

Exercise 11

Write a VBA program to draw the circular arc cam shown in Figure 17-26. The circular arc cam in Figure 17-26 is made up of four arcs. The dwell semi-circle is drawn first. Along the diameter or an extension of the diameter outside the circle, two arc centers equidistant from the dwell circle center are chosen. Since the dwell circle and the arcs drawn will be perpendicular to the diameter, their slopes match. A fillet is then used between the other ends of the arcs to create a lobe or nose radius.

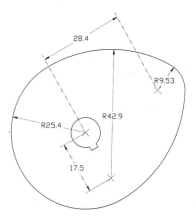

Figure 17-26 *Circular Arc Cam*

Project Exercise

Write an Visual Basic program that will draw the two views of a bushing as shown in Figure 17-27. The program should allow you to enter the starting point P0, lengths L1, L2, and the bushing diameters ID, OD, HD on a user interface form or in the AutoCAD graphical interface. The distance between the front view and the side view of bushing is DIS (DIS = 1.25 * HD). The program should also draw the hidden lines in the HID layer and center lines in the CEN layer. The center lines should extend 0.75 units beyond the object line.

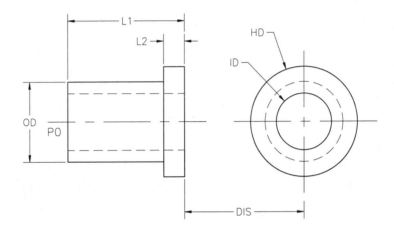

Figure 17-27 *Two views of bushing*

Chapter 18

Accessing External Databases

![Learning Objectives]

After completing this chapter, you will be able to:
- *Understand databases and the database management system (DBMS).*
- *Understand the **AutoCAD database connectivity** feature.*
- ***Configure** external databases.*
- *Access and edit a database using **dbConnect Manager**.*
- *Create **Links** with graphical objects.*
- *Create and display **Labels** in a drawing.*
- *Understand **AutoCAD SQL Environment** (ASE) and create queries using **Query Editor**.*
- *Form selction sets using **Link Select**.*
- *Convert ASE links into AutoCAD 2000 format.*

UNDERSTANDING DATABASES

Database

A **database** is a collection of data that is arranged in a logical order. For example, there are six computers in an office, and we want to keep a record of these computers on a sheet of paper. In a database, the columns are known as **fields**, and the individual rows are called **records**. The entries in the database tables are known as **cells** that store data for a particular variable. One of the ways of recording the computer information is to make a table with rows and columns as shown in Figure 18-1. Each column will have a heading that specifies a certain feature of the computer, such as COMP_CFG, CPU, HDRIVE, or RAM. Once the columns are labeled, the computer data can be placed in the columns for each computer. By doing this, we have created a database on a sheet of paper that contains information on our computers. The same information can be stored in a computer; this is generally known as a computerized database.

COMPUTER

COMP_CFG	CPU	HDRIVE	RAM	GRAPHICS	INPT_DEV
1	PENTIUM350	4300MB	64MB	SUPER VGA	DIGITIZER
2	PENTIUM233	2100MB	32MB	SVGA	MOUSE
3	MACIIC	40MB	2MB	STANDARD	MOUSE
4	386SX/16	80MB	4MB	VGA	MOUSE
5	386/33	300MB	6MB	VGA	MOUSE
6	SPARC2	600MB	16MB	STANDARD	MOUSE

Figure 18-1 A table containing computer information

Most of the database systems are extremely flexible and any modifications or additions of fields or records can be done easily. Database systems also allow you to define relationships between multiple tables, so that if the data of one table is altered, automatically the corresponding values of another table with predefined relationships will change accordingly.

Database Management System

The database management system (DBMS) is a program or a collection of programs (software) used to manage the data in the database. For example, PARADOX, dBASE, INFORMIX, and ORACLE are database management systems.

Components of a Table

A database **table** is a two-dimensional data structure (Figure 18-2) that consists of rows and columns.

COMPUTER

COMP_CFG	CPU	HDRIVE	RAM	GRAPHICS	INPT_DEV
1	PENTIUM350	4300MB	64MB	SUPER VGA	DIGITIZER
2	PENTIUM233	2100MB	32MB	SVGA	MOUSE
3	MACIIC	40MB	2MB	STANDARD	MOUSE
4	386SX/16	80MB	4MB	VGA	MOUSE
5	386/33	300MB	6MB	VGA	MOUSE
6	SPARC2	600MB	16MB	STANDARD	MOUSE

| 3 | MACIIC | 40MB | 2MB | STANDARD | MOUSE |

ROWS

| 6 | SPARC2 | 600MB | 16MB | STANDARD | MOUSE |

Figure 18-2 Rows in a table (the horizontal group of data)

Row. The horizontal group of data is called a **row**. For example, Figure 18-2 shows a table, and just below the full table are two rows of the table. Each value in a row defines an attribute of the item. For example, in Figure 18-2, the attributes assigned to COMP_CFG (1) include PENTIUM350, 4300MB, 64MB, and so on. These attributes are arranged in the first row of the table.

Column. A vertical group of data (attribute) is called a **column**. Three of the columns in Figure 18-2 are shown in Figure 18-3. HDRIVE is the column heading that represents a feature of the computer, and the HDRIVE attributes of each computer are placed vertically in this column.

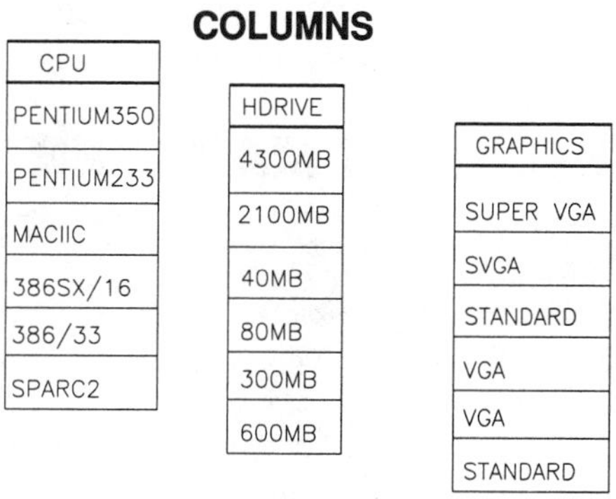

COLUMNS

Figure 18-3 Columns in a table (the vertical group of data)

AUTOCAD DATABASE CONNECTIVITY

AutoCAD can be effectively used in associating data contained in an external database table with the AutoCAD graphical objects by linking. **Links** are the pointers to the database tables from which the data can be referred to. AutoCAD can also be used to attach **Labels** that will display data from the selected tables as text objects. AutoCAD database connectivity offers the following facilities:

1. A **dbConnect Manager** that can be used to associate links, labels, and queries with AutoCAD drawings.
2. An **external configuration utility** that enables AutoCAD to access the data from a database system.
3. A **Data View window** that displays the records of a database table within the AutoCAD session.
4. A **Query Editor** that can be used to construct, store, and execute SQL queries. **SQL** is an acronym for **structured query language**.
5. A **migration tool** that converts links and other displayable attributes of files created by earlier releases to AutoCAD 2000.
6. A **Link Select operation** that creates iterative selection sets based on queries and graphical objects.

DATABASE CONFIGURATION

An external database can be accessed within AutoCAD only after configuring AutoCAD using Microsoft **ODBC** (Open Database Connectivity) and **OLE DB** (Object Linking and Embedding Database) programs. AutoCAD is capable of utilizing data from other applications, regardless of the format and the platform on which the file is stored. Configuration of database involves creating a new **data source** that points to a collection of data and information about the required drivers to access it. A **data source** is an individual table or a collection of tables created and stored in an environment, catalog, or schema. Environments, catalogs, and schemas are the hierarchical database elements in most of the database management systems and they are analogous to Window-based directory structure in many ways. Schemas contain a collection of tables, while Catalogs

Chapter 18

contain subdirectories of schemas and Environment holds subdirectories of catalogs. The external applications supported by AutoCAD 2000 are **dBASE® V** and **III, Oracle®8.0** and **7.3, Microsoft® Access®**, and **PARADOX 7.0, Microsoft Visual FoxPro®6.0, SQL Server 7.0** and **6.5**. The configuration process varies slightly from one database system to other.

DBCONNECT MANAGER*

Menu:	Tools > dbConnect
Command:	DBCONNECT

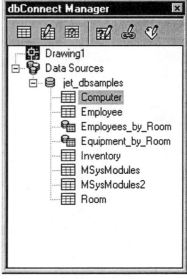

Figure 18-4 dbconnect Manager

This is a new feature with AutoCAD 2000 that enables you to access information from external databases more effectively than previous releases of AutoCAD. The **dbConnect Manager** is a dockable as well as resizable window and contains a set of buttons and a tree view showing all the configured and available databases. You can use **dbConnect Manager** for associating various database objects with an AutoCAD drawing. You can invoke **dbConnect Manager** (Figure 18-4) by entering **DBCONNECT** at the AutoCAD Command prompt or choosing **dbConnect** from the **Tools** menu. The **dbConnect Manager** contains two nodes in the tree view. The **Drawing node** displays all the open drawings and each node will display all the associated database objects with the drawing. **Data Sources node** displays all the configured data on your system.

AutoCAD 2000 contains several Microsoft Access sample database tables and a direct driver (*jet_dbsamples.udl*). The following example shows the procedure to configure a database with your drawing using the **dbConnect Manager**.

Example 1

Configure a data source of Microsoft Access database with your diagram. Update and use the **jet_samples.udl** configuration file with new information.

Step 1. Invoke **dbConnect Manager** either by entering **DBCONNECT** at the Command prompt or choosing from the **Tools** menu. The **dbConnect** menu will be inserted between **Modify** and **Window** menus.

Step 2. From the **dbconnect** menu, choose **Data Sources > Configure** to display the **Configure a Data Source** dialog box (Figure 18-5). You can also invoke the **Configure a Data Source** dialog box by selecting **Data Sources** in **dbConnect Manager** tree view, right-clicking and choosing **Configure Data Source...**.

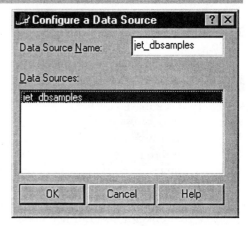

Figure 18-5 Configure a Data Source dialog box

Step 3. From the **Data Sources:** list box in **Configure a Data Source** dialog box, select **jet_dbsamples** and choose the **OK** button. AutoCAD will display **Data Link Properties** (**Connection** tab) dialog box (Figure 18-6).

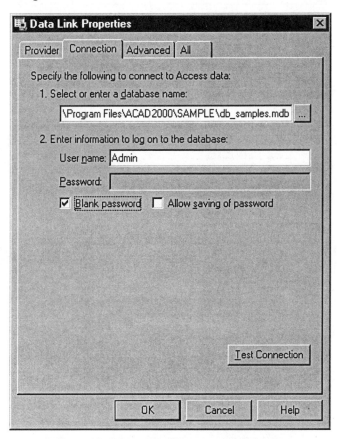

Figure 18-6 Data Link Properties dialog box

Step 4. In the **Data Link Properties** (**Connection** tab) dialog box, choose the [**...**] button adjacent to **Select or enter a database name:** text box. AutoCAD will display the **Select Access Database** dialog box (Figure 18-7). Select **db_samples** (**.mdb file**) and choose **Open** button. You can also type in the text box as **C:\Program Files\ACAD2000\SAMPLE\db_samples.mdb**, if you have selected default during the setup procedure.

Note

*If you are configuring any other driver than **Microsoft Jet** for database linking, you should consult the appropriate documentation for the configuring process.*

*The other tabs in the **Data Links Properties** dialog box are required for configuring different Database Providers supported by AutoCAD 2000.*

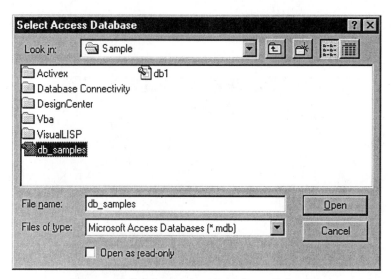

Figure 18-7 Select Access Database dialog box

Step 5. From the **Data Link Properties** (**Connection** tab) dialog box, choose the **Test Connection** button to ensure that the database source has been configured correctly and AutoCAD will display the **Microsoft Data Link** message box (Figure 18-8). Choose **OK** button to end the message and again choose the **OK** button to complete the configuration process.

Figure 18-8 Microsoft Data Link message box

After configuring a data source when you double click on **jet_dbsamples** in **dbConnect Manager,** all the sample tables will be displayed in the tree view and you can connect any table and link its records to your diagram.

VIEWING AND EDITING TABLE DATA FROM AUTOCAD

After you have configured a data source using the **dbConnect Manager**, you can view as well as edit its tables within the AutoCAD session. The **Data View window** can be used for viewing and editing a database table. You can open tables in **Read-only** mode to view its content. But you cannot edit its records in **Read-only** mode. Some database systems may require a valid user-name and password before connecting to AutoCAD drawing files. The records of a database table can be edited by opening in **Edit** mode. The procedures to open a table in various modes follow.

 Read-only mode: Choose **dbConnect** menu > **View Data** > **View External Table** and the table will be opened in **Read-only** mode. You can also view a table in the Read-only mode by selecting the table, right-clicking to invoke the shortcut menu, and choosing **View Table**.

 Edit mode: Choose **dbConnect** menu > **View Data** > **Edit External Table** to open the table in **Edit** mode. You can also open a table in the **Edit** mode by double-clicking on it in the tree view of the **dbConnect Manager** or choosing **Edit Table** from the shortcut menu.

Example 2

From the **jet_dbsamples** data source, select Computer table and edit each row in the table. Add a new computer and replace one by editing the table and save your changes.

Step 1. Select **Computer** table from the **jet_dbsamples** in the **dbConnect Manager** and right-click to invoke the shortcut menu. Choose **Edit Table** from the menu (Figure 18-9). You can also do the same by double clicking on the **Computer** table.

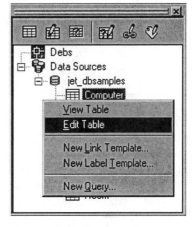

Step 2. AutoCAD will display the **Data View window** with the **Computer** table in it. In the **Data View window**, you can resize, sort, hide, or freeze the columns according to your requirement.

Step 3. To **add a new item** in the table, double-click on the first empty row. In TAG_NUM column, type **24675**. Then type:

MANUFACTURER:	**IBM**
Equipment_Description:	**PIII/450, 4500GL, NETX**
Item_Type:	**CPU**
Room:	**6035**

*Figure 18-9 Choosing the **Edit Table** from the shortcut menu*

Step 4. To **Edit the record** for TAG_NUM **60298**, and display the following in **Data View window** (Figure 18-10), double-click on each cell and type:

MANUFACTURER:	**CREATIVE**
Equipment_Description:	**INFRA 6000, 40XR**
Item_Type:	**CD DRIVE**
Room:	**6996**

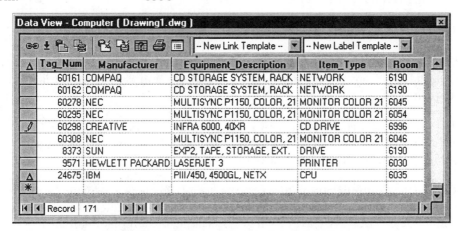

Figure 18-10 Data View window after editing

Step 5. After you have made all the changes in the database table, you have to save the changes for further use. To **save the changes** in the table, right-click on the **Data View grid header,** choose **Commit** from the shortcut menu, and the changes will be saved.

Note

After making changes, if you do not want to save the changes, from the above shortcut menu choose **Restore** *and AutoCAD will restore the orginal values.*

If you quit **Data View window** *without* **Committing**, *AutoCAD will automatically commit all the changes you have made during the editing session.*

CREATING LINKS WITH GRAPHICAL OBJECTS

The main function of the database connectivity feature of AutoCAD is to associate data from external sources with its graphical objects. You can establish the association of the database table with the drawing objects by developing a **link**, which will make reference to one or more record from the table. But you **cannot** link nongraphical objects such as layers or linetypes with the external database. Links are very closely related with graphical objects and change with the change in graphical objects.

To develop links between database tables and graphical objects, you must create a **Link Template**, which will identify the fields of the tables with which links are associated to share the template. For example, you can create a link template that uses the **Tag_number** from the **COMPUTER database table**. Link template also acts as a shortcut that points to the associated database tables. You can associate multiple links to a single graphical object using different link templates. This is useful in associating multiple database tables with a single drawing object. The following example will describe the procedure of linking using link template creation.

Example 3

Create a link template between **COMPUTER** database table from **Jet_dbsamples** and your drawing and use **Tag_Number** as the **key field** for linking.

Step 1. Open your drawing that has to be linked with the **COMPUTER** database table.

Step 2. From the **Tools** menu, choose **dbConnect** to invoke **dbConnect Manager**. Select COMPUTER from the **jet_dbsamples** data source and right click on it to invoke the shortcut menu.

Step 3. In the shortcut menu, choose **New Link Template...** to Invoke the **New Link Template** dialog box (Figure 18-11). You can also invoke the dialog box by selecting the table in the tree view and then choosing the **New Link Template** button.

Step 4. Choose the **Continue** button to accept the default link template name **Computer Link1**. AutoCAD will display the **Link Template** dialog box (Figure 18-12). Select the check box adjacent to **Tag_Number** to accept it as the **Key field** for associating the template with the block reference in the diagram.

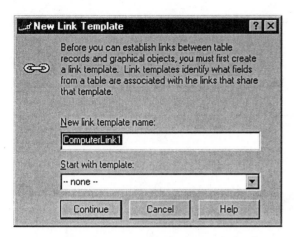

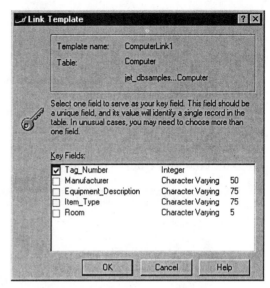

Figure 18-11 New Link Template dialog box

Step 5. Choose the **OK** button to complete the new link template; the name of the link template will be displayed under the Drawing name node of the tree view in **dbConnect Manager**.

Figure 18-12 Link Template dialog box

Step 6. To **link a record with the block reference**, double-click on the COMPUTER table to invoke the **Data View window** (Edit mode). In the table, go to Tag_Number **24675** and highlight the record (row).

 Step 7. Right-click on the row header and choose **Link!** from the shortcut menu (Figure 18-13). You can also link by choosing the **Link!** button from the toolbar in **Data View window**.

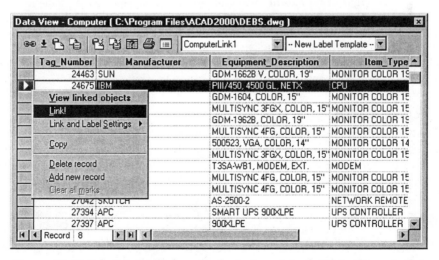

*Figure 18-13 Selecting **Link!** from the shortcut menu in the **Data View window***

Step 8. When AutoCAD prompts you to **Select Objects:**, you can choose the block references you want to link with the record. Repeat the process to link all the records to the corresponding block references in the diagram.

 Step 9. After linking, you can **view the linked objects**, by choosing the **View Linked Objects in Drawing** button in the **Data View** dialog box, and the linked objects will get selected in the drawing area.

Additional Link Viewing Settings

You can set a number of viewing options for linked graphical objects and linked records by using the **Data View and Query Options** dialog box (Figure 18-14).

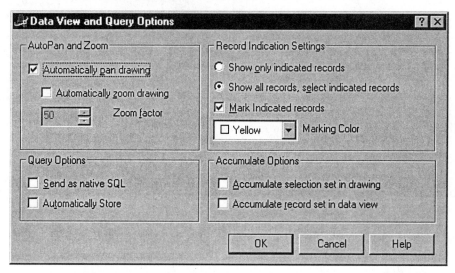

Figure 18-14 Data View and Query Options dialog box

You can set the **Automatically Pan Drawing** option so that AutoCAD will pan the drawing automatically to display objects linked with the current set of selected records in the **Data View** window. You can also set options for **Automatically zoom drawing** and the **Zoom factor**. You can also change the Record Indication settings and the indication marking color.

Editing Link Data

After linking data with the drawing objects, you may need to edit the data or update their **Key field** values. For example, you may need to reallocate the Tag-Number for the computer equipments or Room for each of the linked items. **Link Manager** can be used for changing the Key values. The next example describes the procedure of editing linked data.

Example 4

Use **Link Manager** to edit linked data from the Computer table and change the Key Value from **24352** to **24675**.

Step 1. Open the diagram that was linked with the Computer table and then from the **dbConnect** menu, choose **Links > Link Manager...**. AutoCAD will prompt you to select the linked object.

Step 2. Select the linked object to invoke the **Link Manager** dialog box for the Computer table (Figure 18-15).

Step 3. Choose **Computer link1** from the **Link Template:** drop-down list.

Step 4. Select **24352** field in the **Value** column and choose the [**...**] button to invoke the **Column Values** dialog box (Figure 18-16). Select **24675** from the list and choose the **OK** button.

Figure 18-15 Link Manager *dialog box*

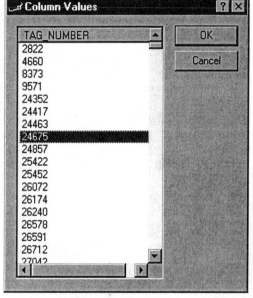

Figure 18-16 Column Values *dialog box*

Step 5. Again choose the **OK** button in **Link Manager** to accept the changes.

CREATING LABELS*

You can use linking as a powerful mechanism to associate drawing objects with external database tables. You can directly access associated records in the database table by selecting linked objects in the drawing. But linking has some limitations. Suppose you want to include the associated external data with the drawing objects. Since during printing, the links are only the pointers to the external database table, they will not appear in the printed drawing. In such situations, AutoCAD provides you a feature called **Labels** that can be used for visible representation of external data in drawing.

Labels are the multiline text objects that display data from the selected fields in the AutoCAD drawing. Labels are generally one of two types:

Freestanding labels. Freestanding labels exists in the AutoCAD drawing independent of the graphical objects. Their properties do not change with any change of graphical objects in the drawing.

Attached labels. Attached labels are very closely connected with the graphical objects they are associated with. If the graphical objects are moved, then the labels also move and if the objects are deleted, then the labels attached with them also get deleted.

Labels associated with the graphical objects in AutoCAD drawing are displayed with a leader. Labels are created and displayed by **Label Templates** and all the properties of label display can be controlled by using the **Label Template** dialog box. The next example will demonstrate the complete procedure of creating and displaying labels in AutoCAD drawing using label template.

Example 5

Create a new label template in the Computer database table and use the following specifications for the display of labels in the drawing:

1. The label includes **Tag_Number, Manufacturer** and **Item_Type** fields.
2. The fields in the label are **0.25** in height, **Times New Roman** font, **black** (**Color 18**) in color and **Middle-left** justified.
3. The label offset starts with **Middle Center** justified and the leader offset is **X=1.5** and **Y=1.5**.

Step 1. Open the drawing where you want to attach a label with the drawing objects.

Step 2. Choose **Tools menu > dbConnect** to invoke **dbConnect Manager**. Select **Computer** table from **jet_dbsamples** data source.

Step 3. Right-click on the table and choose **New Label Template...** from the shortcut menu to invoke the New Label Template dialog box (Figure 18-17). You can also invoke the dialog box by choosing the **New Label Template** button from the **dbConnect Manager** toolbar.

Step 4. In the **New Label Template** dialog box, choose the **Continue** button to accept the default label template name **Computer Label1**. AutoCAD will display the **Label Template** dialog box.

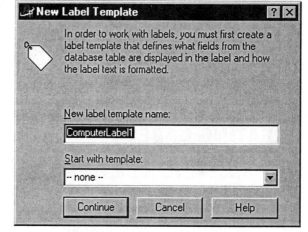

Figure 18-17 New Label Template dialog box

Step 5. In the **Label Template** dialog box, choose the **Label Fields** tab and from the **Field:** drop-down list, add **Tag_Number, Manufacturer**, and **Item_Type** fields by using the **Add** button (Figure 18-18).

Step 6. Highlight the field names in the edit box and choose the **Character** tab. In the **Character** tab, select the **Times New Roman** font, font height to **0.25,** and **Color 18** color (Figure 18-19). Also select **Middle-left** justification in the **Properties** tab.

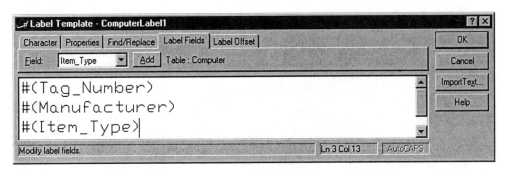

Figure 18-18 Label Template dialog box (Label Fields tab)

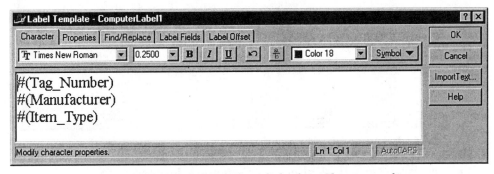

Figure 18-19 Label Template dialog box (Character tab)

Step 7. Choose the **Label Offset** tab and select **Middle-center** in the **Start**: drop-down list. Set the **Leader offset:** value to **X: 1.5** and **Y: 1.5** (Figure 18-20).

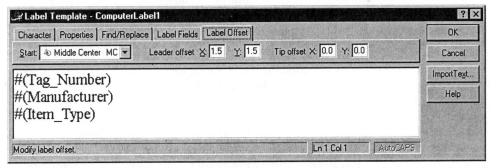

Figure 18-20 Label Template dialog box (Label Offset tab)

Step 8. To display the label in the drawing, select the Computer table in the tree view of **dbConnect Manager**. Right-click on the table and choose **Edit Table...** from the shortcut menu.

Step 9. Select **Computer link1** (created in an earlier example) from the **link name** drop-down list and **Computer label1** from the **label name** drop-down list in the **Data View** window.

Step 10. Select the record (row) you want to use as a label. Then choose the **Link and Label Setting** button and select **Create Attached Label** from the menu (Figure 18-21). Then choose the

Create Attached Label button (adjacent to the **Link and Label setting** button) and select the drawing object you want to label.

Data View - Computer (C:\Program Files\ACAD 2000\DEBS.dwg)			
		ComputerLink1 ▼	**ComputerLabel1** ▼
Create Links		**Equipment_Description**	**Item_Type**
✔ Create Attached Labels		HL6605ATK, COLOR, 16"	MONITOR COLOR 16
Create Freestanding Labels		HL6605ATK, COLOR, 16"	MONITOR COLOR 16
24463	SUN	GDM-1662B V, COLOR, 19"	MONITOR COLOR 19
24857	SUN	GDM-1604, COLOR, 15"	MONITOR COLOR 15
25422	NEC	MULTISYNC 3FGX, COLOR, 15"	MONITOR COLOR 15
25452	SUN	GDM-1962B, COLOR, 19"	MONITOR COLOR 19
26072	NEC	MULTISYNC 4FG, COLOR, 15"	MONITOR COLOR 15
26174	AST	500523, VGA, COLOR, 14"	MONITOR COLOR 14
26240	NEC	MULTISYNC 3FGX, COLOR, 15"	MONITOR COLOR 15
26578	TELEBIT	T3SA-WB1, MODEM, EXT.	MODEM
26591	NEC	MULTISYNC 4FG, COLOR, 15"	MONITOR COLOR 15
26712	NEC	MULTISYNC 4FG, COLOR, 15"	MONITOR COLOR 15
27042	SKUTCH	AS-2500-2	NETWORK REMOTE RESE
27394	APC	SMART UPS 900XLPE	UPS CONTROLLER
Record 3			

*Figure 18-21 Choosing **Create Attached Labels** in the **Data View** window*

AutoCAD will display the Label with given specifications in the drawing.

Note

*You can create **Freestanding Labels** in the same way by selecting **Create Freestanding Labels** from the **Link and Label setting** menu in the **Data View** window.*

Updating Labels with New Database Values

There is a strong possibility that you may have to change the data values in the database table after adding a label in the AutoCAD drawing. So you should update the labels in the drawing after making any alteration in the database table the drawing is linked with. The following is the procedure for updating all label values in the AutoCAD drawing:

1. After editing the database table, open the drawing that has to be updated.

2. From the **dbConnect** menu, choose **Labels > Reload Labels...** to invoke the **Select a Database Object** dialog box (Figure 18-22).

3. In the dialog box, select the label template from the list box and choose the **OK** button for updating with the new database values.

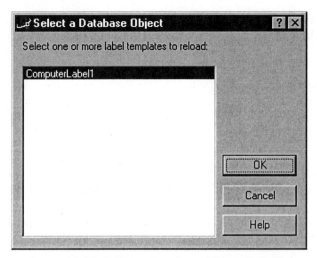

Figure 18-22 *Select a Database Object dialog box*

Importing and Exporting Link and Label Templates

You may want to use the link and label templates that have been developed by some other AutoCAD users. This is very useful when developing a set of common tools to be shared by all of the team members in a project. AutoCAD is capable of importing as well as exporting all the link and label templates that are associated with a drawing. The following is the procedure to **import** a set of templates into the current drawing:

1. From the **dbConnect** menu, choose **Templates > Import Template Set** to invoke the **Import Template Set** dialog box.
2. In the dialog box, select the template set to be imported.
3. Choose **Open** to import the template set into the current drawing.

AutoCAD will display an **alert box** that can be used to provide a unique name for the template if there is a link or label template with the same name already associated with the current drawing.

The following is the procedure to **export** a set of templates from the current drawing:

1. From the **dbConnect** menu, choose **Templates > Export Template Set** to **Export Template Set** dialog box.
2. In the dialog box in the **Save In list**, select the directory where you want to save the template set.
3. Under **File Name**, specify a name for the template set, and then choose the **Save** button to save the template in the specified directory.

AUTOCAD SQL ENVIRONMENT (ASE)

SQL is an acronym for **structured query language**. It is often referred to as **sequel**. SQL is a format in computer programming that lets the user ask questions about a database according to specific rules. The **AutoCAD SQL environment** (ASE) lets you access and manipulate the data that is stored in the external database table and link data from the database to objects in a drawing.

Chapter 18

Once you access the table, you can manipulate the data. The connection is made through a database management system (DBMS). The DBMS programs have their own methodology for working with databases. However, the ASE commands work the same way regardless of the database being used. This is made possible by the ASE drivers that come with AutoCAD software. In AutoCAD 2000, SQL has been incorporated. For example, you want to prepare a report that lists all the computer equipment that costs more then $25. **AutoCAD Query Editor** can be used to easily construct a query that returns a subset of records or linked graphical objects that follow the previously mentioned criterion.

AUTOCAD QUERY EDITOR*

The **AutoCAD Query Editor** consists of four tabs that can be used to create new queries. The tabs are arranged in order of increasing complexities. For example, if you are not familiar with **SQL** (Structured Query Language), you can start with **Quick Query** and **Range Query** initially to get familiar with query syntax.

You can start developing a query in one tab and subsequently add and refine the query conditions in the next tabs. For example, if you have created a query in the **Quick Query** tab and then decide to add an additional query using **Query Builder** tab, when you choose the **Query Builder** tab, all the values initially selected in the previous tabs are displayed in this tab and additional conditions can be added to the query. But it is not possible to go backwards through the tabs once you have created queries with one of the advanced tabs because additional functions are not available in the simpler tabs. AutoCAD will prompt a warning indicating that the query will be reset to its default values if you attempt to move backward through the query tabs.

The AutoCAD **Query Editor** has the following tabs for building queries:

Quick Query: This tab provides an environment where your simple queries can be developed based on a single database field, single operator, and a single value. For example, you can find all records from the current table where the value of the **"Item_type"** field equals **"CPU"**.

Range Query: This tab provides an environment where a query can be developed that returns all records that fall within a given range of values. For example, you can find all records from the current table where the value of the **"Room"** field is **greater than or equal to 6050** and **less than or equal to 6150**.

Query Builder: This tab provides an environment where more complicated queries can be developed based on multiple search criteria. For example, you can find all records from the current table where the **"Item_type"** equals **"CPU"** and **"Room"** number is greater than **6050**.

SQL Query: This tab provides an environment where sophisticated queries can be developed that conform with the SQL 92 protocol. For example, you can select * from **Item_type.Room. Tag_Number** where:

> **Item_type = 'CPU' ; Room > = 6050 and < = 6150**
> and
> **Tag_number > 26072**

The following example describes the complete procedure of creating a new query using all the tabs of the **Query Editor**.

Example 6

Create a new Query for the **Computer** database table and use all the tabs of the **Query Editor** to prepare a SQL query.

Step 1. To create a new query, from the **dbconnect** menu choose **Query > New Query...** to invoke the **New Query** dialog box (Figure 18-23). You can also invoke the dialog box by right clicking on the table (Computer) and choosing **New Query...** from the short-cut menu.

Step 2. In the **New Query** dialog box, choose the **Continue** button to accept the default query name **ComputerQuery1**. AutoCAD will display **Query Editor**.

Step 3. In the **Quick Query** tab of **Query Editor**, select **'Item_Type'** from the **Field:** list box, **'=Equal'** from the **Operator:** drop-down list, and type **'CPU'** in the **Value:** text

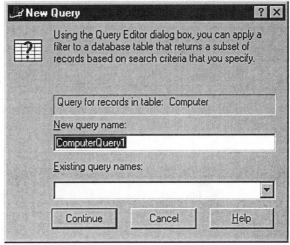

Figure 18-23 New Query dialog box

box. You can also select **'CPU'** from the **Column Values** dialog box by choosing the **Look up values** button. Choose the **Store** button to save the query (Figure 18-24).

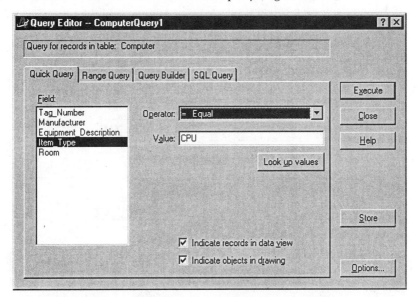

Figure 18-24 Query Editor (Quick Query tab)

Step 4. Choose the **Range Query** tab, and select **Room** number from the **Field:** list box. Enter **6050** in the **From:** text box and **6150** in the **Through:** text box. You can also select the values from the **Column Values** dialog box by choosing the **Look up values** button (Figure 18-25). Choose the **Store** button to save the query.

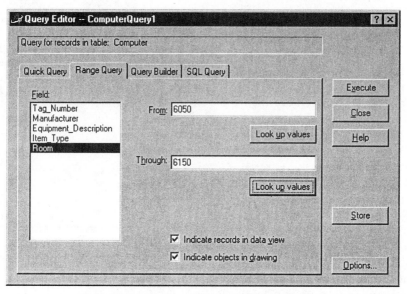

Figure 18-25 Query Editor (Range Query tab)

Step 5. Choose the **Query Builder** tab, and add **Item_Type** in **Show fields:** list box by Selecting **Item_Type** from the **Fields in table:** list box and choosing the **Add** button. In the table area, change the entries of fields, Operator, Value, Logical, and Parenthetical grouping criteria as shown in Figure 18-26, by selecting each cell and selecting the values from the respective drop-down list or options lists. Choose the **Store** button to save the query.

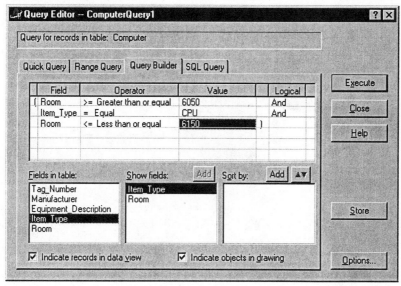

Figure 18-26 Query Editor (Query Builder tab)

Step 6. Choose the **SQL Query** tab; the query conditions specified earlier will be carried to the tab automatically. Here you can create a query using multiple tables. But select **'Computer'** from the **Table:** list box, **'Tag_Number'** from the **Fields:** list box, **'>= Greater than or equal to'** from the **Operator:** drop-down list and enter **26072** in the **Values:** edit box. You can also select from the available values by choosing the [...] button (Figure 18-27). Choose the **Store** button to save the query.

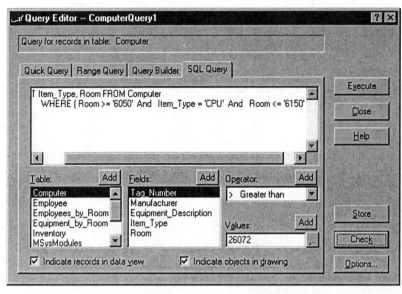

Figure 18-27 Query Editor (SQL Query tab)

Step 7. To check whether the SQL syntax is correct, choose the **Check** button and AutoCAD will display the **Information box** (Figure 18-28) to determine whether the syntax is correct.

Step 8. Choose the **Execute** button to display the **Data View** window showing a subset of records matching the specified query criteria.

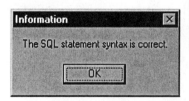

Figure 18-28 Information box

Note

You can view the subset of records conforming to your criterion with any of the four tabs during the creation of queries.

Importing and Exporting SQL Queries

You may occasionally be required to use the queries made by some other user in your drawing or vice-versa. AutoCAD allows you to import or export stored queries. Sharing queries is very useful when developing common tools used by all team members on a project. Following is the procedure of **importing a set of queries** into the current drawing:

1. From the **dbConnect** menu, choose **Queries > Import Query Set** to invoke **Import Query Set** dialog box.
2. In the dialog box, select the query set to be imported.

3. Choose the **Open** button to import the query set into the current drawing.

AutoCAD displays an alert box that you can use to provide a unique name for the query, if there is a query with the same name that is already associated with the current drawing. Following is the procedure of **exporting a set of queries** from the current drawing:

1. From the **dbConnect** menu, choose **Queries > Export Query Set** to invoke the **Export Query Set** dialog box.
2. In the dialog box, select the directory you want to save the query set to from the **Save In** list.
3. Under File Name box, specify a name for the query set, and then choose the **Save** button.

FORMING SELECTION SETS USING LINK SELECT*

It is possible to locate objects on the drawing on the basis of the linked nongraphic information. For example, you can locate the object that is linked to the first row of the COMPUTER table or to the first and second rows of the COMPUTER table. You can highlight specified objects or form a selection set of the selected objects. **Link Select** is an advanced feature of the **Query Editor** that can be used to construct iterative selection sets of AutoCAD graphical objects and the database records. You can start constructing a query or selecting AutoCAD graphical objects for an iterative selection process. The initial selection set is referred to as **set A**. Now you can select an additional set of queries or graphical objects to further refine your selection set. The second selection set is referred to as **set B**. To refine your final selection set you must establish a relation between set A and set B. The following is the list of available relationships or set operations (Figure 18-29):

Select. This creates an initial query or graphical objects selection set. This selection set can be further refined or modified using subsequent Link Select operations.

Union. This operation adds the outcome of a new selection set or query to the existing selection set. Union returns all the records that are member of set A **or** set B.

Intersect. This operation returns the intersection of the new selection set and the existing or running selection set. Intersection returns only the records that are common to both set A **and** set B.

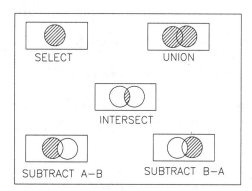

Figure 18-29 *Available relationships or Set operations*

Subtract A - B. This operation subtracts the result of the new selection set or query from the existing selection set.

Subtract B - A. This operation subtracts the result of the existing selection set or query from the new selection set.

After any of the Link Selection operation is executed, the result of the operation becomes the new running selection set and assigned as set A. You can extend refining your selection set by creating

a new set B and then continuing with the iterative process.
Following is the procedure to use **Link Select** for refining selection set:

1. Choose **Links > Link Select...** from the **dbConnect** menu to invoke the **Link Select** dialog box (Figure 18-30).

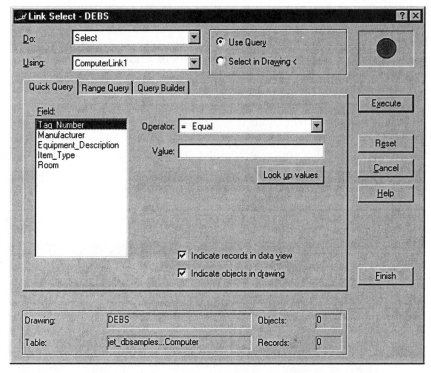

Figure 18-30 Link Select dialog box

2. Select the **Select** option from the **Do:** drop-down list for creating a new selection set. Also select a link template from the **Using:** drop-down list.

3. Choose either the **Use Query** or **Select in Drawing <** option for creating a new query or drawing object selection set.

4. Choose the **Execute** button to execute the refining operation or **Select** to add your query or graphical object selection set.

5. Again choose any **Link Selection** operation from the **Do:** drop-down list.

6. Repeat steps 2 through 4 for creating a set B to the **Link Select** operation and then choose the **Finish:** button to complete the operation.

Note
*You can choose the **Use Query** option to construct a new query or **Select in Drawing <** to select a graphical object from the drawing as a selection set.*

*If you select **Indicate Records in Data View**, then the Link Select operation result will be displayed in the **Data View** window and if you select **Indicate objects in drawing**, then AutoCAD displays a set of linked graphics objects in the drawing.*

CONVERSION OF ASE LINKS TO AUTOCAD 2000 FORMAT*

AutoCAD 2000 stores links in a different format than the previous releases. So if you want to work with your links in AutoCAD 2000, you need to convert them to the AutoCAD 2000 format. You will also have to create a AutoCAD 2000 configuration file that will point to the data source referred by the legacy links. When you open a drawing in AutoCAD 2000 where links were created by earlier releases, AutoCAD attempts to perform an automatic conversion of legacy information. But in some cases, where automatic conversion is not possible or a legacy data source might not match exactly the data source in the AutoCAD 2000 format, you can use the **Link Conversion** dialog box (Figure 18-31). During the conversion process, AutoCAD writes the data source mapping information in **asi.ini** file and applies the information for further usage of the same data source. You can use the following steps to Convert AutoCAD R13 and R14 links to the AutoCAD 2000 format:

1.   To invoke the **Link Conversion** dialog box, choose **dbConnect** menu > **Link Conversion....**

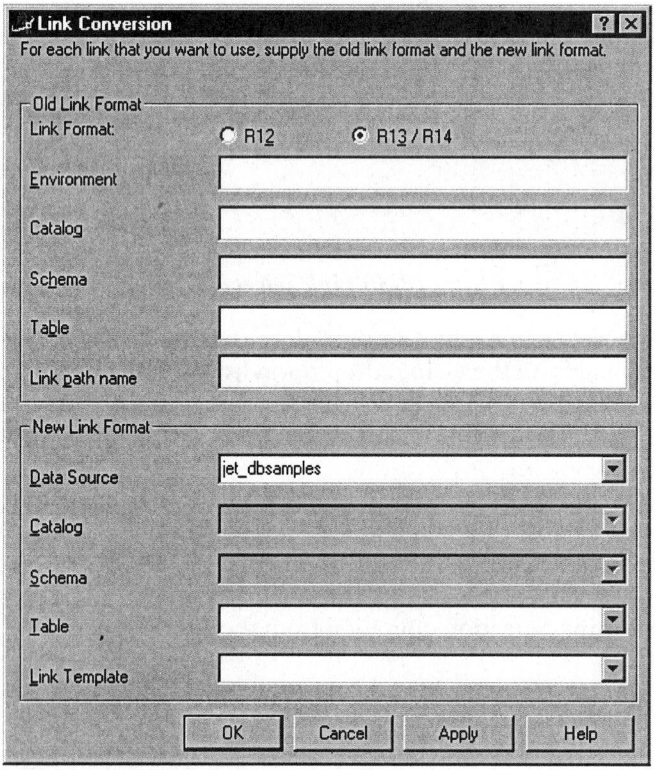

Figure 18-31 Link Conversion dialog box

2.   In **Old Link Format**, enter the following specifications:

Select R13/R14 link format to be converted.
Name of the Environment, Catalog, Schema, and Table.
Link path name of the link that has to be formatted.

3. In the **New Link Format**, enter the following specifications:

Name of the AutoCAD 2000 data source (for example, **jet_dbsamples**).
Name of the new AutoCAD 2000 Environment, Catalog, Schema, and Table.
Name of the AutoCAD 2000 link template.

4. Choose **Apply** and then the **OK** button to complete the conversion procedure.

5. Open the drawing you want to convert and save it in AutoCAD 2000 format drawing.

Saving AutoCAD 2000 Links to R13/R14 Formats[*]

You can convert Links created in AutoCAD 2000 format as R13 or R14 formats. But you cannot save AutoCAD 2000 links in R12 format. When you select SAVE AS for saving the links with drawings, select links to R13 or R14 formats and AutoCAD will automatically convert into earlier releases format.

Review Questions

1. The horizontal group of data is called a _____ .

2. A vertical group of data (attribute) is called a _____ .

3. AutoCAD can access data contained in a particular database system using _____
   _____ .

4. You can connect to an external database table by using _____ manager.

5. You can edit an external database table within an AutoCAD session. (T/F)

6. You can resize, dock, as well as hide the **Data View** window. (T/F)

7. Once you define a link, AutoCAD does not store that information with the drawing. (T/F)

8. To display the associated records with the drawing objects, AutoCAD provides _____ .

9. Labels are of two types, _____ and _____ .

10. It is possible to import as well as export Link and label templates. (T/F)

11. The SQL statements let you search through the database and retrieve the information as specified in the SQL statements. (T/F)

12. What does ASE stand for? _____

13. What is a database? _____

14. What is a database management system (DBMS)? _____
    _____ .

15. A row is also referred as a _____ .

16. A column is sometimes called a _____ .

17. The _____ acts like an identification tag for locating and linking a row.

18. Only one row can be manipulated at a time. This row is called the _____ .

19. **AutoCAD Query Editor** has four tabs, namely _____, _____,
    _____, and _____.

20. It is possible to go back to the previous query tab without resetting the values. (T/F)

21. You can execute a query after any tab in **Query Editor**.(T/F)

22. Link select is an _____ implementation of AutoCAD **Query Editor**, which constructs selection sets of graphical objects or database records.

23. AutoCAD writes data source mapping information in the _____ file during the conversion process of links.

24. AutoCAD 2000 links _____ be converted into AutoCAD R12 format.

Exercises

Exercise 1   *General*

In this exercise, you will select **Employee** table from **Jet_dbsamples** data source in **dbConnect Manager**, then edit the sixth row of the table (EMP_ID=1006, Keyser). You will also add a new row to the table, set the new row current, and then view it.

The row to be added has the following values:

EMP_ID	**1064**
LAST_NAME	**Joel**
FIRST_NAME	**Billy**
Gender	**M**
TITLE	**Marketing Executive**
Department	**Marketing**
ROOM	**6071**

Exercise 2   *General*

From the **Inventory** table in the **Jet_dbsamples** data source, build a SQL query step-by step using all the tabs of the AutoCAD Query Editor. The conditions to be implemented are as follows:

1. Type of Item = **Furniture**   and
2. Range of Cost = **200 to 650**   and
3. Manufacturer = **Office master**

Chapter **19**

Defining Block Attributes

Learning Objectives

After completing this chapter, you will be able to:
- *Understand what attributes are and how to define attributes with a block.*
- *Edit attribute tag names.*
- *Insert blocks with attributes and assign values to attributes.*
- *Extract attribute values from the inserted blocks.*
- *Control attribute visibility.*
- *Perform global and individual editing of attributes.*
- *Insert a text file in a drawing to create a bill of material.*

ATTRIBUTES

AutoCAD has provided a facility that allows the user to attach information to blocks. This information can then be retrieved and processed by other programs for various purposes. For example, you can use this information to create a bill of material, find the total number of computers in a building, or determine the location of each block in a drawing. Attributes can also be used to create blocks (such as title blocks) with prompted or preformatted text, to control text placement. The information associated with a block is known as **attribute value** or simply **attribute**. AutoCAD references the attributes with a block through tag names.

Before you can assign attributes to a block, you must create an attribute definition by using the **ATTDEF** command. The attribute definition describes the characteristics of the attribute. You can define several attribute definitions (tags) and include them in the block definition. Each time you insert the block, AutoCAD will prompt you to enter the value of the attribute. The attribute value automatically replaces the attribute tag name. The information (attribute values) assigned to a block can be extracted and written to a file by using AutoCAD's **ATTEXT** command. This file can then be inserted in the drawing as a table or processed by other programs to analyze the data. The attribute values can be edited by using the **ATTEDIT** command. The display of attributes can be controlled with **ATTDISP** command.

DEFINING ATTRIBUTES
ATTDEF Command

Menu:	Draw > Block > Define Attributes
Command:	ATTDEF

When you invoke the **ATTDEF** command, the **Attribute Definition** dialog box (Figure 19-1) is displayed. The block attributes can be defined through this dialog box. When you create an attribute definition, you must define the mode, attributes, insertion point, and text information for each attribute. All this information can be entered in the dialog box. Following is a description of each area of the **Attribute Definition** dialog box.

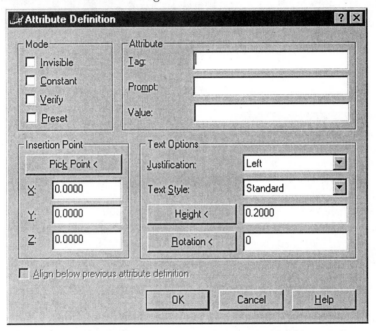

Figure 19-1 Attribute Definition dialog box

Mode. The **Mode** area of the Attribute Definition dialog box (Figure 19-2) has four options: Invisible, Constant, Verify, and Preset. These options determine the display and edit features of the block attributes. For example, if an attribute is invisible, the attribute is not displayed on the screen. Similarly, if an attribute is constant, its value is predefined and cannot be changed. These options are described below:

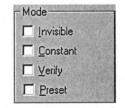

Figure 19-2 Mode area of the **Attribute Definition** *dialog box*

Invisible. This option lets you create an attribute that is not visible on the screen. This mode is useful when you do not want the attribute values to be displayed on the screen to avoid cluttering the drawing. Also, if the attributes are invisible, it takes less time to regenerate the drawing. If you want to make the invisible attribute visible, use the **ATTDISP** command discussed later in this chapter [section "Controlling Attribute Visibility (**ATTDISP** COMMAND)"].

Constant. This option lets you create an attribute that has a fixed value and cannot be changed after block insertion. When you select this mode, the Prompt: edit box and the Verify and Preset check boxes are disabled.

Verify. This option allows you to verify the attribute value you have entered when inserting a block, by asking you twice for the data. If the value is incorrect, you can correct it by entering the new value.

Preset. This option allows you to create an attribute that is automatically set to default value. The prompt is not displayed and an attribute value is not requested when you insert a block with attributes using this option to define a block attribute. Unlike a constant attribute, the preset attribute value can later be edited.

Attribute. The **Attribute** area (Figure 19-3) of the **Attribute Definition** dialog box has three edit boxes: Tag, Prompt, and Value. To enter a value, you must first select the corresponding edit box and then enter the value. You can enter up to 256 characters in these edit boxes.

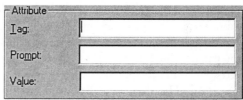

Figure 19-3 Attribute area of the Attribute Definition dialog box

Tag. This is like a label that is used to identify an attribute. For example, the tag name COM-PUTER can be used to identify an item. The tag names can be uppercase, lowercase, or both. Any lowercase letters are automatically converted into uppercase. The tag name cannot be null. Also, the tag name must not contain any blank spaces. You should select a tag name that reflects the contents of the item being tagged. For example, the tag name COMP or COM-PUTER is an appropriate name for labeling computers.

Prompt. The text that you enter in the **Prompt:** edit box is used as a prompt when you insert a block that contains the defined attribute. If you have selected the Constant option in the Mode area, the **Prompt:** edit box is disabled because no prompt is required if the attribute is constant. If you do enter nothing in the **Prompt:** edit box, the entry made in the **Tag:** edit box is used as the prompt.

Value. The entry in the **Value:** edit box defines the default value of the specified attribute; that is, if you do not enter a value, it is used as the value for the attribute. The entry of a value is optional.

Insertion Point. The **Insertion Point** area of the **Attribute Definition** dialog box (Figure 19-4) lets you define the insertion point of block attribute text. You can define the insertion point by entering the values in the **X:**, **Y:**, and **Z:** edit boxes or by selecting **Pick Point <** button. If you select this button, the dialog box clears, and you can enter the X, Y, and Z values of the insertion point at the command line or specify the point by selecting a point on the screen. When you are

Chapter 19

done specifying the insertion point, the **Attribute Definition** dialog box reappears.

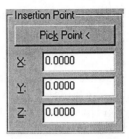

Just under the Insertion Point area of the dialog box is a
check box labeled Align below previous attribute. You can
use this box to place the subsequent attribute text just below the previously defined attribute automatically. This check
box is disabled if no attribute has been defined. When you
select this check box, the Insertion Point area and the Text
Options areas are disabled because AutoCAD assumes previously defined values for text such as text height, text style,

*Figure 19-4 Insertion Point area of
the **Attribute Definition** dialog box*

text justification, and text rotation. Also, the text is automatically placed on the following line.
After insertion, the attribute text is responsive to the setting of the **MIRRTEXT** system variable.

Text Options. The **Text Options** area of the **Attribute
Definition** dialog box (Figure 19-5) lets you define the
justification, text style, height, and rotation of the attribute text. To set the text justification, select justification type in the **Justification:** drop-down list. Similarly,
you can use the **Text Style:** drop-down list to define the
text style. You can specify the text height and text rotation in the **Height <** and **Rotation <** edit boxes. You
can also define the text height by selecting the **Height
<** button. If you select this button, AutoCAD tempo-
rarily exits the dialog box and lets you enter the value

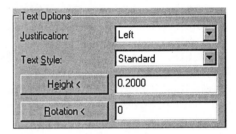

*Figure 19-5 **Text Options** area of
Attribute Definition dialog box*

from the command line. Similarly, you can define the text rotation by selecting the **Rotation <**
button and then entering the rotation angle at the command line.

Note

The text style must be defined before it can be used to specify the text style.

*If you select a style that has the height predefined, AutoCAD automatically disables the **Height <** edit
box.*

*If you have selected the Align option for the text justification, the **Height <** and **Rotation <** edit box
are disabled.*

*If you have selected the **Fit** option for the text justification, the **Rotation <** edit box is disabled.*

After you complete the settings in the **Attribute Definition** dialog box and choose **OK**, the attribute tag text is inserted in the drawing at the specified insertion point.

Using Command Line

You can define block attributes through the command line by entering the **-ATTDEF** command.

Command: **-ATTDEF**
Current attribute modes: Invisible=N Constant=N Verify=N Preset=N

Enter an option to change [Invisible/Constant/Verify/Preset] <done>:
Enter attribute tag name:

The default value of all attribute modes is **N** (No). To reverse the default mode, enter **I**, **C**, **V**, or **P**. For example, if you enter **I**, AutoCAD will change the Invisible mode from **N** to **Y** (Yes). This will make the attribute visible. After setting the modes, press ENTER to go to the next prompts where you can enter the attribute tag, attribute prompt, and the attribute values.

Enter attribute tag name:
Enter attribute prompt:
Enter default attribute value:

The entry at this prompt defines the default value of the specified attribute; that is, if you do not enter a value, it is used as the value for the attribute. The entry of a value is optional. If you have selected the Constant mode, AutoCAD displays the following prompt:

Enter attribute tag name:
Enter attribute value:

Next, AutoCAD displays the following text prompts:

Current text style: "Standard" Text height: *current*
Specify start point of text or [Justify/Style]:
Specify height <*current*>:
Specify rotation angle of text <*current*>:

After you respond to these prompts, the attribute tag text will be placed at the specified location. If you press ENTER at **Specify start point of text or [Justify/Style]:**, AutoCAD will automatically place the subsequent attribute text just below the previously defined attribute, and it assumes previously defined text values such as text height, text style, text justification, and text rotation. Also, the text is automatically placed on the following line.

Example 1

In this example, you will define the following attributes for a computer and then create a block using the **BLOCK** command. The name of the block is COMP.

Mode	Tag name	Prompt	Default value
Constant	ITEM		Computer
Preset, Verify	MAKE	Enter make:	CAD-CIM
Verify	PROCESSOR	Enter processor type:	Unknown
Verify	HD	Enter Hard-Drive size:	100MB
Invisible,Verify	RAM	Enter RAM:	4MB

1. Draw the computer as shown in Figure 19-6. Assume the dimensions, or measure the dimensions of the computer you are using for AutoCAD.

2. Invoke the **ATTDEF** command. The **Attribute Definition** dialog box is displayed (Figure 19-7).

3. Define the first attribute as shown in the preceding table. Select Constant in the Mode area because the mode of the first attribute is constant. In the Tag: edit box, enter the tag name, **ITEM**. Similarly, enter Computer in the **Value:** edit box. The **Prompt:** edit box is disabled because the variable is constant.

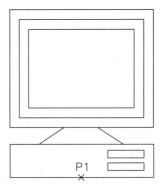

Figure 19-6 Drawing for Example 1

4. In the Insertion Point area, choose the **Pick Point <** button to define the text insertion point. Select a point below the drawing of the computer.

5. In the Text Options area, select the justification, style, height, and rotation of the text.

6. Choose the **OK** button when you are done entering information in the **Attribute Definition** dialog box.

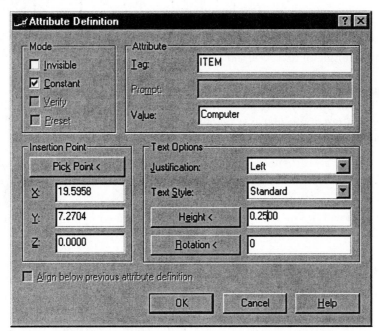

*Figure 19-7 Enter information in the **Attribute Definition** dialog box*

7. Enter **ATTDEF** at the Command prompt to invoke the **Attribute Definition** dialog box. Enter the Mode and Attribute information for the second attribute shown in the table at the beginning of Example 1. You need not define the Insertion Point and Text Options. Check the

Align below previous attribute box that
is located just below the Insertion Point
area. When you check this box, the Insertion Point and Text Justification areas are
disabled. AutoCAD places the attribute
text just below the previous attribute text.

8. Define the remaining attributes
    (Figure 19-8).

9. Use the **BLOCK** command to create a
    block. The name of the block is COMP,
    and the insertion point of the block is P1,
    midpoint of the base. When you select the
    objects for the block, make sure you also

ITEM
MAKE
PROCESSOR
HD
RAM

Figure 19-8 *Define attributes below the computer*
drawing

select the attributes. The order of attribute selection controls the order of prompts.

EDITING ATTRIBUTE TAGS
Using the DDEDIT Command

Toolbar:	Modify II > Edit Text
Menu:	Modify > Text
Command:	DDEDIT

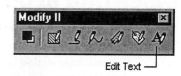

Figure 19-9 *Edit Text button*
in the ModifyII toolbar

The **DDEDIT** command lets you edit text and attribute definitions. After invoking this command, AutoCAD will prompt to **select an annotation object or [Undo]:**. If you select an attribute
definition created by attribute definition, the **Edit Attribute Definition** dialog box is displayed
and lists the tag name, prompt, and default value of the attribute, (Figure 19-10).

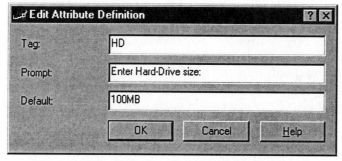

Figure 19-10 *Edit Attribute Definition dialog box*

You can select the edit boxes and enter the changes. Once you are done making the required
changes, choose the **OK** button in the dialog box. After you exit the dialog box, AutoCAD will
prompt you to select text or attribute object (Attribute tag). If you are done editing, press ENTER
to return to the command line.

Command: **DDEDIT**

Select an annotation object or [Undo]: *Select the attribute tag.*

Using the CHANGE Command

You can also use the **CHANGE** command to edit text or attribute objects. The following is the Command prompt sequence for the **CHANGE** command:

Command: **CHANGE**
Select objects: *Select attribute objects.*
Select objects: [Enter]
Specify change point or [Properties]: [Enter]
Specify new text insertion point <no change>: [Enter]
Enter new text style <current>: [Enter]
Specify new height <current>: [Enter]
Specify new rotation angle <0>: [Enter]
Enter new tag <current>: *Enter new tag name or* [Enter]
Enter new prompt <current>: *Enter new prompt or* [Enter]
Enter new default value <current>: *Enter new default or* [Enter]

INSERTING BLOCKS WITH ATTRIBUTES
Using the Dialog Box

The value of the attributes can be specified during block insertion, either at the command line or in the **Edit Attributes** dialog box. The **Edit Attributes** dialog box is invoked by setting the system variable **ATTDIA** to **1** and then using the **INSERT** command to invoke **Insert** dialog box (discussed earlier in Chapter 12). The default value for **ATTDIA** is **0**, which disables the dialog box. The command sequence for the **INSERT** command is as follows:

Command: **ATTDIA** [Enter]
Enter new value for ATTDIA <0>: **1**

Command: **-INSERT** [Enter]
Enter block name or [?] <current>: *Enter block name that has attributes in it.*
Specify insertion point or [Scale/X/Y/Z/Rotate/PScale/PX/PY/PZ/PRotate]: *Specify a point.*
Enter X scale factor, specify opposite corner, or [Corner/XYZ] <1>: *Enter X scale factor.*
Enter Y scale factor <use X scale factor>: *Enter Y scale factor.*
Specify rotation angle <0>: *Enter rotation angle.*

After you respond to these prompts, AutoCAD will display the **Edit Attributes** dialog box (Figure 19-11). This displays the prompts and their default values that have been entered at the time of attribute definition. If there are more attributes, they can be accessed by using the **Next** or **Previous** buttons. You can enter the attribute values in the edit box located next to the attribute prompt. The block name is displayed at the top of the dialog box. After entering the new attribute values, choose the **OK** button; AutoCAD will place these attribute values at the specified location.

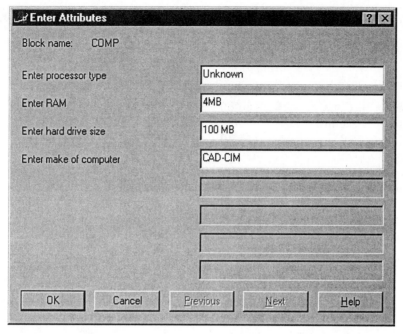

Figure 19-11 Enter attribute values in the **Enter Attributes** dialog box

Note

*If you use the dialog box to define the attribute values, the Verify mode is ignored because the **Enter** Attributes dialog box allows the user to examine and edit the attribute values.*

Using the Command Line

You can also define attributes from the command line by setting the system variable **ATTDIA** to **0** (default value). When you use the **-INSERT** command with **ATTDIA** set to **0**, AutoCAD does not display the **Enter Attributes** dialog box. Instead, AutoCAD will prompt you to enter the attribute values for various attributes that have been defined in the block. To define the attributes from the command line, enter the **-INSERT** command at the Command: prompt. After you define the insertion point, scale, and rotation, AutoCAD will display the following prompt:

Enter attribute values

It will be followed by the prompts that have been defined with the block using the ATTDEF command. For example:

Enter processor type <Unknown>:
Enter RAM <4MB>:
Enter Hard-Drive size <100MB>:

Example 2

In this example, you will use the **-INSERT** command to insert the block (COMP) that was defined in Example 1. The following is the list of the attribute values for computers.

ITEM	MAKE	PROCESSOR	HD	RAM
Computer	Gateway	486-60	150MB	16MB
Computer	Zenith	486-30	100MB	32MB
Computer	IBM	386-30	80MB	8MB
Computer	Del	586-60	450MB	64MB
Computer	CAD-CIM	Pentium-90	100 Min	32MB
Computer	CAD-CIM	Unknown	600MB	Standard

1. Make the floor plan drawing as shown in Figure 19-12 (assume the dimensions).

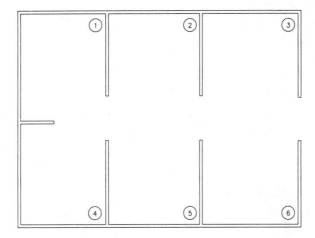

Figure 19-12 Floor plan drawing for Example 2

2. Set the system variable **ATTDIA** to 1. Use the **-INSERT** command to insert the blocks, and define the attribute values in the **Enter Attributes** dialog box (Figure 19-13).

    Command: **-INSERT**
    Enter block name or [?] <current>: **COMP**
    Specify insertion point or [Scale/X/Y/Z/Rotate/PScale/PX/PY/PZ/PRotate]: *Specify a point.*
    Enter X scale factor, specify opposite corner, or [Corner/XYZ] <1>: Enter
    Enter Y scale factor <use X scale factor>: Enter
    Specify rotation angle <0>: Enter

3. Repeat the **-INSERT** command to insert other blocks, and define their attribute values as shown in Figure 19-14

4. Save the drawing for further use.

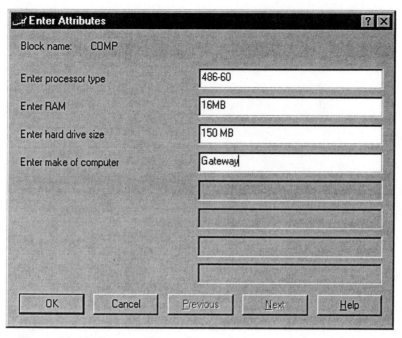

Figure 19-13 *Enter attribute values in the **Enter Attributes** dialog box*

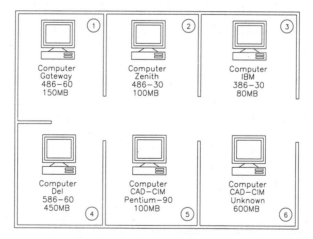

Figure 19-14 *The floor plan after inserting blocks and defining their attributes*

EXTRACTING ATTRIBUTES
Using the Dialog Box (ATTEXT)

Command: ATTEXT

To use the **Attribute Extraction** dialog box (Figure 19-15) for extracting the attributes, enter **ATTEXT** at the Command prompt. The information about the file format, template file, and output file must be entered in the dialog box to extract the defined attribute. Also, you must select the blocks whose attribute values you want to extract.

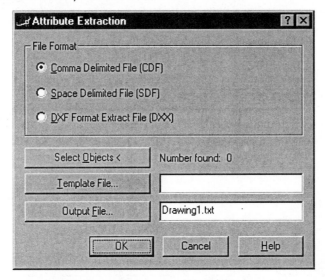

Figure 19-15 Attribute Extraction dialog box

File Format. This area of the dialog box lets you select the file format of the extracted data. You can select Comma Delimited File, Space Delimited File, or DXF Format Extract File (DXX). The format selection is determined by the application that you plan to use to process the data.

 Comma Delimited File (CDF). In CDF format, each character field is enclosed in single quotes, and the records are separated by a delimiter (comma by default). CDF file is a text file with the extension **.TXT.**

 Space Delimited File (SDF). In SDF format, the records are of fixed width as specified in the template file. The records are not separated by a comma, and the character fields are not enclosed in single quotes. The SDF file is a text file with the extension **.TXT.**

 DXF Format Extract File (DXX). If you select the Drawing Interchange File format, the template file name and the Template File edit box in the **Attribute Extraction** dialog box are automatically disabled. DXF™ -format extraction do not requires any template. The file created by this option contains only block references, attribute values, and end-of-sequence objects. The extension of these files is **.DXX.**

Select Objects<. This button closes the dialog box so that you can use the pointing device to select blocks with attributes. When the **Attribute Extraction** dialog box reopens, Number Found displays the number of objects you have selected.

Template File. The **template file** allows you to specify the attribute values you want to extract and the information you want to retrieve about the block. It also lets you format the display of the extracted data. The file can be created by using any text editor, such as Notepad, Windows Write, or WordPad. You can also use a word processor or a database program to write the file. The template file must be saved as an **ASCII** file and the extension of the file must be **.TXT**. The following are the fields that you can specify in a template file (the comments given on the right are for explanation only; they must not be entered with the field description):

BL:LEVEL	Nwww000	(Block nesting level)
BL:NAME	Cwww000	(Block name)
BL:X	Nwwwddd	(X coordinate of block insertion point)
BL:Y	Nwwwddd	(Y coordinate of block insertion point)
BL:Z	Nwwwddd	(Z coordinate of block insertion point)
BL:NUMBER	Nwww000	(Block counter)
BL:HANDLE	Cwww000	(Block's handle)
BL:LAYER	Cwww000	(Block insertion layer name)
BL:ORIENT	Nwwwddd	(Block rotation angle)
BL:XSCALE	Nwwwddd	(X scale factor of block)
BL:YSCALE	Nwwwddd	(Y scale factor of block)
BL:ZSCALE	Nwwwddd	(Z scale factor of block)
BL:XEXTRUDE	Nwwwddd	(X component of block's extrusion direction)
BL:YEXTRUDE	Nwwwddd	(Y component of block's extrusion direction)
BL:ZEXTRUDE	Nwwwddd	(Z component of block's extrusion direction)
Attribute tag		(The tag name of the block attribute)

The extract file may contain several fields. For example, the first field might be the item name and the second field might be the price of the item. Each line in the template file specifies one field in the extract file. Any line in a template file consists of the name of the field, the width of the field in characters, and its numerical precision (if applicable). For example:

ITEM　　　　N015<u>002</u>
<u>BL:NAME</u>　C<u>015</u>000
　　　　　　Where　**BL:NAME** ------ Field name
　　　　　　　　　Blankspaces --- Blank spaces (must not include the tab character)
　　　　　　　　　C ----------------- Designates a character field
　　　　　　　　　N ----------------- Designates a numerical field
　　　　　　　　　015 -------------- Width of field in characters
　　　　　　　　　002 -------------- Numerical precision

BL:NAME
or **ITEM**　　　　　Indicates the field names; can be of any length.

C　　　　　　　　Designates a character field; that is, the field contains characters or it starts with characters. If the file contains numbers or starts with numbers, then C will be replaced by N. For example, **N015002**.

015　　　　　　Designates a field that is 15 characters long.

002　　　　　　Designates the numerical precision. In this example, the numerical precision is 2, two places following the decimal. The decimal point and the two digits following decimal are **included in the field width**. In the next example, (000), the numerical precision, is not applicable because the field does not have any numerical value (the field contains letters only).

After creating a template file in AutoCAD 2000, if you select the **Template File...** button, the **Select Template File** dialog box (Figure 19-16) is displayed, where you can browse and select the required template file.

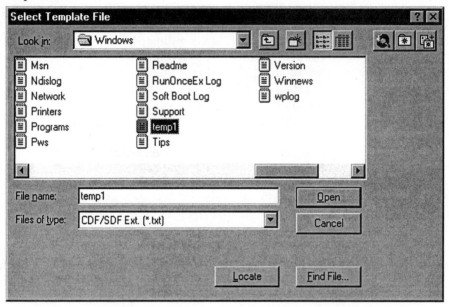

Figure 19-16 Select Template File dialog box

Note

You can put any number of spaces between the field name and the character C or N (ITEM N015002). However, you must not use the tab characters. Any alignment in the fields must be done by inserting spaces after the field name.

In the template file, a field name must not appear more than once. The template file name and the output file name must be different.

The template file must contain at least one field with an attribute tag name because the tag names determine which attribute values are to be extracted and from which blocks. If several blocks have different block names but the same attribute tag, AutoCAD will extract attribute values from all selected blocks. For example, if there are two blocks in the drawing with the attribute tag PRICE, then when you extract the attribute values, AutoCAD will extract the value from both blocks (if both blocks were selected). To extract the value of an attribute, the tag name must match the field name specified in the template file. AutoCAD automatically converts the tag names and the field names to uppercase letters before making a comparison.

Output File. This button specifies the name of the output file. Enter the file name in the box, or choose the **Output File...** button to search for existing template files from **Output File** dialog box (Figure 19-17). AutoCAD appends the **.TXT** file extension for CDF or SDF files and the **.DXF** file extension for DXF files.

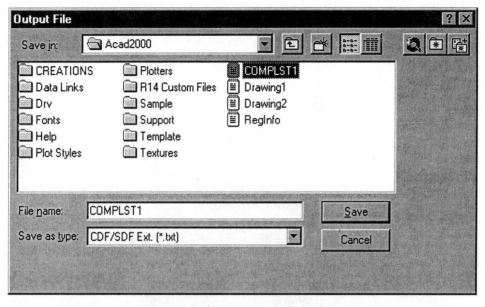

Figure 19-17 Output File dialog box

Example 3

In this example, you will write a template file for extracting the attribute values as defined in Example 2. These attribute values must be written to a file **COMPLST1.TXT** and the values arranged as shown in the following table:

Field width in characters

< 10 >	< 12 >	< 10 >	< 12 >	< 10 >	< 10 >
COMP	Computer	Gateway	486-60	150MB	**16MB**
COMP	Computer	Zenith	486-30	100MB	**32MB**
COMP	Computer	IBM	386-30	80MB	**8MB**
COMP	Computer	Del	586-60	450MB	**64MB**
COMP	Computer	**CAD-CIM**	Pentium-90	100 Min	**32MB**
COMP	Computer	**CAD-CIM**	**Unknown**	600MB	Standard

1. Load the drawing you saved in Example 2.

2. Use the Windows Notepad to write the following template file. You can use any text editor or word processor to write the file. After writing the file, save it as an ASCII file under the file name **TEMP1.TXT**. Exit the Notepad and access AutoCAD.

BL:NAME	C010000	(Block name, 10 spaces)
Item	C012000	(Item, 12 spaces)

Chapter 19

Make	C010000	(Computer make, 10 spaces)
Processor	C012000	(Processor type, 12 spaces)
HD	C010000	(Hard drive size, 10 spaces)
RAM	C010000	(RAM size, 10 spaces)

3. Use the **ATTEXT** command to invoke the **Attribute Extraction** dialog box (Figure 19-18), and select the Space Delimited File (SDF) radio button.

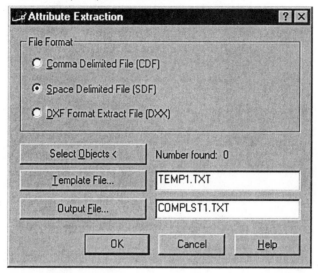

Figure 19-18 *Enter information in the **Attribute Extraction** dialog box*

4. Choose the **Select Objects <** button to select the objects (blocks) present on the screen. You can select the objects by using the Window or Crossing option. After selection is complete, right-click your pointing device to display the dialog box again.

5. In the Template File... edit box, enter the name of the template file, **TEMP1.TXT.**

6. In the Output File... edit box, enter the name of the output file, **COMPLST1.TXT.**

7. Choose the **OK** button in the **Attribute Extraction** dialog box.

8. Use the Notepad again to list the output file, **COMPLST1.TXT.** The output file will be similar to the file shown at the beginning of Example 3.

Using the Command Line (-ATTEXT)

You can also extract the attributes from the command line by entering the **-ATTEXT** command at the Command prompt. When you enter this command, AutoCAD will first prompt you to enter the type of output file. You could select CDF, SDF, DXF, or Objects. If you select the Objects option, AutoCAD will prompt you to select the objects whose attributes you want to extract. After you select the objects, AutoCAD will return the prompt, this time without the Objects option. The following is the command prompt sequence for this command.

Command: **-ATTEXT**
Enter extraction type or enable object selection [Cdf/Sdf/Dxf/Objects] <C>: **O**
Select objects: *Select objects.*
Select objects: Enter
Enter attribute extraction type [Cdf/Sdf/Dxf] <C>: **C** or **S**

If you enter **C** or **S** (CDF or SDF), AutoCAD 2000 will display the **Select Template File** dialog box. You can browse for the required template file. (Template files were discussed earlier in this chapter in the subsection "Template File.")

After you select the template file name, AutoCAD 2000 will display the **Create extract file** dialog box (Figure 19-19). The extract file is the output file where you want to write the attribute values.

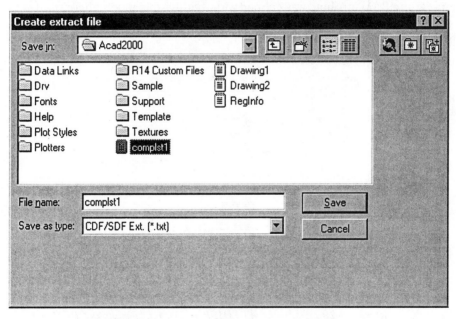

Figure 19-19 Create extract file dialog box

If you do not enter a file name, AutoCAD assumes that the name of the extract file is the same as the drawing file name with the file extension **.TXT** or **.DXF**, depending on the attribute extract format. If you enter the file name and have selected the CDF or SDF file extraction format, the file extension must be **.TXT**; if you have selected the .DXF format, the file extension must be **.DXF**. After you enter the file name, AutoCAD will extract the attribute values from the blocks and write the data to the output file.

CONTROLLING ATTRIBUTE VISIBILITY (ATTDISP COMMAND)

Menu:	View > Display > Attribute Display
Command:	ATTDISP

The **ATTDISP** command allows you to change the visibility of all attribute values. Normally, the attributes are visible unless they are defined invisible by using the Invisible mode. The invisible attributes are not displayed, but they are a part of the block definition. The prompt sequence is:

Command: **ATTDISP**
Enter attribute visibility setting [Normal/ON/OFF] <current>:

If you enter ON and press ENTER, all attribute values will be displayed, including the attributes that are defined with the Invisible mode.

If you select OFF, all attribute values will become invisible. Similarly, if you enter N (Normal), AutoCAD will display the attribute values the way they are defined; that is, the attributes that were defined invisible will stay invisible and the attributes that were defined visible will become visible. In Example 2, the RAM attribute was defined with the Invisible mode; therefore, the RAM values are not displayed with the block. If you want to make the RAM attribute values visible (Figure 19-20), enter the **ATTDISP** command and then turn it ON.

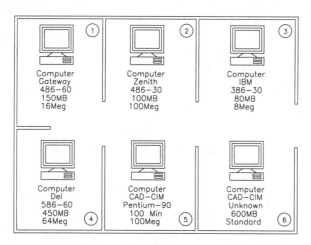

Figure 19-20   *Using the **ATTDISP** command to make the RAM attribute values visible*

EDITING ATTRIBUTES (ATTEDIT COMMAND)
Single Block

Toolbar:	Modify II > Edit Attribute
Menu:	Modify > Attribute > Single...(Option)
Command:	ATTEDIT

Figure 19-21 Modify II toolbar

The **ATTEDIT** command allows you to edit the block attribute values through the **Edit Attributes** dialog box. When you enter this command, AutoCAD prompts you to select the block whose values you want to edit. After selecting the block, the **Edit Attribute** dialog box is displayed. The dialog box shows the prompts and the attribute values of the selected block. If an attribute has been defined with Constant mode, it is not displayed in the dialog box because a constant attribute value cannot be edited. To make any changes, choose the edit box and enter the new value. After you choose the **OK** button, the attribute values are updated in the selected block.

Command: **ATTEDIT**
Select block reference: *Select a block with attributes.*

If the selected block has no attributes, AutoCAD will display the alert message **That block has no editable attributes**. Similarly, if the selected object is not a block, AutoCAD again displays the alert message **That object is not a block.**.

Note
*You cannot use the **ATTEDIT** command to do global editing of attribute values.*

*You cannot use the **ATTEDIT** command to modify position, height, or style of the attribute value.*

Example 4

In this example you will use the **ATTEDIT** command to change the attribute of the first computer (16MB to 16 Meg), which is located in Room-1.

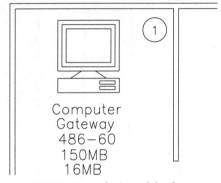

1. Load the drawing that was created in Example 2. The drawing has six blocks with attributes. The name of the block is COMP, and it has six defined attributes, one of them invisible. Zoom in so that the first computer is displayed on the screen (Figure 19-22).

Figure 19-22 Zoomed view of the first computer

2. At the AutoCAD Command: prompt, enter the **ATTEDIT** command. AutoCAD will prompt you to select a block. Select the block, first computer located in Room-1.

Command: **ATTEDIT**
Select block reference: *Select a block.*

AutoCAD will display the **Edit Attributes** dialog box (Figure 19-23), which shows the attribute prompts and the attribute values.

3. Edit the values, and choose the **OK** button in the dialog box. When you exit the dialog box, the attribute values are updated.

Chapter 19

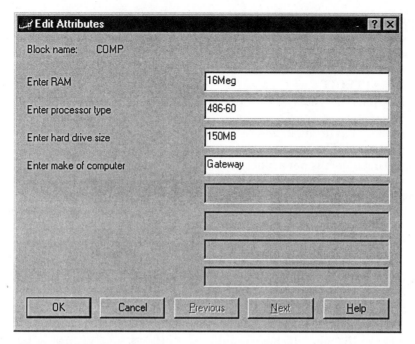

Figure 19-23 *Editing attribute values using the **Edit Attributes** dialog box*

Global Editing of Attributes

Menu: Modify > Attribute > Global
Command: -ATTEDIT

The **-ATTEDIT** command allows you to edit the attribute values independently of the blocks that contain the attribute reference. For example, if there are two blocks, COMPUTER and TABLE, with the attribute value PRICE, you can globally edit this value (PRICE) independently of the block that references these values. You can also edit the attribute values one at a time. For example, you can edit the attribute value (PRICE) of the block TABLE without affecting the value of the other block, COMPUTER. When you enter the **-ATTEDIT** command, AutoCAD displays the following prompt:

    Command: **-ATTEDIT**
    Edit attributes one at a time? [Yes/No] <Y>: **N**
    Performing global editing of attribute values

If you enter **N** at this prompt, it means that you want to do the global editing of the attributes. However, you can restrict the editing of attributes by block names, tag names, attribute values, and visibility of attributes on the screen.

Editing Visible Attributes Only

After you select global editing, AutoCAD will display the following prompt:

    Edit only attributes visible on screen? [Yes/No] <Y>: **Y**

If you enter **Y** at this prompt, AutoCAD will edit only those attributes that are visible and displayed on the screen. The attributes might have been defined with the Visible mode, but if they are not displayed on the screen they are not visible for editing. For example, if you zoom in, some of the attributes may not be displayed on the screen. Since the attributes are not displayed on the screen, they are invisible and cannot be selected for editing.

Editing All Attribute

If you enter **N** at the earlier-mentioned prompt, AutoCAD flips from graphics to text screen and displays the following message on the screen:

Drawing must be regenerated afterwards.

Now, AutoCAD will edit all attributes even if they are not visible or displayed on the screen. Also, changes that you make in the attribute values are not reflected immediately. Instead, the attribute values are updated and the drawing regenerated after you are done with the command.

Editing Specific Blocks

Although you have selected global editing, you can confine the editing of attributes to specific blocks by entering the block name at the prompt. For example:

Enter block name specification <*>: **COMP**

When you enter the name of the block, AutoCAD will edit the attributes that have the given block (COMP) reference. You can also use the wild-card characters to specify the block names. If you want to edit attributes in all blocks that have attributes defined, press ENTER

Editing Attributes with Specific Attribute Tag Names

Like blocks, you can confine attribute editing to those attribute values that have the specified tag name. For example, if you want to edit the attribute values that have the tag name MAKE, enter the tag name at the following AutoCAD prompt:

Enter attribute tag specification <*>: **MAKE**

When you specify the tag name, AutoCAD will not edit attributes that have a different tag name, even if the values being edited are the same. You can also use the wild-card characters to specify the tag names. If you want to edit attributes with any tag name, press ENTER.

Editing Attributes with a Specific Attribute Value

Like blocks and attribute tag names, you can confine attribute editing to a specified attribute value. For example, if you want to edit the attribute values that have the value 100MB, enter the value at the following AutoCAD prompt:

Enter attribute value specification <*>: **100MB**

When you specify the attribute value, AutoCAD will not edit attributes that have a different value, even if the tag name and block specification are the same. You can also use the wild-card characters to specify the attribute value. If you want to edit attributes with any value, press ENTER.

Sometimes the value of an attribute is null, and these values are not visible. If you want to select the null values for editing, make sure you have not restricted the global editing to visible attributes. To edit the null attributes, enter \ at the following prompt:

Enter attribute value specification <*>: \

After you enter this information, AutoCAD will prompt you to select the attributes. You can select the attributes by selecting individual attributes or by using one of the object selection options (Window, Crossing, etc).

Select Attributes: *Select the attribute values parallel to the current UCS only.*

After you select the attributes, AutoCAD will prompt you to enter the string you want to change and the new string. AutoCAD will retrieve the attribute information, edit it, and then update the attribute values.

Enter string to change:
Enter new string:

The following is the complete command prompt sequence of the **-ATTEDIT** command. It is assumed that the editing is global and for visible attributes only.

Command: **-ATTEDIT**
Edit attributes one at a time? [Yes/No] <Y>: **N**
Performing global editing of attribute values.
Edit only attributes visible on screen? [Yes/No] <Y>: **N**
Drawing must be regenerated afterwards.
Enter block name specification <*>:
Enter attribute tag specification <*>:
Enter attribute value specification <*>:
Enter string to change:
Enter new string:

Note

AutoCAD regenerates the drawing at the end of the command automatically unless system variable ***REGENAUTO*** *is off, which controls automatic regeneration of the drawing.*

Example 5

In this example, you will use the drawing from Example 2 to edit the attribute values that are highlighted in the following table. The tag names are given at the top of the table (ITEM, MAKE, PROCESSOR, HD, RAM). The RAM values are invisible in the drawing.

	ITEM	MAKE	PROCESSOR	HD	RAM
COMP	Computer	Gateway	486-60	150MB	**16MB**
COMP	Computer	Zenith	486-30	100MB	**32MB**
COMP	Computer	IBM	386-30	80MB	**8MB**

COMP	Computer	Del	586-60	450MB	**64MB**
COMP	Computer	**CAD-CIM**	Pentium-90	100 Min	**32MB**
COMP	Computer	**CAD-CIM**	**Unknown**	600MB	Standard

Make the following changes in the **highlighted** attribute values (Figure 19-24).

1. Change Unknown to Pentium.
2. Change CAD-CIM to Compaq.
3. Change MB to Meg for all attribute values that have the tag name RAM. (No changes should be made to the values that have the tag name HD.)

The following is the prompt sequence to change the attribute value **Unknown to Pentium**.

1. Enter the **-ATTEDIT** command at the Command prompt. At the next prompt, enter **N** (No).

   Command: **-ATTEDIT**
   Edit attributes one at a time? [Yes/No] <Y>: **N**
   Performing global editing of attribute values.

2. We want to edit only those attributes that are visible on the screen, so press ENTER at the following prompt:

   Edit only attributes visible on screen? [Yes/No] <Y>: *Press ENTER.*

3. As shown in the table, the attributes belong to a single block, COMP. In a drawing, there could be more blocks. To confine the attribute editing to the COMP block only, enter the name of the block (COMP) at the next prompt.

   Enter block name specification <*>: **COMP**

4. At the next two prompts, enter the attribute tag name and the attribute value specification. When you enter these two values, only those attributes that have the specified tag name and attribute value will be edited.

   Enter attribute tag specification<*>: **Processor**
   Enter attribute value specification<*>: **Unknown**

5. Next, AutoCAD will prompt you to select attributes. Use the Crossing option to select all blocks. AutoCAD will search for the attributes that satisfy the given criteria (attributes belong to the block COMP, the attributes have the tag name Processor, and the attribute value is Unknown). Once AutoCAD locates such attributes, they will be highlighted.

   Select Attributes:

6. At the next two prompts, enter the string you want to change, and then enter the new string.

   Enter string to change: **Unknown**

Enter new string: **Pentium**

7. The following is the command prompt sequence to change the make of the computers from
CAD-CIM to **Compaq**.

Command: **-ATTEDIT**
Edit attributes one at a time? [Yes/No] <Y>: **N**
Performing global editing of attribute values.
Edit only attributes visible on screen? [Yes/No] <Y>: *Press ENTER.*
Drawing must be regenerated afterwards.
Enter block name specification <*>: **COMP**
Enter attribute tag specification <*>: **MAKE**
Enter attribute value specification <*>:
Select Attributes:
Enter string to change: **CAD-CIM**
Enter new string: **Compaq**

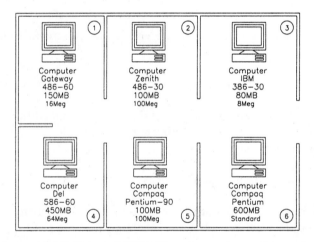

Figure 19-24 Using -ATTEDIT to change the attribute values

8. The following is the command prompt sequence to change **MB** to **Meg**.

Command: **-ATTEDIT**
Edit attributes one at a time? [Yes/No] <Y>: **N**
Performing global editing of attribute values.

At the next prompt, you must enter **N** because the attributes you want to edit (tag name, RAM)
are not visible on the screen.

Edit only attributes visible on screen? [Yes/No] <Y>: **N**
Drawing must be regenerated afterwards.
Enter block name specification <*>: **COMP**

At the next prompt, about the tag specification, you must specify the tag name because the text string MB also appears in the hard drive size (tag name, HD). If you do not enter the tag name, AutoCAD will change all MB attribute values to Meg.

> Enter attribute tag specification <*>: **RAM**
> Enter attribute value specification <*>:
> Select Attributes:
> Enter string to change: **MB**
> Enter new string: **Meg**

9. Use the **ATTDISP** command to display the invisible attributes on the screen.

> Command: **ATTDISP**
> Enter attribute visibility setting [Normal/ON/OFF] <ON>: **ON**

Individual Editing of Attributes

The **-ATTEDIT** command can also be used to edit the attribute values individually. When you enter this command, AutoCAD will prompt **Edit attributes one at a time? [Yes/No] <Y>:**. At this prompt, press ENTER to accept the default or enter **Y.** The next three prompts are about block specification, attribute tag specification, and attribute value specification, which were discussed in the previous section of this chapter. These options let you limit the attributes for editing. For example, if you specify a block name, AutoCAD will limit the editing to those attributes that belong to the specified block. Similarly, if you also specify the tag name, AutoCAD will limit editing to the attributes in the specified block and with the specified tag name.

> Command: **-ATTEDIT**
> Edit attributes one at a time? [Yes/No] <Y>: [Enter]
> Enter block name specification<*>: [Enter]
> Enter attribute tag specification<*>: [Enter]
> Enter attribute value specification<*>: [Enter]
> Select Attributes:

At the **Select Attributes:** prompt, select the objects by choosing the objects or by using an object selection option such as Window, Crossing, WPolygon, CPolygon, or Box. By using these options you can further limit the attribute values selected for editing. After you select the objects, AutoCAD will mark the first attribute it can find with an **X.** The next prompt is:

> Enter an option [Value/Position/Height/Angle/Style/Layer/Color/Next] <N>:

Value. The Value option lets you change the value of an attribute. To change the value, enter **V** at this prompt. AutoCAD will display the following prompt:

> Enter type of value modification [Change/Replace] <R>:

The Change option allows you to change a few characters in the attribute value. To select the Change option, enter Change or **C** at the prompt. AutoCAD will display the next prompt:

Enter string to change:
Enter new string:

At the **Enter string to change**: prompt, enter the characters you want to change and press ENTER. At the next prompt, **Enter new string:**, enter the new string.

 Note
You can use ? and * in the string value. When these characters are used in string values, AutoCAD does not interpret them as wild-card characters.

To use the Replace option, enter **R** or press ENTER at the **Enter type of value modification [Change/Replace] <R>:** prompt. AutoCAD will display the following prompt:

Enter new Attribute value:

At this prompt, enter the new attribute value. AutoCAD will replace the string bearing the **X** mark with the new string. If the new attribute is null, the attribute will be assigned a null value.

Position, Height, Angle. You can change the position, height, or angle of an attribute value by entering, respectively, **P**, **H**, or **A** at the following prompt:

Enter an option [Value/Position/Height/Angle/Style/Layer/Color/Next] <N>:

The Position option lets you define the new position of the attribute value. AutoCAD will prompt you to enter the new starting point, center point, or endpoint of the string. If the string is aligned, AutoCAD will prompt for two points. You can also define the new height or angle of the text string by entering, respectively, **H** or **A** at the prompt.

Layer and Color. The Layer and Color options allow you to change the layer and color of the attribute. For a color change, you can enter the new color by entering a color number (1 through 255), a color name (red, green, etc.), BYLAYER, or BYBLOCK.

Example 6

In this example, you will use the drawing in Example 2 to edit the attributes individually (Figure 19-25). Make the following changes in the attribute values.

a. Change the attribute value 100 Min to 100MB.
b. Change the height of all attributes with the tag name RAM to 0.075 units.

1. Load the drawing that you had saved in Example 2.

2. At the AutoCAD Command prompt, enter the **-ATTEDIT** command. The following is the command prompt sequence to change the value of 100 Min to 100MB.

   Command: **-ATTEDIT**
   Edit attributes one at a time? [Yes/No] <Y>: [Enter]

Enter block name specification <*>: **COMP**
Enter attribute tag specification <*>: [Enter]
Enter attribute value specification <*>: [Enter]
Select Attributes: *Select the attribute.*
Enter an option [Value/Position/Height/Angle/Style/Layer/Color/Next] <N>: **V**
Enter type of value modification [Change/Replace] <R>: **C**
Enter string to change: \ **Min**
Enter new string: **100MB**

When AutoCAD prompts **Enter string to change:**, enter the characters you want to change. In this example, the characters **Min** are preceded by a space. If you enter a space, AutoCAD displays the next prompt, **Enter new string:**. If you need a leading blank space, the character string must start with a backslash (\), followed by the desired number of blank spaces.

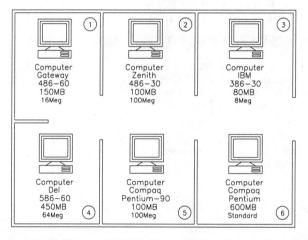

Figure 19-25 *Using* ***-ATTEDIT*** *to change the attribute values individually*

3. To change the height of the attribute text, enter the **-ATTEDIT** command as just shown. When AutoCAD displays the following prompt, enter **H** for height.

Enter an option [Value/Position/Height/Angle/Style/Layer/Color/Next] <N>: **H**
Specify new height <current>: **0.075**

After you enter the new height and press ENTER, AutoCAD will change the height of the text string that has the **X** mark. AutoCAD will then repeat the last prompt. Use the **Next** option to move the **X** mark to the next attribute. To change the height of other attribute values, repeat these steps.

Chapter 19

INSERTING TEXT FILES IN THE DRAWING
Using MTEXT Command

Toolbar:	Draw > Multiline Text
Menu:	Draw > Text > Multiline Text
Command:	MTEXT

You can insert a text file by selecting the Import Text option in the **Multiline Text Editor** dialog box. You can invoke this dialog box by using the **MTEXT** command.

Command: **MTEXT**

Next, AutoCAD prompts you to enter the insertion point and other corner of the paragraph text box. After you enter these points, the **Multiline Text Editor** dialog box appears on screen. To insert the text file **COMPLST1.TXT** (created in Example 3), choose the Import Text button. AutoCAD displays the **Open** dialog box (Figure 19-26).

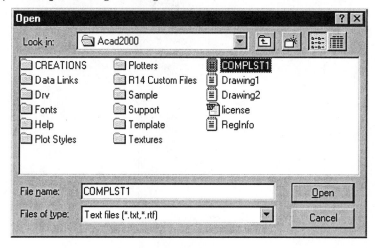

Figure 19-26 Open dialog box

In this dialog box, you can select the text file COMPLST1 and then choose the **Open** button. The imported text is displayed in the text area of the **Multiline Text Editor** dialog box (Figure 19-27). Note that only ASCII files are properly interpreted.

Now choose the **OK** button to get the imported text in the selected area on the screen (Figure 19-28). You can also use the **Multiline Text Editor** dialog box to change the text style, height, direction, width, rotation, line spacing, and attachment.

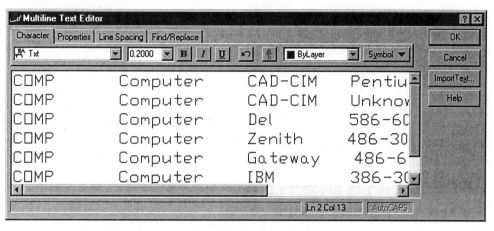

Figure 19-27 Multiline Text Editor dialog box displaying the imported text

```
COMP        Computer      CAD-CIM    Pentium-90    100  Min
COMP        Computer      CAD-CIM    Unknown       600MB
COMP        Computer      Del        586-60        450MB
COMP        Computer      Zenith     486-30        100MB
COMP        Computer      Gateway    486-60        150MB
COMP        Computer      IBM        386-30        80MB
```

Figure 19-28 Imported text file on the screen

Self-Evaluation Test

Answer the following questions, and then compare your answers to the correct answers given at the end of this chapter.

1. The **Verify** option allows you to verify the attribute value you have entered when inserting a block. (T/F)

2. Unlike a constant attribute, the **Preset** attribute cannot be edited. (T/F)

3. For tag names, any lowercase letters are automatically converted to uppercase. (T/F)

4. The entry in the **Value:** edit box of the Attribute Definition dialog box defines the _____ of the specified attribute.

5. If you have selected the **Align** option for the text justification, the Height < and Rotation < edit boxes are _____.

6. You can use the _____ command to edit text or attribute definitions.

7. The default value of the **ATTDIA** variable is _____, which disables the dialog box.

8. In the Space Delimited File, the records are of fixed width as specified in the _____ File.

9. In the _____ File, the records are not separated by a comma and the character fields are not enclosed in single quotes.

10. You must not use the _____character in template files. Any alignment in the fields must be done by inserting spaces after the field name.

11. You cannot use the **ATTEDIT** command to modify the position, height, or style of the attribute value. (T/F)

12. The _____ command allows you to edit the attribute values independently of the blocks that contain the attribute reference.

13. The **-ATTEDIT** command can also be used to edit the attribute values individually. (T/F)

14. You can use ? and * in the string value. When these characters are used in string values, AutoCAD does not interpret them as wild-card characters. (T/F)

15. You can insert the text file in the drawing by entering the _____ command at the AutoCAD Command prompt.

Review Questions

1. Give two major uses of defining block attributes. _____ .

2. The information associated with a block is known as _____ or _____.

3. You can define the block attributes by entering _____ at the Command prompt.

4. The most convenient way to define the block attributes is by using the **Attribute Definition** dialog box, which can be invoked by entering _____.

5. What are the options in the **Mode** area of the **Attribute Definition** dialog box? _____
    _____ .

6. The **Constant** option lets you create an attribute that has a fixed value and cannot be changed later. (T/F)

7. What is the function of the **Preset** option? _____ .

8. The attribute value is requested when you use the **Preset** option to define a block attribute. (T/F)

9. Name the three edit boxes in the **Attribute area** of the **Attribute Definition** dialog box.
    _____ .

10. The **tag** is like a label that is used to identify an attribute. (T/F)

11. The tag names can only be uppercase. (T/F)

12. The tag name cannot be null. (T/F)

13. The tag name can contain a blank space. (T/F)

14. If you select the Constant option in the Mode area of the **Attribute Definition** dialog box, the **Prompt:** edit box is _____ because no prompt is required if the attribute is _____ .

15. If you do not enter anything in the **Prompt:** edit box, the entry made in the **Tag:** edit box is used as the prompt. (T/F)

16. What option, button, or check box should you select in the **Attribute Definition** dialog box to automatically place the subsequent attribute text just below the previously defined attribute? _____ .

17. The text style must be defined before it can be used to specify the text style. (T/F)

18. If you select a style that has the height predefined, AutoCAD automatically disables the Height < edit box in the Attribute Definition dialog box. (T/F)

19. If you have selected the Fit option for the text justification, the _____ edit box is disabled.

20. What is the difference between the **ATTDEF** and the **-ATTDEF** commands? _____ _____ .

21. The _____ command lets you edit both text and attribute definitions.

22. The value of the block attributes can be specified in the **Edit Attribute Definition** dialog box. (T/F)

23. The **Edit Attribute Definition** dialog box is invoked by using the _____ or _____ command with the system variable _____ set to 1.

24. If you use the **Enter Attributes** dialog box to define the attribute values, the Verify mode is _____ because the **Enter Attributes** dialog box allows the user to examine and edit the attribute values.

25. You can also define attributes from the command line by setting the system variable _____ to 0.

26. To use the **Attribute Extraction** dialog box for extracting the attributes, enter _____ at the Command prompt.

Chapter 19

27. You can select the Comma Delimited File, Space Delimited File, or Drawing Interchange File. The format selection is determined by the text editor you use. (T/F)

28. In the Comma Delimited File, each character field is enclosed in _____ and each record is separated by a _____.

Exercises

Exercise 1   *Electronics*

In this exercise, you will define the following attributes for a resistor and then create a block using the **BLOCK** command. The name of the block is RESIS.

Mode	Tag name	Prompt	Default value
Verify	RNAME	Enter name	RX
Verify	RVALUE	Enter resistance	XX
Verify, Invisible	RPRICE	Enter price	00

1. Draw the resistor as shown in Figure 19-29.
2. Enter **ATTDEF** at the AutoCAD Command prompt to invoke the **Attribute Definition** dialog box.
3. Define the attributes as shown in the preceding table, and position the attribute text as shown in Figure 19-29.
4. Use the **BLOCK** command to create a block. The name of the block is RESIS, and the insertion point of the block is at the left end of the resistor. When you select the objects for the block, make sure you also select the attributes.

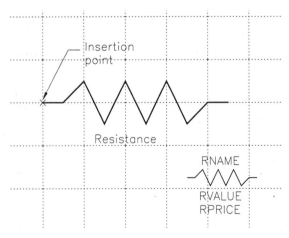

Figure 19-29 Drawing of a resistor for Exercise 1

In this exercise, you will use the **INSERT** command to insert the block that was defined in Exercise 1 (RESIS). The following is the list of the attribute values for the resistances in the electric circuit.

RNAME	RVALUE	RPRICE
R1	35	0.32
R2	27	0.25
R3	52	0.40
R4	8	0.21
RX	10	0.21

1. Draw the electric circuit diagram as shown in Figure 19-30 (assume the dimensions).

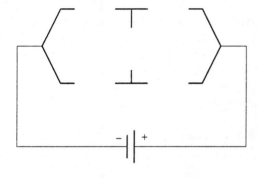
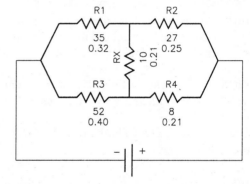

Figure 19-30 *Drawing of the electric circuit diagram without resistors for Exercise 2*

Figure 19-31 *Drawing of the electric circuit diagram with resistors for Exercise 2*

2. Set the system variable **ATTDIA** to 1. Use the **INSERT** command to insert the blocks, and define the attribute values in the **Enter Attributes** dialog box.
3. Repeat the **INSERT** command to insert other blocks, and define their attribute values as given in the table. Save the drawing as **ATTEXR2.DWG** (Figure 19-31).

In this exercise, you will write a template file for extracting the attribute values as defined in Exercise 2. These attribute values must be written to a file **RESISLST.TXT** and arranged as shown in the following table.

Field width in characters			
< 10 >	< 10 >	< 10 >	< 10 >
RESIS	R1	35	0.32
RESIS	R2	27	0.25
RESIS	R3	52	0.40
RESIS	R4	8	0.21
RESIS	RX	10	0.21

Chapter 19

1. Load the drawing **ATTEXR2** that you saved in Exercise 2.
2. Use the Windows Notepad to write the template file. After writing the file, save it as an ASCII file under the file name **TEMP2.TXT.**
3. In the AutoCAD screen, use the **ATTEXT** command to invoke the **Attribute Extraction** dialog box, and select the Space Delimited File (SDF) radio button.
4. Select the objects (blocks). You can also select the objects by using the Window or Crossing option.
5. In the Template File edit box, enter the name of the template file, **TEMP2.TXT.**
6. In the Output File edit box, enter the name of the output file, **RESISLST.TXT.**
7. Choose the **OK** button in the **Attribute Extraction** dialog box.
8. Use the Windows Notepad to list the output file, **RESISLST.TXT.** The output file should be similar to the file shown in the beginning of Exercise 3.

Exercise 4   *Electronics*

In this exercise, you will use the **ATTEDIT** or **-ATTEDIT** command to change the attributes of the resistances that are highlighted in the following table. You will also extract the attribute values and insert the text file in the drawing.

1. Load the drawing **ATTEXR2** that was created in Exercise 2. The drawing has five resistances with attributes. The name of the block is RESIS, and it has three defined attributes, one of them invisible.
2. Use the AutoCAD **ATTEDIT** or **-ATTEDIT** command to edit the values that are **highlighted** in the following table.

RESIS	R1	40	0.32
RESIS	R2	29	0.25
RESIS	R3	52	0.45
RESIS	R4	8	0.25
RESIS	R5	10	0.21

3. Extract the attribute values, and write the values to a text file.
4. Use the **MTEXT** command to insert the text file in the drawing.

Exercise 5   *Electronics*

Use the information given in Exercise 3 to extract the attribute values, and write the data to the output file. The data in the output file should be Comma Delimited CDF. Use the **ATTEXT** and **-ATTEXT** commands to extract the attribute values.

Exercise 6   *Electronics*

In this exercise, you will draw the circuit diagram as shown in Figure 19-32, define the attributes, and then extract the attributes to create a bill of materials.

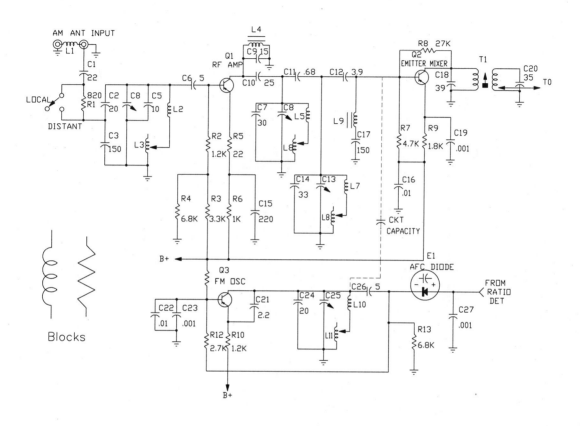

Figure 19-32 *Drawing of the circuit diagram for Exercise 6*

Answers to Self-Evaluation Test
1 - T, **2** - F, **3** - T, **4** - default value, **5** - disabled, **6** - **CHANGE**, **7** - 0, **8** - template, **9** - Space Delimited, **10** - tab, **11** - T, **12** - **ATTEDIT**, **13** - T, **14** - T, **15** - **MTEXT**

Chapter 20

AutoCAD on the Internet

Learning Objectives

The Internet has become the most important way to exchange information in the world. AutoCAD allows you to interact with the Internet in several ways. It can open and save files that are located on the Internet; it can launch a Web browser (AutoCAD 2000 includes a simple Web browser); and it can create drawing Web format (DWF) files for viewing as drawings on Web pages. After completing this chapter, you will be able to:

* *Launch a Web browser from AutoCAD.*
* *Understand the importance of the uniform resource locator (URL).*
* *Open and save drawings to and from the Internet.*
* *Place hyperlinks in a drawing.*
* *Convert drawings to DWF file format.*
* *View DWF files with a Web browser.*
* *Learn about the Whip plug-in and the DWF file format.*

INTRODUCTION

Before you can use the Internet features in AutoCAD 2000, some components of Microsoft's Internet Explorer must be present in your computer. If Internet Explorer v4.0 (or later) is already set up on your computer, then you already have these required components. If not, then the components are installed automatically when during AutoCAD 2000's setup you select: (1) the **Full Install**; or (2) the **Internet Tools** option during the **Custom** installation.

This chapter introduces you to the following Web-related commands:

BROWSER:
1   Launches a Web browser from within AutoCAD.

HYPERLINK*:
1   Attaches and removes a URL to an object or an area in the drawing.

HYPERLINKFWD*:
1   Move to the next hyperlink (an undocumented command).

HYPERLINKBACK*:
1   Moves back to the previous hyperlink (an undocumented command).

HYPERLINKSTOP*:
1   Stops the hyperlink access action (an undocumented command).

PASTEASHYPERLINK*:
1   Attaches a URL to an object in the drawing from text stored in the Windows Clipboard (an undocumented command).

HYPERLINKBASE*:
1   A system variable for setting the path used for relative hyperlinks in the drawing.

In addition, all of AutoCAD 2000's file-related dialog boxes are "Web enabled."

Changes From AutoCAD Release 14

If you used the Internet features in AutoCAD Release 14, then you should be aware of changes in AutoCAD 2000.

Attached URLs from R14

In AutoCAD R14, attached URLs were not active until the drawing was converted to a DWF file; in AutoCAD 2000, URLs are active in the drawing.

If you attached URLs to drawing in Release 14 (as well as LT 97 and LT 98), they are converted to AutoCAD 2000-style hyperlinks the first time you save the drawing in AutoCAD 2000 DWG format.

When you use the **SAVEAS** command to save an AutoCAD 2000 drawing in R14 format, hyperlinks are converted back to R14-style URLs. Hyperlink descriptions are stored as proxy data and not displayed in R14/LT (this ensures that hyperlink descriptions are available again when the drawing is brought back to AutoCAD 2000 format.)

WHIP v3.1 Limitations

AutoCAD 2000 can create DWF files compatible with the WHIP v3.1 plug-in. It is recommended, however, that you download WHIP v4.0 from the Autodesk Web site. Whip v3.1 does not display the following AutoCAD 2000-specific features:

• Hyperlink descriptions are not saved in v3.1 DWF format.
• All objects, except polylines and traces, are displayed at minimum lineweight.
• Linetypes defined by a plot style table are displayed as continuous lines.
• Nonrectangular viewports are displayed as rectangular viewports.

Changed Internet Commands

Many of Release 14's Internet-related commands have been discontinued and replaced. The following summary describes the status of all R14 Internet commands:

BROWSER (launches Web browser from within AutoCAD): continues to work in AutoCAD 2000; the default is now Home.Html, a Web page located on your computer.

ATTACHURL (attaches a URL to an object or an area in the drawing): continues to work in AutoCAD 2000, but has been superceded by the **-HYPERLINK** command's **Insert** option.

SELECTURL (selects all objects with attached URLs): continues to work in AutoCAD 2000, but has been superceded by the **QSelect** dialog box's **Hyperlink** option.

LISTURL (lists URLs embedded in the drawing): removed from AutoCAD 2000. As a replacement, use the **Properties** dialog box's **Hyperlink** option.

DETACHURL (removes the URL from an object): continues to work in AutoCAD 2000, but has been superceded by the **-HYPERLINK** command's **Remove** option.

DWFOUT (exports the drawing and embedded URLs as a DWF file): continues to work in AutoCAD 2000, but has been superceded by the **Plot** dialog box's **ePlot** option.

DWFOUTD (exports drawing in DWF format without a dialog box): removed from AutoCAD 2000. As a replacement, use the **-PLOT** command's **ePlot** option.

INETCFG (configures AutoCAD for Internet access): removed from AutoCAD 2000; it has been replaced by the **Internet** applet of the Windows Control Panel (choose the **Connection** tab).

INSERTURL (inserts a block from the Internet into the drawing): automatically executes the **INSERT** command and displays the **Insert** dialog box. You may type a URL for the block name.

OPENURL (opens a drawing from the Internet): automatically executes the **OPEN** command and displays the **Select File** dialog box. You may type a URL for the filename.

SAVEURL (saves the drawing to the Internet): automatically executes the **SAVE** command and displays the **Save Drawing As** dialog box. You may type a URL for the filename.

You are probably already familiar with the best-known uses for the Internet: email (electronic mail) and the Web (short for "World Wide Web"). Email lets users exchange messages and data at very low cost. The Web brings together text, graphics, audio, and movies in an easy-to-use format. Other uses of the Internet include FTP (file transfer protocol) for effortless binary file transfer, Gopher (presents data in a structured, subdirectory-like format), and USENET, a collection of more than 10,000 news groups.

AutoCAD allows you to interact with the Internet in several ways. AutoCAD is able to launch a Web browser from within AutoCAD with the **BROWSER** command. Hyperlinks can be inserted in drawings with the **HYPERLINK** command, which lets you link the drawing with other documents

Chapter 20

on your computer and the Internet. With the **PLOT** command's **ePlot** option (short for "electronic plot), AutoCAD creates DWF (short for "drawing Web format") files for viewing drawings in 2D format on Web pages. AutoCAD can open, insert, and save drawings to and from the Internet via the **OPEN**, **INSERT**, and **SAVEAS** commands.

Understanding URLs

The uniform resource locator, known as the URL, is the file naming system of the Internet. The URL system allows you to find any resource (a file) on the Internet. Example resources include a text file, a Web page, a program file, an audio or movie clip—in short, anything you might also find on your own computer. The primary difference is that these resources are located on somebody else's computer. A typical URL looks like the following examples:

ExampleURL	Meaning
http://www.autodesk.com	Autodesk Primary Web Site
news://adesknews.autodesk.com	Autodesk News Server
ftp://ftp.autodesk.com	Autodesk FTP Server
http://www.autodeskpress.com	Autodesk Press Web Site
http://users.uniserve.com/~ralphg	Editor Ralph Grabowski's Web site

Note that the **http://** prefix is not required. Most of today's Web browsers automatically add in the *routing* prefix, which saves you a few keystrokes.

URLs can access several different kinds of resources—such as Web sites, email, news groups—but always take on the same general format, as follows:

scheme://netloc

The scheme accesses the specific resource on the Internet, including these:

Scheme	Meaning
file://	File is located on your computer's hard drive or local network
ftp://	File Transfer Protocol (used for downloading files)
http://	Hyper Text Transfer Protocol (the basis of Web sites)
mailto://	Electronic mail (email)
news://	Usenet news (news groups)
telnet://	Telnet protocol
gopher://	Gopher protocol

The **://** characters indicate a network address. Autodesk recommends the following format for specifying URL-style filenames with AutoCAD:

Resource	URL Format
Web Site	**http:**//*servername/pathname/filename*
FTP Site	**ftp:**//*servername/pathname/filename*
Local File	**file:**///*drive:/pathname/filename*
or	*drive:\pathname\filename*
or	**file:**///*drive	/pathname/filename*

	or	**file://\\\localPC\pathname\filename**
	or	**file:////localPC/pathname/filename**
Network File		**file://localhost/drive:/pathname/filename**
	or	**\\\localhost\drive:\pathname\filename**
	or	**file://localhost/drive|/pathname/filename**

The terminology can be confusing. The following definitions will help to clarify these terms.

Term	Meaning
servername	The name or location of a computer on the Internet, for example: *www.autodesk.com*
pathname	The same as a subdirectory or folder name
drive	The driver letter, such as C: or D:
localpc	A file located on your computer
localhost	The name of the network host computer

If you are not sure of the name of the network host computer, use Windows Explorer to check the Network Neighborhood for the network names of computers.

Launching A Web Browser

The **BROWSER** command lets you start a Web browser from within AutoCAD. Commonly used Web browsers include Netscape's Navigator, Microsoft's Internet Explorer, and Operasoft's Opera.

By default, the **BROWSER** command uses whatever brand of Web browser program is registered in your computer's Windows operating system. AutoCAD prompts you for the uniform resource locator (URL), such as *http://www.autodeskpress.com*. The **BROWSER** command can be used in scripts, toolbar or menu macros, and AutoLISP routines to automatically access the Internet.

Command: **BROWSER**
Browse <C:\CAD\ACAD 2000\Home.htm>: *Enter the URL.*

The default URL is an HTML file added to your computer during AutoCAD's installation. After you type the URL and press *, AutoCAD launches the Web browser and contacts the Web site. Figure 20-1 shows the popular Netscape Navigator with the Autodesk Web site at *http://www.autodesk.com*.

Changing Default Web Site

To change the default Web page that your browser starts with from within AutoCAD, change the setting in the **INETLOCATION** system variable. The variable stores the URL used by the **BROWSER** command and the **Browse the Web** dialog box. Make the change, as follows:

Command: **INETLOCATION**
Enter new value for INETLOCATION <"C:\CAD\ACAD 2000\Home.htm">: *(Type URL.)*

Figure 20-1 Netscape Navigator Displaying the Autodesk Web Site

Drawings on the Internet

When a drawing is stored on the Internet, you access it from within AutoCAD 2000 using the standard **OPEN**, **INSERT**, and **SAVE** commands. (In Release 14, these commands were known as **OPENURL**, **INSERTURL**, and **SAVEURL**.) Instead of specifying the file's location with the usual drive-subdirectory-file name format, such as c:\acad14\filename.dwg, use the URL format. (Recall that the URL is the universal file naming system used by the Internet to access any file located on any computer hooked up to the Internet.)

Opening Drawings from the Internet

To open a drawing from the Internet (or your firm's intranet), use the **OPEN** command (**File > Open**). Notice the three buttons at the upper right of the **Select File** dialog box (see Figure 20-2).

These are specifically for use with the Internet. From left to right, they are:

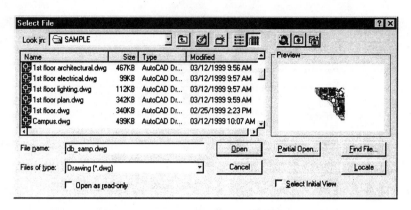

Figure 20-2 Select File dialog box

Search the Web: Displays the **Browse the Web** dialog box, a simplified Web browser.

Look in Favorites: Opens the **Favorites** folder, equivalent to "bookmarks" that store Web addresses.

Add to Favorites: Saves the current URL (hyperlink) to the **Favorites** folder.

When you click the **Search the Web** button, AutoCAD opens the **Browse the Web** dialog box (Figure 20-3). This dialog box is a simplified version of the Microsoft brand of Web browser. The purpose of this dialog box is to allow you to browse files at a Web site.

By default, the **Browse the Web** dialog box displays the contents of the URL stored in the **INETLOCATION** system variable. You can easily change this to another folder or Web site, as noted earlier.

Along the top, the dialog box has six buttons:

Back: Go back to the previous URL.

Forward: Go forward to the next URL.

Stop: Halt displaying the Web page (useful if the connection is slow or the page is very large.)

Refresh: Redisplay the current Web page.

Home: Return to the location specified by the **INETLOCATION** system variable.

Favorites: List stored URLs (hyperlinks) or bookmarks. If you have previously used Internet Explorer, you will find all your favorites listed here. Favorites are stored in the **\Windows\Favorites** folder on your computer.

The **Look in** field allows you to type the URL. Alternatively, click the down arrow to select a previous destination. If you have stored Web site addresses in the Favorites folder, then select a URL from that list.

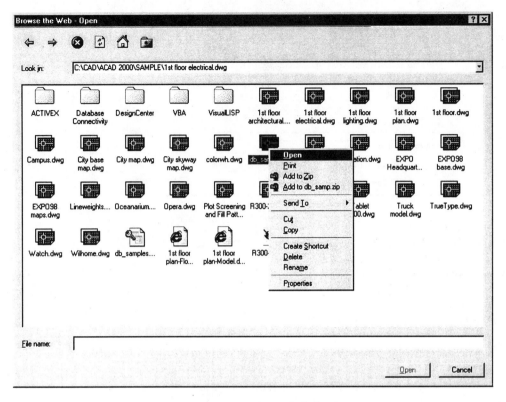

Figure 20-3 Browse the Web dialog box

You can either double-click a filename in the window, or type a URL in the **File name** field. The following table gives templates for typing the URL to open a drawing file:

Drawing Location	Template URL
Web or HTTP Site	*http://servername/pathname/filename.dwg*
	http://practicewrench.autodeskpress.com/wrench.dwg
FTP Site	*ftp://servername/pathname/filename.dwg*
	ftp://ftp.autodesk.com
Local File	*drive:\pathname\filename.dwg*
	c:\acad 2000\sample\tablet2000.dwg
Network File	*\\localhost\drive:\pathname\filename.dwg*
	\\upstairs\e:\install\sample.dwg

When you open a drawing over the Internet, it will probably take much longer than opening a file found on your computer. During the file transfer, AutoCAD displays a dialog box to report the progress (Figure 20-4). If your computer uses a 28.8Kbps modem, you should allow about five to ten minutes per megabyte of drawing file size. If your computer has access to a faster T1 connection to the Internet, you should expect a transfer speed of about one minute per megabyte.

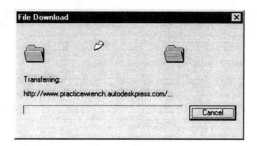

Figure 20-4 File Download *dialog box*

It may be helpful to understand that the **OPEN** command does not copy the file from the Internet location directly into AutoCAD. Instead, it copies the file from the Internet to your computer's designated **Temporary** subdirectory, such as C:\Windows\Temp (and then loads the drawing from the hard drive into AutoCAD). This is known as *caching*. It helps to speed up the processing of the drawing, since the drawing file is now located on your computer's fast hard drive, instead of the relatively slow Internet.

Note that the **Locate** and **Find File** buttons (in the **Select File** dialog box) do not work for locating files on the Internet.

Example 1

The Autodesk Press Web site has an area that allows you to practice using the Internet with AutoCAD. In this example, you open a drawing file located at that Web site.

1. Start AutoCAD.

2. Ensure that you have a live connection to the Internet. If you normally access the Internet via a telephone (modem) connection, dial your Internet service provider now.

3. From the menu bar, select **File > Open.** Or, choose the **Open** button on the toolbar. Notice that AutoCAD displays the **Select File** dialog box.

4. Click the **Search the Web** button. Notice that AutoCAD displays the **Browse the Web** dialog box.

5. In the **Look in** field, type:

    practicewrench.autodeskpress.com

    and press ENTER. After a few seconds, the **Browse the Web** dialog box displays the Web site (Figure 20-5).

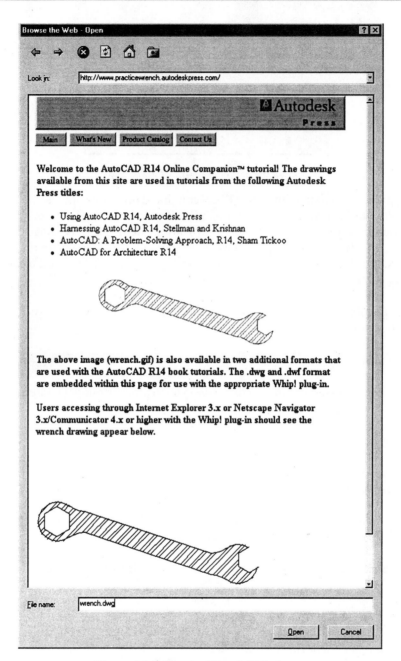

Figure 20-5 Practice Wrench Web site

6. In the Filename field, type:

wrench.dwg
and press ENTER. AutoCAD begins transferring the file. Depending on the speed of your Internet connection, this will take between a couple of seconds and a half-minute. Notice the drawing of the wrench in AutoCAD (Figure 20-6).

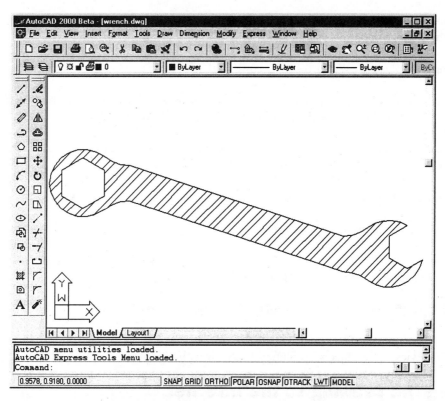

Figure 20-6 Wrench drawing in AutoCAD

Inserting a Block from the Internet

When a block (symbol) is stored on the Internet, you can access it from within AutoCAD using the **INSERT** command. When the **Insert** dialog box appears, choose the **Browse** button to display the **Select Drawing File** dialog box. This is identical to the dialog box discussed previously.

After you choose the select the file, AutoCAD downloads the file and continues with the **INSERT** command's familiar prompts.

The process is identical for accessing external reference (xref) and raster image files. Other files that AutoCAD can access over the Internet include: 3D Studio, SAT (ACIS solid modeling), DXB (drawing exchange binary), WMF (Windows metafile), and EPS (encapsulated PostScript). All of these options are found in the **Insert** menu on the menu bar.

Accessing Other Files on the Internet

Most other file-related dialog boxes allow you to access files from the Internet or intranet. This allows your firm or agency to have a central location that stores drawing standards. When you need to use a linetype or hatch pattern, for example, you access the LIN or PAT file over the Internet. More than likely, you would have the location of those files stored in the Favorites list. Some examples include:

Chapter 20

Linetypes: From the menu bar, choose **Format > Linetype**. In the **Linetype Manager** dialog box, choose the **Load, File**, and **Look in Favorites** buttons.

Hatch Patterns: Use the Web browser to copy .PAT files from a remote location to your computer.

Multiline Styles: From the menu bar, choose **Format > Multiline Style**. In the **Multiline Styles** dialog box, choose the **Load, File**, and **Look in Favorites** buttons.

Layer Name: From the menu bar, choose **Express > Layers > Layer Manager**. In the **Layer Manager** dialog box, choose the **Import** and **Look in Favorites** buttons.

LISP and ARX Applications: From the menu bar, choose **Tools > Load Applications**.

Scripts: From the menu bar, choose **Tools > Run Scripts**.

Menus: From the menu bar, choose **Tools > Customize Menus**. In the **Menu Customization** dialog box, choose the **Browse** and **Look in Favorites** buttons.

Images: From the menu bar, select **Tools > Displays Image > View**.

You cannot access text files, text fonts (SHX and TTF), color settings, lineweights, dimension styles, plot styles, OLE objects, or named UCSs over the Internet.

Saving the Drawing to the Internet

When you are finished editing a drawing in AutoCAD, you can save it to a file server on the Internet with the **SAVE** command. If you inserted the drawing from the Internet (using **INSERT**) into the default Drawing.Dwg drawing, AutoCAD insists you first save the drawing to your computer's hard drive.

When a drawing of the same name already exists at that URL, AutoCAD warns you, just as it does when you use the **SAVEAS** command. Recall from the **OPEN** command that AutoCAD uses your computer system's Temporary subdirectory, hence the reference to it in the dialog box.

Using Hyperlinks with AutoCAD*

AutoCAD 2000 allows you to employ URLs in two ways: (1) directly within an AutoCAD drawing; and (2) indirectly in DWF files displayed by a Web browser. (URLs are also known as *hyperlinks*, the term we use from now on.)

Hyperlinks are created, edited, and removed with the **HYPERLINK** command (displays a dialog box) and the **-HYPERLINKS** command (for prompts at the command line). The **HYPERLINK** command (**Insert > Hyperlink**) prompts you to "Select objects" and then displays the **Insert Hyperlink** dialog box. (As a shortcut, you may press CTRL+K or choose the **Insert Hyperlink** button on the toolbar.)

Figure 27-7(a) Invoking the HYPERLINK command from the Standard toolbar

As an alternative, you may use the **–HYPERLINK** command. This command displays its prompts at the command line, and is useful for scripts and AutoLISP routines. The **–HYPERLINK** command has the following syntax:

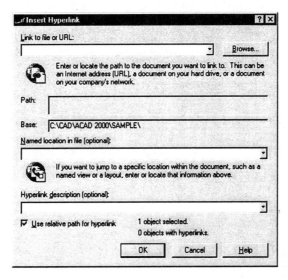

*Figure 20-7(b) The **Insert Hyperlink** dialog box*

Command: **-HYPERLINK**
Enter an option [Remove/Insert] <Insert>: *Press ENTER.*
Enter hyperlink insert option [Area/Object] <Object>: *Press ENTER.*
Select objects: *Select an object.*
1 found Select objects: *Press ENTER.*
Enter hyperlink <current drawing>: *Enter the name of the document or Web site.*
Enter named location <none>: *Enter the name of a bookmark or AutoCAD view.*
Enter description <none>: *Enter a description of the hyperlink.*

Notice that the command also allows you to remove a hyperlink. It does not, however, allow you to edit a hyperlink. To do this, use the **Insert** option and respecify the hyperlink data.

In addition, this command allows you to create hyperlink *areas*, which is a rectangular area that can be thought of as a 2D hyperlink (the dialog box-based **HYPERLINK** command does not create hyperlink areas). When you select the **Area** option, the rectangle is placed automatically on layer URLLAYER and colored red (Figure 20-8).

In the following sections, you learn how to apply and use hyperlinks in an AutoCAD drawing and in a Web browser via the dialog box-based **HYPERLINK** command.

Figure 20-8 A rectangular hyperlink area

Chapter 20

Hyperlinks Inside AutoCAD*

AutoCAD allows you to add a hyperlink to any object in the drawing. An object is permitted just a single hyperlink; on the other hand, a single hyperlink may be applied to a selection set of objects.

You can tell an object has a hyperlink by passing the cursor over it. The cursor displays the "linked Earth" icon, as well as a tooltip describing the link. See Figure 20-9.

Figure 20-9 The cursor reveals a hyperlink

If, for some reason, you do not want to see the hyperlink cursor, you can turn it off. From the **Tools** menu, choose **Options**, then choose the **User Preferences** tab. The **Display hyperlink cursor and shortcut menu** item toggles the display of the hyperlink cursor, as well as the **Hyperlink** tooltip option on the shortcut menu.

Attaching Hyperlinks*

Let's see how this works with an example.

Example 2

In this example, we have the drawing of a floorplan. To this drawing, we will add hyperlinks to another AutoCAD drawing, a Word document, and a Web site. Hyperlinks must be attached to objects. For this reason, we will place some text in the drawing, then attach the hyperlinks to the text.

1. Start AutoCAD 2000.

2. Open the **1st floor plan.dwg** file found in the AutoCAD 2000 **Sample** folder. (If necessary, choose the **Model** tab to display the drawing in model space.)

3. Invoke the **TEXT** command and place the following text in the drawing (Figure 20-10):

    Command: **TEXT**
    Current text style: "Standard" Text height: 0.20000
    Specify start point of text or [Justify/Style]: *Specify a point in the drawing.*
    Specify height <0.2000>: **2**
    Specify rotation angle of text <0>: *Press ENTER.*
    Enter text: **Site Plan**
    Enter text: **Lighting Specs**
    Enter text: **Electrical Bylaw**
    Enter text: *Press ENTER.*

4. Select the "Site Plan" text to select it.

5. Choose the **Insert Hyperlink** button on the toolbar. Notice the **Insert Hyperlinks** dialog box.

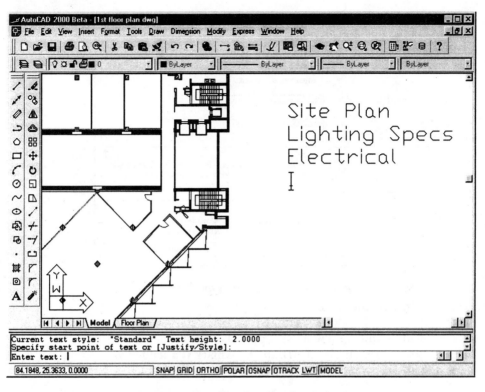

Figure 20-10 Text placed in drawing

6.   Next to the **Link to File or URL** field, choose the **Browse** button. Notice the **Browse the Web – Select Hyperlink** dialog box (Figure 20-11).

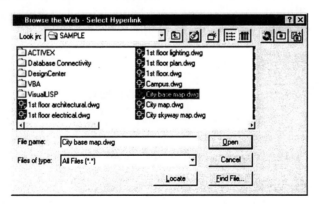

Figure 20-11 Browse the Web – Select Hyperlink dialog box

7.   Go to AutoCAD 2000's **Sample** folder and select the **City base map.dwg** file. Choose **Open**. AutoCAD does not open the drawing; rather, it copies the file's name to the **Insert Hyperlinks** dialog box (Figure 20-12). Notice the name of the drawing you selected in the **Link to File or URL** field.

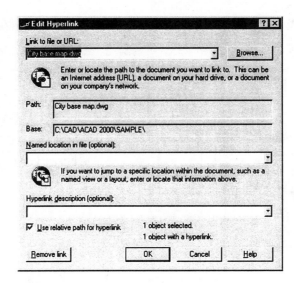

*Figure 20-12 The **Insert Hyperlinks** dialog box*

You can fill in two other fields:

Named Location in File: When the hyperlinked file is opened, it goes to the named location, called a "Bookmark." In AutoCAD, the bookmark is a named view (created with the **VIEW** command).

Hyperlink Description: You may type a description for the hyperlink, which is displayed by the tooltip. If you leave this blank, the URL is displayed by the tooltip.

8. Choose **OK** to dismiss the dialog box. Move the cursor over the Site Plan text. Notice the display of the "linked Earth" icon; a moment later, the tooltip displays "City base map.dwg"(Figure 20-13).

Figure 20-13 The Hyperlink cursor and tooltip

9. Repeat the **HYPERLINK** command twice more, attaching these files to the drawing text:

Text	URL
Lighting Specs	License.Rtf (found in the AutoCAD 2000 folder)
Electrical Bylaw	Home.Htm (also found in the AutoCAD 2000 folder)

You have now attached a drawing, a text document, and a Web document to objects in the drawing. Let's try out the hyperlinks.

10. Click "Site Plan" to select it. Right-click and select **Hyperlink > Open "City base map.dwg"** from the shortcut menu (Figure 20-14).

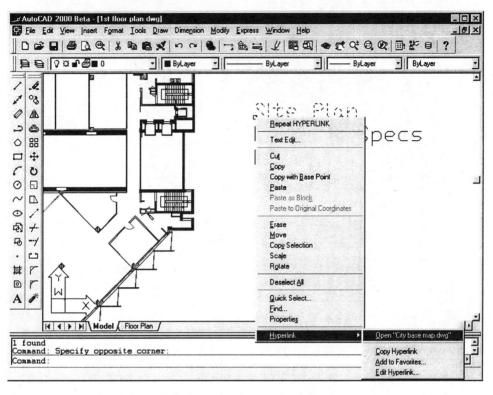

Figure 20-14 *Selecting a hyperlink from the shortcut menu*

Notice that AutoCAD opens "City base map.dwg". To see both drawings, select **Window > Tile Vertically** from the menu bar (Figure 20-15).

11. Select, then right-click "Lighting Specs" hyperlink. Choose **Hyperlink > Open**. Notice that Windows starts a word processor and opens the License.Rtf file.

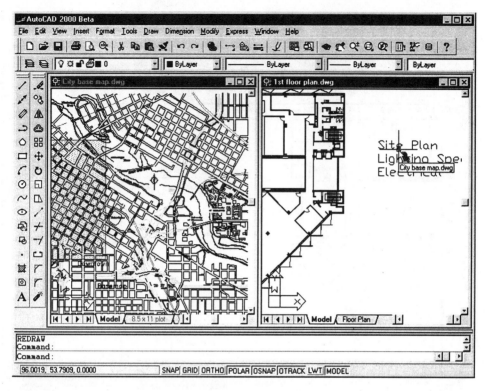

Figure 20-15 *Viewing two drawings*

12. If your word processor has a **Web** toolbar, open it. Choose the **Back** button (the back arrow), as shown in Figure 20-16. Notice how that action sends you back to AutoCAD.

13. Select, then right-click "Electrical Bylaw" hyperlink. Select **Hyperlink > Open**. Notice that Windows starts your Web browser and opens the Home.Html file (Figure 20-17).

14. Back in AutoCAD, right-click any toolbar and choose **Web**. Notice that the **Web** toolbar has four buttons. From left to right, these are:

Button	Command	Meaning
Go Back	**HYPERLINKBACK**	Move back to the previous hyperlink.
Go Forward	**HYPERLINKFWD**	Move to the next hyperlink.
Stop Navigation	**HYPERLINKSTOP**	Stops the hyperlink access action.
Browse the Web	**BROWSER**	Launches the Web browser.

15. Try choosing the **Go Forward** and **Go Back** buttons. Notice how these allow you to navigate between the two drawings, the RTF document, and the Web page.
    When you work with hyperlinks in AutoCAD, you may come across these limitations:

    • AutoCAD does not check that the URL you type is valid.

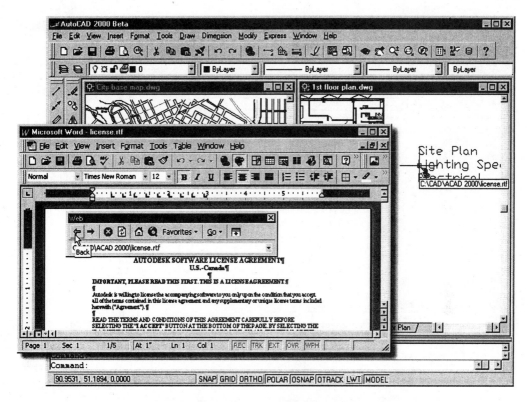

Figure 20-16 *Viewing a Word document*

- If you attach a hyperlink to a block, be aware that the hyperlink data is lost when you scale the block unevenly, stretch the block, or explode it.
- Wide polylines and rectangular hyperlink areas are only "sensitive" on their outline.

Pasting As Hyperlink*

AutoCAD 2000 has a shortcut method for creating hyperlinks in the drawing. The undocumented **PASTEASHYPERLINK** command pastes any text in the Windows Clipboard as a hyperlink to any object in the drawing. Here is how it works:

1. In a word processor, select some text and copy it to the Clipboard (via CTRL+C or choose **Edit > Copy**). The text can be a URL (such as http://www.autodeskpress.com) or any other text.

2. Switch to AutoCAD and select **Edit > Paste As Hyperlink** from the menu bar. Note that this command does not work (is grayed out) if anything else is in the Clipboard, such as a picture.

3. Select one or more objects, as prompted:
   Command: _pasteashyperlink
   Select objects: *Select an object.*
   1 found Select objects: *Press ENTER.*

Figure 20-17 *Viewing a Web Site*

Figure 20-18 *AutoCAD's* **Web** *Toolbar*

4. Pass the cursor over the object and note the hyperlink cursor and tooltip. The tooltip displays the same text that you copied from the document.

If the text you copy to the Clipboard is very long, AutoCAD displays only portions of it in the tooltip, using ellipses (...) to shorten the text. You cannot select text in the AutoCAD drawing to paste as a hyperlink.

You can, however, copy the hyperlink from one object to another. Select the object, right-click, and select **Hyperlink > Copy Hyperlink** from the shortcut menu. The hyperlink is copied to the Clipboard. You can now paste the hyperlink into another document, or use AutoCAD's **Edit > Paste as Hyperlink** command to attach the hyperlink to another object in the drawing.

Note
*The **MATCHPROP** command does not work with hyperlinks.*

Highlighting Objects with URLs*

Although you can see the rectangle of area URLs, the hyperlinks themselves are invisible. For this reason, AutoCAD has the **QSELECT** command, which highlights all objects that match specifications. From the menu bar, choose **Tools > Quick Select**. AutoCAD displays the **Quick Select** dialog box (Figure 20-19).

In the fields, enter these specifications:

      Apply to: **Entire drawing**
      Object type: **Multiple**
      Properties: **Hyperlink**
      Operator: *** Wildcard Match**
      Value: *****

Choose **OK** and AutoCAD highlights all objects that have a hyperlink. Depending on your computer's display system, the highlighting shows up as dashed lines or as another color.

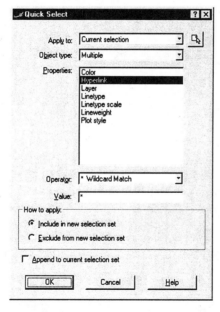

*Figure 20-19 The **Quick Select** dialog box*

Editing Hyperlinks*

Now that you know where the objects with hyperlinks are located, you can use the **HYPERLINK** command to edit the hyperlinks and related data. Select the hyperlinked object and invoke the **HYPERLINK** command. When the **Edit Hyperlink** dialog box appears (it looks identical to the **Insert Hyperlink** dialog box), make the changes and choose **OK**.

Removing Hyperlinks from Objects*

To remove a URL from an object, use the **HYPERLINK** command on the object. When the **Edit Hyperlink** dialog box appears, choose the **Remove Hyperlink** button.

To remove a rectangular area hyperlink, you can simply use the **ERASE** command; select the rectangle and AutoCAD erases the rectangle. (Unlike in Release 14, AutoCAD no longer purges the URLLAYER layer.)

As an alternative, you may use the **–HYPERLINK** command's **Remove** option, as follows:

> Command: **-HYPERLINK**
> Enter an option [Remove/Insert] <Insert>: R
> Select objects: ALL
> 1 found Select objects: *Press ENTER.*
> 1. *www.autodesk.com*
> 2. *www.autodeskpress.com*
> Enter number, hyperlink, or * for all: 1
> Remove, deleting the Area.
> 1 hyperlink deleted.

Hyperlinks Outside AutoCAD*

The hyperlinks you place in the drawing are also available for use outside of AutoCAD. The Web browser makes use of the hyperlink(s) when the drawing is exported in DWF format. To help make the process clearer, here are the steps that you need to go through:

Step 1
1   Open a drawing in AutoCAD.

Step 2
1   Attach hyperlinks to objects in the drawing with the **HYPERLINK** command. To attach hyperlinks to areas, use the **–HYPERLINK** command's **Area** option.

Step 3
1   Export the drawing in DWF format using the **PLOT** command's "DWF ePlot PC2" plotter configuration.

Step 4
1   Copy the DWF file to your Web site.

Step 5
1   Start your Web browser with the **BROWSER** command.

Step 6
1   View the DWF file and click on a hyperlink spot.

THE DRAWING WEB FORMAT

To display AutoCAD drawings on the Internet, Autodesk invented a new file format called drawing Web format (DWF). The DWF file has several benefits and some drawbacks over DWG files.

The DWF file is compressed as much as eight times smaller than the original DWG drawing file so that it takes less time to transmit over the Internet, particularly with the relatively slow telephone modem connections. The DWF format is more secure, since the original drawing is not being displayed; another user cannot tamper with the original DWG file.

However, the DWF format has some drawbacks:

- You must go through the extra step of translating from DWG to DWF.
- DWF files cannot display rendered or shaded drawings.
- DWF is a flat 2D-file format; therefore, it does not preserve 3D data, although you can export a 3D view.
- AutoCAD itself cannot display DWF files.
- DWF files cannot be converted back to DWG format without using file translation software from a third-party vendor.
- Earlier versions of DWF did not handle paperspace objects (version 2.x and earlier), or linewidths and non-rectangular viewports (version 3.x and earlier).

To view a DWF file on the Internet, your Web browser needs a *plug-in*—a software extension that lets a Web browser handle a variety of file formats. Autodesk makes the DWF plug-in freely available from its Web site at *http://www.autodesk.com/whip*. It is a good idea to regularly check for updates to the DWF plug-in, which is updated about twice a year.

Other DWF Viewing Options

In addition to Whip with a Web browser, Autodesk provides two other options for viewing DWF files.

CADViewer Light is designed to be a DWF viewer that works on all operating systems and computer hardware because it is written in Java. As long as your Windows, Mac, or Unix computer has access to Java (included with most Web browsers), you can view DWF files.

Volo View Express is a stand-alone viewer that views and prints DWG, DWF, and DXF files. Both products can be downloaded free from the Autodesk Web site.

Viewing DWG Files

To view AutoCAD DWG and DXF files on the Internet, your Web browser needs a DWG-DXF plug-in. At the time of publication, the plug-ins were available from the following vendors:

Autodesk: *http://www.autodesk.com/whip*
SoftSource: *http://www.softsource.com/*
California Software Labs: *http://www.cswl.com*
Cimmetry systems: *http://www.cimmetry.com*

Creating a DWF File[*]

To create a DWF file from AutoCAD 2000, follow these steps:

1. Type the **PLOT** command or choose **Plot** from the **File** menu. Notice that AutoCAD displays the **Plot** dialog box (Figure 20-20).

2. In the **Name** list box (found in the **Plotter Configuration** area), select **DWF ePlot pc3**.

3. Choose the **Properties** button. Notice the **Plotter Configuration Editor** dialog box (Figure 20-21).

4. Choose **Custom Properties** in the tree view. Notice the **Custom Properties** button, which appears in the lower half of the dialog box.

5. Choose the **Custom Properties** button. Notice the **DWF Properties** dialog box (Figure 20-22).

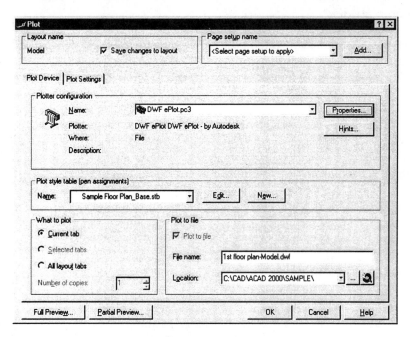

Figure 20-20 Plot Dialog Box

Resolution

Unlike AutoCAD DWG files, which are based on real numbers, DWF files are saved using integer numbers. The **Medium** precision setting saves the drawing using 16-bit integers, which is adequate for all but the most complex drawings. **High** resolution saves the DWF file using 20-bit integers, while **Extreme** resolution saves using 32-bit integers. Figure 20-23 shows an extreme close-up of a drawing exported in DWF format. The portion at left was saved at medium resolution and shows some bumpiness in the arcs. The portion at right was saved at extreme resolution.

More significant is the difference in file size. When a DWF file was created from the "1st floor plan.dwg" file (342KB), the medium resolution DWF file was just 27KB, while the extreme resolution file became 1,100KB – some 40 times larger. That means that the medium resolution DWF file transmits over the Internet 40 times faster, a significant savings in time.

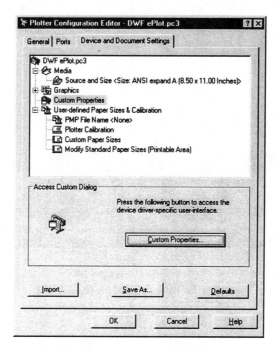

Figure 20-21 Plotter Configuration Editor dialog box

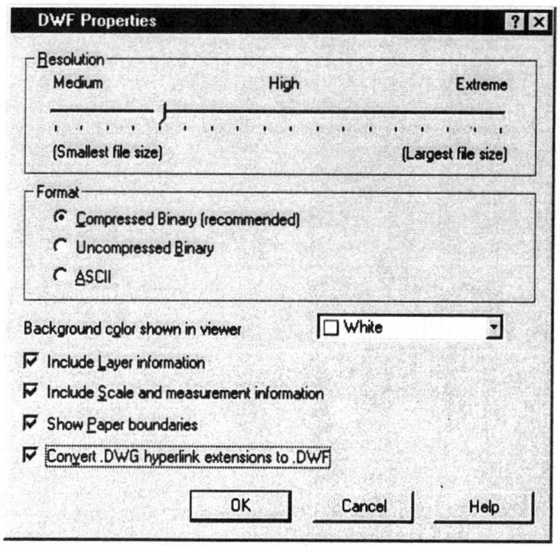

Figure 20-22 DWF Properties Dialog Box.

Format

Compression further reduces the size of the DWF file. You should always use compression, unless you know that another application cannot decompress the DWF file. Compressed binary format is seven times smaller than ASCII format (7,700KB). Again, that means the compressed DWF file transmits over the Internet seven times faster than an ASCII DWF file.

Other Options

Background Color Shown in Viewer: While white is probably the best background color, you may choose any of AutoCAD's 255 colors.

Include Layer Information: This option includes layers, which allows you to toggle layers off and on when the drawing is viewed by the Web browser.

Figure 20-23 DWF output in medium resolution (at left) and extreme resolution (at right).

Include Scale and Measurement Information: This allows you to use the **Location** option in the Web browser's Whip plug-in to show scaled coordinate data.

Show Paper Boundaries: Includes a rectangular boundary at the drawing's extents.

Convert .DWG Hyperlink Extensions to .DWF: Includes hyperlinks in the DWF file.

In most cases, you would turn on all options.

6. Choose **OK** to exit the dialog boxes back to the **Plot** dialog box.

7. Accept the DWF file name listed in the **File name** text box, or type a new name. If necessary, change the location that the file will be stored in. Note the two buttons: one has an ellipsis (…), which displays the **Browse for Folder** dialog box. The second button brings up AutoCAD's internal Web browser, a simplified version of the Microsoft-branded product.

8. Choose **OK** to save the drawing in DWF format.

Viewing DWF Files

To view a DWF file, you need to use a Web browser with a special plug-in that allows the browser to correctly interpret the file. (Remember: you cannot view a DWF file with AutoCAD.) Autodesk has named their DWF plug-in "Whip!," short for "Windows HIgh Performance."

Autodesk updates the DWF plug-in approximately twice a year. Each update includes some new features. In summary, all versions of the DWF plug-in perform the following functions:

- Views DWF files created by AutoCAD within a browser.
- Right-clicking the DWF image displays a shortcut menu with commands.
- Real-time pan and zoom lets you change the view of the DWF file as quickly as a drawing

file in AutoCAD.
- Embedded hyperlinks let you display other documents and files.
- File compression means that a DWF file appears in your Web browser much faster than the equivalent DWG drawing file would.
- Print the DWF file alone or along with the entire Web page.
- Works with Netscape Navigator or Microsoft Internet Explorer. A separate plug-in is required, depending upon which of the two browsers you use.
- Allows you to "drag and drop" a DWG file from a Web site into AutoCAD as a new drawing or as a block.
- Views a named view stored in the DWF file.
- Can specify a view using x,y- coordinates.
- Toggles layers off and on.

If you don't know whether the DWF plug-in is installed in your Web browser, select **Help > About Plug-ins** from the browser's menu bar. You may need to scroll through the list of plug-ins to find something like this (Figure 20-24):

If the plug-in is not installed, or is an older version, then you need to download it from Autodesk's Web site at:

http://www.autodesk.com/whip

For Netscape users, the file is quite large at 3.5MB and takes about a half-hour to download using a typical 28.8Kbaud modem. The file you download from the Autodesk Web site is either a self-installing file (has a .JAR extension) or a self-extracting installation file with a name such as Whip4.Exe. After the download is complete, follow the instructions on the screen.

For Internet Explorer users, the DWF plug-in is an ActiveX control. It is automatically installed into Explorer when you set up AutoCAD 2000.

DWF Plug-in Commands

To display the DWF plug-in's commands, move the cursor over the DWF image and press the mouse's right button. This displays a shortcut menu with options, such as **Pan**, **Zoom**, and **Named Views** (Figure 20-25). To select an option, move the cursor over the option name and press the left mouse button.

Pan is the default option. Press the left mouse button and move the mouse. The cursor changes to an open hand, signaling that you can pan the view around the drawing. This is exactly the same as real-time panning in AutoCAD. Naturally, panning only works when you are zoomed in; it does not work in Full View mode.

Zoom is like the **ZOOM** command in AutoCAD. The cursor changes to a magnifying glass. Hold down the left mouse button and move the cursor up (to zoom in) and down (to zoom out).

Zoom Rectangle is the same as **ZOOM Window** in AutoCAD. The cursor changes to a plus sign. Click the left mouse button at one corner, then drag the rectangle to specify the size of the new view.

Chapter 20

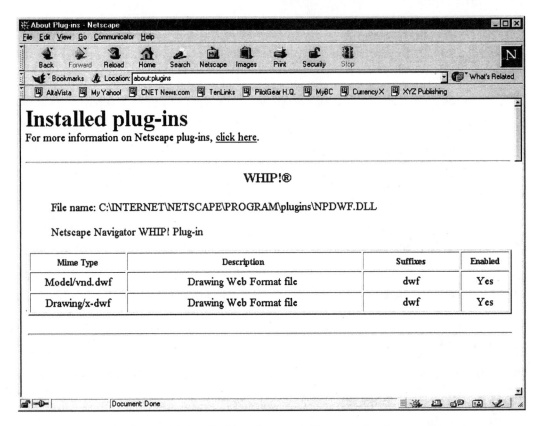

Figure 20-24 *Checking the status of browser plug-ins*

Fit to Window is the same as AutoCAD's **ZOOM Extents** command. You see the entire drawing.
Layers displays a non-modal dialog box (Figure 20-26), which lists all layers in the drawing. (A *non-modal* dialog box remains on the screen; unlike AutoCAD's modal dialog boxes, you do not need to dismiss a non-modal dialog box to continue working.) Click a layer name to toggle its visibility between on (yellow light bulb icon) and off (blue light bulb).
Named Views works only when the original DWG drawing file contained named views created with the **VIEW** command. Selecting **Named Views** displays a non- modal dialog box that allows you to select a named view (Figure 20-27). Click a named view to see it; click the small **x** in the upper-right corner to dismiss the dialog box.
Location displays a non-modal dialog box that shows the x,y,z-coordinates of the cursor location (Figure 20-28). Since DWF files are only 2D, the Z coordinate is always 0.
Full View causes the Web browser to display the DWF image as large as possible all by itself. This is useful for increasing the physical size of a small DWF image. When you have finished viewing the large image, right-click and choose the **Back** or choose the browser's **Back** button to return to the previous screen.

Highlight URLs displays a flashing gray effect on all objects and areas with a hyperlink (Figure 20-29). This helps you see where the hyperlinks are. To read the associated URL, pass the

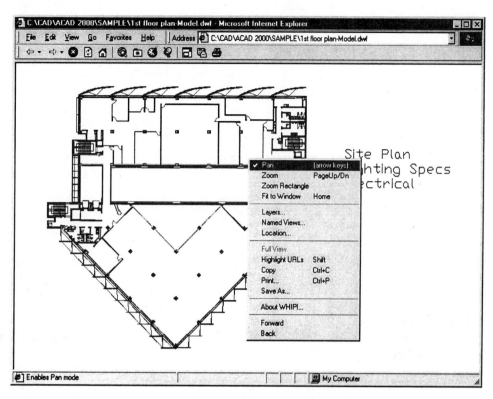

Figure 20-25 *Whip's shortcut menu*

cursor over a highlighted area and look at the URL on the browser's status line. (A shortcut is to hold down the SHIFT key, which causes the URLs to highlight until you release the key). Choosing the **URL** opens the document with its associated application.

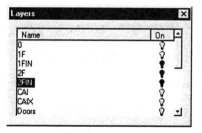

Figure 20-26 *Whip's* **Layer** *dialog box*

Copy copies the DWF image to the Windows Clipboard in EMF (enhanced Metafile or "Picture") format.

Print prints the DWF image alone. To print the entire Web page (including the DWF image), use the browser's **Print** button.

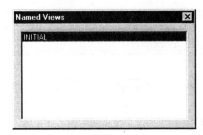

*Figure 20-27 Whip's **Named Views** dialog box*

*Figure 20-28 Whip's **Location** dialog box*

Save As saves the DWF file in three formats to your computer's hard drive:
DWF
BMP (Windows bitmap)
DWG (AutoCAD drawing file) works only when a copy of the DWG file is available at the same subdirectory as the DWF file.

You cannot use the **Save As** option until the entire DWF file has been transmitted to your computer.

About WHIP displays information about the DWF file, including DWF file revision number, description, author, creator, source filename, creation time, modification time, source creation time, source modification time, current view left, current view right, current view bottom, and current view top (Figure 20-30).

Forward is almost the same as choosing the browser's **Forward** button. When the DWF image is in a frame, only that frame goes forward.

Back is almost the same as choosing the browser's **Back** button. It works differently when the DWF image is displayed in a frame.

Drag and Drop from Browser to Drawing

The DWF plug-in allows you to perform several "drag and drop" functions. *Drag and drop* is when you use the mouse to drag an object from one application to another.
Hold down the (key to drag a DWF file from the browser into AutoCAD. Recall that AutoCAD cannot translate a DWF file into DWG drawing format. For this reason, this form of drag and drop only works when the originating DWG file exists in the same subdirectory as the DWF file. AutoCAD executes the **-INSERT** command, as follows:

        Command: _.-insert
        Enter block name or [?]: C:\CAD\ACAD 2000\SAMPLE\1st floor plan.dwg

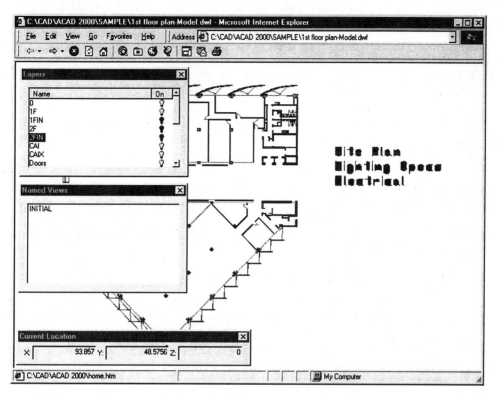

Figure 20-29 *Highlighting hyperlinks*

Specify insertion point or [Scale/X/Y/Z/Rotate/
PScale/PX/PY/PZ/PRotate]: 0,0
Enter X scale factor, specify opposite corner,
or [Corner/XYZ] <1>: *Press ENTER.*
Enter Y scale factor <use X scale factor>:
Press ENTER.
Specify rotation angle <0>: *Press ENTER.*

To see the inserted drawing, you may need to use
the **ZOOM Extents** command.

Another drag and drop function is to drag a DWF
file from the Windows Explorer (or File Manager)
into the Web browser. This causes the Web browser
to load the DWF plug-in, then display the DWF file.
Once displayed, you can execute all of the commands
listed in the previous section.

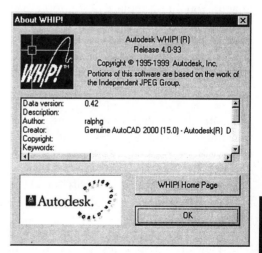

Figure 20-30 About *dialog box*

Finally, you can also drag and drop a DWF file from Windows Explorer (in Windows 95/98) or File
Manager (in Windows NT) into AutoCAD. This causes AutoCAD to launch another program that
is able to view the DWF file. This does not work if you do not have other software on your com-

puter system capable of viewing DWF files.

Embedding a DWF File

To let others view your DWF file over the Internet, you need to embed the DWF file in a Web page. Here are the steps to embedding a DWF file in a Web page.

Step 1. This hype text markup language (HTML) code is the most basic method of placing a DWF file in your Web page:

            <embed src="filename.dwf">

The **<embed>** tag embeds an object in a Web page. The **src** option is short for "source." Replace **filename.dwf** with the URL of the DWF file. Remember to keep the quotation marks in place.

Step 2. HTML normally displays an image as large as possible. To control the size of the DWF file, add the **Width** and **Height** options:

            <embed width=800 height=600 src="filename.dwf">

The **Width** and **Height** values are measured in pixels. Replace 800 and 600 with any appropriate numbers, such as 100 and 75 for a "thumbnail" image, or 300 and 200 for a small image.

Step 3. To speed up a Web page's display speed, some users turn off the display of images. For this reason, it is useful to include a description, which is displayed in place of the image:

            <embed width=800 height=600 name=description src="filename.dwf">

The **Name** option displays a textual description of the image when the browser does not load images. You might replace description with the DWF filename.

Step 4. When the original drawing contains named views created by the **VIEW** command, these are transferred to the DWF file. Specify the initial view for the DWF file:

            <embed width=800 height=600 name=description namedview="viewname"
            src="filename.dwf">

The **namedview** option specifies the name of the view to display upon loading. Replace **viewname** with the name of a valid view name. When the drawing contains named views, the user can right-click on the DWF image to get a list of all named views.

As an alternative, you can specify the 2D coordinates of the initial view:

            <embed width=800 height=600 name=description view="0,0 9,12"
            src="filename.dwf">

The **View** option specifies the x,y-coordinates of the lower-left and upper-right corners of the initial view. Replace 0,0 9,12 with other coordinates. Since DWF is 2D only, you cannot specify a 3D viewpoint. You can use **View** or **Namedview**, but not both.

Step 5. Before a DWF image can be displayed, the Web browser must have the DWF plug-in called "Whip!". For users of Netscape Communicator, you must include a description of where to get the Whip plug-in when the Web browser is lacking it.

```
<embed pluginspage=http://www.autodesk.com/products/autocad/whip/whip.htm
width=800 height=600 name=description view="0,0 9,12" src="filename.dwf">
```

The **pluginspage** option describes the page on the Autodesk Web site where the Whip-DWF plug-in can be downloaded.

The code listed above works for Netscape Navigator. To provide for users of Internet Explorer, the following HTML code must be added:

```
<object classid ="clsid:B2BE75F3-9197-11CF-ABF4-08000996E931" codebase = "ftp://
ftp.autodesk.com/pub/autocad/plugin/whip.cab#version=2,0,0,0" width=800
height=600>
<param name="Filename" value="filename.dwf">
<param name="View" value="0,0 9,12">
<param name="Namedview" value="viewname">
<embed pluginspage=http://www.autodesk.com/products/autocad/whip/
whip.htm width=800 height=600 name=description view="0,0 9,12"
src="filename.dwf">
</object>
```

The two **<object>** and three **<param>** tags are ignored by Netscape Navigator; they are required for compatibility with Internet Explorer. The **classid** and **codebase** options tell Explorer where to find the plug-in. Remember that you can use **View** or **Namedview** but not both.

Step 6: Save the HTML file.

Review Questions

1. Can you launch a Web browser from within AutoCAD? _____

2. What does DWF mean? _____

3. What is the purpose of DWF files? _____

4. URL is an acronym for? _____

5. Which of the following URLs are valid?
    www.autodesk.com
    http://www.autodesk.com
    Both of the above.
    None of the above.

6. FTP is an acronym for: _____

7. What is a "local host"? _____

8. Are hyperlinks active in an AutoCAD 2000 drawing? _____

9. The purpose of URLs is to let you create _____ between files.

10. When you attach a hyperlink to a block, the hyperlink data is _____ when you scale the block unevenly, stretch the block, or explode it.

11. Can you can attach a URL to any object? _____

12. The **-HYPERLINK** command allows you to attach a hyperlink to _____ and_____.

13. To see the location of hyperlinks in a drawing, use the _____ command.

14. Rectangular (area) hyperlinks are stored on layer:_____.

15. Compression in the DWF file causes it to take (less, more, the same) time to transmit over the Internet.

16. A DWF is created from a _____ file using the _____ command.

17. What does a "plug-in" let a Web browser do?

18. Can a Web browser view DWG drawing files over the Internet? _____

19. _____ is an HTML tag for embedding graphics in a Web page.

20. A file being transmitted over the Internet via a 28.8Kpbs modem takes about _____ minutes per megabyte.

Chapter 21

Inquiry Commands, Data Exchange, and Object Linking and Embedding

Learning Objectives

After completing this chapter, you will be able to:
- *Make inquiries about drawings and objects.*
- *Calculate area using the **AREA** command.*
- *Calculate distances using the **DIST** command.*
- *Identify a position on the screen using the **ID** command.*
- *List information about drawings and objects using the **DBLIST** and **LIST** commands.*
- *List drawing-related information using the **STATUS** command.*
- *Import and export .dxf files using the **SAVEAS** and **OPEN** commands.*
- *Convert scanned drawings into the drawing editor using the **DXB** command.*
- *Export raster files using the **SAVEIMG** command.*
- *Understand the embedding and the linking functions of the **OLE** feature of Windows.*

MAKING INQUIRIES ABOUT A DRAWING (INQUIRY COMMANDS)

When you create a drawing or examine an existing one, you often need some information about the drawing. In manual drafting, you inquire about the drawing by performing measurements and calculations manually. Similarly, when drawing in an AutoCAD environment, you will need to make inquiries about data pertaining to your drawing. The inquiries can be about the distance from one location on the drawing to another, the area of an object like a polygon or circle, coordinates of a location on the drawing, and so on. AutoCAD keeps track of all the details pertaining to a drawing. Since inquiry commands are used to obtain information about the selected objects, these commands do not affect the drawings in any way. The following is the list of Inquiry commands:

AREA	**DIST**	**ID**	**LIST**	**DBLIST**
MASSPROP	**STATUS**	**TIME**	**DWGPROPS**	

For most of the Inquiry commands, you are prompted to select objects; once the selection is complete, AutoCAD switches from graphics mode to text mode, and all the relevant information about the selected objects is displayed. For some commands information is displayed in the AutoCAD Text Window. The display of the text screen can be tailored to your requirements with the help of a pointing device; hence, by moving the text screen to one side,

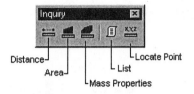

*Figure 21-1 Selecting Inquiry commands from the **Inquiry** toolbar*

you can view the drawing screen and the text screen simultaneously. If you select the **minimize** button or select the close button, you will return to the graphics screen. You can also return to the graphics screen by entering **GRAPHSCR** command at the Command prompt. Similarly, you can return to the AutoCAD Text Window by entering **TEXTSCR** at the Command prompt.

AREA COMMAND

Toolbar:	Inquiry > Area
Menu:	Tools > Inquiry > Area
Command:	AREA

 Finding the area of a shape or an object manually is time-consuming. In AutoCAD, the **AREA** command is used to automatically calculate the area of an object in square units. This command saves time when calculating the area of shapes, especially when the shapes are complicated.

You can use the default option of the **AREA** command to calculate the area and perimeter or circumference of the space enclosed by the sequence of specified points. For example, to find the area of an object (one which is not formed of a single object) you have created with the help of the **LINE** command (Figure 21-2), you need to select all the vertices of that object. By selecting the points, you define the shape of the object whose area is to be found. This is the default method for determining the area of an object. The only restriction is that all the

*Figure 21-2 Using the **AREA** command*

points you specify be in a plane parallel to the XY plane of the current UCS. You can make the best possible use of object snaps such as ENDpoint, INTersect, and TANgent, or even use running Osnaps, to help you select the vertices quickly and accurately. For AutoCAD to find the area of a shape, the shapes need not have been drawn with polylines; nor do the lines need to be closed. However, curves must be approximated with short straight segments. In such cases, AutoCAD computes the area by assuming that the first point and the last point are joined. The prompt sequence in this case is:

Command: **AREA** `Enter`
Specify first corner point or [Object/Add/Subtract]: *Specify first point.*
Specify next corner point or press ENTER for total: *Specify the second point.*
Specify next corner point or press ENTER for total: *Continue selecting until all the points enclosing the area have been selected.*
Specify next corner point or press ENTER for total: `Enter`
Area = X, Perimeter = Y

Here, X represents the numerical value of the area and Y represents the circumference/perimeter. It is not possible to accurately determine the area of a curved object, such as an arc, with the default (Point) option. However, the approximate area under an arc can be calculated by specifying several points on the given arc. If the object whose area you want to find is not closed (formed of independent segments) and has curved lines, you should use the following steps to determine the accurate area of such an object:

1. Convert all the segments in that object into polylines using the **PEDIT** command.
2. Join all the individual polylines into a single polyline. Once you have performed these operations, the object becomes closed and you can then use the **Object** option of the **AREA** command to determine the area.

If you specify two points on the screen, the **AREA** command will display the value of the area as 0.00; the perimeter value is the distance between the two points.

Object option. You can use the **Object** option of the **AREA** command to find the area of objects such as polygons, circles, polylines, regions, solids, and splines. If the selected object is a polyline or polygon, AutoCAD displays the area and perimeter of the polyline. In case of open polylines, the area is calculated assuming that the last point is joined to the first point but the length of this segment is not added to the polyline length unlike the default option. If the selected object is a circle, ellipse, or planar closed spline curve, AutoCAD will provide information about its area and circumference. For a solid, the surface area is displayed. For a 3D polyline, all vertices must lie in a plane parallel to the XY plane of the current UCS. The extrusion direction of a 2D polyline whose area you want to determine should be parallel to the Z axis of the current UCS. In case of polylines which have a width, the area and length of the polyline is calculated using the centerline. If any of these conditions is violated, an error message is displayed on the screen. The prompt sequence is:

Command: **AREA** `Enter`
Specify first corner point or [Object/Add/Subtract]: O `Enter`
Select objects : Select an object `Enter`
Area = (X), Circumference = (Y)

The X represents the numerical value of the area, and Y represents the circumference/perimeter.

Note
*In many cases, the easiest and most accurate way to find the area of an area enclosed by multiple objects is to use the **BOUNDARY** command to create a polyline, then use the **AREA Object** option.*

Add option. Sometimes you want to add areas of different objects to determine a total area. For example, in the plan of a house, you need to add the areas of all the rooms to get the total floor area. In such cases, you can use the **Add** option. Once you invoke this option, AutoCAD activates the **Add** mode. By using the **First corner point** option at the **Specify first corner point or [Object/Subtract]:** prompt, you can calculate the area and perimeter by selecting points on the screen. Pressing ENTER after you have selected points defining the area which is to be added, calculates the total area, since the **Add** mode is on. The command prompt is as follows:

Specify next corner point or press ENTER for total [ADD mode]:

If the polygon whose area is to be added, is not closed, the area and perimeter are calculated assuming a line is added that connects the first point to the last point to close the polygon. The length of this area is added in the perimeter. The **Object** option adds the areas and perimeters of selected objects. While using this option, if you select an open polyline, the area is calculated considering the last point is joined to the first point but the perimeter does not consider the length of this assumed segment, unlike the **First corner point** option. When you select an object, the area of the selected object is displayed on the screen. At this time the total area is equal to the area of the selected object. When you select another object, AutoCAD displays the area of the selected object as well as the combined area (total area) of the previous object and the currently selected object. In this manner you can add areas of different objects. Until the **Add** mode is active, the string ADD mode is displayed along with all subsequent object selection prompts to remind you that the **Add** mode is active. When the **AREA** command is invoked, the total area is initialized to zero.

Subtract option. The action of the **Subtract** option is the reverse of that of the **Add** option. Once you invoke this option, AutoCAD activates the **Subtract** mode. The **First corner point** and **Object** options work similar to the way they work in the ADD mode. When you select an object, the area of the selected object is displayed on the screen. At this time, the total area is equal to the area of the selected object. When you select another object, AutoCAD displays the area of the selected object as well as the area obtained by subtracting the area of the currently

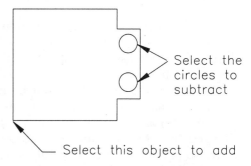

*Figure 21-3 Using the **Add** and **Subtract** options*

selected object from the area of the previous object. In this manner, you can subtract areas of objects from the total area. Until the **Subtract** mode is active, the string SUBTRACT mode is displayed along with all subsequent object selection prompts, to remind you that the SUBTRACT mode is active. To exit the **AREA** command, press ENTER (null response) at the **Specify first corner point or [Object/Add/Subtract]:** prompt. The prompt sequence for these two modes for Figure 21-3 is:

Command: **AREA** [Enter]
Specify first corner point or [Object/Add/Subtract]: A [Enter]
Specify first corner point or [Object/Subtract]: O [Enter]
(ADD mode) Select objects: *Select the polyline.*
Area = 2.4438, Perimeter = 6.4999
Total area = 2.4438
(ADD mode) Select objects: [Enter]
Specify first corner point or [Object/Subtract]: S [Enter]
Specify first corner point or [Object/Add]: O [Enter]
(SUBTRACT mode) Select object: *Select the circle.*
Area = 0.0495, Circumference = 0.7890
Total area = 2.3943
(SUBTRACT mode) Select objects: *Select the second circle.*
Area = 0.0495, Circumference = 0.7890
Total area = 2.3448
(SUBTRACT mode) Select object: [Enter]
Specify first corner point or [Object/Add]: [Enter]

The **AREA** and **PERIMETER** system variables hold the area and perimeter (or circumference in the case of circles) of the previously selected polyline (or circle). Whenever you use the **AREA** command, the **AREA** variable is reset to zero.

Calculating Distance Between Two Points (DIST Command)

Toolbar:	Inquiry > Distance
Menu:	Tools > Inquiry > Distance
Command:	DIST

The **DIST** command is used to measure the distance between two selected points (Figure 21-4). The angles that the selected points make with the X axis and the XY plane are also displayed. The measurements are displayed in current units. Delta X (horizontal displacement), delta Y (vertical displacement), and delta Z are also displayed. The distance computed by the **DIST** command is saved in the **DISTANCE** variable. The prompt sequence is:

Command: **DIST** [Enter]
Specify first point: *Specify a point.*
Specify second point: *Specify a point.*

AutoCAD returns the following information:

Distance = *Calculated distance between the two points.*
Angle in XY plane = *Angle between the two points in the XY plane.*
Angle from XY plane = *Angle the specified points make with the XY plane.*
Delta X = *Change in X,* Delta Y = *Change in Y,* Delta Z = *Change in Z.*

If you enter a single number or fraction at the **Specify first point:** prompt, AutoCAD will display that number in the current unit of measurement.

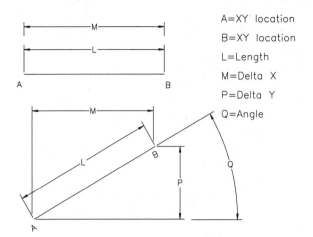

A=XY location
B=XY location
L=Length
M=Delta X
P=Delta Y
Q=Angle

Figure 21-4 *Using the **DIST** command*

Command: **DIST**
First point: 3-3/4   *(Enter a number or a fraction.)*
Distance = 3.7500

Note
The Z coordinate is used in 3D distances. If you do not specify the Z coordinates of the two points between which you want to know the distance, AutoCAD takes the current elevation as the Z coordinate value.

Identifying a Position on the Screen (ID Command)

Toolbar:	Inquiry > Locate Point
Menu:	Tools > Inquiry > ID Point
Command:	ID

    The **ID** command identifies the position of a point you specify and tells you its coordinates.

Command: **ID** [Enter]
Specify point: *Specify the point to be identified.*
X = X coordinate   Y = Y coordinate   Z = Z coordinate

AutoCAD takes the current elevation as the Z coordinate value. If an **Osnap** mode is used to snap to a 3D object in response to the **Specify point:** prompt, the Z coordinate displayed will be that of the selected feature of the 3D object. You can also use the **ID** command to identify the location on the screen. This can be realized by entering the coordinate values you want to locate on the screen. AutoCAD identifies the point by drawing a blip mark at that location (**BLIPMODE** should be On). For example, say you want to find the position on the screen where X = 2.345, Y = 3.674, and Z = 1.0000 is located. This can be achieved with the following prompt sequence:

Command: **ID** [Enter]

Specify point: 2.345,3.674,1.00 [Enter]
X = 2.345   Y = 3.674   Z = 1.0000

The coordinates of the point specified in the **ID** command are saved in the **LASTPOINT** system variable. You can locate a point with respect to the **ID** point by using the relative or polar coordinate system. You can also snap to this point by typing @ when AutoCAD prompts for a point.

Command: **ID**
Specify point: *Select a point.*
Command: **LINE**
Specify first point: @ *(The line will snap to the ID point.)*

You can also use the **X/Y/Z point filters** to specify points on the screen. You can respond to the **Specify point:** prompt with any desired combination of .X, .Y, .Z, .XY, .XZ, .YZ. The 2D or 3D point is specified by specifying individual (intermediate) points and forming the desired point from the selected X, Y, and Z coordinates of the intermediate points. More simply, you can specify a 2D or 3D point by supplying separate information for the X, Y, and Z coordinates. For example, say you want to identify a point whose X, Y, and Z coordinates are marked on the screen separately. This can be achieved in the following manner:

Command: **ID** [Enter]
Specify point: .X [Enter]
of   *Specify the location whose X coordinate is the X coordinate of the final desired point to be identified.*
of (need YZ):   *Specify the location whose YZ coordinate is the YZ coordinate of the final desired point to be identified.*
X = X coordinate   Y = Y coordinate   Z = Z coordinate

A blip mark is formed at the point of intersection of the specified X, Y, and Z coordinates. This blip mark identifies the desired point.

Listing Information About Objects (LIST Command)

Toolbar:	Inquiry > List
Menu:	Tools > Inquiry > List
Command:	LIST

The **LIST** command displays all the data pertaining to the selected objects. The prompt sequence is:

Command: **LIST** [Enter]
Select objects: *Select objects whose data you want to list.*
Select objects: [Enter]

Once you select the objects to be listed, AutoCAD shifts you from the graphics screen to the AutoCAD Text Window. The information displayed (listed) varies from object to object. Information on an object's type, its coordinate position with respect to the current UCS (user coordinate system), the name of the layer on which it is drawn, and whether the object is in model space or paper space is listed for all types of objects. If the color, lineweight, and the linetype are not BYLAYER, they are also listed. Also, if the thickness of the object is greater than 0, that is also displayed. The elevation value is displayed in the form of a Z coordinate (in the case of 3D objects). If an object has an extrusion direction different from the Z axis of the current UCS, the object's extrusion direction is also provided.

More information based on the objects in the drawing is also provided. For example, for a line the following information is be displayed:

1. The coordinates of the endpoints of the line.
2. Its length (in 3D).
3. The angle made by the line with respect to the X axis of the current UCS.
4. The angle made by the line with respect to the XY plane of the current UCS.
5. Delta X, delta Y, delta Z: this is the change in each of the three coordinates from the start point to the endpoint.
6. The name of the layer in which the line was created.
7. Whether the line is drawn in Paper space or Model space.

The center point, radius, true area, and circumference of circles is displayed. For polylines, this command displays the coordinates. In addition, for a closed polyline, its true area and perimeter are also given. If the polyline is open, AutoCAD lists its length and also calculates the area by assuming a segment connecting the start point and endpoint of the polyline. In the case of wide polylines, all computation is done based on the centerlines of the wide segments. For a selected viewport, the **LIST** command displays whether the viewport is on and active, on and inactive, or off. Information is also displayed about the status of Hideplot and the scale relative to paper space. If you use the **LIST** command on a polygon mesh, the size of the mesh (in terms of M, X, N), the coordinate values of all the vertices in the mesh, and whether the mesh is closed or open in M and N directions are all displayed. As mentioned before, if all the information does not fit on a single screen, AutoCAD pauses to allow you to press ENTER to continue the listing.

Listing Information About All Objects in a Drawing (DBLIST Command)

Command:	DBLIST

The **DBLIST** command displays information pertaining to all the objects in the drawing. Once you invoke this command, information is displayed in the AutoCAD Text Window. If the information does not fit on a single screen, AutoCAD pauses to allow you to press ENTER to continue the listing. To terminate the command, press ESCAPE. To return to the graphics screen, close the AutoCAD Text Window. This command can be invoked by entering **DBLIST** at the Command prompt.

Exercise 1   *Mechanical*

Draw Figure 21-5 and save it as EX1. Using INQUIRY commands, determine the following values:

a. Area of the hexagon.
b. Perimeter of the hexagon.
c. Perimeter of the inner rectangle.
d. Circumference of each circle.
e. Area of the hexagon minus the inner rectangle.
f. Use the **DBLIST** command to get the database listing.

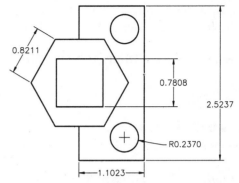

Figure 21-5 Drawing for Exercise 1

Listing Information About Solids and Regions (MASSPROP Command)

Toolbar:	Inquiry > Mass properties
Menu:	Tools > Inquiry > Mass Properties
Command:	MASSPROP

With the **MASSPROP** command, you can determine volumetric information, such as principle axes, center of gravity, and moment of inertia of 2D or 3D objects.

Command: **MASSPROP** [Enter]
Select objects: *Select an object (solid object).*

CHECKING TIME-RELATED INFORMATION (TIME COMMAND)

The time and date maintained by your system are used by AutoCAD to provide information about several time factors related to the drawings. Hence, you should be careful about setting the current

Menu:	Tools > Inquiry > Time
Command:	TIME

date and time in your computer. The **TIME** command can be used to display information pertaining to time related to a drawing and the drawing session. The display obtained by invoking the **TIME** command is similar to the following:

Command: **TIME**

Current time:	Monday, April 26 1999 at 11:22:35.069 AM
Times for this drawing:	
Created:	Thursday, April 15 1999 at 09:34:42.157 AM
Last updated:	Friday, April 16 1999 at 02:54:25.700 PM

Total editing time:	0 days 07:52:16.205
Elapsed timer (on)	0 days 00:43:07.304
Next automatic save in:	0 days 00:54:27:153

Enter an option [Display/ON/OFF/Reset]:

The foregoing display gives you information on the following.

Current Time

Provides today's date and the current time.

Drawing Creation Time

Provides the date and time that the current drawing was created. The creation time for a drawing is set to the system time when the **NEW**, **-WBLOCK**, or **SAVE** command is used to create that drawing file.

Last Updated Time

Provides the most recent date and time you saved the current drawing. In the beginning, it is set to the drawing creation time, and it is modified every time you use the **QUIT** or **SAVE** command to save the drawing.

Total Editing Time

This tells you the total time spent on editing the current drawing since it was created. If you terminate the editing session without saving the drawing, the time you have spent on that editing session is not added to the total time spent on editing the drawing. Also, the last update time is not revised.

Elapsed Timer

This timer operates while you are in AutoCAD. You can stop this timer by entering OFF at the **Enter an option [Display/ON/OFF/Reset]:** prompt. To activate the timer, enter ON. If you want to know how much time you have spent on the current drawing or part of the drawing in the current editing session, use the **Reset** option as soon as you start working on the drawing or part of the drawing. This resets the user-elapsed timer to zero. By default, this timer is ON. If you turn this timer OFF, the time accumulated in this timer up to the time you turned it OFF will be displayed.

Next Automatic Save in Time

This tells you when the next automatic save will be performed. The automatic save time interval can be set in the **Options** dialog box (**Open and Save** tab) or with the **SAVETIME** system variable. If the time interval has been set to zero, the **TIME** command displays the following message:

Next automatic time save in: <disabled>

If the time interval is not set to zero, and no editing has taken place since the previous save, the **TIME** command displays the following message:

Next automatic time save in: <no modification yet>

If the time interval is not set to zero, and editing has taken place since the previous save, the **TIME** command displays the following message:

Next automatic time save in: 0 days hh:mm:ss.msec

hh stands for hours
mm stands for minutes
ss stands for seconds
msec stands for milliseconds

At the end of the display of the **TIME** command, AutoCAD prompts:

Enter an option [Display/ON/OFF/Reset]:

The information displayed by the **TIME** command is static; that is, the information is not updated dynamically on the screen. If you respond to the last prompt with DISPLAY (or D), the display obtained by invoking the **TIME** command is repeated. This display contains updated time values.

With the ON response, the user-elapsed timer is started, if it was off. As mentioned earlier, when you enter the drawing editor, by default the timer is on. The OFF response is just the opposite of the ON response and it stops the user-elapsed time, if it is on. With the **Reset** option, you can set the user-elapsed time to zero.

OBTAINING DRAWING STATUS INFORMATION (STATUS COMMAND)

Menu:	Tools > Inquiry > Status
Command:	STATUS

The **STATUS** command displays information about the prevalent settings of various drawing parameters, such as snap spacing, grid spacing, limits, current space, current layer, current color, and various memory parameters.

Command: **STATUS**

Once you enter this command, AutoCAD displays information similar to the following:

106 objects in Drawing.dwg

Model space limits are	X: 0.0000	Y: 0.0000(On)	
	X: 6.0000	Y: 4.4000	
Model space uses	X:0.6335	Y:-0.2459	**Over
	X:8.0497	Y: 4.9710	**Over
Display shows	X: 0.0000	Y:-0.2459	
	X: 8.0088	Y: 5.5266	

Insertion base is	X: 0.0000	Y: 0.0000	Z: 0.0000
Snap resolution is	X: 0.2500	Y: 0.2500	
Grid spacing is	X: 0.2500	Y: 0.2500	

Current space: Model space
Current layout: Model
Current layer: OBJ
Current color: BYLAYER .... 7 (white)
Current linetype: BYLAYER .... CONTINUOUS
Current lineweight: BYLAYER
Current Plot Style: By layer
Current elevation: 0.0000   thickness: 0.0000
Fill on Grid on Ortho off Qtext off Snap off Tablet on
Object snap modes: None
Free dwg disk space: 2047.7 MBytes
Free temp disk: 2047.7 MBytes
Free physical memory: 15.0 MBytes
Free swap file space: 1987.5 MBytes

All the values (coordinates and distances) on this screen are given in the format declared in the **UNITS** command. You will also notice **Over in the **Model space uses** or **Paper space uses** line. This signifies that the drawing is not confined within the drawing limits. The amount of memory available on the disk is given in the **Free dwg disk space:** line. Information on the name of the current layer, current color, current space, current linetype, current lineweight, current Plot Style, current elevation, snap spacing (snap resolution), grid spacing, various tools that are on or off (such as Ortho, Snap, Fill, Tablet, Qtext), and which object snap modes are active is also provided by the display obtained by invoking the **STATUS** command.

Exercise 2   *General*

Use the **STATUS** command to check the disk space available on the hard drive.

Displaying Drawing Properties* (DWGPROPS Command*)

Menu:	File > Drawing Properties
Command:	DWGPROPS

On choosing **Drawing Properties** from the **File** Menu, the **Drawing Properties** dialog box is displayed (Figure 21-6). This dialog box has four tabs under which information about the drawing is displayed. The information displayed in this dialog box helps you look for the drawing more easily. The tabs are as follows:

General. This tab displays general properties about the drawing like the **Type**, **Size** and **Location**.

Summary. The **Summary** tab displays predefined properties like the Author, title, and subject.

Custom. This tab displays custom file properties including values assigned by you.

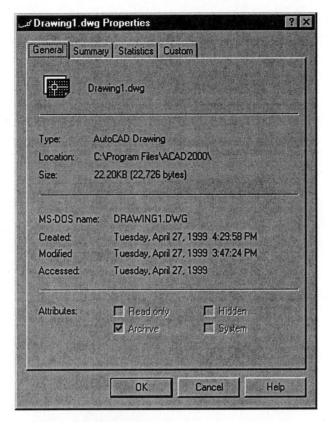

*Figure 21-6 The **Drawing Properties** dialog box*

Statistics. This tab stores and displays data such as the file size and data such as the date when the drawing was last saved on or modified on.

DATA EXCHANGE IN AUTOCAD

Different companies have developed different software for applications such as CAD, desktop publishing, and rendering. This nonstandardization of software has led to the development of various data exchange formats that enable transfer (translation) of data from one data processing software to another. We will now discuss various data exchange formats provided in AutoCAD. AutoCAD uses the .DWG format to store drawing files. This format is not recognized by most other CAD software, such as Intergraph, CADKEY, and MicroStation. To solve this problem so that files created in AutoCAD can be transferred to other CAD software for further use, AutoCAD provides various data exchange formats, such as DXF (data interchange file) and DXB (binary drawing interchange).

DXF FILE FORMAT (DATA INTERCHANGE FILE)

The DXF file format generates a text file in ASCII code from the original drawing. This allows any computer system to manipulate (read/write) data in a DXF file. Usually, DXF format is used for CAD packages based on microcomputers. For example, packages like SmartCAM use DXF files. Some desktop publishing packages, such as Pagemaker and Ventura Publisher, also use DXF files.

Creating a Data Interchange File

The **SAVE** or the **SAVEAS** command is used to create an ASCII file with a .DXF extension from an AutoCAD drawing file. Once you invoke any of these commands, the **Save Drawing As** dialog box (Figure 21-7) is displayed. By default, the DXF file to be created assumes the name of the drawing file from which it will be created. However, you can specify a file name of your choice for the DXF file by typing the desired file name in the **File Name:** edit box. Select the extension as **DXF [*dxf]**. This can be observed in the **Save as type:** drop-down list where you can select the output file format. You can also use the **-WBLOCK** command. The **Create Drawing** dialog box is displayed, where you can enter the name of the file in the **File Name:** edit box and select **DXF [*dxf]** from the **Save as type:** drop-down list.

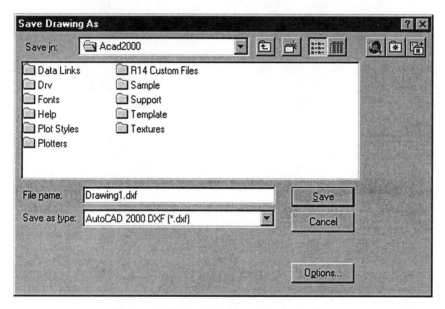

Figure 21-7 Save Drawing As dialog box

Choose the **Options** button to display the **Saveas Options** dialog box (Figure 21-8). In this dialog box, choose the **DXF options** tab and enter the degree of accuracy for the numeric values. The default value for the degree of accuracy is six decimal places, however, this results in less accuracy than the original drawing, which is accurate to 16 places. You can enter a value between 0 and 16 decimal places.

In this dialog box, you can also select the **Select Objects** check box, which allows you to specify objects you want to include in the DXF file. In this case, the definitions of named objects such as block definitions, text styles, and so on, are not exported. Selecting the **Save thumbnail preview image** check box, saves a preview image with the file that can be previewed in the **Preview** window of the **Select File** dialog box.

Select the **ASCII** radio button. Choose **OK** to return to the **Save Drawing As** dialog box. Choose the **Save** button here. Now an ASCII file with a .DXF extension has been created, and this file can be accessed by other CAD systems. This file contains data on the objects specified. By default, DXF files are created in ASCII format. However, you can also create binary format files by selecting the

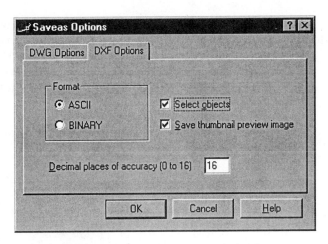

Figure 21-8 Saveas Options dialog box

Binary radio button in the **Saveas Options** dialog box. Binary DXF files are more efficient and occupy only 75 percent of the ASCII DXF file. You can access a file in binary format more quickly than the same file in ASCII format.

Information in a DXF File

The DXF file contains data on the objects specified using the **Select objects** option in the **Save Drawing As** dialog box. You can change the data in this file to your requirement. To examine the data in this file, load the ASCII file in word processing software. A DXF file is composed of the following parts.

Header. In this part of the drawing database, all the variables in the drawing and their values are displayed.

Classes. This section deals with the internal database.

Tables. All the named objects, such as linetypes, layers, blocks, text styles, dimension styles, and views, are listed in this part.

Blocks. The objects that define blocks and their respective values are displayed in this part.

Entities. All the entities in the drawing, like Circle, and so forth, are listed in this part.

Objects. Objects in the drawing are listed in this part.

Converting DXF Files into a Drawing File (OPEN Command)

You can import a DXF file into a new AutoCAD drawing file with the **OPEN** command. From the **File** menu choose **Open**.

Command: **OPEN**

After you invoke the **OPEN** command, the **Select File** dialog box (Figure 21-9) is displayed. From the **Files of Type:** drop-down list, select **DXF [*dxf]**. In the **File Name:** edit box enter the name of the file you want to import into AutoCAD or select the file from the list. Choose the **Open** button. Once this is done, the specified DXF file is converted into a standard DWG file, regeneration is carried out, and the file is inserted into the new drawing. Now you can perform different operations on this file just as with other drawing files. You can also use the **INSERT** command to insert a DXF file into the current drawing.

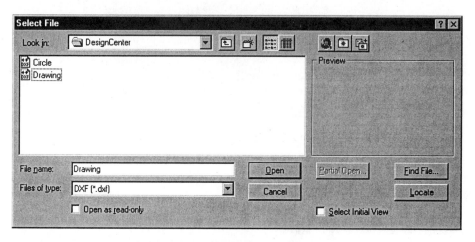

Figure 21-9 Select File dialog box

Importing Scanned Files into the Drawing Editor DXB File Format

Menu:	Insert > Drawing Exchange Binary
Command:	DXBIN

AutoCAD offers another file format for data interchange: DXB. This format is much more compressed than the binary DXF format and is used when you want to translate large amounts of data from one CAD system to another. For example, when a drawing is scanned, a DXB file is created. This file has huge amounts of data in it. The **DXBIN** command is used to create drawing files out of DXB format files.

Command: **DXBIN**

After you invoke the **DXBIN** command, the **Select DXB File** dialog box is displayed. In the **File Name:** edit box, enter the name of the file (in DXB format) you want to import into AutoCAD. Choose the **Open** button. Once this is done, the specified DXB file is converted into a standard DWG file and is inserted into the current drawing.

Note

Before importing a DXB file, you must create a new drawing file. No editing or drawing setup (limits, units, and so on) can be performed in this file. This is because if you import a DXB file into an old drawing, the settings (definitions) of layers, blocks, and so on, of the file being imported are overruled by the settings of the file into which you are importing the DXB file.

Creating and Using an ACIS file

Menu:	Tools > Export...
Command:	EXPORT, ACISOUT

Trimmed Nurbs surfaces, regions, and solids can be exported to an ACIS file with an ASCII format. Choose **Export** from the **File** menu or enter **EXPORT** at the Command prompt. The **Export Data** dialog box is displayed, enter a file name in the **File name:** edit box. From the **Save as type:** drop-down list select **ACIS [*.sat]** and then choose the **Save** button. At the **Select Objects:** prompt, select the objects you wish to save as an ACIS file and press ENTER. AutoCAD appends the file extension .sat to the file name. You can also use the **ACISOUT** command. It prompts you to select objects before the **Create ACIS File** dialog box appears on the screen.

The command **ACISIN** displays the **Select ACIS File** dialog box. Select a file to import and choose the **Open** button. This command can also be invoked by choosing **ACIS Solid** from the **Insert** menu.

Creating and Using a 3D Studio File

Menu:	Tools > Export...
Command:	EXPORT, 3DSOUT

3D geometry, views, lights and objects with surface characteristics can be saved in the 3D Studio format. In the **Export Data** dialog box, enter the file name in the **File Name:** edit box and select **3D Studio [*.3ds]** from the **Save as type:** drop-down list. Choose **Save**. At the **Select objects:** prompt, select the objects you want to save in the 3D Studio format. Press ENTER; the file extension .3ds is automatically added to the file name. You can also use the **3DSOUT** command. AutoCAD prompts you to select the objects first. When you press ENTER, the **Create 3D Studio Output File** dialog box appears. Enter the file name in the edit box and choose the **Save** button.

The **3DSIN** command displays the **3D Studio File Import** dialog box. Select the file you wish to import and choose the **Open** button. You can also invoke this command by choosing **3DStudio** from the **Insert** menu.

Creating and Using a Windows WMF File

Menu:	File > Export
Command:	EXPORT, WMFOUT

The Windows Metafile File format file contains screen vector and raster graphics format. In the **Export Data** dialog box or the **Create WMF file** dialog box, enter the file name. Select the objects you want to save in this file format. Select **Metafile [*wmf]** from the **Save as type:** drop-down list. The extension .wmf is appended to the file name.

The **WMFIN** command displays the **Import WMF** dialog box. Window metafiles are imported as blocks in AutoCAD. Select the wmf file you want to import and choose the **Open** button. Specify an insertion point, rotation angle and scale factor. Specify scaling by entering a **scale factor,** using the **corner** option to specify an imaginary box whose dimensions correspond to the scale factor or entering **xyz** to specify 3D scale factors. You can also invoke the **WMFIN** command by choosing

Windows Metafile from the **Insert** menu.

Creating a BMP File

Menu:	File > Export
Command:	EXPORT, BMPOUT

This is used to create bitmap images of the objects in your drawing. In the **Export Data** dialog box, enter the name of the file, select **Bitmap [*bmp]** from the **Save as type:** drop-down list and then choose **Save**. Select the objects you wish to save as bitmap and press ENTER. Entering **BMPOUT** displays the **Create BMP File** dialog box. Enter the file name and choose **Save**. Select the objects to save as bitmap. The file extension .bmp is appended to the file name.

DATA INTERCHANGE THROUGH RASTER FILES

Until now we have discussed importing and exporting files in the DXF file format. To uphold the accuracy of the drawing, the DXF file includes almost all the information about the original drawing file. The accuracy is maintained at the expense of DXF file size and degree of complexity of these files. There are many applications in which accuracy is not very important, like desktop publishing. In such applications, you are concerned primarily with image presentation. A very simple and effective way of storing an image for import/export is in the form of **raster files**. In a raster file, information is stored in the form of a dot pattern on the screen. This bit pattern is also known as a **bit map**. For example, in a raster file a picture is stored in the form of information about the position and color of the screen pixels. AutoCAD allows you to add raster images to the vector base AutoCAD drawings, and view and plot the resulting file. We will be discussing three types of raster files: TIFF files, TGA files, and BMP files. These formats make it possible to transfer a file from AutoCAD to other software.

TIFF (tagged image file format)

TGA (targa format)

BMP (bitmap format)

Exporting the Raster Files (SAVEIMG Command)

Menu:	Tools > Display Image > Save
Command:	SAVEIMG

If there is a picture on the screen that you want to save as a raster file, use the **SAVEIMG** command. With this command, you can save a rendered image to TIFF, TGA, or BMP types of raster files. These raster files can be used by most software. Sometimes, the raster files need to be converted into some other format before they can be used by certain software. This operation is performed with a file conversion program such as Hijaak or Pizazz Plus.

Command: **SAVEIMG**

Once you invoke this command, the **Save Image** dialog box is displayed (Figures 21-10). Depending on the **Destination** option you have selected in the **Rendering Preferences** dialog box, there is a difference in the **Portion** area.

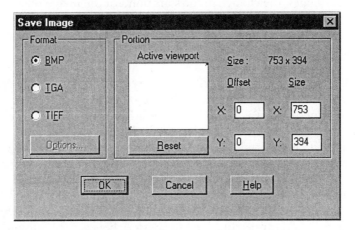

Figure 21-10 Save Image dialog box

Format. This area of the **Save Image** dialog box provides three raster formats: TGA, TIFF, and BMP. You can select any one of the formats by selecting the radio button next to the desired format. A brief description of each format follows.

> **TGA**. (default format). With this format, you can create compressed or uncompressed 32-bit RGBA Truevision v2.0 format files. These files have the extension **.tga**.

> **TIFF**. (tagged image file format). This file format has been developed by Aldus and Microsoft. With this format, you can create compressed or uncompressed 32-bit RGBA tagged image file format. These files have the extension **.tif**.

> **BMP**. (bitmap format). These files have the extension **.bmp**.

Portion. In the **Portion** area of the **Save Image** dialog box, you can specify the portion of the screen you want to save in the specified file format and the portion of the image you wish to render. In case AutoCAD is configured for rendering to a viewport (in the **Render Preferences** dialog box), the **Portion** area provides options to save either the active viewport, the drawing area, or the full screen. The **Portion** area contains an image tile that graphically represents the section of the screen to be saved. You will notice that by default the entire screen is selected for saving. If you want to save only a section of the screen, you need to specify the desired section. This can be achieved by specifying two diagonally opposite points on the image tile. Specify the lower left corner of the desired section first, then specify the upper right corner. After specifying the two corner points, AutoCAD creates a box whose lower left corner point and upper right corner point coincide with the respective points you specified. Also, the values for the X and Y Offsets and Sizes are automatically updated.

Offset. This includes **X** and **Y** edit boxes. The values you enter in these edit boxes are treated as the X and Y offsets. The start point (lower left corner) of the area to be saved as a raster file can be specified by X and Y offset values. The default value for the offset is (0,0). The values entered are taken as the pixel values. The offset values should be specified within the screen size value. When you change the offset values, the change in the start point of the image selection area is reflected by a margin formed by two intersecting lines in the image tile. The point of intersection of these two lines is the start point of the image selection area. Values of X and Y offsets greater than the screen size are not accepted by AutoCAD.

Size. With the **Offset** option, you can specify the lower left corner of the area to be selected for conversion into a raster file format. To specify the area completely, however, you need to specify the upper right corner of this area as well. This can be achieved with the **Size** option. In the case of the **Offset** option, the values entered in the **X** and **Y** edit boxes are the pixel values. The default for the **Size** option is the upper right corner of the display area. The change in location of the upper right corner of the image selection area is reflected by two intersecting lines.

Options. The display provided by choosing the **Options** button varies from one file format to another. The image compression options are displayed only for TGA and TIFF formats. For the BMP format, the image file compression is not supported, hence, the **Options** button is disabled.

    None. This is the default option. If this option is selected, compression does not take place.

    PACK. With this option, run-length encoded image compression can be realized only for TIFF files (Figure 21-11). It uses Macintosh packbits.

    RLE. This option works only for TGA format files (Figure 21-12). If you select this option, run-length encoded image compression can be realized.

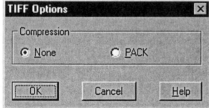 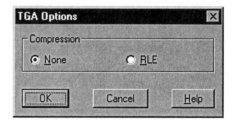

Figure 21-11 TIFF Options dialog box   *Figure 21-12 TGA Options dialog box*

Reset. When you choose this button in the **Save Image** dialog box, the Offset values and the Size values are set to their default values.

Example 1

In this example, you will load a drawing or draw a figure and then use the **SAVEIMG** command to save the image.

Creating an image using the **SAVEIMG** command involves the following steps. Assume that the drawing is already loaded or drawn on the screen.

1.   Invoke the **SAVEIMG** command by any one of the methods explained to display the **Save**

Image dialog box.

2. Specify the file type in the **Format** area by selecting the radio button with the desired file type format.

3. Specify the area you want to save in the image file to be created by selecting the desired offset and size.

4. Choose the **OK** button.

5. The **Image File** dialog box is displayed. Enter the name you want to give the image file.

After this, AutoCAD starts saving the selected area to the file whose name you have specified. The saving operation is acknowledged by the following message: Writing file.

RASTER IMAGES

A raster image consists of small square-shaped dots known as pixels. In a colored image, the color is determined by the color of pixels. The raster images can be moved, copied, or clipped, and used as a cutting edge with the **TRIM** command. They can also be modified by using grips. You can also control the image contrast, transparency, and quality of the image. AutoCAD stores images in a special temporary image swap file whose default location is the Windows Temp directory. You can change the location of this file modifying it under **Temporary File Location** in the **Files** tab of the **Options** dialog box.

The images can be 8-bit gray, 8-bit color, 24-bit color, or bitonal. When image transparency is set to On, the image file formats with transparent pixels is recognized by AutoCAD and transparency is allowed. The transparent images can be in color or gray scale. AutoCAD supports the following file formats.

Image Type	File Extension	Description
BMP	.bmp, .dib, .rle	Windows and OS/2 Bitmap Format
CALS-I	.gp4, .mil, .rst	Mil-R-Raster I
FLIC	.flc, .fli	Flic Autodesk Animator Animation
GEOSPOT	.bil	GeoSPOT (BIL files must be accompanied with HDR and PAL files with connection data in the same directory.)
IG4	.ig4	Image Systems group 4
IGS	.igs	Image Systems Grayscale
JFIF or JPEG	.jpg, .jpeg	Joint Photographics Expert group
PCX	.pcx	Picture PC Paintbrush Picture
PICT	.pct	Picture Macintosh Picture
PNG	.png	Portable Network Graphic
RLC	.rlc	Run-length Compressed
TARGA	.tga	True Vision Raster based Data format
TIFF/LZW	.tif	Taffed image file format

When you store images as Tiled Images, that is, in the Tagged Image File Format [TIFF], you can edit or modify any portion of the image; only the modified portion is regenerated thus saving time. Tiled images load much faster compared to nontiled images.

Attaching Raster Images

Toolbar:	Reference > Image Attach
Menu:	Insert > Raster Image...
Command:	IMAGEATTACH

*Figure 21-13 Invoking the **IMAGEATTACH** command from the **Reference** toolbar*

Attaching Raster images do not make them part of the drawing. You can also drag and drop images from other files into the current drawing using the **AutoCAD DesignCenter** window. When you invoke the **IMAGEATTACH** command, AutoCAD displays the **Select Image File** dialog box (Figure 21-14). Enter the name of the image file you want to attach to the current drawing in the **File name:** edit box or select the image from the list box. The name can be up to 255 characters long and can include spaces, numbers and special characters that are not being used by AutoCAD or Windows. A preview image of the selected image is displayed in the **Preview** window of the dialog box. If you choose the **Hide Preview** button, the preview image of the selected image is not displayed in the **Preview** window. Choose the **Open** button.

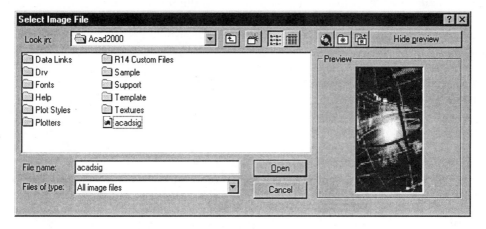

*Figure 21-14 **Select Image File** dialog box*

AutoCAD displays the **Image** dialog box (Figure 21-15). The name of the selected image is displayed in the **Name:** edit box. You can select another file by using the **Browse** button. The **Name** drop-down list displays names of all the images in the current drawing. The **Path** is also displayed just below the **Name:** edit box. Select the **Specify On-Screen** check boxes to specify the **Insertion point**, **Rotation Angle**, and **Scale** on screen. Alternately you can clear these check boxes and enter values in the respective edit boxes. Choosing the **Details** button gives Image information like the Horizontal and Vertical resolution values, current AutoCAD unit, and Image size in pixels and units. Choose **OK** to return to the drawing screen.

You can attach and scale an image from the Internet. In the **Select Image File** dialog box, choose the **Search the Web** button. Once you have located the image file you wish to attach to the current drawing, enter the URL address in the **Look In:** edit box and the Image file name in the **File name:** edit box. Choose the **Open** button. Specify the **Insertion point**, **scale**, and **Rotation** angle in the **Image** dialog box. You can also right-click the image you wish to attach to the current drawing and choose **Properties** from the shortcut menu to display the **Object Properties** window.

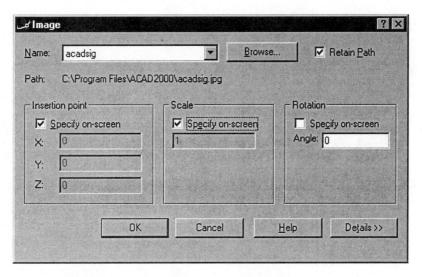

Figure 21-15 Image dialog box

From here you can select , cut, and paste the URL address to the **File name:** edit box in the **Select File** dialog box.

Managing Raster Images* (Image Manager Dialog Box*)

Toolbar:	Reference > Image
	Insert > Image
Menu:	Insert > Image Manager...
Command:	IMAGE

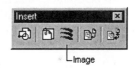

When you invoke the **IMAGE** command, AutoCAD displays the **Image Manager** dialog box (Figure 21-17). You can also invoke the **Image Man-** ager dialog box by selecting an image and right-clicking to display a shortcut menu. Choose

Figure 21-16 Invoking the IMAGE command from the Insert toolbar

Image > Image Manager. You can view image information either as a list or as a tree view by choosing the respective buttons located in the upper-left corner of the **Image Manager** dialog box. The **List View** displays the names of all the images in the drawing, its loading status, size, date last modified on, and its search path. The **Tree View** displays the images in a hierarchy that show its nesting levels within blocks and Xrefs. The **Tree View** does not display the Status, size, or any other information about the image file. You can rename an image file in this dialog box. The different options in the **Image Manager** dialog box are:

Attach. If you want to attach an image file, choose the **Attach** button; the **Select Image File** dialog box is displayed. Select the file that you want to attach to the drawing; the selected image is displayed in the **Preview** box. Choose the **Open** button. The **Image** dialog box is displayed. The name of the file and its path and extension are displayed in the **Image** dialog box. You can use the **Browse** button to select another file. Choose the **OK** button in the **Image** dialog box and specify a point where you want to insert the image. Next, AutoCAD will prompt you to enter the scale factor. You can enter the scale factor or specify a point on the screen.

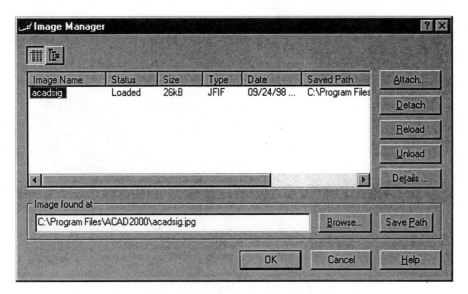

Figure 21-17 Image Manager dialog box

Detach. Detaches the selected and all associated raster images from the current drawing. Once the image is detached, no information about the image is retained in the drawing.

Reload. Reloads an image. The changes made to the image since the last insert will be loaded on the screen. You can change the status of the image file in the **Image Manager** dialog box by double-clicking on the current status. It changes from **Unload** to **Reload** and vice versa.

Unload. Unloads an image. An image is unloaded when it is not needed in the current drawing session. When you unload an image, AutoCAD retains information about the location and size of the image and the image boundary is displayed. If you reload the image, the image will appear at the same point and in the same size as the image was before unloading. Unloading the raster images enhances AutoCAD performance. Also, the unloaded images are not plotted, but unloading does not unlink the file from the drawing. If multiple images are to be loaded and the memory is insufficient, AutoCAD automatically unloads them.

Details. Displays the **Image File Details** box, which lists information about the image like file name, saved path, file creation date, file size, file type, color and color path, pixel width and height, resolution, and default size. It also displays the image in the preview box.

Image found at. This edit box displays the path of the selected image. You can edit this path and choose the **Save path** button to save the new path. If you have changed the path of an image, choose the **Browse** button to display the **Select Image File** dialog box. In this dialog box, locate the image file and then choose the **Open** button. The new path is displayed in the **Image found at** edit box. Choose the **Save Path** button to save this new path. This path is also displayed in the **Saved path** column of the dialog box.

Save Path. Saves the current path of the image.

Browse. Displays the **Select Image File** dialog box.

Using the -IMAGE Command

The options in the **Image Manager** dialog box can also be used from the command line when you use the **-IMAGE** command. The Command prompts are as follows:

Command: **-IMAGE**
Enter image option [?/Detach/Path/Reload/Unload/Attach] <Attach>: *Enter an option or press ENTER to attach an image.*

Entering ? at the above prompt gives the next prompt as follows:
　　Images to list<*>: *Press ENTER.*

The AutoCAD Text window lists information about the image.

EDITING RASTER IMAGE FILES
Clipping Raster Images

Toolbar:	Reference > Image Clip
Menu:	Modify > Clip > Image
Command:	IMAGECLIP

You can clip a raster image by invoking the **IMAGECLIP** command. You can select an image to clip and right-click to display a shortcut menu. Choose **Image > Clip** to invoke the **IMAGECLIP** command. When you invoke the command, AutoCAD will prompt you to select the image to clip. Select the raster image by selecting the boundary edge of the image. The image boundary must be visible to select the image. You can turn off or on the clipping boundary. To improve performance you can also make sure that on selecting an image, only the boundary is highlighted. You can turn On or Off the image highlighting that is visible when we select an image by selecting or clearing the **Highlight Raster Image Frame Only** check box in the **Display** tab of the **Options** dialog box. You can also use the **IMAGEHLT** system variable. By default, **IMAGEHLT** is set to 0, that is, only the raster image frame will be highlighted. At the **Enter image clipping option [ON/OFF/Delete/New boundary] <New Boundary>:** prompt, select the **New boundary** option and then define the clip boundary either by defining a rectangular or polygonal boundary. The clipping boundary must be specified in the same plane as the image or a plane that is parallel to it (see Figure 7-18). The following is the command prompt sequence for the **IMAGECLIP** command:

Command: **IMAGECLIP**
Select image to clip: *Select the image at the edge.*
Enter image clipping option [ON/OFF/Delete/New boundary] <New boundary>: *Press ENTER to select New boundary.*
Enter clipping type [Polygonal/Rectangular] <Rectangular>: *Type P for Polygonal and then define the polygon shape.*

ON/OF. The **ON/OFF** option allows you to turn the clipping boundary on or off. When the clipping boundary is off, you can see the complete image; when it is on, the image that is within the

Figure 21-18 *Object before and after clipping*

clipping polygon is displayed. In the **Object properties** window, under the **Misc** area , you can select options from the **Show Clipped** drop-down list. Selecting **Yes**, shows the clipped image and selecting **No**, displays the complete image (the clipping boundary is off).

Delete. The **Delete** option deletes the clipping boundary.

Adjusting Raster Image

Toolbar:	Reference > Image Adjust
Menu:	Modify > Object > Image > Adjust
Command:	IMAGEADJUST

The **IMAGEADJUST** command allows you to adjust the brightness, contrast, and fade of the raster image. When you choose an image and right-click, a shortcut menu is displayed. Choose **Image > Adjust** to invoke the **IMAGEADJUST** command. When you invoke this command, AutoCAD prompts you to select image(s). After selecting the image(s), AutoCAD displays the **Image Adjust** dialog box (Figure 21-19) that you can use to adjust the image. As you

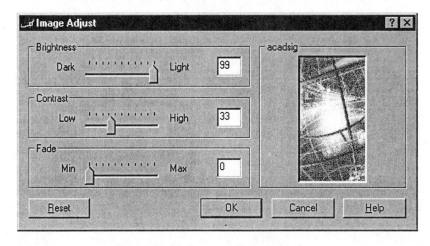

Figure 21-19 Image Adjust *dialog box*

adjust the brightness, contrast, or fade, the image box displays the effect on the image. If you choose the **Reset** button, the values are returned to default values (Brightness= 50, Contrast= 50, and Fade= 0). You can also choose **Properties** in the shortcut menu to display the **Object Properties** window, where you can enter new values of **Brightness**, **Contrast**, and **Fade**.

Image Quality

Toolbar:	Reference > Image Quality
Menu:	Modify > Object > Image > Quality
Command:	IMAGEQUALITY

The **IMAGEQUALITY** command allows you to control the quality of the image that affects the display performance. A high quality image takes longer time to display. When you change the quality, the display changes immediately without causing a **REGEN**. The images are always plotted using high quality display. Although draft quality images appear more grainy, they are displayed more quickly. The command prompts are as follows:

Command: **IMAGEQUALITY**
Enter image quality setting [High/Draft] <High>: *Enter h or d.*

Transparency

Toolbar:	Reference > Image Transparency
Menu:	Modify > Object > Image >Transparency
Command:	TRANSPARENCY

The **TRANSPARENCY** command is used with bitonal images to turn the transparency of the image background on or off. A bitonal image is one that consists of only a foreground and a background color. When you attach a bitonal image, the image assumes the color of the current layer. Also, the bitonal image and the bitonal image boundaries are always in the same color. You can also right-click an image and choose **Properties** from the shortcut menu that is displayed. In the **Object Properties** window under the **Misc** list, select an option from the **Transparency** drop-down list. The command prompts for the **TRANSPARENCY** command are:

Command: **TRANSPARENCY**
Select image(s): *Select the image.*
Enter transparency mode [ON/OFF] <OFF>:

Image Frame

Toolbar:	Reference > Image Frame
Menu:	Modify > Object > Image > Frame
Command:	IMAGEFRAME

The **IMAGEFRAME** command is used to turn the image boundary on or off. If the image boundary is off, the image cannot be selected with the pointing device and hence, cannot be accidently moved or modified. The command prompt is as follows:

Command: **IMAGEFRAME**

Enter image frame setting [ON/OFF] <OFF>:

Other Editing Commands

You can use other editing commands like **COPY**, **MOVE**, and **STRETCH** to edit the raster image. You can also use the image as the trimming edge for trimming objects. However, you cannot trim an image. You can insert the raster image several times or make multiple copies of it. Each copy could have a different clipping boundary. You can also edit the image using grips. You can also use the **Object Properties** window to change the image layer, boundary linetype, and color, and perform other editing commands like changing the scale, rotation, width, and height by entering new values in the respective edit boxes. You can also display an image or turn off the display by selecting options from the **Show Image** drop-down list in the **Misc** field of the **Object Properties** window. If you select **Yes**, the image is displayed and if you select **No**, the display of the image is turned off. You can turn off the display of the image when you do not need it thus improving the performance.

Scaling Raster Images

The scale of the inserted image is determined by the actual size of the image and the unit of measurement (inches, feet, and so on). For example, if the image is 1" x 1.26" and you insert this image with a scale factor of 1, the size of the image on the screen will be 1 AutoCAD unit by 1.26 AutoCAD units. If the scale factor is 5, the image will be five times larger. The image that you want to insert must contain the resolution information (DPI). If the image does not contain this information, AutoCAD treats the width of the image as one unit.

POSTSCRIPT FILES

PostScript is a page description language developed by Adobe Systems. It is used mostly in DTP (desktop publishing) applications. AutoCAD allows you to work with PostScript files. You can create and export PostScript files as well as convert PostScript files into regular AutoCAD drawing files (import). PostScript images have higher resolution than raster images. The extension for these files is .EPS (Encapsulated PostScript).

PSOUT Command

Menu:	File > Export...
Command:	PSOUT, EXPORT

As just mentioned, any AutoCAD drawing file can be converted into a PostScript file. This can be accomplished with the **PSOUT** command. Once the **PSOUT** command is invoked, AutoCAD displays the **Create PostScript File** dialog box (Figure 21-20). You can also use the **EXPORT** command to display the **Export Data** dialog box. The options are similar in both the dialog boxes.

In the **File Name:** edit box, enter the name of the PostScript (EPS) file you want to create. When you use the **EXPORT** command, in the **Export Data** dialog box, select **Encapsulated PS (*.eps)** from the **Save as type:** drop-down list. Then you can choose the **Save** button to accept the default setting and create the PostScript file. You can also choose the **Options** button to change the settings through the **PostScript Out Options** dialog box (Figure 21-21) and then save the file. The **PostScript Out Options** dialog box has the following options:

Figure 21-20 Create PostScript File dialog box

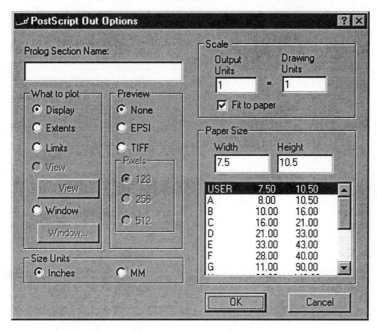

Figure 21-21 PostScript Out Options dialog box

Prolog Section Name

In this edit box, you can assign a name for a prolog section to be read from the acad.psf file.

What to plot

The **What to plot** area of the dialog box has the following options:

Display. If you specify this option when you are in model space, the image in the current viewport is saved in the specified EPS file. Similarly, if you are in paper space, the current view is saved in the specified EPS file.

Extents. If you use this option, the PostScript file created will contain the section of the AutoCAD drawing that currently holds objects. In this way, this option resembles the **ZOOM Extents** option. If you add objects to the drawing, they are also included in the PostScript file to be created because the extents of the drawing are also altered. If you reduce the drawing extents by erasing, moving, or scaling objects, then you must use the **ZOOM Extents** or **ZOOM All** option. Only then does the **Extents** option of the **PSOUT** command understand the extents of the drawing to be exported. If you are in model space, the PostScript file is created in relation to the model space extents; if you are in paper space, the PostScript file is created in relation to the paper space extents. If you invoke the **PSOUT** command **Extents** option when perspective view is on and the position of camera is not out of the drawing extents, the following message is displayed:

　　　PLOT Extents incalculable, using display

In such cases, the EPS file is created as it would be created with the **Display** option.

Limits. With this option, you can export the whole area specified by the drawing limits. If the current view is not the plan view [viewpoint (0,0,1)], the **Limits** option exports the area just as the **Extents** option would.

View. Any view created with the **VIEW** command can be exported with this option. When this radio button is activated, the **View** button is available. Choose the **View** button to display the **View Name** dialog box from where you can select the view.

Window. In this option, you need to specify the area to be exported with the help of a window. When this radio button is selected, the **Window** button is also available. Choose the **Window** button to display the **Window Selection** dialog box where you can select the **Pick** button and then specify the two corners of the window on the screen. You can also enter the coordinates of the two corners in the **Window Selection** dialog box.

Preview

The **Preview** area of the dialog box has two types of formats for preview images: **EPSI** and **TIFF**. If you want a preview image with no format, select the **None** radio button. If you select **TIFF** or **EPSI**, you are required to enter the pixel resolution of the screen preview in the **Pixels** area. You can select a preview image size of 128 x 128, 256 x 256, or 512 x 512.

Size Units

In this area, you can set the paper size units to **Inches** or **Millimeters** by selecting their corresponding radio buttons.

Scale

In this area, you can set an explicit scale by specifying how many drawing units are to be output per unit. You can select the **Fit to paper** check box so that the view to be exported is made as large as possible for the specified paper size.

Paper Size

You can select a size from the list or enter a new size in the **Width** and **Height** edit boxes to specify a paper size for the exported PostScript image.

PSIN Command

Menu:	Insert > Encapsulated PostScript
Command:	PSIN

AutoCAD allows you to import PostScript files. The **PSIN** (PostScript IN) command can be used to import a PostScript file. The PostScript file is imported into an AutoCAD drawing as a block.

Command: **PSIN**

The **Select PostScript File** dialog box (Figure 21-22) is displayed.

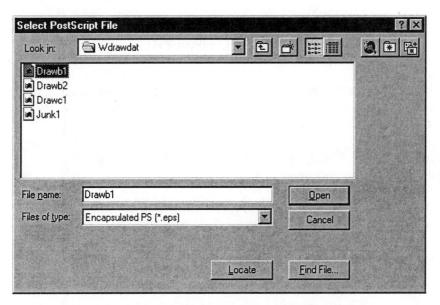

***Figure 21-22 Select PostScript File** dialog box*

In this dialog box, all the EPS files in the current directory are listed. Select the file you want to import into AutoCAD, and then choose the **Open** button. The dialog box is cleared from the screen, and a rectangular box is displayed. You can move this rectangular box anywhere on the screen. The next prompts ask for the insertion point and the scale factor for the rectangular box (imported image), respectively:

Specify insertion point <0,0,0>: *Specify the location for the insertion point.*
Specify scale factor: *Specify the scale factor.*

The scale factor can also be specified by dynamically dragging the box. After you specify the scale factor, whatever is in the specified EPS file is inserted into the AutoCAD screen. From the prompts associated with **PSIN** command, it is obvious that the image is inserted into the AutoCAD drawing file as a block. By default, a box appears on the screen when you import an EPS file. This box contains the imported image, which you are unable to see until you specify the insertion point and the scale factor. If you want to see the image immediately instead of the box, set the **PSDRAG** variable to 1. By default, the value of this variable is 0; that is why, by default, you get the box. The quality of the rendering of the imported PostScript file depends on the **PSQUALITY** variable.

Command: **PSQUALITY**
New value for PSQUALITY <current>:

Different settings for **PSQUALITY** and their effects are as follows:

PSQUALITY = 0. When you insert an EPS file, the rectangular box representing the imported image is displayed. The file name of the imported EPS file is displayed in this rectangular box, and the size of this box corresponds to the size of the imported file.

PSQUALITY = A Positive Value. In this case, the imported image is displayed with the **PSQUALITY** value number of pixels per AutoCAD drawing unit. In other words, this is the resolution of the imported image. By default, the value of **PSQUALITY** is 75; hence, the imported image is displayed with 75 pixels per AutoCAD drawing unit. PostScript objects are filled.

PSQUALITY = A Negative Value. In such situations, the absolute value of the **PSQUALITY** variable controls the resolution of the imported image. The difference lies in the fact that the PostScript objects are not filled, but are displayed as outlines.

PostScript Fill Patterns (PSFILL Command)

Command: PSFILL

With the **PSFILL** command, you can fill closed 2D polylines with PostScript fill patterns. These fill patterns are declared in the acad.psf file. The declarations for a PostScript fill pattern are various parameters and arguments needed to influence the definition of a fill pattern on the screen. Some of the PostScript fill patterns are:

GRAYSCALE	LINEARGRAY	RADIALGRAY	SQUARE
WAFFLE	BRICK	ZIGZAG	STARS
AILOGO	SPECKS	RGBCOLOR	

You can also define your own fill patterns in the acad.psf file with the PostScript procedures. When you assign a fill pattern to a closed polyline, the pattern is not displayed on the screen. However, the fill pattern is recognized by the **PSOUT** command. If you import the PostScript file with a fill pattern using the **PSIN** command, the fill pattern is displayed on the screen as determined by the setting of **PSQUALITY, PSDRAG**, and **FILL** mode. Also, if you print such a file on a PostScript device, the fill pattern is printed. Depending on the fill pattern you specify, AutoCAD issues prompts concerning various parameters and arguments needed by the fill pattern. For example, if you want to use the STARS fill pattern, the prompt sequence is:

Command: **PSFILL**
Select polyline: *Select the polyline to be filled.*

Enter postScript fill pattern name (. = none) or [?]: STARS
Enter scale <1.0000>: *Specify the scale factor for the fill pattern.*
Enter lineWidth <1>: *Specify the line width for the fill pattern.*
Enter ForegroundGray <100>: *Press ENTER.*
Enter BackgroundGray <0>: *Press ENTER.*

To view a list of patterns, enter ? at the **Enter PostScript fill pattern name (.=none) or [?]:** prompt.
If you precede the pattern name with *, the polyline outline does not appear.

Example 2

Draw a rectangle, and use the **PSFILL** command to assign the waffle pattern to the rectangle [Figure 21-23(a)]. Convert the rectangle
to a PostScript file by using the **PSOUT** command, and then import the PostScript file using the **PSIN** command [Figure 21-23(b)].

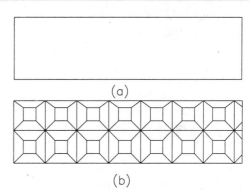

1. Set **PLINEWID** to 0 and use the
   RECTANG command to draw a rectangle
   [Figure 21-23(a)].

2. Use the **PSFILL** command to assign the
   waffle pattern to the rectangle.

Figure 21-23 Assigning the waffle pattern to a polyline

Command: **PSFILL**
Select polyline: *Select the rectangle.*
Enter postScript fill pattern name (. = none) or [?] <current>: Waffle
Enter scale <1.0000>: [Enter]
Enter proportion <30>: [Enter]
Enter lineWidth <1>: [Enter]
Enter UpLeftGray <100>: [Enter]
Enter BotRightGray <0>: [Enter]
Enter TopGray <0>: [Enter]

3. Use the **PSOUT** command to convert the drawing into a PostScript file. Once the **PSOUT**
   command is invoked, AutoCAD displays the **Create PostScript File** dialog box. In the **File
   Name:** edit box, enter the name of the PostScript (EPS) file you want to create (Waffle). Then
   choose the **Options** button. AutoCAD displays **PostScript Out Options** dialog box.

4. In the **Preview** area of the **PostScript Out Options** dialog box, choose the **EPSI** radio button;
   then, choose the **128** radio button in the **Pixels** area.

5. In the **Scale** area, select the **Fit to paper** check box.

6. In the **Paper Size** area, enter 7.5 in the **Width** edit box and 10.2 in the **Height** edit box.

7. Select the **Window** radio button in the **What to plot** area.

8. Choose **OK** in the **PostScript Out Options** dialog box, and then choose **Save** in the **Create PostScript File** dialog box.

9. In the command line area, the following prompts are displayed to select the window.

    Specify window for export
    Specify first corner: *Specify a point close to the lower left corner of the rectangle.*
    Specify second corner: *Specify a point close to the upper right corner of the rectangle.*

10. Set **PSQUALITY** to -75 and **PSDRAG** to 0. Use the **PSIN** command to import and insert the PostScript file.

    Command: **PSIN**   *(Select the file Waffle from the dialog box.)*
    Specify insertion point <0,0,0>: *Specify the location for the insertion point.*
    Specify scale factor: *Specify the scale factor.*

OBJECT LINKING AND EMBEDDING

With Windows, it is possible to work with different Windows-based applications by transferring information between them. You can edit and modify the information in the original Windows application, and then update this information in other applications. This is made possible by creating links between the different applications and then updating those links, which in turn updates or modifies the information in the corresponding applications. This linking is a function of the OLE feature of Microsoft Windows. The OLE feature can also join together separate pieces of information from different applications into a single document. AutoCAD and other Windows-based applications, such as Microsoft Word, Notepad, and Windows WordPad support the Windows OLE feature.

For the OLE feature, you should have a source document where the actual object is created in the form of a drawing or a document. This document is created in an application called a **server** application. AutoCAD for Windows and Paintbrush can be used as server applications. Now this source document is to be linked to (or embedded in) the **compound** (destination) document, which is created in a different application, known as the **container** application. AutoCAD for Windows, Microsoft Word, and Windows WordPad can be used as container applications.

Clipboard

The transfer of a drawing from one Windows application to another is performed by copying the drawing or the document from the server application to the Clipboard. The drawing or document is then pasted in the container application from the Clipboard; hence, a Clipboard is used as a medium for storing the documents while transferring them from one Windows application to another. The drawing or the document on the Clipboard stays there until you copy a new drawing, which overwrites the previous one, or until you exit Windows. You can save the information present on the Clipboard with the .CLP extension.

Object Embedding

You can use the embedding function of the OLE feature when you want to ensure that there is no effect on the source document even if the destination document has been changed through the server application. Once a document is embedded, it has no connection with the source. Although editing is always done in the server application, the source document remains unchanged. Embedding can be accomplished by means of the following steps. In this example, AutoCAD for Windows is the server application and Windows WordPad is the container application.

1. Create a drawing in the server application (AutoCAD).

2. Open Windows WordPad (container application) from the Accessories group in the Program.

3. It is preferable to arrange both the container and the server windows so that both are visible (Figure 21-24).

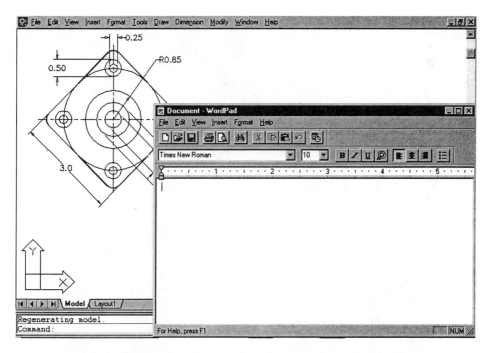

Figure 21-24 AutoCAD graphics screen with the WordPad window

4. In the AutoCAD graphics screen, use the **COPYCLIP** command. This command can be used in AutoCAD for embedding the drawings. This command can be invoked from the **Standard** toolbar by choosing the **Copy to Clipboard** button, from the **Edit** menu (Choose **Copy**), or by entering **COPYCLIP** at the command line.

Command: **COPYCLIP**

The next prompt, **Select objects:**, allows you to select the entities you want to transfer. You can either select the full drawing by entering ALL or select some of the entities by selecting them.

You can use any of the selection set options for selecting the objects. With this command the selected objects are automatically copied to the Windows Clipboard.

5. After the objects are copied to the Clipboard, make the WordPad window active. To get the drawing from the Clipboard to the WordPad application (client), select the **Paste** in the WordPad application. Choose **Paste** from the **Edit** menu in Windows WordPad (Figure 21-25). You can also use **Paste Special** from the **Edit** menu, which will display the **Paste Special** dialog box (Figure 21-26). In this dialog box, select the **Paste** radio button (default) for embedding, and then choose **OK**. The drawing is now embedded in the WordPad window.

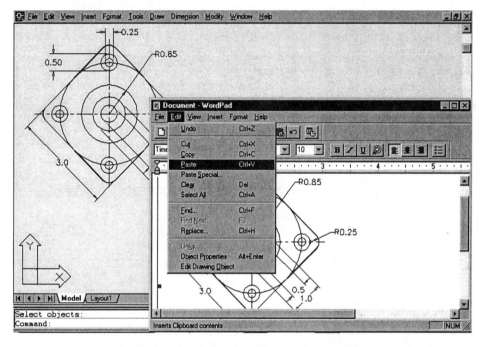

*Figure 21-25 Pasting a drawing to the WordPad application by selecting **Paste** from the **Edit** menu*

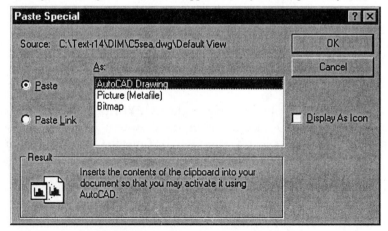

Figure 21-26 Paste Special dialog box

6. Your drawing is now displayed in the Write window, but it may not be displayed at the proper position. You can get the drawing in the current viewport by moving the scroll button up or down in the WordPad window. You can also save your embedded drawing by choosing **Save** from the **File** menu. It displays a **Save As** dialog box where you can enter a file name. You can now exit AutoCAD.

7. You can now edit your embedded drawing. Editing is performed in the server application, which in this case is AutoCAD for Windows. You can get the embedded drawing into the server application (AutoCAD) directly from the container application (WordPad) by double-clicking on the drawing in WordPad. The other method is by choosing **Edit Drawing Object** in the **Edit** menu. (This menu item has replaced **Object**, which was present before pasting the drawing.)

8. Now you are in AutoCAD, with your embedded drawing displayed on the screen, but as a temporary file with a file name, such as [Drawing in Document]. Here you can edit the drawing by changing the color and linetype or by adding and deleting text, entities, and so on. In Figure 21-27 the two upper circles and their dimensions (diameter 25 and 50) have been erased, and a hexagon (diameter 50) has been drawn in its place.

9. After you have finished modifying your drawing, choose **Update WordPad** from the **File** menu in the server (AutoCAD). This menu item has replaced the previous **Save** menu item. When you choose **Update**, AutoCAD automatically updates the drawing in Wordpad (container application). Now you can exit this temporary file in AutoCAD.

10. This completes the embedding function so you can exit the container application. While exiting, a dialog box that asks whether or not to save changes in WordPad is displayed.

Linking Objects

The linking function of OLE is similar to the embedding function. The only difference is that here a link is created between the source document and the destination document. If you edit the source, you can simply update the link, which automatically updates the client. This allows you to place the same document in a number of applications, and if you make a change in the source document, the clients will also change by simply updating the corresponding links. Consider AutoCAD for Windows to be the server application and Windows WordPad to be the container application. Linking can be performed by means of the following.

1. Open a drawing in the server application (AutoCAD). If you have created a new drawing, then you must save the drawing before you can link it with the container application.

2. Open Windows WordPad (the container application) from **Accessories** in the **Program** directory.

3. It is preferable to arrange both the container and the server windows so that both are visible.

4. In the AutoCAD graphics screen, use the **COPYLINK** command. This command can be used in AutoCAD for linking the drawing. This command can be invoked from the **Edit** menu (Choose **Copy Link**) or by entering **COPYLINK** at the Command prompt. The prompt

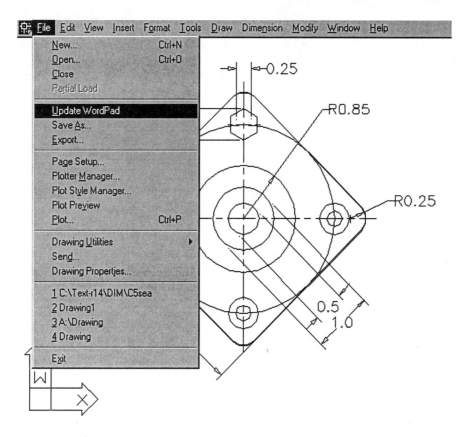

*Figure 21-27 Selecting **Update WordPad** from the **File** menu*

sequence is:

Command: **COPYLINK**

The **COPYLINK** copies the whole drawing in the current viewport directly to the Clipboard. Here you cannot select the objects for linking. If you want only a portion of the drawing to be linked, you can zoom into that view so that it is displayed in the current viewport prior to invoking the **COPYLINK** command. This command also creates a new view of the drawing having a name OLE1. Now you can exit AutoCAD.

5. Make the WordPad window active. To get the drawing from the Clipboard to the Write (container) application, choose **Paste Special** from the **Edit** menu, which will display the **Paste Special** dialog box. In this dialog box, select the **Paste Link** radio button for linking. Choose **OK**. The drawing is now linked to the WordPad window.

6. Your drawing is now displayed in the WordPad window. You can also save your linked drawing by choosing **Save** from the **File** menu. It displays a **Save As** dialog box where you can enter a file name.

7. You can now edit your linked drawing. Editing can be performed in the server application,

which in this case is AutoCAD for Windows. You can get the linked drawing in the server (AutoCAD) directly from the client (WordPad) by double-clicking on the drawing in WordPad. The other method is by choosing **Edit Linked Drawing Object** in the **Edit** menu. (This menu item has replaced **Object**, which was present before pasting the drawing.)

8. Now you are in AutoCAD, with the original drawing displayed on the screen. You can edit the drawing by changing the color and linetype or by adding and deleting text, entities, etc. Then save your drawing in AutoCAD by using the **SAVE** command. You can now exit AutoCAD.

9. You will notice that the drawing is automatically updated, and the changes made in the source drawing are present in the destination drawing also. This automatic updating is dependent on the selection of the **Automatic** radio button (default) in the **Links** dialog box (Figure 21-28). The **Links** dialog box can be invoked by choosing **Links** from the **Edit** menu. For updating manually, you can select the **Manual** radio button in the dialog box. In the manual case, after making changes in the source document and saving it, you need to invoke the **Links** dialog box and then choose the **Update Now** button; then, choose **Cancel**. This will update the drawing in the container application and display the updated drawing on the WordPad.

10. Exit the container application after saving the updated file.

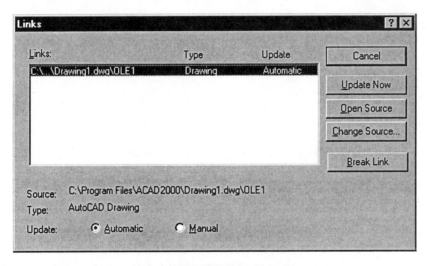

Figure 21-28 Links dialog box

Linking Information into AutoCAD*

Similarly, you can also embed and link information from a server application into an AutoCAD drawing. You can also drag selected OLE objects from another application into AutoCAD, provided this application supports Microsoft Activex and the application is running and visible on the screen. Dragging and dropping is like cutting and pasting. If you press the CTRL key while you drag the object it is copied to AutoCAD. Dragging and dropping an OLE object into AutoCAD embeds it into AutoCAD.

Linking Objects into AutoCAD*

Start any server application like the Windows Wordpad and open a document in it. Select the information you wish to use in AutoCAD with your pointing device and choose **Copy** from the **Edit** menu or choose the **Copy** button in the toolbar to copy this data to the Clipboard. Open the AutoCAD drawing you wish to link this data to. Choose **Paste Special** from the **Edit** menu or use the **PASTESPEC** command. The **Paste Special** dialog box is displayed (Figure 21-29). In the **As:** list box, select the data format you wish to use. For example, for a word pad document, select **WordPad document**. Picture format uses a Metafile format. Select the **Paste Link** radio button to paste the contents of the Clipboard to the current drawing. If you select the **Paste** radio button, the data is embedded and not linked. Choose **OK** to exit the dialog box. The data is displayed in the drawing and can be positioned as needed.

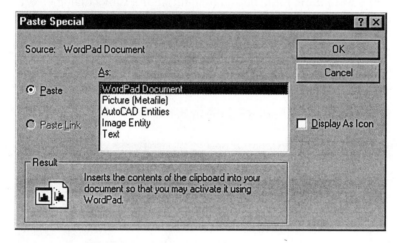

Figure 21-29 Paste Special dialog box

You can also use the **INSERTOBJ** command by entering INSERTOBJ at the Command prompt, by choosing **Ole Object...** from the **Insert** menu, or by choosing the **OLE Object** button in the **Insert** toolbar. This command links an entire file to a drawing from within AutoCAD. Using this command displays the **Insert Object** dialog box (Figure 21-30).

Select the **Create from File** radio button. Also select the **Link** check box. Choosing the **Browse** button, displays the **Browse** dialog box. Select a file you want to link from the list box or enter a name in the File name: edit box and choose the **Insert** button. The path of the file is displayed in the **File:** edit box. If you select the **Display as icon** check box, an icon is also displayed in the dialog box. Choose **OK** to exit the dialog box, the selected file is linked to the AutoCAD drawing.

AutoCAD updates the links automatically by default, whenever the server document changes, but you can use the **OLELINKS** command to display the **Links** dialog box (Figure 21-31) where you can change these settings. This dialog box can also be displayed by choosing **OLE Links...** from the **Edit** menu. In the **Links** dialog box, select the link you want to update and then choose the **Update Now** button. Then choose the **Close** button. If the server file location changes or if it is renamed, you can choose the **Change Source** button in the **Links** dialog box to display the **Change Source** dialog box. In this dialog box, locate the server file name and location and choose the **Open** button. You can also choose the **Break Link** button in the **Links** dialog box to disconnect

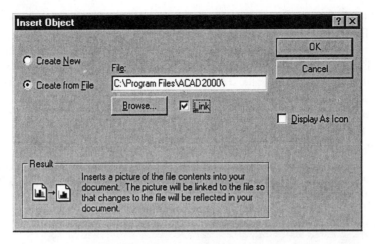

Figure 21-30 Insert Object dialog box

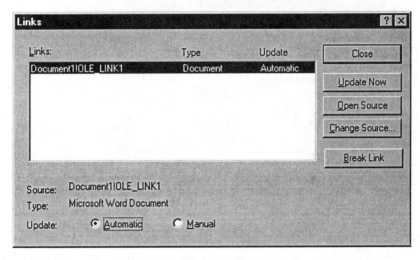

Figure 21-31 Links dialog box

the inserted information from the server application. This is done when the linking between the two is not required anymore.

Embedding Objects into AutoCAD*

Open the server application and select the data you want to embed into the AutoCAD drawing. Copy this data to the Clipboard by choosing the **Copy** button in the toolbar or choosing **Copy** from the **Edit** menu. Open the AutoCAD drawing and choose **Paste** from the AutoCAD **Edit** menu. You can also use the command **PASTECLIP**. The selected information is embedded into the AutoCAD drawing.

You can also create and embed an object into an AutoCAD drawing starting from AutoCAD itself using the **INSETOBJ** command. You can also choose **OLE Object** from the **Insert** menu or choose the **OLE Object** button from the **Insert** toolbar. The **Insert Object** dialog box is displayed

(Figure 21-32). In this dialog box, select the **Create New** radio button and select the application you wish to use from the **Object Type:** list box. Choose **OK**. The selected application opens, now, you can create the information you wish to insert in the AutoCAD drawing and save it before closing the application. The **OLE Properties** dialog box is displayed (Figure 21-33) where you can resize and rescale the inserted objects. This dialog is displayed by default. If you do not want to display this dialog box, clear the **Display dialog when pasting OLE objects** check box in the **OLE Properties** dialog box. You can also clear the **Display OLE Properties Dialog** check box in the **Systems** tab of the **Options** dialog box. You can edit information embedded in the AutoCAD drawing by opening the server application by double-clicking on the inserted OLE object. You can also select the object and right-click to display a shortcut menu. Choose **Object > Edit**. After editing the data in the server application, choose **Update** from the **File** menu to reflect the modifications in the AutoCAD drawing.

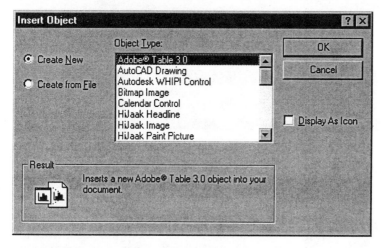

Figure 21-32 Insert Object dialog box

Working with OLE Objects*

Select an OLE Object and right-click to display a shortcut menu; choose **Properties**. The **OLE Properties** dialog box is displayed (Figure 21-33). You can also select the OLE Object and use the **OLESCALE*** command. Specify a new height and a new width in the **Height:** and **Width:** edit boxes in the **Size** area or Under **Scale**, enter a value in percentage of the current values in the **Height:** and **Width:** edit boxes. Here, if you select the **Lock aspect Ratio** check box, whenever you change either the height or the width under **Scale** or **Size**, the respective width or height changes automatically to maintain the aspect ratio. If you want to change only the height or only the Width, clear this check box. Choose **OK** to apply changes.

Choosing the **Reset** button restores the selected OLE objects to their original size, that is, the size they were when inserted. If the AutoCAD drawing contains an OLE object with text with different fonts and you wish to select and modify specific text. You can select a particular font and point size from the drop-down lists under the **Text size** area and enter in the box after the = sign the value in drawing units. For example if you wish to select text in the Times Roman font, of point size 10 and modify it to size 0.5 drawing units, select Times Roman and 10 point size from the drop-down lists and in the text box after the = sign enter .5. All the text that is in Times Roman and is of point

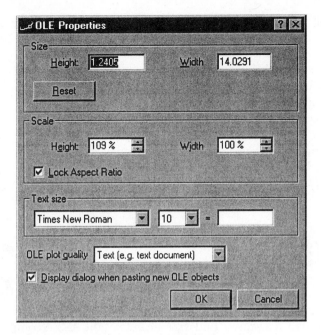

Figure 21-33 OLE Properties dialog box

size 10 will change to 0.5 drawing units in height.

The pointing device can also be used to modify and scale an OLE Object in the AutoCAD drawing. Selecting the object displays the object frame and the move cursor. The move cursor allows you to select and drag the object to a new location. The middle handle allows you to select the frame and stretch it. It does not scale objects proportionately. The corner handle scales the object proportionately.

Select an OLE object and right-click to display the shortcut menu. Choosing **Cut** removes the object from the drawing and pastes it on the Clipboard, **Copy** places a copy of the selected object on the clipboard; and **Clear** removes the object from the drawing and does not place it on the clipboard. Choosing **Object** displays the **Convert, Open**, and **Edit** options. Choosing **Convert** displays the **Convert** dialog box where you can convert objects from one type to another and **Edit** opens the object in the Server application where you can edit it and update it in the current drawing. **Undo** cancels the last action. **Bring to Front** and **Send to Back** options place the OLE objects in the front of or back of the AutoCAD objects. The **Selectable** option, turns the selection of the OLE object on or off. If the **Selectable** option is on, the object frame is visible and the object is selected.

If you want to change the layer of an OLE object, select the object and right-click to display the shortcut menu; choose **Cut**. The selected object is placed on the clipboard. Change the current layer to the one you want to change the OLE object's layer to using the **Layer Properties Manager** dialog box. Now, choose **Paste** from the **Edit** menu to paste the contents of the clipboard in the AutoCAD drawing. The OLE object is pasted in the New layer, in its original size.

The **OLEHIDE** system variable controls the display of OLE objects in AutoCAD. The default value is 0, that is, all the OLE objects are visible. The different values and their effects are as follows:

0   All OLE objects are visible
1   OLE objects are visible in paper space only
2   OLE objects are visible in model space only
3   No OLE objects are visible

The **OLEHIDE** system variable affects both screen display and printing.

Self-Evaluation Test

Answer the following questions, and then compare your answers to the answers given at the end of this chapter.

1. The object whose area you want to find with the help of the **AREA** command must be a closed object. (T/F)

2. If you quit the editing session without saving the drawing, the time you spent on that editing session is added to the total time spent on editing the drawing. Also, the last update time is revised. (T/F)

3. The angle between two points can be measured with the help of the _____ command.

4. The _____ command displays all the information pertaining to the selected objects.

5. _____ time provides the most recent date and time you edited the current drawing.

6. You can set the automatic save time interval with the **Options** dialog box or with the _____ system variable.

7. You can import a DXF file into an AutoCAD drawing file with the _____ command.

8. The _____ command is used to create drawing files out of DXB format files.

9. If you import a DXB file into an old drawing, the settings of layers, blocks, and so on, of the file being imported are _____ by the settings of the file into which you are importing the DXB file.

10. A block can be imported into a drawing file only if the block definition exists in the _____ drawing file.

11. The _____ system variable holds the value of distance computed.

12. The _____ command is used to identify the coordinate values of a point on the screen.

13. If you import a PostScript file with a fill pattern using the **PSIN** command, the fill pattern is not displayed on the screen. (T/F)

14. The _____ command displays all the information pertaining to all the objects in the drawing.

15. The _____ is used as a medium for storing the documents while transferring them from one Windows application to another.

16. The _____ command can be used in AutoCAD for embedding drawings.

17. You can edit your embedded drawing in the _____ application.

18. You can get an embedded drawing into the server application directly from the container application by _____ on the drawing.

19. The _____ command can be used in AutoCAD for linking a drawing.

20. The **COPYLINK** command copies a drawing in the _____ to the Clipboard.

Review Questions

Inquiry

1. Inquiry commands are used to obtain information about the drawn figures. (T/F)

2. The default method of specifying the object whose area you want to find is by specifying all the vertices of the object. (T/F)

3. The **AREA** and **PERIMETER** system variables are reset to zero whenever you invoke the **AREA** command. (T/F)

4. To find the circumference or perimeter of an object, you can use the _____ command.

5. You can use the _____ option to add subsequently measured areas to the running total.

6. You can use the _____ option to subtract subsequently measured areas from the running total.

7. The distance between two points can be measured with the _____ command.

8. The _____ command can be used to display the data pertaining to time related to a drawing and the drawing session.

9. Drawing _____ provides the date and time the current drawing was created.

10. The drawing creation time for a drawing is set to the system time when either the _____ command, the _____ command, or the _____ command was used to create that drawing file.

Data Exchange

11. The _____ command is used to create an ASCII format file with the .DXF extension from AutoCAD drawing files.

12. With the **Binary** option of the **Save As Options** dialog box, you can also create binary format files. Binary DXF files are _____ efficient and occupy only 75 percent of the ASCII DXF file. File access for files in binary format is _____ than for the same file in ASCII format.

13. In a _____ file, information is stored in the form of a dot pattern on the screen. This bit pattern is also known as _____ .

14. If there is a picture on the screen that you want to save as a raster file, use the _____ command.

15. With the **SAVEIMG** command you can create _____ , _____ , or _____ type of raster files from the current drawing.

16. The section of screen to be saved can be specified by specifying _____ on the image tile.

17. If you are rendering to a viewport and you select the _____ option, then whatever is in the current viewport is saved to the raster file of the selected format.

18. If you are rendering to a viewport and you select the _____ option, the entire screen, including the menu area and the command line area, is saved to the raster file.

Exercises

Exercise 3   *Mechanical*

Draw Figure 21-34 without dimensions, and determine the indicated parameters using the required commands. Select the **DBLIST** command. Select **TIME**, and note the time in the drawing editor. Save the drawing.

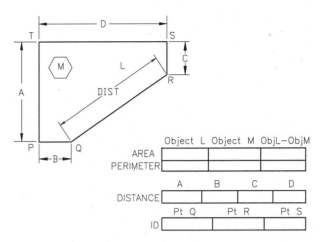

Figure 21-34 Drawing for Exercise 3

Exercise 4   *Graphics*

Draw two rectangles, and use the **PSFILL** command to assign stars and brick patterns to these rectangles (Figure 21-35). Convert the rectangles to the PostScript file using the **PSOUT** command, and then import the PostScript file using the **PSIN** command.

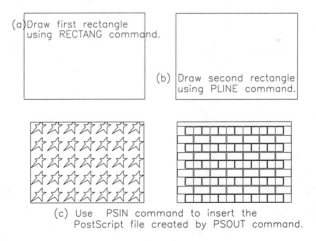

Figure 21-35 Drawing for Exercise 4

Answers to Self-Evaluation Test
1 - F, **2** - F, **3** - **DIST**, **4** - **LIST**, **5** - Last Update, **6** - **SAVETIME**, **7** - **OPEN**, **8** - **DXBIN**, **9** - overruled, **10** - importing, **11** - **DISTANCE**, **12** - **ID**, **13** - F, **14** - **DBLIST**, **15** - clipboard, **16** - **COPYCLIP**, **17** - server, **18** - double-clicking, **19** - **COPYLINK**, **20** - current viewport

Chapter 22

Rendering

Learning Objectives

After completing this chapter, you will be able to:
- *Understand rendering and why to render the objects.*
- *Configure, load, and unload AutoCAD Render.*
- *Insert and modify light sources.*
- *Select rendering type and render objects.*
- *Define and render scenes.*
- *Modify scenes.*
- *Use AutoCAD Render light sources.*
- *Attach materials to objects.*
- *Define materials and modify existing materials.*
- *Replay and print renderings.*

RENDERING

A rendered image makes it easier to visualize the shape and size of a three-dimensional (3D) object, compared to a wireframe image or a shaded image. A rendered object also makes it easier to express your design ideas to other people. For example, if you want to make a presentation of your project or a design, you do not need to build a prototype. You can use the rendered image to explain your design much more clearly because you have complete control over the shape, size, color, and surface material of the rendered image. Additionally, any required changes can be incorporated into the object, and the object can be rendered to check or demonstrate the effect of those changes. Thus, rendering is a very effective tool for communicating ideas and demonstrating the shape of an object. You can create a rendered image of a 3D object by using AutoCAD's **RENDER** command. It allows you to control the appearance of the object by defining the surface material and reflective quality of the surface and by adding lights to get the desired effects.

Determining Which Sides Are to Be Rendered in a Model

Some of the faces of a 3D model, such as the back faces and the hidden faces, need not be rendered since it would amount to an unnecessary waste of time. In the process of rendering, AutoCAD determines the front faces and the back faces of the 3D model with the use of normals on each face. A **vector** perpendicular to each face on a 3D model and whose direction is outward toward space is known as **normal**. If a face has been drawn in the clockwise direction, then the normal points inward; if a face has been drawn counterclockwise, the normal points outward. Now, depending on the location of your viewpoint, if the normal of a face points away from the viewpoint, the face is a back face. As mentioned before, rendering of such faces is not advisable (since those faces are not visible from the viewpoint) and can be avoided by invoking the Discard Back Faces option (explained later in this chapter). The faces concealed by other faces are discarded. In this manner, the time required to render objects can be decreased by discarding faces that need not be rendered.

Points to Be Remembered While Defining a Model

1. To make the rendering process as time-efficient as possible, you must use the fewest possible faces to define a plane.

2. There should be consistency in your drafting technique. Models formed of a complex mixture of faces, extruded lines, and wireframe meshes should be avoided.

3. If you are rendering circles, ellipses, or arcs, set the **VIEWRES** command to a high number. This way, the circles, arcs, or ellipses will appear smoother and the rendering of such objects will be better. But increasing the value of **VIEWRES** increases the time taken to render the objects. The smoothness of rendered curved solids depends on **FACETRES** variable.

4. If you are using the Smooth Shading option (available in the **Render** and **Rendering Preferences** dialog box, explained later), then you should specify the mesh density in such manner that the angle described by normals of any two adjoining faces is less than 45 degrees. This is because if the angle is greater than 45 degrees, then after rendering, an edge is displayed between the faces even when the Smooth Shading option is active.

LOADING AND UNLOADING AUTOCAD RENDER

When you select any AutoCAD **RENDER** command, AutoCAD Render is loaded automatically. AutoCAD will display the following message in the command prompt area:

Command: **RENDER**
Loading Landscape Object module.
Initializing Render...
Initializing preferences... done

If you do not need AutoCAD Render, you can unload it by entering the **ARX** command at the Command prompt. AutoCAD acknowledges the unloading of Render by issuing the message "Render successfully unloaded." After unloading Render, you can reload AutoCAD Render by invoking the **RENDER** command or any other command associated with rendering (such as **SCENE** or **LIGHT**).

Command: **ARX**
Enter an option [?/Load/Unload/Commands/Options]: **U**
Enter ARX/DBX file name to unload: **ACRENDER**
ACRENDER successfully unloaded.

ELEMENTARY RENDERING

Toolbar:	Render > Render
Menu:	View > Render > Render
Command:	RENDER

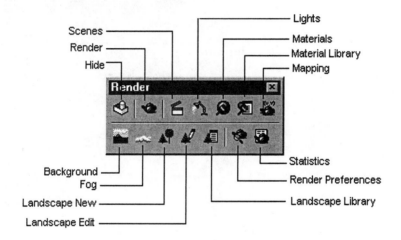

Figure 22-1 Render toolbar

In this section you are going to perform an ordinary rendering. You will encounter many terms of which you will not be aware, but you need not bother about them at this stage. All the terms are explained later in this chapter. In this rendering, lights, materials, and other advanced features of rendering will not be applied. Only a default distant light is used. This will give you a feel of rendering in the simplest possible manner. Perform the following steps:

1. Using the **SPHERE** command, create four spheres on the screen as shown in Figure 22-2.
2. Invoke the **Render** dialog box (Figure 22-3).
3. In the **Render** dialog box, choose the **Render** button. AutoCAD will render all objects that are on the screen.

After some time, all the spheres are rendered and displayed on the screen (Figure 22-4). Press any key to exit the rendered image screen and return to the AutoCAD drawing screen.

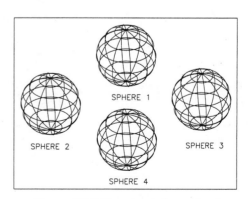

Figure 22-2 Drawing to be rendered

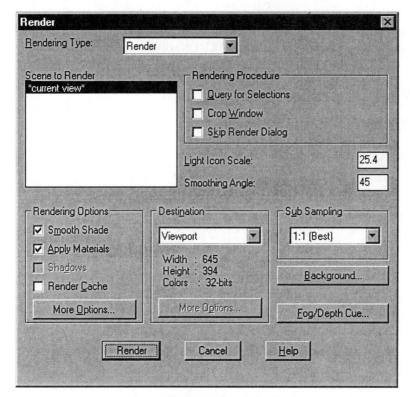

Figure 22-3 Render dialog box

If you want to render only some spheres (for example, SPHERE 2 and SPHERE 3), select the **Query for Selections** check box in the **Render** dialog box. When you choose the **Render** button, AutoCAD will prompt you to select the objects that you want to render. Select the objects and press ENTER; you will notice that only the selected spheres are rendered and displayed on the screen.

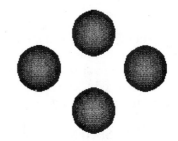

Figure 22-4 Rendering all spheres

Exercise 1   *Mechanical*

In this exercise you will render the drawing that was created in Example 2 of Chapter 23, Solid Modeling. Load and render the drawing. Figure 22-5 shows the object before rendering; Figure 22-6 shows the object after rendering.

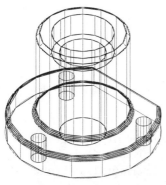

Figure 22-5 Drawing for Exercise 1   *Figure 22-6* After rendering

SELECTING DIFFERENT PROPERTIES FOR RENDERING

Toolbar:	Render > Preferences
Menu:	View > Render > Preferences
Command:	RPREF

AutoCAD Render allows you to select various properties for the rendering. This can be achieved through the **Rendering Preferences** dialog box (Figure 22-7). The various sections of the dialog box are described next.

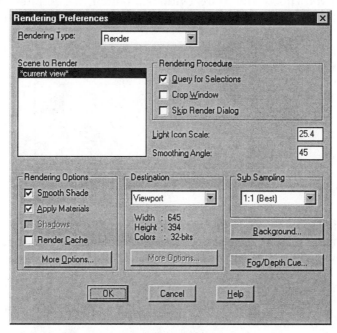

Figure 22-7 Rendering Preferences dialog box

Rendering Type

In the **Rendering Preferences** dialog box you can select the type of rendering (Render, Photo Real, Photo Raytrace) you want from the **Rendering Type** drop-down list.

Rendering Options

Various rendering options are provided in the Rendering Options area of the dialog box. These are described next.

Smooth Shade. The **Smooth Shade** allows you to smooth the rough edges. If this option is enabled, the rough-edged appearance of a multifaceted surface is smoothed. Only polygon meshes are affected by Smooth Shading. The surface normals are determined, and colors across two or more adjoining faces are blended.

Apply Materials. The Apply Materials option allows you to assign surface materials to objects. If this option is disabled, then the objects in the drawing are assigned the ***GLOBAL*** material.

Shadows. When you select this option, AutoCAD generates shadows. This option applies only to Photo Real and Photo Raytrace rendering.

Render Cache. Selecting this option results in writing rendering information to a cache file on the hard disk that can be used for subsequent renderings. This eliminates the need for AutoCAD to recalculate (tessellate) the objects for rendering. This saves time, especially when rendering solids.

More Options. If you choose the **More Options...** button, the **Render Options** dialog box (Figure 22-8) is displayed. The following is a brief description of these additional options.

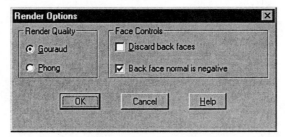

Figure 22-8 Render Options dialog box

Render Quality Area. In the **Render Quality** area of this dialog box, you can choose either the Gouraud or the Phong option. These options determine the quality of shading that will be used if you have activated the Smooth Shading option.

Phong. The **Phong** option creates a high-quality rendered image. The light intensity is calculated for all the pixels, and hence, more accurate highlights are generated.

Gouraud. On the other hand, the **Gouraud** option produces a lower-quality rendered image. Light intensity at each vertex is determined, and intermediate intensities are interpolated. The advantage of the Gouraud render option is that the rendering occurs faster as compared with the Phong render option.

Face Controls Area. The **Face Controls** area governs the definition of faces of a 3D solid.

Discard Back Faces. If you select the **Discard Back Faces** option, the back faces of a 3D solid object are not taken into consideration (that is, they are made invisible), and hence, they are omitted from the calculations for the rendering. Rendering time can be reduced by enabling this option.

Back Face Normal is Negative. The **Back Face Normal** is **Negative** option can be used to define the back faces in a drawing. If this option is enabled, the faces with negative normal vectors are treated as back faces and discarded. When this option is disabled, the selection faces treated as back faces by AutoCAD are just reversed.

Rendering Procedures

The Rendering Procedure of the **Rendering Preferences** dialog box has the following options:

Query for Selections. If you select the **Query for Selections** option, AutoCAD prompts you to select objects to render.

Crop Window. When you select this option, AutoCAD will prompt you to select a window. Only the objects inside the specified window are rendered.

Skip Render Dialog. In the Rendering Procedure area, if the **Skip Render Dialog** option is invoked, the current view is rendered without displaying the Render dialog box.

Light Icon Scale. The **Icon Scale** edit box can be used to set the size of the light blocks in the drawing.

Smoothing Angle. The **Smoothing Angle** option lets you specify the angle defined by two edges. The default value for smoothing an angle is 45 degrees. Angles of less than 45 degrees are smoothed. Angles greater than 45 degrees are taken as edges.

Destination

The **Destination** area of the **Rendering Preferences** dialog box allows you to specify the destination for the rendered image output.

Viewport Option. If you select the **Viewport** option, AutoCAD renders to the current viewport.

Render Window Option. This option renders to the AutoCAD for Windows Render window.

File Option. The **File** option lets you output the rendered image to a file.

More Options. The **More Options...** button can be used to set the configuration for the output file through the **File Output Configuration** dialog box (Figure 22-9).

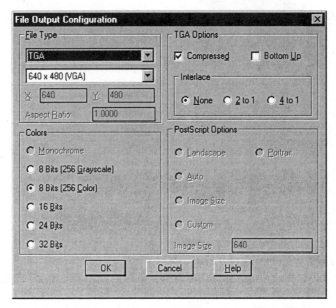

Figure 22-9 File Output Configuration dialog box

File Type Area. In this area you can specify the output file type and rendering resolution. The file formats allowed are BMP, PCX, Postscript, TGA, and TIFF. The screen resolution can also be specified in this area. The aspect ratio of the output file can be specified in the **Aspect Ratio** edit box.

Colors Area. The colors in the output file can be specified in this area.

TGA Options Area. The **Compressed** option lets you specify compression for those file types that allow compression. The **Bottom Up** option lets you specify the scan line start point as bottom left instead of top left.

Interlace Area. Selecting the None option turns off line interlacing. The **2** to **1** and **4** to **1** options turn on interlacing.

PostScript Options Area. The **Landscape** and **Portrait** options in this area specify the orientation of the file. The **Auto** option automatically scales the image. The Custom option sets the image size in pixels. The **Image Size** edit box uses the explicit image size.

Sub Sampling

The Sub Sampling edit box controls the quality and the rendering time by reducing the number of pixels to be rendered. A **1:1** ratio (default) produces the best quality rendering. If you are just testing a rendering, you can change the value to reduce rendering time.

Background

The background option allows you to add a background to a rendering. You can use solid colors, gradient colors, or an image file as a background.

Fog/Depth Cue

You can use the Fog option to add a misty effect to a rendering. You can also assign a color to the fog.

AUTOCAD RENDER LIGHT SOURCE

Lights are vital to rendering a realistic image of an object. Without proper lighting, the rendered image may not show the features the way you expect. Colors and surface reflection can be set for the lights with RGB or HLS color systems. AutoCAD Render supports the following four light sources: ambient light, point light, distant light, and spotlight.

Ambient Light

You can visualize **ambient light** as the natural light in a room that equally illuminates all surfaces of the objects. Ambient light does not have a source and hence has no location or direction. However, you can increase or decrease the intensity of ambient light or completely turn it off. Normally, you should set the ambient light to a low value because high values give a washed-out look to the image. If you want to create a dark room or a night scene, turn off the ambient light. With ambient light alone you cannot render a realistic image. Figure 22-10 shows an object illuminated by ambient light.

Point Light

A **point light** source emits light in all directions, and the intensity of the emitted light is uniform. You can visualize an electric bulb as a point light source. In AutoCAD Render, a point source does not cast a shadow because the light is assumed to be passing **through** the object. The intensity of light radiated by a point source decreases over distance. This phenomenon is called attenuation. Figure 22-11 shows a light source that radiates light in all directions.

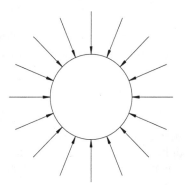

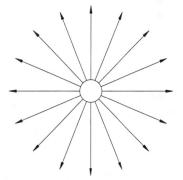

Figure 22-10 *Ambient light provides constant illumination*

Figure 22-11 *Point light emits light in all directions*

Spotlight

A **spotlight** emits light in the defined direction with a cone-shaped light beam (Figure 22-12). The direction of the light and the size of the cone can be specified. The phenomenon of **attenuation** (falloff) also applies to spotlights. This light is used mostly to highlight particular features and portions of the model. If you want to simulate a soft lighting effect, set the falloff cone angle a few degrees larger than the hot-spot cone angle.

Attenuation

Light intensity is the amount of light falling per unit area. The intensity of light is directly proportional to the degree of brightness of the object. The intensity of light decreases as the distance increases. This phenomenon, called **attenuation**, occurs only with spotlights and point light. In Figure 22-13, the light is emitted by a point source. Assume that the amount of light incident on Area-1 is I. Therefore, the intensity of light on Area-1 = I/Area. As the light travels farther from the source, it covers a larger area. The amount of light falling on Area-2 is the same as on Area-1, but the area is larger. Therefore, the intensity of light for Area-2 is smaller (Intensity of light for Area-2 = I/Area). Area-1 will be brighter than Area-2 because of higher light intensity. AutoCAD Render has three options for controlling the light falloff: None, Inverse Linear, and Inverse Square.

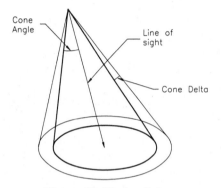

Figure 22-12 Spotlight

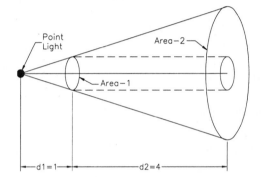

Figure 22-13 Light intensity decreases with distance

None

If you select the **None** option for light falloff, the brightness of objects is independent of distance. This means that objects that are far away from the point light source will be as bright as objects close to the light source.

Inverse Linear

In this option, the light falling on the object (brightness) is inversely proportional to the distance of the object from the light source (Brightness = 1/Distance). As distance increases, brightness decreases. For example, let us assume the intensity of the light source is **I** and the object is located at a distance of **2** units from the light source. Brightness or intensity = **I/2**. If the distance is 8 units, the intensity (light falling on the object per unit area) = **I/8**. The brightness is a linear function of the distance of the object from the light source.

Inverse Square

In this option, the light falling on the object (brightness) is inversely proportional to the square of the distance of the object from the light source (Brightness = 1/Distance $^\wedge$ 2). For example, let us assume the intensity of the light source is **I** and the object is located at a distance of **2** units from the light source. Brightness or intensity = **I/(2) $^\wedge$ 2** = **I/4**. If the distance is 8 units, the intensity (light falling on the object per unit area) = **I/(8) $^\wedge$ 2** = **I/64**.

Distant Light

A **distant light** source emits a uniform parallel beam of light in a single direction only
(Figure 22-14). The intensity of the light beam
does not decrease with distance; it remains
constant. For example, the sun's rays can be
assumed to be a distant light source because
the light rays are parallel. When you use a distant light source in a drawing, the location of
the light source does not matter; only the direction is critical. Distant light is used mostly
to light objects or a backdrop uniformly and
for getting the effect of sunlight.

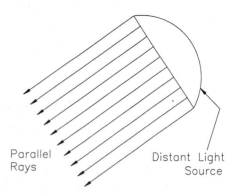

Figure 22-14 Distant light source

INSERTING AND MODIFYING LIGHTS

Toolbar:	Render > Lights
Menu:	View > Render > Lights
Command:	LIGHT

In a rendering, the lights and lighting effects are important to create a realistic representation of an object. The sides of an object that face the light must appear brighter and the
sides that are on the other side of the object must be darker. This smooth gradation of
light produces a realistic image of the object. If the light intensity is uniform over the
entire surface, the rendered object probably will not look realistic. For example, if you use the
SHADE command to shade an object, that object does not look realistic because the displayed
model lacks any gradation of light. Any number of lights can be installed in a drawing. The color,
location, and direction of all the lights can be specified individually. As mentioned before, you can
specify attenuation for point lights and spotlights. AutoCAD also allows you to change the color,
position, and intensity of any light source. The only limitation is that light types cannot be changed.
For example, you cannot change a distant light into a point light. The following sections describe
how to insert, position, and modify lights.

Inserting Distant Light

As mentioned before, the light plays an important role in producing a realistic representation of a
3D model. The gradation effect produced by the lights makes the object look realistic. In this
example, you will insert a distant light source and then render the object.

1. Use the **VPOINT** command to change the viewpoint to (1,-1,1).
    Command: **VPOINT**
    Specify a view point or [Rotate] <current>: **1,-1,1** *Press ENTER.*

2. Choose **Render** from the **View** menu.
3. Select the **Lights...** option from the cascading menu. The **Lights** dialog box is displayed
    (Figure 22-15). You can use any one of the methods just described to invoke the **Lights** dialog
    box.
4. Select the **Distant Light** option from the drop-down list, and then choose **New...** to insert a

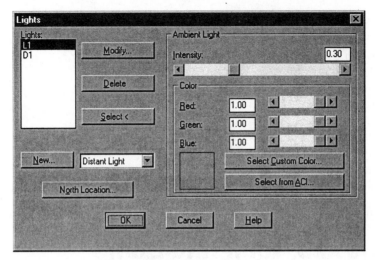

Figure 22-15 Lights dialog box

new distant light. The **New Distant Light** dialog box is displayed [Figure 22-16(a)].

5. Enter the name of the distant light (D1) in the **New Distant Light** dialog box. Leave the intensity at its default value (1.00).

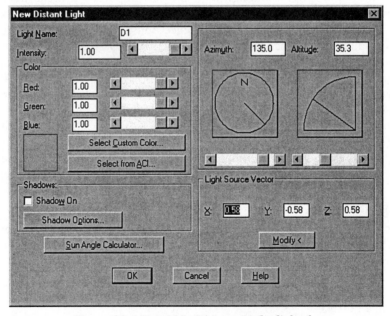

Figure 22-16(a) New Distant Light dialog box

6. Choose the **OK** button to exit this dialog box, then choose **OK** button again to exit the **Lights** dialog box.
7. Now render the image.

Modifying Distant Light

In this rendering, notice that the light falls on the object from the viewpoint direction and the right front corner is brighter than the rest of the object. We would like to modify the light source so that the light falls on the object from the right at an angle. When the light source is at an angle, the top surface also receives some light. Notice that as the oblique angle increases, the amount of light reflected by the oblique surface decreases.

1. Choose the **Lights** button from the **Render** toolbar.
2. Select the light (D1) from the **Lights** dialog box. If there is only one light, the light is automatically highlighted.
3. Choose the **Modify...** button. The **Modify Distant Light** dialog box is displayed [Figure 22-16(b)].

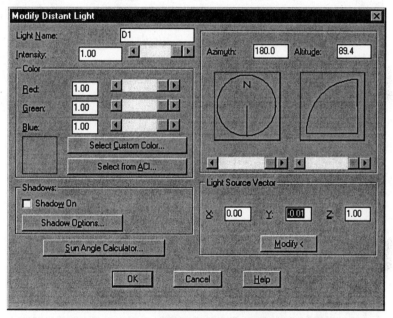

Figure 22-16(b) Modify Distant Light dialog box

Note

Double-clicking on the light name is equivalent to selecting the light name and the Modify... option.

4. Choose the **Sun Angle Calculator** button to display the **Sun Angle Calculator** dialog box (Figure 22-17). Change the date and clock time, if needed.
5. Choose the **Geographic Location** button to display the **Geographic Location** dialog box (Figure 22-18). You can specify the location by choosing the name of the city in the City list box or selecting a point in the map.
6. Choose **OK** to exit the **Geographic Location** dialog box and choose **OK** again to exit the **Sun Angle Calculator** dialog box.
7. Choose **OK** to exit the **Modify Distant Lights** dialog box and choose **OK** again to exit the **Lights** dialog box.

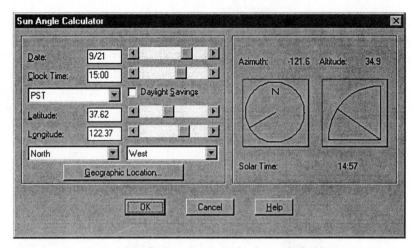

Figure 22-17 Sun Angle Calculator dialog box

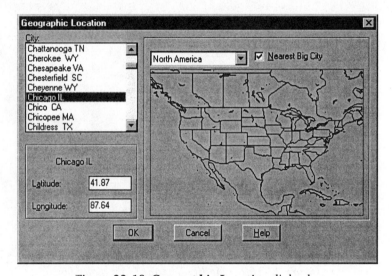

Figure 22-18 Geographic Location dialog box

8.   Render the object.

Notice that the sides (vertical faces) of the object and the horizontal surfaces have different degrees of brightness. This is because the oblique angle that the vertical surfaces make with the direction of the light rays is smaller than the oblique angle that the horizontal surfaces make. Maximum brightness occurs when the object is perpendicular to the direction of the light rays.

Inserting Point Light
In Example 1 you will insert a point light source. Then you will render the object.

Example 1

Insert a point light source and render the object of Exercise 1 with Inverse Linear and Inverse Square options.

1. Load the drawing of Exercise 1.
2. Choose the **Lights** button from the **Render** toolbar.
3. Select the Point Light option from the **Lights** dialog box.
4. Choose the **New...** button. The **New Point Light** dialog box is displayed.
5. Enter the name of the point light (P1) and set the intensity (6.0).
6. Choose the **Modify<** option from the **New Point Light** dialog box.
7. Enter the light location (-2,0,5).
8. In the **New Point Light** dialog box, select Inverse Linear, and then choose the **OK** button to exit the dialog box. Choose **OK** to exit the **Lights** dialog box.
9. Change the viewpoint to (1,-1,1), and then render the object.

Notice that the top surface of the cylindrical part and the top surface of the flange are equally bright. Next, change the falloff to Inverse Square:

10. Choose the **Lights** button from the **Render** toolbar.
11. Select the point light (P1) and then choose the **Modify...** button.
12. The **Modify Point Light** dialog box will be displayed. Select **Inverse Square**, and then choose **OK** to exit the dialog box. Choose **OK** to exit the Lights dialog box.
13. Render the object again. Now, notice that the top surface of the cylindrical part is brighter than the top surface of the flange.

DEFINING AND RENDERING A SCENE

Toolbar:	Render > Scenes
Menu:	View > Render > Scene
Command:	SCENE

 The rendering depends on the view that is current and the lights that are defined in the drawing. Sometimes the current view or the lighting setup may not be enough to show all features of an object. You might need different views of the object with a certain light configuration to show different features of the object. When you change the view or define the lights for a rendering, the previous setup is lost. You can save the rendering information by defining a **scene**. For each scene, you can assign a view and the lights. When you render a particular view, AutoCAD Render uses the view information and the lights that were assigned to that scene. It ignores the lights that were not defined in the scene. Defining scenes makes it convenient to render different views with the required lighting arrangement. The following example describes the process of defining scenes and assigning views and lights to the scenes.

Example 2

In this example, you will draw a rectangular box and a sphere that is positioned at the top of the box. Next, you will insert lights, define views and scenes, and then render the scenes. You may start a new drawing or continue with the current drawing. Before defining a scene, perform the following steps in a new drawing:

1. Draw a rectangular box with length = 3, width = 3, and height = 1.5.
2. Use the **UCS** command to define the new origin at the center of the top face.
3. Draw a sphere of 1.5 radius.
4. Move the sphere so that the bottom of the sphere is resting at the top face of the box, as shown in Figure 22-19.

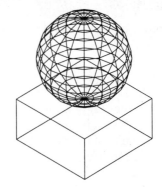

Next, insert the lights. In this example, you will insert a point light and a distant light. Let us first insert a distant light source.

Figure 22-19 Drawing for Example 2

1. Select the **Render** option from the **View** menu and then select the **Lights...** option from the cascading menu. The **Lights** dialog box is displayed.
2. Select the **Distant** option, and then choose **New...** to insert a new distant light. The **New Distant Light** dialog box is displayed.
3. Enter the name of the distant light (D1) in the **Light Name:** edit box, and press ENTER. Now choose **Modify<** from the **New Distant Light** dialog box.
4. Enter the light target point (direction TO) (0,0,0) and the light location (direction FROM) (3,-3,4). Leave the intensity at 1.0.
5. Choose the **OK** button to exit the dialog box. Choose **OK** to exit the Lights dialog box.
6. Change the viewpoint to **1,-1,1,** and then render the object.

After inserting a distant light source, you need to insert a point light.

1. Choose the **Lights** button from the **Render** toolbar.
2. Select the **Point Light** option from the **Lights** dialog box and choose the **New...** button. The **New Point Light** dialog box is displayed.
3. Enter the name of the point light (P1) and set the intensity (6.0).
4. Choose the **Modify<** button from the dialog box and enter the light location (0,0,4).
5. In the **New Point Light** dialog box, select **Inverse Linear**, and then choose **OK** to exit the dialog box. Choose **OK** to exit the **Lights** dialog box.
6. Change the viewpoint to **1,-1,1,** and then render the object.

Now you will create three scenes. The first scene, SCENE1, contains distant light D1; the second scene, SCENE2, contains point light P1. The third scene, SCENE3, will contain the view VIEW1, distant light D1, and point light P1. Create the first scene by performing the following steps:

1. Choose the **Scenes** button from the **Render** toolbar.
2. Choose **New...** from the **Scenes** dialog box (Figure 22-20). The **New Scene** dialog box (Figure 22-21) will be displayed.
3. Enter the name of the scene (SCENE1) in the **Scene Name:** edit box. Select Distant Light D1 in the lights area, and then choose the **OK** button to exit the dialog box.
4. Choose the **OK** button in the **Scenes** dialog box.

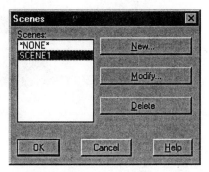

Figure 22-20 Scenes dialog box

Figure 22-21 New Scene dialog box

Similarly, create the second scene, SCENE2, and assign the point light P1 to the scene. Choose the **OK** button to close the **Scenes** dialog box.

Next, create the third scene, SCENE3:

1. Use the **VPOINT** command to change the viewpoint to **1,-0.5,1** and use the **VIEW** command to save the view as **VIEW1**. Choose **OK**.
2. Invoke the **Scenes** dialog box and choose **New...** from the **Scenes** dialog box to display the **New Scene** dialog box.
3. Enter the name of the scene (SCENE3) in the **Scene Name:** edit box. Select ***ALL*** to assign Distant Light D1 and Point Light P1 to the scene. Now select VIEW1.
4. Choose the **OK** button from the **Scenes** dialog box.

You have created three scenes, and each scene has been assigned some lights. SCENE3 has been assigned a view in addition to the lights. You can render any scene, after making the scene current, as follows:

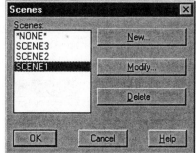

1. Invoke the **Scenes** dialog box (Figure 22-22).
2. Select on the scene name you want to render, and then choose **OK** to exit the dialog box.
3. Render the object.

Figure 22-22 Scenes dialog box
displaying the scene names

Modifying a Scene

You can modify a scene by changing the view and the lights that are assigned to that scene. When you render the object, AutoCAD Render will use the newly assigned view and lights to render the object.

1. Choose the **Scenes** button from the **Render** toolbar; the **Scenes** dialog box with the scene names is displayed on the screen.

2. Select the scene, SCENE3, and then choose **Modify....**
   The **Modify Scene** dialog box (Figure 22-23) is displayed.

3. Select on CURRENT, and then choose **OK** to exit the **Modify Scene** dialog box. This will assign the current view to SCENE3.

4. Choose the **OK** button to exit the **Scenes** dialog box.

5. Render the object.

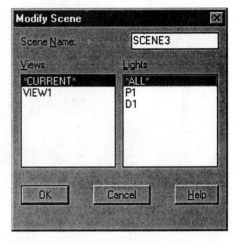

Figure 22-23 Modify Scene dialog box displaying the views and lights

OBTAINING RENDERING INFORMATION

Toolbar:	Render > Statistics
Menu:	View > Render > Statistics
Command:	STATS

You can obtain information about the last rendering. When you enter the **STATS** command, AutoCAD will display the **Statistics** dialog box (Figure 22-24). Information is provided about the name of the current scene, the last rendering type used, the time taken to produce the last rendering, the number of faces processed by the last rendering, and the number of triangles processed by the last rendering. The information contained in the dialog box cannot be edited. However, the information can be saved to a file by checking the Save Statistics to

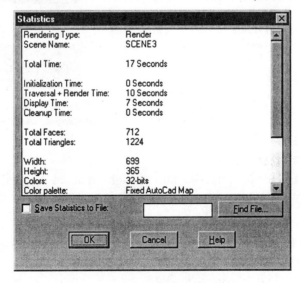

Figure 22-24 Statistics dialog box

File box and then entering the name of the file in the edit box. The file is saved as an **ASCII** file. In case a file by the specified name already exists, AutoCAD adds the present information to that file. You can use the EDIT function of any text editor to read the file.

ATTACHING MATERIALS

You can assign materials to objects, to blocks, to layers, and to an **AutoCAD color index** (ACI). Descriptions of each of them follows.

Attaching Material to an Object

Toolbar:	Render > Materials Library
Menu:	View > Render > Materials Library
Command:	MATLIB

 Any object is made up of some material; hence, to get the actual (realistic) rendered image of the object, it is necessary to assign materials to the surface of the object. AutoCAD supports different materials such as bronze, copper, brass, steel, and plastic. These materials are located in material libraries such as **RENDER.MLI**. You can assign materials to objects, to blocks, to layers, and to an AutoCAD color index (ACI). By default, only ***GLOBAL*** material is assigned to a new drawing. However, if you have an object on the screen to which you want to attach a material that is already defined in the library of materials provided by AutoCAD (**RENDER.MLI**), you need to import the desired material to the drawing and then attach it to the object. For example, if you have a sphere in the current drawing and you want to attach Aqua Glaze to it, the following steps need to be executed.

1. Invoke the **Materials Library** dialog box (Figure 22-25).
2. Once the **Materials Library** dialog box is displayed, select the material you want from the Library List box. In our case, we will select AQUA GLAZE from the list. (In case the material

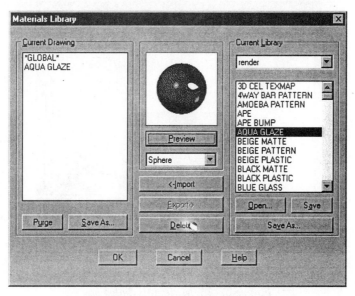

Figure 22-25 Materials Library dialog box

you want to import is in a different library (other than **RENDER.MLI**), choose the **Open...** button. The **Library File** dialog box is displayed. Now you can specify the library file from which you want to select the material.) If you want to check how this material will appear after rendering, choose the **Preview** button. A rendered sphere with an Aqua Glaze material surface appears in the preview image tile in the dialog box.

3. Next, choose the **<-Import** button to import the Aqua Glaze material into the drawing. Once you do so, the entry Aqua Glaze is automatically displayed in the Materials List column. After this, you can save the materials list in a file by choosing the **Save...** button and then specifying the **.mli** file in the **Library File** dialog box. Next, choose the **OK** button.

Note

*When you open a new drawing, the Materials List column contains only the ***GLOBAL*** entry, which is the default material assigned to a new drawing.*

When you import a material from the library to the drawing, the material and its properties are copied to the list of materials in the drawing, and in no case is the material deleted from the library list. Unattached materials can be deleted from the Materials List by choosing the **Purge** button in the **Materials Library** dialog box.

4. Invoke the Materials dialog box (Figure 22-26). You can invoke the Materials dialog box by selecting Materials from the **Render** toolbar or entering **RMAT** at the Command prompt. Notice that in the **Materials:** column, AQUA GLAZE is now listed along with ***GLOBAL***.

5. Select AQUA GLAZE from the list of materials, and then choose the **Attach <** button. AutoCAD clears the dialog box and issues the following prompt:

Select objects to attach "AQUA GLAZE" to:

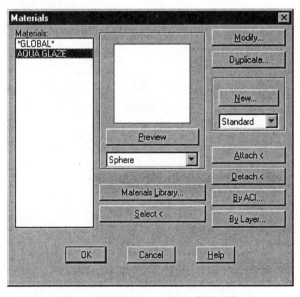

Figure 22-26 Materials dialog box

6. Select the sphere. The Materials dialog box is again displayed. Choose the **OK** button. Now if you render the drawing, you will see that the Aqua Glaze material has been attached to the sphere.

Assigning Materials to the AutoCAD Color Index (ACI)

It is possible to attach materials to the AutoCAD color index (ACI). This can be realized in the following manner:

1. Invoke the **Materials** dialog box (Figure 22-26).
2. Choose the **By ACI...** button. The **Attach by AutoCAD Color Index** dialog box (Figure 22-27) is displayed.

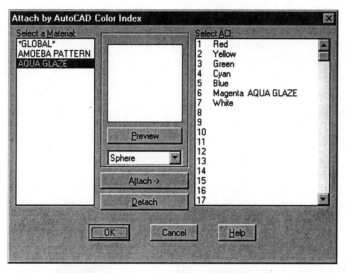

Figure 22-27 Attach by AutoCAD Color Index dialog box

3. Select the material to be attached to an ACI from the Materials: column of the **Materials** dialog box.
4. Select the desired ACI from the **Select ACI:** list and choose the **Attach->** button.
5. Choose the **OK** button to complete the process of attaching the specified material to objects with the specified ACI.

Assigning Materials to Layers

Materials can be associated with layers. This option is useful when you want all the objects on a specified layer to have the assigned material. This can be realized in the following manner:

1. Invoke the **Materials** dialog box.
2. Choose the **By Layer...** button. The **Attach by Layer** dialog box (Figure 22-28) is displayed.
3. Select the material to be attached to all objects on a particular layer.
4. Select the desired layer and choose the **Attach->** button.
5. Choose the **OK** button.

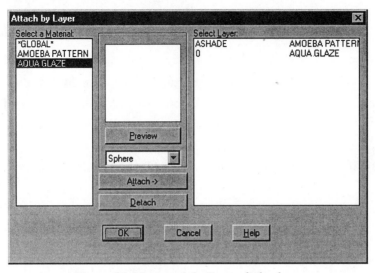

Figure 22-28 Attach by Layer dialog box

With this, the specified material is attached to all objects on the specified layer. AutoCAD attaches a material to an object while rendering according to the following order:

1. Materials explicitly attached to objects have highest priority.
2. Next, materials attached by ACI are considered.
3. Finally, materials attached by layer are considered.
4. GLOBAL material is used if no material is attached to an object.

If you assign different materials to different objects and then combine those objects into a single block, then, upon rendering the block, the different components of the block are rendered according to the materials individually assigned to them. Now, if different components of a block have been created on different layers to which different materials have been attached, and you attach a material to the layer on which the block exists, then the block still will be rendered according to the materials attached to different layers on which the individual parts of the block were created. However, if there is some component in the block to which no material has been attached, then the material attached to the block will be attached to this component. In such a case, if no material has been attached to the block, then the component in the block to which no material has been attached will be assigned the ***GLOBAL*** material.

DETACHING MATERIALS

If you want to assign a different material to an object, it is important first to dissociate the material previously attached to it; only then can you attach another material to it. Also, sometimes it might be required that no material be attached to an object. To dissociate the material attached to an object, choose the **Detach <** button in the **Materials** dialog box. If you have attached a material by **ACI** (AutoCAD Color Index), then to detach this material, choose the **By ACI...** button in the **Materials** dialog box. The **Attach by ACI** dialog box is displayed. Choose the **Detach** button. Similarly, if you want to detach a material attached by layer, choose the **By Layer...** button in the **Materials** dialog box. The **Attach by Layer** dialog box is displayed. Choose the **Detach** button.

CHANGING THE PARAMETERS OF A MATERIAL

It is possible to change the parameters of a material, such as color and reflection properties, to meet your requirements. The **Modify Standard Material** dialog box can be effectively used for changing the parameters. The following steps need to be performed:

1. Invoke the **Materials** dialog box.
2. Select the name of the material you want to modify from the list of material names.
3. Choose the **Modify...** button. The **Modify Standard Material** dialog box appears (Figure 22-29). If you have specified the wrong material name, you can rectify this problem by entering the name of the material you want to modify in the **Material Name:** edit box.

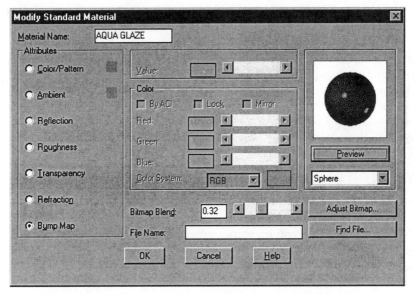

Figure 22-29 Modify Standard Material dialog box

The Modify Standard Material dialog box is comprised of the following options.

Attributes Area

The Attributes area specifies the attributes of a material. You can modify an attribute of a material by selecting it in the Attributes area and then making the alterations to it. The following attributes are listed in this area.

Color Pattern. You can change the diffuse (base) color of the material with this attribute. The intensity of the color can be changed by changing the value in the Value control with the help of the slider bar. The color can be changed by using the RGB or HLS slider bar in the Color area in this dialog box. You can also change the color with the help of the color wheel or the AutoCAD Color Index (ACI) options. The change made to the color is reflected in the color swatch next to the Color option in the Attributes area.

Ambient. By selecting this attribute, you can change the ambient color (shadow) of the material.

Changes to the intensity can be made by changing the value in the Value control box with the help of the slider bar. The color can be changed by using the RGB or HLS slider bar in the Color area in this dialog box. You can also change the color with the help of the color wheel or the AutoCAD Color Index (ACI) options. The change made to the color is reflected in the color swatch next to the Ambient option in the Attributes area.

Reflection. With this attribute, you can modify the reflective (highlight or specular) color of the material. Changes to intensity can be made by changing the value in the Value control box by using the scroll bar to increase or decrease the value. Changes to the reflective color can be made by changing the RGB or HLS slider bar in the Color area in this dialog box. You can also change the color with the help of the color wheel or the AutoCAD Color Index (ACI) options. The change made to the color is reflected in the color swatch next to the Reflection option in the Attributes area.

Roughness. With this attribute you can modify the shininess or roughness level of the material. Changes can be made by changing the value in the Value control. The size of the material's reflective highlight alters by changing the Roughness level. As you increase the level of roughness, the highlight also increases.

Transparency. With this attribute, you can change the transparency of an object. The degree of transparency can be controlled by changing the value in the **Value**: edit box.

Refraction. With this attribute, you can change how refractive a material is. The degree of refraction can be controlled by changing the value in the **Value**: edit box.

Bump Map. When you select this option, you can use the **File Name**: edit box to enter the name of a bump map that you want to assign to the object.

Note

*If you want to see the effect the changes will have on the properties of an object after rendering, choose the **Preview** button. A rendered sphere is displayed in the Preview image tile that reflects the effect of the present properties of the material.*

DEFINING NEW MATERIALS

Sometimes you may want to define a new material. To define a new material in AutoCAD, you have to perform the following steps:

1. Invoke the **Materials** dialog box.
2. Choose the **New...** button. The **New Standard Material** dialog box (Figure 22-30) appears.
3. Enter the name you want to assign to the new material in the **Material Name:** edit box. The material name cannot be more than 16 characters long, and there should be no other material by this name.
4. Just as you modified the attributes of a previously defined material in the **Modify Standard Material** dialog box, set the color and values for the Color, Ambient, Reflection, and Roughness attributes of the new material to your requirement.
5. To examine the effect of rendering with the newly defined material, choose the **Preview** button.

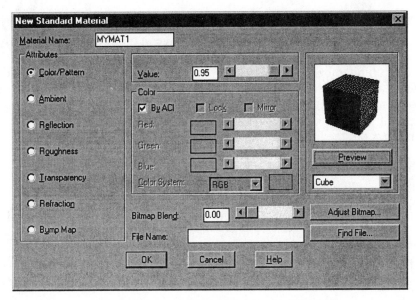

Figure 22-30 New Standard Material dialog box

A rendered sphere depicting the effect of rendering with the new material is displayed in the image tile.

6.   After defining the attributes, choose the **OK** button. In this manner, a new material has been defined.

EXPORTING A MATERIAL FROM A DRAWING TO THE LIBRARY OF MATERIALS

Sometimes you may have a material attached to a drawing, but the material may not be present in the library of materials. One way this situation can arise is when you have transferred a drawing that has materials assigned to it from one system to another system that does not have the materials in its library of materials. It is possible to export a material from a drawing to the library of materials. The following steps need to be performed:

1.   Open the drawing that has the material you want to export.
2.   Invoke the **Materials** Library dialog box.
3.   Select the material you want to export from the **Materials List:** column.
4.   Choose the **Export->** button. You will notice that the material name is appended to the list of materials in the **Library List**: column.
5.   With the **Save...** button just below the **Library List**: column, save the material in the present drawing to a library.

SAVING A RENDERING

A rendered image can be saved by rendering to a file or by rendering to the screen and then saving the image. Redisplaying a saved rendered image requires very little time compared to the time involved in rendering.

Saving a Rendering to a File

You can save a rendering directly to a file. One advantage of doing so is that you can replay the rendering in a short time. Another advantage is that when you render to the screen, the resolution of the rendering is limited by the resolution of your current display. Now if you render to a file, you can render to a higher resolution than that of your current display. Later on you can play this rendered image on a computer with a higher-resolution display. The rendered images can be saved in different formats, such as TGA, TIFF, BMP, PostScript, BMP, PCX, and IFF. To perform this operation, carry out the following steps:

1. Invoke the **Render** dialog box from the menu (Tools >Render > Render) or by entering **RENDER** at the Command prompt.
2. In the Destination area, select File in the pop-up window.
3. Choose the **More Options...** button. The **File Output Configuration** dialog box is displayed.
4. Specify the file type, rendering resolution, colors, and other options. Then choose the **OK** button. The **File Output Configuration** dialog box is cleared from the screen.
5. Choose the **Render** button in the **Render** dialog box. The **Rendering File** dialog box is displayed. Specify the name of the file to which you want to save the rendering, then choose the **OK** button.

Saving a Viewport Rendering

Command:	SAVEIMG

A rendered image in the viewport can be saved with the **SAVEIMG** command. In this case, the file formats that can be used are TGA, TIFF, and BMP. To perform this operation, carry out the following steps:

1. Invoke the **Render** dialog box and choose the **Render** button to render the object.
2. Invoke the **Save Image** dialog box.
3. Specify the file format (TGA, TIFF, BMP), size, and offsets for the image, then choose the **OK** button.
4. The **Image File** dialog box is displayed.
5. Specify the name of the file to which you want to save the rendering, and then choose the **OK** button. In this way, a rendered image in the viewport can be saved in the specified file format.

Saving a Render-Window Rendered Image

A rendered image in the Render window can be saved with the **SAVE** command, which is available in the **Render** window. In this case, the file format in which the Render-window rendered image can be saved is bitmap (.BMP). To perform this operation carry out the following steps:

1. Invoke the **Render** dialog box, and select the **Render Window** option in the **Destination** area. Choose the **Render** button to render the object. In our case, we have a single sphere to be rendered. Shortly after you select the object (sphere), the **Render window** is displayed. The rendered object is displayed in this window.
2. From the **File** menu, select **Options**. The **Window Render Options** dialog box is displayed. In this dialog box, you can set the size and color depth of the rendered image. In our case, the

size is 640 x 480 and the color depth is 8 bit.

3.   Next, choose the **Save** button in **Render window** to save the rendered image. The **Save BMP** dialog box is displayed. Enter the file name in the **File Name:** edit box, and then choose the **OK** button.

The problem with bitmap images is that when you scale them up, the images become nonuniform (blocky) and cannot be printed properly. On the other hand, if you scale down these images, they lose information.

REPLAYING A RENDERED IMAGE

In the previous section, saving a rendered image was explained. In this section, we will explain how to replay the saved rendered image.

Replaying a Rendered Image to a Viewport

| **Menu:** | Tools > Display Image > View |
| **Command:** | REPLAY |

If you have saved the rendered image in a TGA, TIFF, or BMP format, you can use the **REPLAY** command to display the image on the screen. This operation can be performed in the following manner.

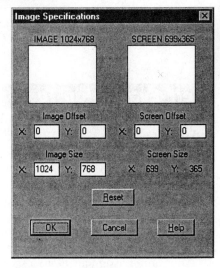

Figure 22-31 Image Specification dialog box

1.   Invoke the **REPLAY** command to display the **Replay** dialog box.
2.   Select the file you want to replay and choose the **OK** button.
3.   Next, the **Image Specification** dialog box (Figure 22-31) is displayed. You can define the size and offsets, or take the default full screen size for displaying the rendered image.
4.   After specifying the size and offsets, choose the **OK** button. The selected image is displayed on the screen per the specified size and offsets.

Replaying a Rendered Image to the Windows Render Window

1.   Select the **Open** option in the **Render Window File** menu.
2.   The **Render Open** dialog box (Figure 22-32) is displayed. Select the file you want AutoCAD to replay, and then choose the **OK** button. The selected image is displayed.

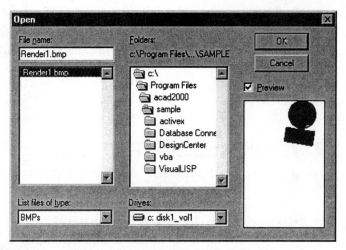

Figure 22-32 Render Open dialog box

Self-Evaluation Test

Answer the following questions, and then compare your answers to the answers given at the end of this chapter.

1. A rendered image makes it easier to visualize the shape and size of a 3D object. (T/F)

2. You can unload AutoCAD Render by invoking the **ARX** command. (T/F)

3. Falloff occurs only with a distant source of light. (T/F)

4. You cannot assign a view to a scene. (T/F)

5. You can modify a scene by changing the lights that are assigned to that scene. (T/F)

6. You can use a spotlight to create a light source that emits light in a conical fashion in the defined direction. (T/F)

7. The Phong render option produces a _____ quality image, but it is computationally _____ and takes more time to display the image.

8. Name the three methods that AutoCAD Render has provided for controlling light falloff:
    1._____ 2._____ 3._____

9. A _____ light source emits light in all directions, and the intensity of the emitted light is uniform.

10. A rendered image in the viewport can be saved with the _____ command.

11. Materials can be imported from a(n) _____ file.

12. A predefined material cannot be modified. (T/F)

13. By default, only _____ material is assigned to a new drawing.

14. Materials once attached to an object cannot be detached. (T/F)

15. In case you have saved the rendered image in a TGA, TIFF, or BMP format, you can use the _____ command to display the image.

Review Questions

1. Rendering is not an effective tool for communicating design ideas. (T/F)

2. When you select any AutoCAD Render command, AutoCAD Render is loaded automatically. (T/F)

3. You cannot render a drawing without inserting any lights or defining reflective characteristics of the surface. (T/F)

4. The Gouraud render option saves computation time, but does not generate a good quality image. (T/F)

5. You can increase or decrease the intensity of ambient light, but you cannot turn it completely off. (T/F)

6. AutoCAD Render allows you to control the appearance of an object by defining the reflective quality of the surface and by adding lights to get the desired effects. (T/F)

7. The gradation effect produced by the lights does not make any difference in the appearance of the object. (T/F)

8. By default, AutoCAD Render uses smooth shading to render the image. (T/F)

9. In the _____ option, the light falling on the object (brightness) is inversely proportional to the distance of the object from the light source (Brightness = 1/Distance).

10. _____ light does not have a source, and hence, has no location or direction.

11. The intensity of the light radiated by a point source _____ with the distance.

12. You can create a rendered image of a 3D object by using the _____ command.

13. You can load AutoCAD Render from the Command prompt by entering _____.

14. The intensity of light _____ as distance increases.

15. In the _____ option, the light falling on the object (brightness) is inversely proportional to the square of the distance of the object from the light source (Brightness = 1/Distance $^\wedge$ 2).

16. A distant light source emits _____ beam of light in one direction only.

17. The intensity of a light beam from a(n) _____ light source does not change with the distance.

18. What are the names of the two rendering shading types available in AutoCAD Render?
    1._____ 2._____

19. As the oblique angle _____, the amount of light reflected by the oblique surface decreases.

20. By defining _____, it is convenient to render different views with the required lighting arrangement.

21. Materials can only be assigned explicitly to objects. (T/F)

22. When you import a material from the library of materials to a drawing, the material and its properties are copied to the list of materials in the drawing. (T/F)

23. When you import a material from the library of materials to a drawing, it gets deleted from the library of materials. (T/F)

24. New materials can be defined. (T/F)

25. A material existing in a drawing can be exported to the library of materials (.mli file). (T/F)

26. A rendered image can be saved by _____.

27. Redisplaying a saved rendered image takes _____ time as compared to the time involved in rendering.

28. You can output a _____ directly to a hard-copy device if the hard-copy device driver software is loaded and the AutoCAD renderer is configured to output to hard copy.

29. Name the light sources that AutoCAD Render supports. _____

30. Define **intensity**. _____

31. What is **attenuation**? _____

Exercises

Exercise 2   *Mechanical*

Make the drawing shown in Figure 22-33. This drawing will then be used for rendering. You can also load this drawing from the accompanying disk. The name of the drawing is REND. The

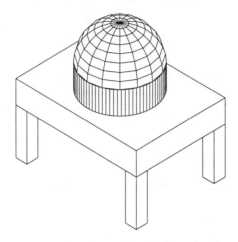

Figure 22-33 Drawing for Exercise 2

following is the description of one of the methods of making this drawing.
1.   Draw a rectangle 2.75 x 2.0.
2.   Use the **CHANGE** command to change the thickness to 0.25.
3.   Use the **3DFACE** command to draw a 3dface at the top of the rectangular box.
4.   Draw another rectangle 0.25 x 0.25.
5.   Use the **CHANGE** command to change the thickness to 1.0.
6.   Copy the leg of the table to four corners.
7.   Define the center point of the table as the new origin.
8.   Draw a circle at the top surface of the table. The center of the circle must be at the center of the table.
9.   Use the **CHANGE** command to change the thickness to 0.5.
10. Draw a dome that has a diameter of 1.5. The bottom edge of the dome must coincide with the top edge of the cylinder, as shown in Figure 22-33.
11. To render the drawing, first position a distant light source at (3,-2,5) and the target at (0,0,0). Now render the drawing.
12. Position a point light source at (-1,0,5) with the falloff set to Inverse Square. Render the object again.

Exercise 3

13. Modify the distant and the point light sources to study the effect of lighting on the rendering. Generate the 3D drawings as shown in Figures 22-34 and 22-35. Next, render the 3D view of the drawing after inserting the lights at appropriate locations so as to get a realistic 3D image of the

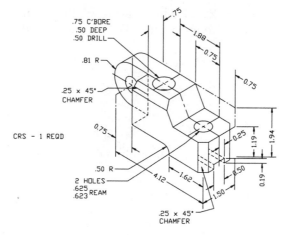

Figure 22-34 Drawing for Exercise 3

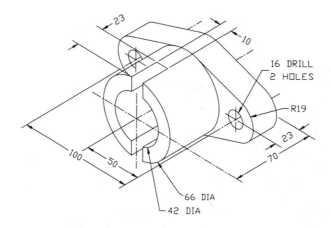

Figure 22-35 Drawing for Exercise 3

Exercise 4

object.
a. Draw different objects on the screen, and assign different materials explicitly to them. Then render the drawing.

Exercise 5   *General*

b.   Detach the materials previously assigned to the objects and assign new materials to them. Attach copper material to some layer on your system. Draw objects on this layer, and then render

Exercise 6   *General*

the drawing. Observe the results carefully.

a.   Define a new material named NEWMAT. Assign desired attribute values to it in the New Standard Material dialog box, and then attach this material to an object and render it to a viewport.

b.   Save the rendered image in the desired size and offset. Give it any name.

Exercise 7   *Mechanical*

c.   Redisplay the saved image on the screen.

Generate the 3D drawing shown in Figure 22-36. (Assume a value for the missing dimensions.). Next, render the 3D view of the drawing after inserting the lights at appropriate locations so as to

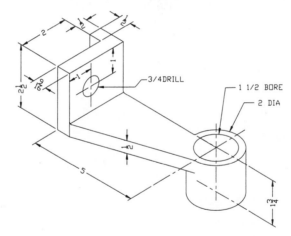

Figure 22-36 *Drawing for Exercise 7*

get a realistic 3D image of the object.

Appendix A

System Requirements and AutoCAD Installation

SYSTEM REQUIREMENTS

The following are the minimum system requirements for running AutoCAD 2000:

1. Operating systems: Windows® 95, Windows NT® 4.0, and Windows 98.
2. 32 MB of RAM
3. Pentium 133, or better/compatible processor
4. 130MB of hard disk space
5. 64MB of disk swap space (minimum)
6. 10MB of additional memory (RAM) for each concurrent session of AutoCAD
7. 2.5 MB of free disk space required at the time of installation for temporary files that are removed when installation is complete
8. 4X speed or faster CD-ROM for initial installation
9. 800 by 600 VGA video display
10. Mouse or digitizer (with Wintab driver) pointing device
11. IBM compatible parallel port and hardware lock for international single user and educational versions only
12. Serial port (for digitizers and some plotters)
13. Sound card for multimedia learning

The following hardware is optional:

1. Printer
2. Plotter
3. Digitizing tablet

INSTALLING AUTOCAD

The installation process has now been simplified that has made it easier and faster to install AutoCAD. Also, the installation procedure is more consistent with other Windows installations. The following steps describe the installation process for AutoCAD 2000:

1. The first step is to insert the AutoCAD 2000 CD in the CD-ROM drive of your computer.

2. If AutoPlay is enabled, the **Setup** dialog box is automatically displayed and the **InstallShield Wizard** is installed on your system. (AutoPlay is a Windows 95, Windows 98, and Windows NT 4.0 feature that automatically runs an executable file. The AutoPlay feature can be turned off by holding down the Shift key when you insert the CD in the CD-ROM drive, or by changing the play setting of the CD-ROM.) If your system matches the minimum hardware requirements, the **Welcome** dialog box is displayed on the screen.

3. Choose the **Next** button in the **Welcome** dialog box; the **Software License Agreement** dialog box is displayed on the screen.

4. After reading the Software License Agreement, select **Accept** to continue installation; the **Serial Number** dialog box appears on the screen.

5. Enter the Serial number and CD Key and then choose the **Next** button. (The Serial Number and the CD Key can be found at the back of the box containing the AutoCAD 2000 CD.). If the numbers you entered are correct, the **Personal Information** dialog box is displayed.

6. Enter the requested information and then choose the **Next** button. Each field must contain at least one character to display the next screen that shows the personal information you just entered. You can edit the information or choose the **Next** button to confirm the data. The **Destination Location** dialog box is displayed on the screen.

7. In the **Destination Location** dialog box, use the **Browse** button to select the directory where you want to install AutoCAD and then choose the **Next** button to display the **Setup Type** dialog box.

8. In the **Setup Type** dialog box select the type of setup you prefer (Typical, Full, Compact, or Custom). If you select Custom, the **Custom Components** dialog box is displayed. From this dialog box clear the check boxes for the components that you do not want to install and then choose the **Next** button. If your computer has enough hard disk space, the **Folder Name** dialog box is displayed.

9. In the **Folder Name** dialog box, accept the AutoCAD 2000 (default) folder or select any other folder. Choose the **Next** button; the **Setup Confirmation** dialog box is displayed.

10. If the selection you made is OK, choose the **Next** button in the **Setup Confirmation** dialog box. The installation program installs the program files on your system and when the installation is complete, the **Setup Complete** dialog box is displayed on the screen.

Appendix B

Express Tools

EXPRESS TOOLS

Express tools consist of programs, sample drawings, driver files, and font files that enhance the existing AutoCAD. The **Express tools** are productivity tools and are in the form of AutoLISP and ARX applications that help the user increase the speed of drafting. The **Express Tools** are Layers, Blocks, Text, Dimension, Modify, and Draw Tools.

LAYER TOOLS

Toolbar:	Express Layer Tools
Menu:	Express > Layers

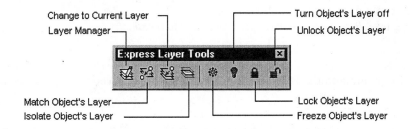

Figure C-1 Express Layer Tools toolbar

The different **layer tools** have been added to help in the manipulation of layers.

LMAN

 When you invoke this command the **Layer Manager: Save and Restore Layer Settings** dialog box is displayed. Through this dialog box you can save, edit, and restore the layers. The layers can also be exported to or imported from a LAY file.

LAYCUR

 This command changes the layer of the selected objects to the current layer. The prompt sequence is:

Command: **LAYCUR**
Select objects to be changed to the current layer
Select objects: *Select the objects whose layer you want to change*.
Select objects: *Press ENTER*.
1 object(s) changed to layer LAYER1 (the current layer).

LAYFRZ

 This command **freezes the layer** of the selected objects. The prompt sequence is:

Command: **LAYFRZ**
Select an object on the layer to be frozen or [Options/Undo]: *Select the objects whose layer you want to freeze or enter O for Options*.
If you enter **O** then the next prompt issued is:
[No nesting/Entity level nesting/]<Block level nesting>: *Select an option*.

The **No nesting** option freezes the layer of the object selected. If a block or an xref is selected, the layer in which they are inserted are also frozen. The **Entity level nesting** freezes the layers of the objects selected even if the selected object is nested in xref or block. In the **Block level nesting**, if a block is selected then the layer in which the block is inserted is frozen. But if an xref is selected, then the layer of the object is frozen and the xref layer is not frozen.

LAYISO

 With this command the layer of the selected objects is isolated by turning all other layers off. The prompt sequence is:

Command: **LAYISO**
Select object(s) on the layer(s) to be isolated.
Select objects: *Select the objects whose layer you want to isolate*.

LAYLCK

 With this command the layer of the selected objects is locked. The prompt sequence is:

Command: **LAYLCK**
Select an object on the layer to be locked: *Select the object whose layer you want to lock*.

LAYMCH

 With this command the layer(s) of the selected object(s) is changed such that it matches the layer of the selected destination object. The prompt sequence is:

Command: **LAYMCH**
Select objects to be changed.

Select objects: *Select the object whose layer you want to change.*
Select object on destination layer or [Type-it]:

LAYOFF

The layer of the selected object is turned off. The prompt sequence is:

Command: **LAYOFF**
Select an object on the layer to be turned off or [Options/Undo]: *Select an object or specify an option.*
If you enter **O** for Options the next prompt issued is:
[No nesting/Entity level nesting/]<Block level nesting>:
If you select an object on the current layer you are prompted:
Really want layer <layer name> (the CURRENT layer) off? <N>:

TEXT TOOLS

Toolbar:	Express Text Tools
Menu:	Express > Text

The Express text commands are used to manipulate the text and attributes.

Figure C-2 Express Text Tools toolbar

ARCTEXT

This command is used to place the text along an arc. The prompt sequence is:

Command: **ARCTEXT**
Select an Arc or an Arc Aligned Text: *Select the arc or a text.*

The **Arc Aligned Text Workshop-Create** dialog box is displayed. You can enter the text in the **Text** edit box and also choose a text style and a font for the text from the respective drop-down lists in the dialog box. You can also edit an arc aligned text by selecting it and then changing the different properties through the dialog box. In the Properties area of the dialog box you can change the height, width, and offset from the arc. After making the changes, choose the **OK** button.

TEXTEXP

This command explodes text or Mtext into geometry (lines and arcs) and enables you to assign a thickness or elevation. The prompt sequence is:

Command: **TEXTEXP**
Select text to be EXPLODED: *Select the text using the object selection process.*

RTEXT

Remote Text (RTEXT) objects are displayed as other text or Mtext objects are, but the source for the Rtext is either an ASCII text file or the value of a DIESEL expression. Rtext objects can be edited by the **RTEDIT** command. The command sequence for **RTEXT** is:

Command: **RTEXT**
Current text style: STANDARD Text height: 0.2000 Text rotation: 0
Enter an option [Style/Height/Rotation/File/Diesel] <File>: *Specify an option or source file.*

RTEDIT

This command is used to edit existing Remote Text (Rtext) objects. The prompt sequence is:

Command: **RTEDIT**
Select objects: Select
Style: STANDARD Height: 0.1000, Rotation: 0
Enter an option [Style/Height/Rotation/Edit]: *Select an option.*

The **Style** option specifies a new style for the Rtext object. The **Height** option allows you to change the height of the Rtext entity and also you can list the objects' current height. With the **Rotation** option you can specify a new rotation angle for the Rtext objects. With the **Edit** option, you can edit the Rtext objects.

TEXTFIT

 This command shrinks or stretches text objects such that they fit between the specified startpoint and endpoints. The prompt sequence is:

Command: **TEXTFIT**
Select text to stretch or shrink: *Select the text.*
Specify end point or [Start point]: *Enter* **S** *for new start point or select an ending point.*
If you enter **S** for a new starting point then the next prompt is:
Pick new starting point: *Specify a new starting point.*
Ending point: *Specify the ending point.*

TEXTMASK

 This command hides the objects behind the selected text. **TEXTMASK** works together with the **WIPEOUT** bonus routine. The prompt sequence is:

Command: **TEXTMASK**
Current settings: Offset factor = <default>, Mask type = <default>
Select text objects to mask or [Masktype/Offset]: *Select the text or an option.*

If you want to remove mask from a text object that has been masked by the **TEXTMASK** command, you can use the **TEXTUNMASK** command. The command sequence is:

Select text or MText object from which mask is to be removed.
Select objects: *Select a masked text object.*

TXT2MTXT

This command converts one or more lines of text into Mtext. The command sequence is:

Command: **TXT2MTXT**
Select text objects or press ENTER to set Options:
Select objects: *Select the text object to be converted.*

STANDARD TOOLS

Toolbar:	Express Standard Toolbar

The Express standard commands are based on the standard AutoCAD modify commands.

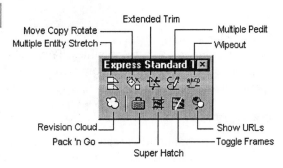

Figure C-3 Express Standard Tools toolbar

MSTRETCH

 This command is used to stretch multiple objects. If the complete object is selected, this command works like the **MOVE** command. The prompt sequence is:

Command: **MSTRETCH**
Define crossing windows or crossing polygons...
Options: Crossing Polygon or Crossing first point
Specify an option [CP/C] <Crossing first point>: *Specify first point of a crossing window.*
Specify an option [CP/C/Undo] <Crossing first point>: *Specify second point of a crossing window.*
Specify an option [CP/C/Undo] <Crossing first point>: *Press ENTER.*
Specify an option [Remove objects] <Base point>: *Select a base point.*
Second base point: *Select a new base point.*

MOCORO

This single command can be used to move, copy, rotate, and scale the selected object or objects. The prompt sequence is:

Command: **MOCORO**
Select objects: *Select objects for modifying.*
Base point: *Select a base point.*
[Move/Copy/Rotate/Scale/Base/Undo]<eXit>: *Specify an option.*

EXTRIM

This command allows you to trim the objects by specifying a cutting edge that can be a polyline, line, circle, or arc. The prompt sequence is:

Command: **EXTRIM**
Pick a POLYLINE, LINE, CIRCLE, ARC, ELLIPSE, IMAGE or TEXT for cutting edge...
Select objects: *Select a cutting edge.*
Pick the side to trim on: *Select a point to specify the side of trimming.*

The portion of the objects in the selected side are trimmed.

MPEDIT

 This command edits multiple polylines and also converts lines and arcs to polylines. The prompt sequence is:

Command: **MPEDIT**
Select objects: *Select objects.*
Convert Lines and Arcs to polylines? [Yes/No] <Yes>: *Press ENTER.*
Enter an option [Open/Close/Join/Width/Fit/Spline/Decurve/Ltype gen/eXit] <eXit>: *Specify an option.*

WIPEOUT

 With this command the selected area is covered with the background color so as to hide it. The area you want to select must be defined by a closed polyline. The prompt sequence is:

Command: **WIPEOUT**
Select first point or [Frame/New from Polyline] <New>: *Press ENTER to create a new wipeout.*
Select a polyline: *Select the closed polyline.*
Erase polyline? Yes/No <No>: *Enter your choice.*

The area inside the polyline boundary is wiped out. You can also enter **F** at the **Select first point or [Frame/New from Polyline] <New>:** prompt to display the following prompt:

Select an option [OFF/ON] <ON>:

The **On** option displays the wipeout frame on all viewport entities. The **Off** options turns off the wipeout for all wipeout entities.

REVCLOUD

 This command is used to create a revision cloud shape from a polyline made up of sequential arcs. The prompt sequence is:

Command: **REVCLOUD**
Arc Length = <current>, Arc style = <current>l
Pick cloud starting point or [eXit/Options] <eXit>: *Specify start point or enter **O**.*

AutoCAD will display the **Revcloud Options** dialog box. As you move the crosshairs, continuous arcs are created having the set arc lengths. When the crosshairs are close to the start point, the last arc joins the start point to close the Revcloud and you come out of the command.

PACK

 With this command all the files associated with a particular drawing are copied to a specified location.

Command: **PACK**

The **Pack & Go** dialog box is displayed. All the filenames with their size, date, and drawing version (if any) are listed in the dialog box. You can choose the Tree View tab to display the files as a directory tree. If you choose the **Browse** button the **Choose Directory** dialog box is displayed where you can specify a path to copy files. The **Copy to** button copies the files to the specified path. The **Print** button allows you to print a report of information displayed in the dialog box. The **Report** button displays the **Report** dialog box specifying information about associated files.

SUPERHATCH

 The Superhatch option is much like the AutoCAD **HATCH** command, but you can use images, xrefs, block, or Wipeout objects as hatch pattern. The command sequence varies with the type of hatch patterns.

For **images**:

Insertion point <0,0>: *Specify an insertion point.*
Base image size: Width: 1.00, Height: 0.24 <Unitless>
Scale factor <1>: *Specify the scale factor for insertion.*

For **blocks and xrefs**:

Insertion point: *Specify an insertion point.*
X scale factor <1> / Corner / XYZ: *Specify the X scale factor.*
Y scale factor (default=X): *Specify the Y scale factor or press ENTER.*
Rotation angle <0>: *Specify the rotation angle.*
Select a window around the block to define column and row tile distances.
Specify block [Extents] First corner <magenta rectang>: *Specify a frame for tiling and press ENTER when done.*

After the required hatch pattern is selected, the next command sequence is:

Selecting visible objects for boundary detection...Done.
Specify an option [Advanced options] <Internal point>: *Specify an internal point for hatching* boundary or *A for advanced options and press ENTER.*
Preparing hatch objects for display...
Done.

TFRAMES

 TFRAMES command can be used for toggling the state of frames for Wipeout and image objects. For example, if frames are on, this command will turn them off, and vice versa.

Command: **TFRAMES**.

SHOWURLS

 This command shows all the embedded URLs (Uniform Resource Locators) in the drawing. **SHOWURLS** also allows you to edit the associated URLs.

BLOCK TOOLS

Toolbar:	Express Block Tools

The Block Tools are used to edit the blocks
and xrefs in the drawing.

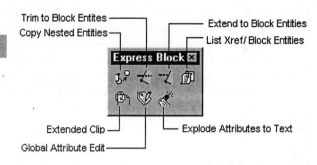

Trim to Block Entites — Copy Nested Entities — Extend to Block Entities — List Xref/ Block Entities — Extended Clip — Global Attribute Edit — Explode Attributes to Text

Figure C-3 Express Block Tools toolbar

NCOPY

This command is used to copy
objects that are nested in an xref
or block. The prompt sequence is:

Command: **NCOPY**
Select nested objects to copy: *Select objects.*
Specify base point or displacement, or [Multiple]: *Select the base point or M for multiple copies.*
Second point of displacement: *Select the second point.*

BTRIM

With this command the selected objects are trimmed to a selected cutting edge. The cutting edge can be a block or an xref object. The prompt sequence is:

Command: **BTRIM**
Select cutting edges: *Select objects for cutting edge.*
Select object to trim or [Project/Edge/Undo]: *Select the objects to trim or select an option.*

BEXTEND

With this command the selected objects are extended to a selected boundary. The boundary can be a block or an xref object. The prompt sequence is:

Command: **BEXTEND**
Select edges for extend: Select objects for boundary edge.
Select object to extend or [Project/Edge/Undo]: *Select the objects to extend or select an option.*

XLIST

This command lists the different properties such as object, block name, layer, color, and
linetype of a nested object in a block.

Command: **XLIST**
Select nested xref or block object to list: *Select the object.*

An **Xref/Block Nested Object List** dialog box is displayed that lists all the information about the
selected object.

CLIPIT

This command allows you to separate some portion of block, xref drawing, or image by
removing the display of the remaining objects using a polyline, circle, or arc. The prompt

sequence is:

> Command: **CLIPIT**
> Pick a POLYLINE, CIRCLE, ARC, ELLIPSE, or TEXT object for clipping edge...
> Select objects: *Select a clipping edge.*
> Pick an IMAGE, a WIPEOUT, or an XREF/BLOCK to clip: *Select the object.*
> Enter max error distance for resolution of arcs <0.0200>: *Select a resolution value.*

The curved boundary edge is traversed with a series of short straight segments, and the resolution of arc is the distance between the midpoint of the given segment and the arc. Hence a large error value creates less segments and faster speed but the arcs become less smooth.

GATTE

 With this command the attribute values are changed globally for all insertion of specific block. The prompt sequence is:

> Command: **GATTE**
> Select block or attribute [Block name]: *Enter **B** to use a block name or select the block.*
> Known tag names for block: MDLN.
> Select attribute or type attribute name: *Select attribute or enter the tag name.*
> Number of inserts in drawing = 2   Process all of them? [Yes/No] <Yes>: *Select Yes to process all and No to process the selected attributes.*

BURST

 This command explodes a block so that the attribute values are changed to text entities. The prompt sequence is:

> Command: **BURST**
> Select objects: *Select the block containing attributes.*

DIMENSION TOOLS
DIMEX

With this command the named dimension styles and all their settings are exported to an exported file.

> Command: **DIMEX**

The **Dimension Style Export** dialog box is displayed. You can enter the desired **DIM** filename in the **Export filename:** edit box. The list of dimension styles available in the current drawing are listed in the Available **Dimension Style** list box. You can select the dimension style you wish to write to the ASCII file. In the Text Style Options area you can choose the **Full Text Style Information** radio button or **Text Style Name Only** radio button.

DIMIM

With this command the named dimension styles are imported from a DIM file into the current drawing.

> Command: **DIMIM**

The **Dimension Style Import** dialog box is displayed. You can enter the desired dimension style name in the **Import filename** edit box, or from the Open dialog box that can be invoked by choosing the **Browse** button. You can retain the dimension style in the current drawing or overwrite it by selecting the appropriate radio button in the Import Options area.

QLATTACH

This command is used to attach the leader line to an annotation object such as Mtext, Tolerance, or Block Reference Object. The prompt sequence is:

> Command: **QLATTACH**
> Select Leader: *Select the leader.*
> Select Annotation: *Select the annotation object.*

QLATTACHSET

This command is used to globally attach the leader line to an annotation object such as Mtext, Tolerance, or Block Reference Object. The prompt sequence is:

> Command: **QLATTACHSET**
> Select objects: *Select the leader and the annotation objects.*
> Select objects: *Press ENTER.*
> Number of Leaders=n
> Number with annotations detached=n

QLDETACHSET

This command is used to detach the leader line from an annotation object such as Mtext, Tolerance, or Block Reference Object. The prompt sequence is:

> Command: **QLDETACHSET**
> Select objects: *Select the leader and the annotation objects.*
> Select objects: *Press ENTER.*
> Number of Leaders=n
> Number of annotations detached=n

SELECTION TOOLS
GETSEL

This command creates a temporary selection set.

> Command: **GETSEL**
> Select an object on the Source layer <*>: *Select object or press ENTER for all layers.*
> Select an object of the Type you want <*>: *Select entity type or press ENTER for all.*

SSTOOLS

This command creates "anti" or exclusionary selection sets. The selection set method is preceeded by **EX**, and all the drawing objects except the ones in the selection area are selected. For example, if **W** is for window selection method, then **EXW** is for the exclusionary window selection method.

Command: **'EXW, 'EXC, 'EXP, 'EXF, 'EXWP, 'EXCP**

TOOLS COMMANDS

Menu:	Express > Tools

The Express tool commands help you to manage the AutoCAD environment.

FULLSCREEN

The **FULLSCREEN** command resizes the drawing screen to a maximum area. It also hides the title bar and the menu bar in the AutoCAD window. The **FULLSCREEN** command toggles the screen size between normal and full screen.

MKLTYPE

The **MKLTYPE** command creates a new linetype based on the selected objects. This is a quick and useful method of making new ready-to-use linetype. The command line sequence is:

Command: **MKLTYPE**
Select a ".LIN" file: <acad.lin>: *Enter the name of the linetype filename.*
Enter linetype name: *Specify a new linetype name.*
Linetype description: *Specify description of the new linetype.*
Starting point for line definition: *Specify a starting point of the line.*
Ending point for line definition: *Specify an ending point.*
Select objects: *Select a drawing object to be used to define the linetype.*

MKSHAPE

The **MKSHAPE** command creates a new shape definition based on the selected drawing objects. The command sequence is:

Command: **MKSHAPE**
Select a ".SHP" file: <drawing.shp>: *Enter a shape file name to save your shape.*
Reading shape file: drawing.shp...Done.
Enter the name of the shape: *Specify a shape name.*
Enter resolution <n>: *Enter a resolution value (between 8 - 32767)*
Insertion base point: *Specify a basepoint.*
Select objects: *Select drawing objects for definition purposes, which will be treated as though they are of a continuous linetype.*
Determining geometry extents...Done.
Building coord lists...Done.
Formatting coords...Done.
Writing new shape...Done.

Compiling shape/font description file
Compilation successful. Output file C:\drawing.shx contains **n** bytes.
Use the **SHAPE** command to place shapes in your drawing.

REDIR

The **REDIR** command redefines hard-coded paths in xrefs, images, shapes, styles, and rtext. The command line sequence is:

Command: **REDIR**
Current REDIRMODE: Styles,Xrefs,Images,Rtext
Find and replace directory names
Enter old directory (use '*' for all), or ? <options>: *Enter the name of directory to be replaced or press ENTER for* **options**.
Replace "*" with: *Enter the name of the directory to be replaced with the old one.*

REDIRMODE

The **REDIRMODE** command sets the options for the **REDIR** command by specifying which objects are to be redefined. If you enter the **REDIRMODE** in the command line, the **REDIRMODE** dialog box appears where you can select object types to be redefined. The command line sequence is:

Command: **REDIRMODE**
Current REDIRMODE: Styles,Xrefs,Images,Rtext
Replace directories in Xrefs,Styles,Images,Rtext. <current>: *Enter the options (object types).*
Current REDIRMODE: *Selected object types.*

XDATA

This command allows you to attach extended entity data (xdata) to a selected entity.

Command: **XDATA**
Select object: *Select the objects.*
Application name: *Enter the name of application.*
[3Real/DIR/DISP/DIST/Hand/Int/LAyer/LOng/Pos/Real/SCale/STr/eXit]<eXit>: *Enter an option to attach the extended entity data.*

XDLIST

This command lists all the xdata associated with an object.

Command: **XDLIST**
Select object: *Select the objects.*
Application name <*>: *Enter the name of the application or press ENTER to list all.*

ADDITIONAL COMMANDS

These commands are accessed through the command line only.

BLOCK?

This command lists the different entities in a block definition.

> Command: **BLOCK?**
> Enter block name <Return to select>: *Press ENTER.*
> Select a block: *Select the block using a selection method.*
> Enter an entity type <Return for all>: *Press ENTER for all entities or enter a type of entity.*

BCOUNT

This command counts the number of insertions of each block in the drawing or in the selected objects. It then displays it in a tabular form.

> Command: **BCOUNT**
> Press Enter to select all or...
> Select objects: *Press ENTER or select objects.*

EXPRESSMENU

This command loads the Express tools menu and displays **Express** menu on the AutoCAD menu bar. Since AutoCAD can display only 24 menus at a time, make sure that you have 23 or fewer menu items before running the **EXPRESSMENU** command.

> Command: **EXPRESSMENU**

EXPRESSTOOLS

This command loads the **AutoCAD Express Tools** libraries, places the **Express** directory on the search path, and also loads and places the **Express** menu on the **AutoCAD** menu bar. **EXPESSTOOLS** adds the **Express menu group** in the **Toolbars** dialog box from where you can view **Express** toolbars.

> Command: **EXPRESSTOOLS**

JULIAN

This command contains the AutoLISP functions for conversion of Julian date.

> Command: **DATE**
> Tue 1999/5/25 10:25:52.391

LSP

This command displays a list of all the AutoLISP commands that are available at the Command prompt. The command sequence is:

> Command: **LSP**
> Enter an option [?/Commands/Functions/Variables/Load]: *Specify an option.*

SSX

This command creates a selection set.

> Command: **SSX**
> Select object/<None>: *Select a template object.*
> >>Block name/Color/Entity/Flag/LAyer/LType/Pick/Style/Thickness/Vector: *Enter an option.*

It creates a selection set exactly like the selected entity or very similar to the filter list that you have adjusted.

OPERATING SYSTEM COMMANDS
TTC2TFF

The **TTC2TFF** command converts a **TTC** file (which is a TrueType font collection file that contains one or more TrueType fonts) to **TTF** files. This command is valid only in **DOS command syntax**.

> DOS command line: **TTC2TFF**.

Appendix C

AutoCAD Commands

Command	Description and Options

3D Draws a 3D polygon entities with surfaces. Options:

B - Draws a 3D box, C - Draws a wire frame having a cone shape, DI - Draws a 3D dish shaped (lower half of a sphere) polygon mesh by specifying the center and then the diameter or radius, DO - Draws the 3D upper half of the spherical polygon mesh by specifying the center and then the diameter or radius, M - Draws a polygon mesh by specifying the corners and the M and N sizes, P - Draws a 3D tetrahedron or a pyramid by specifying the relevant number of base points and the apex point or the top points, S - Draws a spherical polygon mesh by specifying the center and then the diameter or radius, T - Draws a polygon mesh having a toroidal shape. It is drawn by specifying the center and then the diameter or the radius, W - Draws a polygon wire frame having the shape of a wedge. It is drawn by specifying a corner, height, length, and width.

3DARRAY Draws a 3D rectangular or polar array. Options:

R - Rectangular 3D array, P - Polar array

3DCLIP Starts the interactive 3D viewing and opens the **Adjust Clipping Planes** window.

3DCORBIT Starts the interactive 3D viewing and enables you to set the drawing objects in the 3D view into continuous motion.

3DDISTANCE Simulates the effect of moving the camera closer or farther to the drawing objects.

3DFACE Draws a 3D surface with three or four sides. Options:

First point, I- Invisible.

3DMESH Draws a polygon mesh by specifying the size of M and N and the location of the vertices.

3DORBIT Controls the interactive viewing of 3D objects in the drawing area.

3DPAN Allows you to drag the view horizontally, vertically or diagonally.

3DPOLY Draws a 3D polyline having segments of a straight line. Options:
Point - Draws the 3D polyline to the specified point, C - Closes the 3D polyline by joining the last point with the first point, U - Deletes the last segment.

3DSIN Displays a dialog box that allows you to import the specified objects in a 3D Studio file.

3DSOUT Displays a dialog box that exports the AutoCAD objects with surface characteristics to a 3D studio file.

3DSWIVEL Simulates the effects of turning the camera on a tripod.

3DZOOM Allows you to zoom in and out on the view.

ABOUT Displays AutoCAD version and serial numbers and other information.

ACISIN Imports an ASCII ACIS file into the AutoCAD drawing.

ACISOUT AutoCAD exports the selected 3D solid objects to an ASCII ACIS file.

ADCCLOSE Closes the AutoCAD Design Center.

ADCENTER Starts and displays AutoCAD Design Center.

ADCNAVIGATE Directs the Desktop in AutoCAD DesignCenter.

ALIGN Allows specified objects to align with other 2D or 3D objects by moving and rotating them.

AMECONVERT Changes the regions and solids of AME to region and solids of AutoCAD.

APERTURE Controls the size of the object snap target box.

APPLOAD Displays a dialog box that loads certain applications such as AutoLISP, ADS, and ARX at startup.

ARC Draws an arc of any size. The default method is to specify two endpoints and a point along the arc. Options:
A - Included angle, C - Center point, D - Starting direction, E - Endpoint, L - Length of chord, R - Radius.

AREA Computes the area and perimeter of different objects and of a region formed by specifying a sequence of points. Options:
A - Add mode, F - First point (Area by specifying points), O - Area of the object, S - Subtract mode.

ARRAY Creates specified number of copies of a selected object. Options:
P - Polar array, R - Rectangular array.

ARX Loads, unloads, and provides information about ARX applications.

ATTDEF Creates an attribute definition (characteristics of an attribute). Options:
I - Invisible mode: Attribute remains invisible, C - Constant mode: Constant value of attribute, V - Verify mode: Verifies attribute value is correct, P - Preset mode: Default value to attribute.

ATTDISP Controls the visibility of the attribute globally. Options:
ON - Attributes made visible, OFF - Attributes made invisible, N - Current visibility kept.

ATTEDIT Edits attributes within a block or blocks. Options:
Y - Yes (one attribute at a time), N - No (all attributes at a time).

ATTEXT Attribute information is extracted from the drawing. Options:
C - CDF: Comma-Delimited File, D - DXF:

Drawing Interchange File, S - SDF: Space-Delimited File, O - Drawing objects.

ATTREDEF An existing block is redefined and its attributes are updated.

AUDIT Identifies errors in a drawing. Options: Y - Corrects the errors, N - Informs about the error without correcting it

BACKGROUND Creates a background for your scene.

BASE Sets the point of origin for inserting a drawing into another drawing.

BHATCH A specified enclosed area is filled with an associative hatch pattern through a dialog box. Previewing a hatch and adjusting the boundary is also possible. Options:
I - Internal point, P - Properties, S - Select, R - Remove islands, A - Advanced.

'BLIPMODE Controls the appearance of marker blip that is displayed on the screen when a point is picked. Options:
ON - Marker blip displayed, OFF - Marker blip not displayed.

BLOCK Creates a compound object as a block definition from a set of entities. Options:
? - Lists names of previously defined blocks.

BLOCKICON Generates preview images for the blocks created in Release 14 or earlier.

BMPOUT Creates a bitmap image of the drawing and saves the screen to a file having a .bmp extension.

BOUNDARY Creates a polyline or region of a boundary that defines an enclosed area. Options:
A - Advanced Options.

BOX Creates a solid box that is 3D in nature with its base parallel to the XY plane. Options:

C - Corner of box, CE - Center of box.

BREAK Removes specified portions of an object or splits the object. Options:
F - Respecifies first point.

BROWSER Launches the default Web browser defined in the registry of the system.

CAL Calculates expressions that can be mathematical as well as geometrical.

CAMERA Sets a different camera location and target locations.

CHAMFER Connects two nonparallel objects with a beveled line. Options:
A - Chamfer distance is set using angle and distance, D - Sets chamfer distance, P - Chamfers entire polyline, T - Controls the trimming of the edges to chamfer line endpoints, M - Trim method.

CHANGE Alters the properties of selected objects. Options:
C - Change point - Changes lines, circles, Text, Attribute Definitions, Blocks, P - Changes properties like Color, Elev, LAyer, LType, Thickness.

CHPROP Alters the drawing properties of selected objects. Options:
C - Changes color, LA - Changes layer, LT - Changes linetype, S - Changes linetype scale factor, LW - Changes lineweight, T - Changes thickness, PL - Changes the plotstyle of the selected objects.

CIRCLE Draws a circle using any of the four methods available. Options:
C - Circle drawn on the basis of center point and diameter or radius, 3P - Drawn on the basis of 3 points on circumference, 2P - Drawn on the basis of 2 endpoints of the diameter, TTR - Circle drawn tangent to two objects with a specified radius.

Appendix-C

CLOSE Closes the current drawing if no change has occurred since last save or will prompt you to save the drawing.

COLOR Sets color for subsequent objects drawn. Options:
value - Sets color by number (1-255), name - Sets color by name, Byblock - The current color setting is inherited by the block at the time of insertion, Bylayer - Objects inherit the color of the layer in which they are drawn.

COMPILE Shape files and PostScript font files are compiled.

CONE Draws a 3D solid cone. Options:
Center - Cone having circular base, E - Cone having elliptical base.

COPY Draws a copy of the selected object leaving the original object intact. The default method is to specify the base point. Options:
M - Multiple copies of object in single **COPY** command.

COPYBASE Copies the drawing objects with a specified base point.

COPYCLIP Copies the selected objects to the Clipboard.

COPYHIST Copies the text in the command line history to the Clipboard.

COPYLINK Copies the current view to the Clipboard to link with other OLE applications.

CUTCLIP Copies objects to the Clipboard but erases them from the drawing.

CYLINDER Draws a 3D solid cylinder. Options:
Center - Specifies the center of the circular base, E - Forms an elliptical base.

DBCCLOSE Closes the **dbConnect Manager**.

DBCONNECT Provides an AutoCAD interface to external database tables by starting **dbConnect Manager**.

DBLIST Lists all the database information for each object in the drawing.

DDEDIT Displays a dialog box that allows the user to edit text and attribute definitions.

DDPTYPE Displays a dialog box that sets the point style and also the size of the point object.

DDVPOINT Controls the direction of 3D views through a dialog box.

DELAY The execution of the next command is postponed for a specified time duration. In other words a specified pause is provided within the script.

DIM and DIM1 Dimensioning mode is invoked and permits the use of dimension subcommands from AutoCAD's previous releases.

DIMALIGNED Creates a linear dimension aligned to the specified points or object.

DIMANGULAR Creates an angular dimension.

DIMBASELINE Starts drawing from the baseline of the previous dimension. The new dimension can be linear, angular, or ordinate.

DIMCENTER Draws center mark or center lines in circles and arcs.

DIMCONTINUE Starts drawing a new dimension from the second extension line of the previous or selected dimension. The new dimension can be a linear, angular, or ordinate dimension.

DIMDIAMETER Draws diameter dimensions for different circles and arcs.

DIMEDIT Edits dimension text and extension

lines.Options:

> H - Dimension text moved back to default position, N - Dimension text is replaced, R - Dimension text is rotated, O - Extension lines placed at obliquing angle.

DIMLINEAR Draws linear dimensions.

DIMORDINATE Creates ordinate dimensions.

DIMOVERRIDE The settings of the dimensioning system variables concerning the dimension object are overridden. The current dimension style is not affected.

DIMRADIUS Draws radial dimensions for different circles and arcs.

DIMSTYLE New dimension styles are created and the existing ones are modified. Options:

> R - Dimensioning system variable setting changed, S - Current settings of dimensioning variables saved, ST - Current values of dimensioning variables displayed, V - Dimensioning variable setting of a style is listed, A - Selected dimension objects are updated, ? - Named dimension styles are listed.

DIMTEDIT Dimension text is moved and rotated. Options:

> A - Angle of dimension text changed, H - Dimension text moved to default position, C - Dimension text center justified, L - Dimension text left justified, R - Dimension text right justified.

DIST Distance and angle between two points is measured.

DIVIDE Places blocks or points as markers at equal distance along the length or perimeter of an entity, thus dividing it into a specified number of equal parts. Options:

> B - Places blocks as markers.

DONUT Draws wide polyline with specified in-

side and outside diameters, thus forming a ring.

DRAGMODE Controls the dragging feature for appropriate commands. Options:

> ON - Permits dragging, OFF - Ignores dragging, A - Permits dragging wherever possible.

DRAWORDER Changes the display order of images and other objects.

DSETTINGS Specifies the settings for Snap mode, grid, polar, and object snap tracking.

DSVIEWER Displays the Arial View window.

DVIEW Parallel projection or perspective views are defined. Options:

> CA - Sets camera position by rotating about the target, TA - Sets target position by rotating about the camera, D - Camera to target distance is set, PO - Locates target and camera points, PA - Pans image, Z - Zooms In/Out, TW - Tilts view around line of sight, CL - The view is clipped in front and back, H - Hidden lines removed on selected objects, OFF - Perspective viewing turned off, U - Last DVIEW operation reversed.

DWGPROPS Sets and displays all the properties of the current drawing.

DXBIN Specially coded binary files are imported into a drawing.

EDGE The visibility of 3D sides is altered. By default the selected edge is hidden. Options:

> D - Display mode invisible edges are highlighted.

EDGESURF A 3D polygon mesh is created with four adjoining edges that define a Coons surface patch.

ELEV The elevation and extrusion thickness is set for the new objects.

Appendix-C

ELLIPSE Draws ellipses or elliptical arcs using different options. Specifying the axis endpoint is the default method. Options:
A - Draws elliptical arc, C - Specifies the center point of the ellipse, I - Draws isometric circle in current isometric plane.

ERASE Erases the selected objects from the drawing.

EXPLODE Compound objects (blocks, dimensions, polylines, 3D solids, regions, polygon meshes, multilines) are broken into their constituent parts.

EXPORT Objects are saved to other file formats via a dialog box.

EXPRESSTOOLS Activates the installed AutoCAD Express Tools if currently not available.

EXTEND Lengthens a selected entity to meet another entity. Options:
P - Specifies projection mode like UCS and View, E - Controls the extension to implied or actual edge, U - Latest extension is undone.

EXTRUDE Solids are created by extruding 2D entities along a selected path. By default, height of extrusion is to be specified which extrudes the object along the positive Z axis. Options:
P - Extrusion path is selected.

FILL Controls whether multilines, traces, solids, or wide polylines are filled or not filled. Options:
ON - Fill mode is enabled, OFF - Fill mode is disabled.

FILLET The edges of two specified lines, arcs, or circles are filleted by construction of an arc of specified radius. The default method is to specify the two objects. Options:
P - Entire polyline is filleted, R - The radius of the fillet arc is specified, T - Controls the trimming of the edges to fillet arc endpoints.

FILTER Creates a list of properties on the basis of which the objects are selected.

FIND Finds, replaces, selects, or zooms to specified text through a dialog box.

FOG Provides visual clues for the apparent distance of objects.

GRAPHSCR Flips to the graphics window from the text window.

GRID A grid of dots at specified spacing is displayed. Options:
Grid spacing(X) - Grid set to specified value, ON - Grid turned on at current spacing, OFF - Grid turned off, S - Grid spacing set to current Snap interval, A - Grid set to different spacing in X and Y.

GROUP Creates and changes object selection groups, which are sets of objects having specific names. Options:
? - Lists names and descriptions of all the group, O - Changes the numerical order of objects in the group, A - Adds objects to the group, R - Remove objects from the group, E - Explodes the group into its component objects, REN - Assigns a new name to an existing group, S - Specifies whether a group is selectable, C - Creates a new group.

HATCH A specified area is filled with a selected pattern. Options:

? - Lists the hatch patterns in acad.pat file, name - A pattern name as defined in acad.pat file is specified, U - User-defined hatch pattern is specified. U can be followed by a comma and a hatch style, S - Specifies a solid fill.

HATCHEDIT Edits a hatch block through a dialog box. It sets the pattern type and properties and then applies it to a block. Options:

D - Removes the associative quality from an associative hatch, S - Changes the hatch style type, P - Specifies new hatch properties.

HELP(F1) Displays help for a specific command and also lists the commands and data entry options.

HIDE Regenerating of a 3D object is performed with the removal of hidden lines.

HYPERLINK Attaches a hyperlink to a graphical object or modifies an existing hyperlink through a dialog box.

HYPERLINKOPTIONS Controls the visibility of the hyperlink cursor and the display of hyperlink tooltips. Options:
Y - Shows the hyperlink cursor and tooltip,
N - Does not display.

ID The UCS coordinates of a specified point are displayed.

IMAGE Inserts images into an AutoCAD drawing file.

IMAGEADJUST Controls the brightness, contrast, and fade values of the selected image.

IMAGEATTACH Attaches a new image object and definition.

IMAGECLIP Creates new clipping boundaries for single image objects.

IMAGEFRAME Controls the display of image frame on the screen. Options:
On - Displays image frames, Off- Hides image frames.

IMAGEQUALITY Controls the display quality of images. Options:
H - Produces high quality display, D - Produces low quality display.

IMPORT Imports the different file formats into an AutoCAD drawing via a dialog box.

INSERT Places a previously defined named block or drawing into the current drawing.

INSERTOBJ Inserts a previously linked or embedded object.

INTERFERE Highlights all of the interfering solids and then creates new solids from the intersections of the interfering pairs of solids.

INTERSECT A new composite solid is created from the intersecting region of two or more solids.

ISOPLANE An isometric plane is selected to be the current plane for an orthogonal drawing. Options:
L - Left-hand plane, T - Top plane,
R - Right-hand plane.

LAYER Creates layers and sets different properties for the specified layers. Options:
? - Lists defined layers, M - Creates a layer and makes it the current layer, S - Makes a specified already existing layer current, N - Creates one or more new layers, ON - Turns on the specified layers, OFF - Turns off the specified layers, C - Sets the color of the specified layer, L - Sets the linetype of the specified layer, LW -Changes the lineweight, P - Controls whether visible layers are plotted, PS - Sets the plot style assigned to a layer, F - Makes a layer invisible by freezing it, T - The frozen layer is thawed, LO - Locks layers, thus prevents editing on them, U - Unlocks specified locked layers.

LAYOUT Creates a new layout and renames, copies, saves, or deletes an existing layout. Options:
C - Copies a layout, D - Deletes a layout,
N - Creates a new layout tab, T - Creates a new template, R - Renames a layout, SA - Saves a layout, S - Makes a layout current,
? - Lists all the available layouts.

LAYOUTWIZARD Starts the Layout wizard to designate page and plot settings for a new layout.

LEADER Creates a line segment with an arrowhead that connects the text to a feature. The leader is created from a specified point to another point depending upon the options. Options:
> A - Annotation is inserted at the end of leader line, F - Controls the type of leader (Spline, Straight, Arrow), U - The last vertex point is removed.

LENGTHEN Alters the length of specified entities and the included angle of arcs. Options:
> DE - Lengthens the object by a specified incremental distance, P - Alters the length by a specified percentage of its total length, T - Alters the length by specified total absolute length, DY - The object is lengthened to where its endpoint is dragged.

LIGHT Controls the lighting effects in the model space via a dialog box. It creates, modifies, deletes the lights and controls the color system in a drawing. It manages different lights (Point light, Distant light, Spotlight) through a series of dialog boxes.

LIMITS Sets the drawing boundaries (and WCS grid extents) for the current space. Options:
> Lower - Specifies 2 points-lower left corner and the left-upper right corner, ON - Limits checking is enabled, OFF - Limits checking is disabled.

LINE Draws straight line segments of any length by specifying the endpoints. Options:
> ENTER - Continues from end of previous line or arc, U - Removes the most recent segment, C - Closes polygon.

LINETYPE Defines line characteristics, loads linetypes and sets them for new entities. It also creates new linetype definitions to a library file. Options:

? - Lists linetypes in a file, C - Creates new linetype definition, L - Loads an already existing linetype definition, Sets linetype for new entities, S - Set the current linetype.

LIST Lists database information (type, layer, X,Y,Z position, thickness, and so on) about the specified entity.

LOAD Loads the shapes from the shape file to be used by the **SHAPE** command.

LOGFILEOFF The log file already opened is closed by this command.

LOGFILEON The subsequent contents of the text window are recorded into the log file.

LSEDIT Lets you edit a landscape object.

LSLIB Lets you maintain libraries of landscape objects.

LSNEW Lets you add realistic landscape items, such as trees and bushes, to your drawing.

LTSCALE Sets the global scale factor of the linetype so as to alter the relative length of dashes and dots.

LWEIGHT Sets the current lineweight, lineweight display options, and lineweight units.

MASSPROP Calculates and lists the mass characteristics of 2D and 3D objects. The properties displayed are Area, Perimeter, Bounding box, Centroid. For Coplanar regions, additional properties displayed are Moments of Inertia, Products of Inertia, Radii of Gyration, and Principal Moments. The properties displayed for solids are Mass, Volume, and the properties of the Coplanar region.

MATCHPROP Copies the properties from one object to one or more objects.

MATLIB Displays a dialog box that lists all the

predefined materials (material list) and lists the materials in the selected library (library list). It also imports and exports materials between those two lists.

MEASURE Places blocks or points as markers at measured intervals along the length or perimeter of an entity. Options:
    B - Places blocks as markers.

MENU Loads a customized menu file into the menu area. The menu file contains the command strings and menu syntax.

MENULOAD Displays a dialog box that loads and permits you to add partial menu files to an already present base menu file.

MENUUNLOAD Displays the same dialog box as in the case of the **MENULOAD** command that can also be used to unload the partial menu files.

MINSERT Places multiple copies of a previously drawn named block or drawing into the current drawing in a rectangular array. Options:
    ? - Lists the defined block definitions, ~ - Displays a dialog box.

MIRROR Reflects objects so as to create their mirror images about a specified line.

MIRROR3D Reflects objects so as to create their mirror images about a specified plane. Options:
    3points - 3points specify the mirroring plane, O - The plane of a planar object specifies the mirroring plane, L - The previous mirroring plane is taken as the present one, Z - The point on the plane and another point on the Z axis (normal) of the plane specifies the mirroring plane, V - A point on the viewing plane specifies the mirroring plane, XY/YZ/ZX - The mirroring plane is aligned to any one of the standard planes.

MLEDIT Displays a dialog box that controls intersection between multiple parallel lines and

edits them. Different types of cross, tee, corner joints, and vertices can be created between multilines via the dialog box. It is also possible to cut and weld segments of a single multiline.

MLINE Draws multiple parallel lines between two points. Options:
    J - Justification- How multiline is drawn between two points, S - Scale- Sets the width of the multiline, ST - Sets the multiline style.

MLSTYLE Displays a dialog box that creates a multiline style, makes a specific style current, saves, adds a style to the current list, renames a style, adds a description to a style, and loads a style from the library file. It also controls the element properties (number, offset, color, linetype) and the multiline properties (start and end caps, angle, background color).

MODEL Switches from a layout tab to the Model tab and makes it current.

MOVE Moves objects from one location to another by specifying a displacement.

MSLIDE Creates a slide file from the current display.

MSPACE Switches to model space in a floating viewport from paper space.

MTEXT Creates paragraph text within a specified text boundary. Displays a dialog box where you can specify the different options for the multiline text.

MULTIPLE Causes the repetition of the next command until it is cancelled.

MVIEW Creates viewports and controls the number and layout of paper space viewports. You specify diagonal corners of new viewport as the default option. Options:
    ON - Viewport is turned on, OFF - Viewport is turned off, H - Hideplot- Hidden lines removed during plotting, F - Fit- Single

viewport created that fills the display area completely, L - Locks the selected viewport, O - Specifies a closed polygon to convert it into viewport, P - Creates an irregularly shaped viewport using specified points, 2 - The specified area is divided into two viewports either horizontally or vertically, 3 - The specified area is divided into 3 viewports, 4 - The specified area is divided into 4 viewports, R - Restore - Viewport configurations changed into individual viewports.

MVSETUP The specifications of a drawing are set. Depending upon the system variable TILEMODE, the working of MVSETUP is different. When TILEMODE is On, drawing scale factor, units type, and paper size is set and lastly a bounding box is drawn. When TILEMODE is Off a set of floating viewports is created. Options (TILEMODE Off):

A - Aligns the view in a viewport with another viewport. The view can be panned in a specified direction, aligned horizontally, vertically, or rotated, C - The viewport can be created, S - Sets the scale factor of objects in the viewport, O - Options - The layer can be set, reset limits, set units, Xref attach, T - Creates a title block and drawing border, U - Reverses the previous operation

NEW Displays a dialog box that creates a new drawing.

OFFSET Creates offset curves, concentric circles, and parallel lines at a specified distance from the original object. Options:

value - Specify the offset distance, T - Through- The offset object passes through the specified point.

OLELINKS Updates, changes, and cancels existing OLE links.

OLESCALE Displays the OLE Properties through a dialog box.

OOPS Restores those entities that have been erased by the last **ERASE** command.

OPEN Displays a dialog box through which an existing drawing can be opened. The dialog box also displays the directory, files, preview, name of the file, and the pattern.

OPTIONS Customizes the AutoCAD settings through a dialog box.

ORTHO The movement of the cursor is restrained to only vertical or horizontal directions and aligned with the grid. Options:

ON - Constrains cursor movement, OFF - Does not constrain cursor movement.

OSNAP Specifies a point at an exact location on an object by setting the Object Snap modes. Options:

END - Closest endpoint of arc (arcs and lines include polyline segments), elliptical arc, ray, mline, line and closest corner of trace, solid, 3D face, MID - Midpoint of arc, elliptical arc, spline, ellipse, ray, solid, xline, mline, or line, INT - Intersection of line, arc, spline, elliptical arc, ellipse, ray, xline, mline, or circle, EXT - Snaps to the extension point of an object, APP - Apparent or extended (projected) intersection (which may not actually intersect in 3D space) of line, arc, spline, elliptical arc, ellipse, ray, xline, mline, or circle, CEN - Center of arc, elliptical arc, ellipse, or circle, QUA - Quadrant point of arc, elliptical arc, ellipse, solid, or circle, PER - Point perpendicular to arc, elliptical arc, ellipse, spline, ray, xline, mline, line, solid, or arc, TAN - Tangent to arc, elliptical arc, ellipse, or circle, NOD - Point object, INS - Insertion point of text, block, shape, or attribute, NEA - Nearest point of arc, elliptical arc, ellipse, spline, ray, xline, mline, line, circle, or point, QUI - First snap point, NON - Turns Object Snap mode off.

PAGESETUP Specifies the layout page, paper size, plotting device, and settings for each new layout through a dialog box.

PAN   Moves the drawing display by a specified displacement.

PARTIALOAD   Loads additional geometry into a partially opened drawing.

PARTIALOPEN   Loads additional geometry from a selected view or layer into a drawing.

PASTEBLOCK   Pastes a copied block in a new drawing.

PASTECLIP   Inserts data from the Clipboard.

PASTEORIG   Pastes a copied object in the new drawing using the geometric coordinates of the original drawing.

PASTESPEC   Inserts data from the Clipboard and controls the format of the data.

PCINWIZARD   Starts a wizard to import PCP and PC2 configuration file plot settings in current layout or model tab.

PEDIT   Editing of 2D polyline, 3D polyline, or 3D mesh. Options:
  2D polyline C - Closes polyline segment, O - Closing segment removed, J - Joins to polyline, W - Specifies uniform width, E - Edits the vertices. The first vertex is marked by placing an X. Editing includes moving the X to next or previous vertex, adding a new vertex, setting the first vertex for break, moving the vertex, regenerating, straightening and attaching a tangent direction to the current vertex, F - Fits arc curves smoothly to the polyline, replacing each line segment with a pair of arcs, S - Vertices are used as a frame for spline curve, L - Linetype generation in a continuous pattern, U - Reverses the previous operation, 3D polyline C - Closes polyline segment, O - Closing segment removed, E - Edits the vertices. Same suboptions as in

2D Edit except the tangent suboption, S - Vertices are used as a frame for spline curve, D - Removes a spline curve to its control frame, U - reverses the previous option, X - Exits **PEDIT**. 3D polygon mesh E - Edits vertices. The first vertex is marked by placing an X. Editing includes moving the X to next or previous vertex, moving the X marker to the next vertex or the previous vertex in the N direction, moving the marker to the next or previous vertex in the M direction, regenerating the mesh, S - Fits a smooth surface, D - The control point polygon mesh is restored, Mclose - M-direction polylines are closed, Mopen - M-direction polylines are opened, Nclose - N-direction polylines are closed, Nopen - N-direction polylines are opened, U - Reverses editing operations as far back as the beginning of the **PEDIT** session.

PFACE   A 3D polyface mesh is created vertex by vertex.

PLAN   Allows you to view the drawing from plan view of a User Coordinate System. Options:
  C - Plan view of the current UCS, U - Plan view of the specified UCS, W - Plan view of the World Coordinate System.

PLINE   Draws 2D polylines. The default is to draw a polyline between two specified points. Options:
  A - Arc mode- Arc segments can be added to polyline. The arc segment starts from the endpoint of the previous polyline segment and can be drawn by specifying the endpoint of the arc, the included angle, center of the arc, starting direction of the arc, halfwidth of the arc, radius of the arc, Width. You can also close the polyline with the arc segment, or reverse the previous operation or you can shift to the Line mode, C - Closes the polyline, H - Sets the halfwidth, L - Draws polyline of specified length, U - Last polyline segment is removed, W - The width of the next segment is specified.

PLOT Displays a dialog box that allows you to plot the drawing to the plotting device or file. Through a series of dialog boxes you can set the different parameters, device information, drawing extents and limits, plot size, paper size, orientation, plot scale, rotation, and origin. You can also plot a view or a specific portion of the drawing and also preview the plot.

PLOTSTYLE Sets the current plot style for new drawing objects, or the assigned plot style for selected objects through a dialog box.

PLOTTERMANAGER Opens the **Plotter Manager** and enables you to launch Add-a-plotter wizard and Plotter Configuration Editor.

POINT Draws a point object at a specified location.

POLYGON Draws a polygon (closed polyline object) having specified number of sides. Options:
    C - Specifies the center of polygon. Suboptions:
    I - Inscribed in the circle, C - Circumscribed about the circle.
    E - Defines one edge of the polygon.

PREVIEW Shows how the drawing will look when it is printed or plotted.

PROPERTIES Controls properties of a drawing object and displays Properties window.

PROPERTIESCLOSE Closes the Properties window.

PSDRAG An imported PostScript file is dragged into place by the **PSIN** command and as it is dragged, the **PSDRAG** command controls its appearance. Options:
    0 - Only the bounding box and the file name of the image is displayed as the image is being dragged, 1 - The rendered PostScript image is displayed.

PSETUPIN Imports a user-defined page setup into a new drawing layout.

PSFILL A 2D polyline boundary is filled with a PostScript fill pattern. Options:
    name - Fills the polyline with the specified pattern, ? - Lists all the previously defined PostScript fill patterns.

PSIN A PostScript file is inserted into a drawing.

PSOUT Creates an Encapsulated PostScript file into which the current view of the drawing is exported through a dialog box.

PSPACE Switches from a model space viewport to paper space.

PURGE Removes those references from the database that are not being used. Options:
    B - Removes unused blocks, D - Removes unused dimstyles, LA - Removes unused layers, LT - Removes unused linetypes, P - Removes unused plotstyle, SH - Removes unused shape files, ST - Removes unused text styles, M - Removes unused mline styles, A - Removes all unused objects.

QDIM Quickly creates a dimension on the drawing objects.

QLEADER Quickly creates a leader and its annotation.

QSAVE Saves and backs up the drawing without asking for a filename.

QLEADER Quickly creates selection sets based on filtering criteria.

QTEXT Sets the text and the attribute objects to be displayed without drawing the text detail. Options:
    ON - Text displayed as a bounding box.
    OFF - Quick text mode off.

QUIT Exits AutoCAD without saving.

RAY Draws a semi-infinite line used as a construction line.

RECOVER Recovers a damaged and corrupted drawing.

RECTANG Creates a polyline rectangle by specifying the diagonally opposite corners.

REDEFINE Restores an AutoCAD built-in command that has been previously overridden by **UNDEFINE**.

REDO The effect of the previous command if it was **UNDO** is reversed.

REDRAW Cleans up the current viewport by removing the blip marks and other stray pixels and redrawing missing portions of objects.

REDRAWALL Refreshes or cleans up all the viewports.

REFCLOSE Saves back or discards changes made during in-place editing of a reference.

REFEDIT Selects an external reference for editing.

REFSET Adds or removes objects from a working set during in-place editing of a reference.

REGEN Regenerates the current viewport.

REGENALL Regenerates all the viewports.

REGENAUTO Controls automatic regeneration of the drawing. Options:
    ON - Permits automatic regeneration.
    OFF - Does not permit automatic regeneration.

REGION Region entities (2D enclosed areas) are created from a selection set.

REINIT Reinitializes the I/O ports, digitizer, display, or parameters file through a dialog box.

RENAME Alters the name of entities. Options: B - Renames block, D - Renames dimstyle, LA - Renames layers, LT - Renames linetype, S - Renames style, U - Renames UCS, VI - Renames view, VP - Renames viewport configuration.

RENDER Displays a dialog box that shades a 3D wireframe or solid, so that a realistically shaded image is created. It is possible to render the current scene or just the specified objects. You can also control the color map and the shading of different materials.

REPLAY The BMP, TGA, or TIFF images are displayed via a dialog box.

RESUME Resumes an interrupted script.

REVOLVE By revolving a 2D entity (polygon, closed polyline, circle, ellipse, donuts, and so on), a solid is formed. Options:
    point - The axis of revolution is specified by two points, O - The axis of revolution is specified by selecting an existing line or a segment polyline, X - The positive X axis used as the axis direction, Y - The positive Y axis used as the axis direction.

REVSURF A polygon mesh is constructed by rotating a curve or profile around a specified axis.

RMAT Displays a dialog box that manages the materials used for rendering. A new material can be created or the existing ones can be modified through a series of dialog boxes. It is possible to adjust the value and color of the materials. AutoCAD's color index can also be attached by layers or by using a color wheel.

ROTATE Rotates specified entities about a base point. Options:
    angle - Rotates object through a specified

angle, R - Rotates object with respect to the reference angle.

ROTATE3D   Rotates object about a 3D axis. Options:

2points - The axis of rotation is given by specifying 2 points, A - Axis by object- The axis of rotation is aligned with an object, L - The previous rotation axis is considered, V - The axis of rotation is aligned with the viewing direction, X/Y/Z - The axis of rotation is aligned with any one of the axes (X-axis, Y-axis, Z-axis).

RPREF   Displays a dialog box that controls the rendering preferences. It controls the color map, the behavior of the **RENDER** command by default, rendering display, and the image output setting. Through a series of subdialog boxes, the type of shading used and 3D solid faces can be controlled. You can also set the color and the aspect ratio of the output file.

RSCRIPT   Repeats a script continuously.

RULESURF   Creates a polygon mesh representing a ruled surface between two curves.

SAVE   A name is requested under which the drawing is saved. If the drawing is already named, then it is saved under the current filename.

SAVEAS   An unnamed drawing is saved with a filename or the current drawing is renamed.

SAVEIMG   Displays a dialog box that saves a rendered image to a file. Through the subdialog boxes, image compression for TGA and TIFF formats is possible.

SCALE   The size of the existing objects is changed. The default is to specify a scale factor. Options:

R - The object is scaled according to the reference length and a new length.

SCENE   Controls different scenes (particular view) in model space. Through a series of dialog boxes all the scenes in the current drawing are listed, new scenes can be added, scene names can be modified, and the lights can be controlled in the scene.

SCRIPT   Executes a command script.

SECTION   Creates regions from the intersection of a plane and solids. Options:

3points - Specifying 3 points on sectioning plane, O - Sectioning plane is aligned with the object, Z - Sectioning plane is aligned with the plane's normal direction, V - Sectioning plane is aligned with the viewing plane of current viewport, XY - Sectioning plane aligned with XY plane of UCS, YZ - Sectioning plane aligned with YZ plane of UCS, ZX - Sectioning plane aligned with ZX plane of UCS.

SELECT   Creates a selection set of specified group of objects. Options:

AU - Automatic selection, A - Add mode - Objects are added to the selection set, ALL - Selects all objects, BOX - Objects inside or crossing a rectangle are selected, C - Objects are selected that lie inside and crossing an area specified by two points, CP - Those objects are selected that lie inside and crossing the polygon created by specifying points around the objects, F - Those objects are selected that are crossing the specified fence, G - Objects within a group are selected, L - Recently created object is selected, M - Objects are picked without highlighting them, P - Recent selection set is selected, R - Remove mode - Objects can be removed from the selection set, SI - Selects first object or a set of objects, U - Removes the most recently added object from the selection set, W - Selects those objects that lie completely inside an area specified by two points, WP - Selects those objects that lie completely inside an area specified by picking points around the objects.

SETUV Lets you map materials onto geometry.

SETVAR Sets the values of the system variables. Options:

? - Lists the variables with their current values.

SHADEMODE Displays a shaded picture of the drawing in the current viewport. Options:

2D - Displays the objects using lines and curves to represent the boundaries,

3D - Displays the objects using lines and curves to represent the boundaries, 3D wireframe - Displays the objects using 3D wireframe representation, H - Hides lines representing back faces, F - Shades the objects between the polygon faces, G - Shades the objects and smooths the edges, L - Combines the Flat Shaded and Wireframe options, O - Combines the Gouraud Shaded and Wireframe options.

SHAPE Predefined shapes are inserted. Options:

? - Lists the shape names.

SHELL Permits the access to the commands in the operating system while in AutoCAD.

SHOWMAT Lists the material type and attachment method for a selected object.

SKETCH Allows you to draw freehand drawings. Options:

P - Pen- sketching pen raised and lowered, X - Reports the number of temporary lines drawn and then exits SKETCH Q - Temporary lines discarded and then exits SKETCH R - Temporary lines recorded as permanent, E - Removes portion of the temporary line, C - Pen lowered for sketching, . - Draws a straight line from endpoint of sketched line to current position of pen.

SLICE Solid is cut with a plane. Options:

3points - Cutting plane specified by defining 3 points, O - Cutting plane aligned with an object (Circle, ellipse, elliptical arc, 2D spline, or polyline), Z - Cutting plane specified by locating a point on Z-axis, V - Cutting plane aligned to the viewing plane of the current viewport, XY - Cutting plane aligned with the XY plane, YZ - Cutting plane aligned with the YZ plane, ZX - Cutting plane aligned with the ZX plane.

'SNAP The movement of the cursor is constrained to the snap spacing. Options:

ON - Snap mode is turned on, OFF - Snap mode is turned off, A - Sets different X and Y spacings, R - Snap grid is rotated, S - Sets the style (Standard or Isometric) of the snap grid, T - Specifies the snap type (Polar or Grid).

SOLDRAW Generates profiles and sections in viewports created with **SOLVIEW**.

SOLID Draws polygons that are solid-filled.

SOLIDEDIT Edits faces and edges of 3D solid objects.

SOLPROF Creates profile images of three-dimensional solids.

SOLVIEW Creates floating viewports using orthographic projection to lay out multi and sectional view drawings of 3D solid and body objects.

SPELL Allows spellcheck of text objects in a drawing. If an ambiguous word is found, then the dialog box is displayed that lists the alternatives for the word, or permits you to replace the current word with another one, or add the word to the dictionary.

SPHERE A 3D solid sphere is drawn. Options:

R - Radius of the sphere, D - Diameter of the sphere.

SPLINE Draws smooth spline curves between points. Options:

Point - Specify points to define the spline curve. Suboptions:

Point - Adds spline curve segments by specifying points, C - Spline curve is closed, F - Fit, Tolerance - The tolerance for fitting is changed, O - 2D or 3D spline-fit polylines are changed to splines.

SPLINEDIT   Allows you to edit a spline entity. Options:

F - Fit data is edited. Suboptions:

A - Fit points are added, C - An open spline is closed, O - A closed spline is opened, D - Fit points are removed, M - Fit points are moved, P - A spline fit data is removed from database, T - Beginning and end tangents are edited, L - Tolerance value for spline fit are changed, X - Exits fit data option, C - An open spline is closed, O - A closed spline is opened, M - Move Vertex - The position of the control vertices is changed, R - Refines a spline by adding control points, or by increasing its order, or by changing the weight, E - Spline direction is reversed, U - Reverses the previous operation of SPLINEDIT.

STATS   Displays a dialog box that provides the rendering statistics. It also saves the statistics to a file.

STATUS   Lists the drawing statistics, modes, and extents.

STLOUT   Creates a binary or ASCII file and stores the solid in the specified file.

STRETCH   Stretches lines, arcs, and polylines by moving the endpoints to another specified location, and moves the objects.

STYLE   Creates new text styles or modifies the existing ones through a dialog box.

STYLESMANAGER   Displays Plot style Manager dialog box.

SUBTRACT   Subtracts the area of one set of regions from another and subtracts the volume of one set of solids from another, thus creating a new composite region or solid.

SYSWINDOWS   Arranges windows and is equivalent to standard Window menu options in Windows applications.

TABLET   Aligns the tablet with the coordinate system of a paper drawing. Options:

ON - Tablet mode is turned on, OFF - Tablet mode is turned off, CAL - Calibrates the tablet, CFG - Configures tablet menu area and screen pointing area.

TABSURF   Creates a polygon mesh that represents a tabulated surface formed from a path curve and direction vector.

TEXT   Writes text using a variety of character patterns. Displays text on screen as it is entered. Options:

Start Point - Specifies a start point for the text object, J - Controls justification of the text, Suboptions:

A - Specifies both text height and text orientation, F - Specifies that text fits within an area, C - Center aligned, M - Horizontally aligned, R - Right justified, TL - Top left, TC - Top centered, TR - Top right, ML - Middle left, MC - Middle centered, MR - Middle right, BL - Bottom left, BC - Bottom centered, BR - Bottom right. S - Specifies the text style, which determines the appearance of the text characters.

TEXTSCR   Flips to the text window from the graphics window.

TIME   The date and time of drawing creation is displayed. It also displays the time and the date when the current drawing was last updated and controls an elapsed timer. Options:

D - Displays the updated times, O - Elapsed timer is turned on, OFF - Elapsed timer is turned off, R - Resets the user elapsed timer.

TOLERANCE   Creates and adds geometric tolerances to a drawing through a dialog box.

TOOLBAR   Displays, hides, and customizes toolbars.

TORUS   Draws a solid having the shape of a donut. Options:
R - Radius of the tube, D - Diameter of the tube.

TRACE   Draws filled lines having a specified width.

TRANSPARENCY   Controls whether background pixels in an image are transparent or opaque.

TREESTAT   Displays the current spatial index (position of objects in space) of a drawing. The information includes the number of nodes, number of objects, depth of branch, and so forth.

TRIM   Removes the extra portion of an entity which extends beyond a specified boundary. Options:
P - Sets projection mode, E - Controls trimming of objects until the implied edge, U - Reverses the previous operation of the **TRIM** command, O - Specifies the object to trim.

U   Reverses the most recent operation.

UCS   Sets and modifies the user coordinate system. Options:
W - Current UCS set to World Coordinate System, N - New UCS, M - Moves the origin, G - Specifies one of the six orthographic UCSs, P - Restores the previous UCS, R - Restores a saved UCS so that it becomes the current, S - Saves the current UCS to a specified name, D - Removes the specified UCS from the list, A - Applies the current UCS setting, ? - Lists names of user coordinate systems.

UCSICON   Manages the location and the visibility of the UCS icon. Options:
ON - Coordinate system icon is enabled, OFF - Coordinate system icon is disabled, A - Icon is changed in all active viewports, N - Icon displayed at the lower left corner, OR - Icon displayed at the origin of current coordinate system.

UCSMAN   Manages defined user coordinate systems through a dialog box.

UNDEFINE   A built-in AutoCAD command is disabled.

UNDO   Reverses the effect of commands. Options:
N - The effect of a specified number of previous commands used is reversed, A - The effect of the menu items is reversed by a single U command, C - The **UNDO** command is limited or is turned off, BE - A number of operations are grouped together and are treated as a single operation, E - The group is terminated, M - Mark - A marker is placed in the undo information, B - Back - Undoes all work until the marker is encountered.

UNION   Combines the area of two or more regions, or the volume of two or more solids to create a composite region or solid.

UNITS   Sets the coordinate and angle display formats and precision.

VBAIDE   Displays the Visual Basic Editor.

VBALOAD   Loads a global VBA project into the current AutoCAD session.

VBAMAN   Loads, unloads, saves, creates, embeds, and extracts VBA projects.

VBARUN   Runs a VBA macro applications.

VBASTMT   Executes a VBA statement on the

AutoCAD command line.

VBAUNLOAD Unloads a VBA global project.

VIEW The graphics display is saved and restored as a view with a specified name through a dialog box.

VLISP Displays the Visual LISP interactive development environment (IDE).

VIEWRES Controls the appearance of objects by setting their resolution in the current viewport.

VPCLIP Clips the specified viewport objects.

VPLAYER Controls the visibility of layers in different viewports. Options:
? - Lists the frozen layers in a specified viewport, F - Layers are frozen in current, or all, or specified viewport, T - Layers are thawed in current, or all, or specified viewport, R - Rests the layers default visibility, N - New layers that are frozen in all viewports are created, V - Viewport Visibility Default - Controls thawing and freezing of layers.

VPOINT The viewing direction for 3D visualization. Options:
ENTER - Displays compass and axis tripod for controlling viewing direction, V - Specifies a point from which drawing can be viewed, R - New direction using two angles is specified.

VPORTS Divides the graphics display into a number of viewports through a dialog box.

VSLIDE Displays an existing raster image slide file in the current viewport.

WBLOCK Writes a block definition or specified objects to a new disk file through a dialog box.

WEDGE Creates a 3D solid in the shape of a wedge having its one of the faces as tapered and sloping. Options:
point - Specifies the first corner of the wedge. Suboptions:
point - Specifies the other corner of the wedge, C - Wedge having sides of equal length, L - Wedge with specified length, width, and height, CE - Creates wedge with specified center point. Suboptions:
point - Specifies the other corner of the wedge, C - Creates wedge having all sides equal, L - Creates wedge with specified length, width, and height.

WMFIN Imports a Windows metafile.

WMFOPTS Sets options for WMFIN.

WMFOUT Saves objects to a Windows metafile.

XATTACH Attaches an external reference to the current drawing.

XBIND Adds Xref's dependent symbols to a drawing through a dialog box.

XCLIP Defines an xref clipping boundary and sets the front or back clipping planes. Options:
ON - Displays the clipped portion, OFF - Displays all of the geometry of the xref ignoring the clipping boundary, C - Sets the front and back clipping planes, D - Removes a clipping boundary, P - Automatically draws a polyline coincident with the clipping boundary, N - Defines a rectangular or polygonal clipping boundary.

XLINE Creates a line of infinite length. Options:
point - Specifies the point through which the xline passes, H - Creates a horizontal xline, V - Creates a vertical xline, A - Creates a xline at an angle, B - Creates an xline through the vertex of two lines so that it bisects the angle between those two lines, O - Creates an xline parallel to another linear object.

XPLODE   Breaks a compound object into its individual objects. Options:

   G - Changes selected objects. Suboptions:

   E - Explodes the entire compound object,

   A - Sets color, linetype, layer of the component entities, C - Sets the color, LA - Sets the layer, LT - Sets the linetype, I - Sets all the properties to that of the original compound object, I - Changes selected objects one by one.

XREF   Manages external references to a drawing through Xref Manager.

ZOOM   Changes the display of the entities in the current drawing. Options:

   value - Scale(X/XP) - Changes the display by a specified scale factor, Scale X - Zoom relative to current scale, Scale XP - Scale relative to paper space, A - Zooms the entire drawing in current viewport, C - Displays at a specified center point, D - Displays the portion of the drawing with a view box, E - Displays the drawing extents, S - Zooms the display at a specified scale factor, W - Displays an area specified by two corners of the window, Realtime - Using the pointing device, zooms interactively to a logical extent.

Appendix **D**

AutoCAD System Variables

Variable Name	Type and Description

ACADLSPASDOC   Integer
Controls whether AutoCAD loads the LISP file into every drawing or just the first drawing opened in an AutoCAD session.

ACADPREFIX   String
The ACADPREFIX variable contains the direction path for support files specified by the ACAD environment variable. Path separators are attached if needed. This is a read-only variable.

ACADVER   String
The ACADVER variable contains the AutoCAD version number, which can have values. This variable is different from the DXF file $ACADVER header variable, which stores the drawing database level number. This is a read-only variable.

ACISOUTVER   Integer
The ACISOUTVER variable controls the ACIS version of SAT files created using the ACISOUT command. Initial value is 40.

AFLAGS   Integer
The AFLAGS variable establishes the attribute flags for the **ATTDEF** command bitcode. The initial value for this variable is 0. Basically the value of this variable is the addition of the following:
0 - No attribute mode selected, 1 - Invisible, 2 - Constant, 4 - Verify, 8 - Preset

ANGBASE   Real
The ANGBASE variable establishes the base angle 0 in relation to the current UCS. This variable is saved in the drawing and has an initial value of 0.0000.

ANGDIR   Integer
The ANGDIR variable establishes the angle from angle 0 in relation to the prevailing UCS. This variable is saved in the drawing and has an initial value of 0.

0 - Direction is counterclockwise, 1 - Direction is clockwise.

APBOX   Integer
The APBOX variable turns the AutoSnap aperture box on or off. Initial value is 1.
0 - AutoSnap aperture box is not displayed, 1 - AutoSnap aperture box is displayed.

APERTURE   Integer
The APERTURE variable defines the object snap target height in pixels. This variable is saved in registry and has an initial value of 10.

AREA   Real
The most recently calculated area with commands such as AREA, LIST, or DBLIST is stored in this variable. This is a read-only variable. You can examine this variable through the **SETVAR** command.

ATTDIA   Integer
With the ATTDIA variable you can specify whether you want to enter the attribute value through the INSERT dialog box or from the command line. This variable is saved in the drawing and has an initial value of 0.
0 - Attribute values can be specified on the command line, 1 - Attribute values can be specified in the dialog box.

ATTMODE   Integer
The ATTMODE variable controls the Attribute Display mode and is saved in the drawing and its initial value is 1.
0 - Attribute Display mode is off, 1 - Normal, 2 - On

ATTREQ   Integer
The value contained in the ATTREQ variable determines whether **INSERT** command uses the default attribute settings when the blocks are being inserted.
0 - The default values for all the attributes are used, 1 - This is also the initial value and enables prompts or dialog box for at-

tribute values (depending on the value of ATTDIA variable).

AUDITCTL   Integer
The AUDITCTL variable determines whether an .adt file (audit report file) will be created by AutoCAD. This variable is saved in registry.
0 - Does not allow writing of .adt files. This is also the initial value, 1 - Allows writing of .adt files.

AUNITS   Integer
The AUNITS variable establishes the Angular Units mode and is saved in the drawing.
0 - Decimal degrees (initial value), 1 - Degrees/ minutes/ seconds, 2 - Gradians, 3 - Radians, 4 - Surveyor's units

AUPREC   Integer
The AUPREC variable establishes the angular units decimal places. This variable is saved in the drawing and has an initial value of 0.

AUTOSNAP   Integer
The AUTOSNAP variable controls the display of AutoSnap marker, and Snap Tips and turns the AutoSnap magnet on or off. Initial value is 7.
0 - Turns off marker, Snap Tip and magnet, 1 - Turns on marker, 2 - Turns on Snap tooltips, 4 - Turns on magnet

BACKZ   Real
The BACKZ variable contains the back clipping plane offset (in current drawing units) from the target plane for the current viewport. You can determine the distance between the back clipping plane and the camera point by subtracting the BACKZ value from the camera to target distance. This variable is saved in the drawing.

BINDTYPE   Integer
Controls how xref names are handled when

binding xrefs or editing xrefs in place. The value is not saved and its initial value is 0.
0 - Traditional binding behavior, 1 - Insert-like behavior.

BLIPMODE Integer

The visibility of the blip marks is controlled by BLIPMODE variable. This variable is saved in the drawing and its initial value is 0.
0 - Blip marks are not visible, 1 - Blip marks are visible.

CDATE Real

The calendar date and time is stored in this variable. This is a read-only variable.

CECOLOR String

The CECOLOR variable defines the color of new objects. This variable is saved in the drawing and its initial value is "BYLAYER" (256)

CELTSCALE Real

The CELTSCALE variable defines the current global linetype scale factor for objects. This variable is saved in the drawing and its initial value is 1.0000.

CELTYPE String

The CELTYPE variable defines the linetype that will be used in the new objects. This variable is saved in the drawing and its initial value is "BYLAYER."

CELWEIGHT Integer

The CELWEIGHT variable is stored in the drawing. It set the lineweight of the new objects. Its initial value is 1.
1 - Sets the lineweight to "ByLayer", 2 - Sets the lineweight to "ByBlock", 3 - Sets the lineweight to "Default" (controlled by LWDEFAULT system variable).

CHAMFERA Real

The CHAMFERA variable defines the first chamfer distance. This variable is saved in the drawing and its initial value is 0.5000.

CHAMFERB Real

The CHAMFERB variable defines the second chamfer distance. This variable is saved in the drawing and its initial value is 0.5000.

CHAMFERC Real

The CHAMFERC variable sets the chamfer length. This variable is saved in the drawing and its initial value is 1.0000.

CHAMFERD Real

The CHAMFERD variable sets the chamfer angle. This variable is saved in the drawing and its initial value is 0.0000.

CHAMMODE Integer

With the CHAMMODE variable you can specify the method that will be used to create chamfers.
0 - This is the initial value and in this case two chamfer distances are required, 1 - One chamfer length and an angle are required.

CIRCLERAD Real

The CIRCLERAD variable defines the default circle radius. The initial value of this variable is 0.0000.

CLAYER String

The CLAYER variable sets the current layer. This variable is saved in the drawing and its initial value is "0."

CMDACTIVE Integer

The CMDACTIVE variable contains the bit-code that signifies whether an ordinary command, transparent command, dialog box, or script is active. Basically the value of this variable is the addition of the following:
1 - Only ordinary command is active, 2 - Ordinary command as well as transparent command are active, 4 - Script is active, 8 - If this bit is set then Dialog box is active, 16 - AutoLISP is active.

CMDECHO   Integer

The CMDECHO variable determines whether the prompts and input of a AutoLISP (command) function are echoed. This variable is saved in registry and its initial value is 1.

0 - Echoing is disabled, 1 - Echoing enabled.

CMDNAMES   String

The CMDNAMES variable displays the name of the presently active command and transparent command. This variable is read-only.

CMLJUST   Integer

The CMLJUST variable determines the justification of a multiline. This variable is saved in registry and its initial value is 1.

0 - Sets Top justification, 1 - Sets Middle justification, 2 - Sets Bottom justification.

CMLSCALE   Real

The CMLSCALE variable determines the width of a multiline. For example, a scale factor of 3.0 generates a multiline that is thrice as wide as specified in the style definition. If the value is set to 0, the multiline takes the form of a single line. By specifying a negative scale factor, the order of offset lines is flipped. This variable is saved in registry and has an initial value of 1.0000.

CMLSTYLE   String

The CMLSTYLE variable sets the name of the multiline style that is used to draw multilines. This variable is saved in registry and has an initial value "STANDARD".

COMPASS   Integer

The COMPASS variable determines whether the 3D compass is on or off in the current viewport.

0 - Turns off the 3D compass, 1 - Turns on the 3D compass.

COORDS   Integer

The COORDS variable determines when the

coordinates are updated. This variable is saved in the drawing and its initial value is 1.

0 - Coordinates are updated only upon picking points, 1 - Absolute coordinates are continuously updated, 2 - Absolute continuously plus, when a distance or angle are requested, then the distance and angle from the last point are displayed.

CPLOTSTYLE   String

Controls the current plot style for new objects. The AutoCAD defined values are ByLayer, ByBlock, Normal, User Defined.

CPROFILE   String

The CPROFILE variable displays the name of the current profile. The value is saved in registry and the initial value is < <Unnamed Profile> >. This is a read only variable.

CTAB   String

The CTAB variable returns the name of the current (model or layout) tab in the drawing. The value is saved in drawing and it is a read only variable.

CURSORSIZE   Integer

This variable determines the size of the crosshairs as a percentage of the screen size. Initial value is 5.

CVPORT   Integer

The CVPORT variable establishes the identification number of the current viewport. When this value is changed, the current viewport is also changed in case the following conditions hold good:

1 - The specified identification number belongs to an active viewport, 2 - The cursor movement to the specified viewport is not locked by the command being executed, 3 - Tablet mode is off. The variable is saved in the drawing and its initial value is 2.

DATE   Real

The DATE variable contains the current date

and time as a Julian date and fraction in a real number. This variable is read-only.

DBMOD Integer

The DBMOD variable expresses the drawing modification status using bit-code. This variable is read-only. Basically the value of this variable is the addition of the following:
0 - No changes, 1 - The object database is changed, 2 - The symbol table is changed, 4 - The database variable is changed, 8 - The window is changed, 16 - The view is changed.

DCTCUST String

The DCTCUST variable shows the current custom spelling dictionary path and filename. This variable is saved in registry and its initial value is "".

DCTMAIN String

The DCTMAIN variable shows the current main spelling dictionary filename. Normally this file is located in the \support directory. The default main spelling dictionary can be specified using the **SETVAR** command. This variable is saved in registry and its initial value is "".

DEFLPSTYLE String

The DEFLPSTYLE variable specifies the default plot style for new layers. This variable is saved in registry and its initial value is "".

DEFPLSTYLE String

The DEFPLSTYLE variable specifies the default plot style for new objects. This variable is saved in registry and its initial value is "By Layer".

DELOBJ Integer

Thw DELOBJ variable determines whether objects used to draw other objects are kept or deleted from the drawing database. This variable is saved in the drawing and its ini-

tial value is 1.
1 - Objects are deleted from the drawing database, 0 - Objects are kept in the drawing database.

DEMANDLOAD Integer

This variable specifies if and when AutoCAD demand loads a third-party application if a drawing contains custom objects created in that application. Initial value is 3.

DIASTAT Integer

The method of exiting from the most recently used dialog box is held in the DIASTAT variable. This variable is read-only.
0 - Cancel, 1 - OK.

DIMALT Switch

The DIMALT variable controls the dimensioning in alternate units system. If the DIMALT variable is on (1), alternate unit dimensioning is facilitated. This variable is saved in the drawing and its initial value is Off (0).

DIMALTD Integer

The DIMALTD (DIMension ALTernate units Decimal places) variable controls the number of decimal places (decimal precision) of the dimension text in the alternate units if DIMALT variable is on. This variable is saved in the drawing and its initial value is 2.

DIMALTF Real

The DIMALTF variable (DIMension ALTernate units scale Factor) controls alternate units scale factor. In case DIMALT variable is enabled, all the linear dimensions will be multiplied with this factor to generate a value in an alternate units system. The initial value for DIMALTF is 25.4. This variable is saved in the drawing.

DIMALTRND Real

The DIMALTRND variable rounds off the

alternate dimension units. The value is saved in drawing and initial value is 0.00.

DIMALTTD   Integer
The DIMALTTD variable establishes the number of decimal places for the tolerance values of an alternate units dimension. This variable is saved in the drawing and has an initial value of 2.

DIMALTTZ   Integer
The DIMALTTZ variable controls the suppression of zeros for alternate tolerance values. With this variable, the real-to-string transformation carried out by AutoLISP functions rtos and angtos is also influenced. This variable is saved in the drawing and has an initial value of 0.
0 - Suppresses zero feet and precisely zero inches, 1 - Includes zero feet and precisely zero inches, 2 - Includes zero feet and suppresses zero inches, 3 - Includes zero inches and suppresses zero feet. Value in the range of 0 and 3 influence only the feet and inches dimensions. However, you can add 4 to the above values to omit the leading zeroes in all decimal dimensions. If you add 8, the trailing zeroes are omitted. If 12 (both 4 and 8) is added, the leading and the trailing zeroes are omitted.

DIMALTU   Integer
The DIMALTU variable establishes the units format for alternate units of all dimensions except angular. This variable is saved in the drawing and has an initial value of 2.
1 - Scientific, 2 - Decimal, 3 - Engineering, 4 - Architectural(Stacked), 5 - Fractional (Stacked), 6 - Architectural, 7 - Fractional, 8 - Windows® Desktop settings.

DIMALTZ   Integer
The DIMALTZ variable controls the suppression of zeros for alternate units dimension values. With this variable, the real-to-string transformation carried out by AutoLISP functions rtos and angtos is also

influenced. This variable is saved in the drawing and has an initial value of 0.
0 - Suppresses zero feet and precisely zero inches, 1 - Includes zero feet and precisely zero inches, 2 - Includes zero feet and suppresses zero inches, 3 - Includes zero inches and suppresses zero feet. Value in the range of 0 and 3 influence only the feet and inches dimensions. However, you can add 4 to the above values to omit the leading zeroes in all decimal dimensions. If you add 8, the trailing zeroes are omitted. If 12 (both 4 and 8) is added, the leading and the trailing zeroes are omitted.

DIMAPOST   String
With the help of DIMAPOST variable, you can append a text prefix, suffix, or both to an alternate dimensioning measurement. This can be done in case of all the dimensions except angular dimensions. The variable is saved in the drawing and has an initial value of "". In order to disable an existing suffix or prefix, set the value of this variable to a single period.

DIMASO   Switch
The DIMASO variable governs the creation of associative dimensions. This variable is saved in the drawing (not in the dimension style) and its initial value is set to on.
Off (0) - The dimension created are not associative in nature and hence in such dimensions no association exists between the dimension and the points on the object. All the dimensioning entities such as arrowheads, dimension lines, extension lines, dimension text, etc. are drawn as separate entities, On (1)- The dimension created are associative in nature and hence in such dimensions there exists an association between the dimension and the definition points. If you edit the object, (Editing like trimming or stretching) the dimensions associated with that object also change. Also, the appearance of associative dimensions can be preserved when they are edited by commands

such as STRETCH or TEDIT. For example, a vertical associative dimension is retained as a vertical dimension even after an editing operation. The associative dimension is always generated with the same dimension variable settings as defined in the dimension style.

DIMASZ   Real
The DIMASZ (Dimension arrowhead size) variable specifies the size of dimension line and leader line arrowheads when DIMTSZ is set to zero. The size of arrowhead blocks set by DIMBLK is also controlled by DIMASZ variable. Multiples of this variable determine whether the dimension line and text will be located between the extension lines. This variable is saved in the drawing and has an initial value of 0.18 units.

DIMATFIT   Integer
The DIMATFIT variable determines how dimension text and arrows are arranged when space is not sufficient to place both within the extension lines. The initial value is 3 and is saved in drawing.
0 - Places both text and arrows outside extension lines, 1 - Moves arrows first, then text, 2 - Moves text first, then arrows,
3 - Moves either text or arrows, whichever fits best.

DIMAUNIT   Integer
The DIMAUNIT variable establishes the angle format for angular dimensions. This variable is saved in the drawing and its initial value is 0.
0 - Decimal degrees format, 1 - Degrees/minutes/seconds format, 2 - Gradians format, 3 - Radians format, 4 - Surveyor's units format.

DIMAZIN   Integer
The DIMAZIN variable suppresses zeros for angular dimensions. The initial value is 0 and saved in drawing.
0 - Displays all leading and trailing zeros,

1 - Suppresses leading zeros in decimal dimensions, 2 - Suppresses trailing zeros in decimal dimensions, 3 - Suppresses leading and trailing zeros.

DIMBLK   String
DIMBLK variable replaces the default arrowheads at the end of the dimension lines with a user defined block. The user defined block that may replace the standard arrowhead can be a custom designed arrow or some other symbol. DIMBLK (DIMension BLocK) takes the name of the block as its string value. This variable is saved in the drawing and its initial value is no block (""). To discard an existing block name, set its value to a single period (.).

DIMBLK1   String
DIMBLK1 variable designates user defined arrow block for the first end of the dimension line. This option can be used only if the DIMSAH (DIMension Separate Arrow blocks) variable is on. The value of this variable is the name of earlier formulated block as in the case of DIMBLK. You can discard an existing block name by setting its value to a single period (.). This variable is saved in the drawing and its initial value is no block ("").

DIMBLK2   String
DIMBLK2 variable designates a user defined arrow block for the second end of the dimension line. This option can be used only if DIMSAH (DIMension Separate Arrow blocks) variable is on. The value of this variable is the name of earlier formulated block as in the case of DIMBLK. You can discard an existing block name, by setting its value to a single period (.). This variable is saved in the drawing and its initial value is no block ("").

DIMCEN   Real
The DIMCEN (DIMension CENter) variable governs the drawing of center marks and

the center lines of circles and the arcs by the DIMCENTER, DIMDIAMETER, and DIMRADIUS commands. DIMCEN takes a distance as its argument. The value of the DIMCEN variable determines the result. This variable is saved in the drawing and its initial value is 0.0900.

0 - Center marks or center lines are not drawn, >0 - Center marks are drawn and their size is governed by the value of the DIMCEN. For example, a value of 0.250 displays center dashes which are 0.2500 units long, <0 - Center lines in addition to center marks are drawn and again the size of the mark portion is governed by the absolute value of the DIMCEN. The center lines extend beyond the circle or arc by the value entered. For example a value of -0.2500 for DIMCEN variable will draw a center dashes 0.25 units long ant also the center lines will be extended beyond the circle/arc by a distance of 0.25 units. With the DIMRADIUS and DIMDIAMETER commands, center mark or center line is generated only when the dimension line is located outside the circle or arc.

DIMCLRD   Integer
The DIMCLRD variable is used to assign colors to dimension lines, arrowheads, and the dimension leader lines. This variable can take any permissible color number or the special color labels BYBLOCK (0) or BYLAYER (256) as its value. If you use the SETVAR command, then you have to enter the integer number of the color you want to assign to the DIMCLRD variable. This variable is saved in the drawing and its initial value is 0.

DIMCLRE   Integer
DIMCLRE variable is used to assign color to the dimension extension lines. Just as DIMCLRD, DIMCLRE (DIMension CoLOr Extension) can take any permissible color number or the special color labels BYBLOCK or BYLAYER. This variable is saved in the drawing and its initial value is 0.

DIMCLRT   Integer
The DIMCLRT (DIMension CoLoR Text) variable is used to assign a color to the dimension text. DIMCLRT can take any permissible color number or the special color labels BYBLOCK (0) or BYLAYER (256). This variable is saved in the drawing and its initial value is 0.

DIMDEC   Integer
The DIMDEC variable establishes the number for decimal places of a primary units dimension. This variable is saved in the drawing and its initial value is 4.

DIMDLE   Real
By default the dimension lines meet the extension lines. But if you want that the dimension line to continue past the extension lines, DIMDLE (Dimension Line Extension) variable can be used for this function. DIMDLE is used only when DIMTSZ variable is nonzero (When DIMTSZ variable is nonzero, ticks are drawn instead of arrows). The dimension line will extend past the extension line by the value of DIMDLE. This variable is saved in the drawing and its initial value is 0.0000.

DIMDLI   Real
The DIMDLI variable governs the spacing between the successive dimension lines when dimensions are created with the DIMCONTINUE and DIMBASELINE commands. Successive dimension lines are offset by the DIMDLI value, if needed, to avert drawing over the previous dimension. This variable is saved in the drawing and its initial value is 0.38 units.

DIMDSEP   Single character
The DIMDSEP variable Specifies a single-character decimal separator to use when creating dimensions whose unit format is

decimal. The initial value is a Decimal point and it is saved in drawing.

DIMEXE   Real

The extension of the extension line past the dimension line is governed by the DIMEXE (Dimension EXtension line Extension) variable. This variable is saved in the drawing and has an initial value of 0.18 units.

DIMEXO   Real

There exists a small space between the origin points you specify and the start of the extension lines. The size of this gap is controlled by the DIMEXO (DIMension EXtension line Offset) variable. The offset distance is equal to the value of the DIMEXO variable. This variable is saved in the drawing and has an initial value of 0.0625 units.

DIMFIT   Integer

Obsolete. Has no effect in AutoCAD 2000 except to preserve the integrity of pre-AutoCAD 2000 scripts and AutoLISP routines. This variable is saved in the drawing and its initial value is 3. DIMFIT is replaced by DIMATFIT and DIMTMOVE.

DIMFRAC   Integer

The DIMFRAC variable controls the fraction format when DIMLUNIT is set to 4 (Architectural) or 5 (Fractional). It is saved in drawing and initial value is 0.
0 - Horizontal, 1 - Diagonal, 2 - Not stacked (for example, 3/5).

DIMGAP   Real

The DIMGAP variable controls the space between the dimension line and the dimension text (distance maintained around the dimension text), when the dimension line is split into two for the placement of dimension text. The gap between the leader and annotation created with the LEADER command is also governed by DIMGAP variable.

This variable is saved in the drawing and its initial value for DIMGAP is 0.0900 units. By entering a negative DIMGAP value, you can create a reference dimension, in which case you get the dimension text with a box drawn around it. DIMGAP value is also used by AutoCAD as the measure of minimum length needed for the segments of the dimension line. AutoCAD places the dimension text inside the extension lines only if the dimension line is split into two segments each of which is at least as long as DIMGAP. In case the text is positioned over or under the dimension line, it is placed inside the dimension line only if there is space for the arrows, dimension text, and a margin between them has a minimum value at least as much as DIMGAP: 2*(DIMGAP + DIMASZ).

DIMJUST   Integer

The DIMJUST variable governs the horizontal dimension text position. This variable is saved in the drawing and its initial value is 0.
0 - The text is center justified between the extension lines, 1 - The text is placed next to the first extension line, 2 - The text is placed next to the second extension line, 3 - The text is placed above and aligned with the first extension line, 4 - The text is placed above and aligned with the second extension line.

DIMLDRBLK   String

The DIMLDRBLK variable controls the arrow type for leaders. To turn off arrowhead display, enter a single period (.).

DIMLFAC   Real

The DIMLFAC (DIMension Length FACtor) variable acts as a global scale factor for all linear dimensioning measurements. The linear distances measured by dimensioning include coordinates, diameter, and radii. These linear distances are multiplied by the prevailing DIMLFAC value before they are

projected as dimension text. In this manner DIMLFAC scales the contents of the default text. The angular dimensions are not scaled. Also DIMLFAC does not apply to the values held in DIMTM, DIMTP, or DIMRND. For example, if you want to scale the default dimension measurement by a value of 2, set the value of DIMLFAC to 2. When dimensioning in the paper space, if the value of DIMLFAC variable is not zero, then the distance measured is multiplied by the absolute value of DIMLFAC. In case of dimensioning in the model space, values less than zero are neglected, instead the value of DIMLFAC is taken as 1.0. If in paper space you select the Viewport option and try to change DIMLFAC from the Dim: prompt, AutoCAD will compute a value for the DIMLFAC for you. This is illustrated as follows:
Dim: DIMLFAC, Current value <1.0000>
New value (Viewport): V Select viewport to set scale: The scaling of model space to paper space is computed by AutoCAD and the negative of the computed value is assigned to DIMLFAC. This variable is saved in the drawing and its initial value is 1.0000.

DIMLIM   Switch
The DIMLIM (DIMension LIMits) variable acts as a switch and creates the dimension limits as the default text if it is on (1). Also DIMTOL is forced to be off. This variable is saved in the drawing and its initial value is off.

DIMLUNIT   Integer
The DIMLUNIT variable controls units for all dimension types except Angular. The initial value is Off (0) and it is saved in drawing.
1 - Scientific, 2 - Decimal, 3 - Engineering,
4 - Architectural, 5 - Fractional,
6 - Windows desktop settings.

DIMLWD   Enum
The DIMLWD variable assigns lineweight

to dimension lines. Values are standard lineweights. The initial value is "By Block" and it is saved in drawing.

DIMLWE   Enum
The DIMLWD variable assigns lineweight to extension lines. Values are standard lineweights. The initial value is "By Block" and it is saved in drawing.

DIMPOST   String
The DIMPOST variable is used to define prefix or suffix to the dimension measurement. The variable is saved in the drawing and has an initial value "" (empty string). DIMPOST takes a string value as its argument. For example if you want to have a suffix for centimeters, set DIMPOST to "cm". A distance of 4.0 units will be displayed as 4.0cm. In case tolerances are enabled, the suffix you have defined gets applied to the tolerances as well as to the main dimension. To establish a prefix to a dimension text, type "<>" and then the prefix at the same prompt.

DIMRND   Real
The DIMRND (DIMension RouND) variable is used for rounding all the dimension measurements to the specified value. For example if the DIMRND is set to 0.10, then all the measurements are rounded to the nearest 0.10 unit. Like wise a value of 1 for this variable will result in the rounding of all the measurements to the nearest integer. The angular measurements cannot be rounded. The variable is saved in the drawing and has an initial value of 0.0000.

DIMSAH   Switch
The DIMSAH (DIMension Separate custom Arrow Head) variable governs the placement of user-defined arrow blocks instead of the standard arrows at the end of the dimension line. As explained before, DIMBLK1 variable places a user defined arrow block

at the first end of the dimension line and DIMBLK2 places a user defined arrow block at the other end of the dimension line. This variable is saved in the drawing and its initial value is off.

On - DIMBLK1 and DIMBLK2 specify different user-defined arrow blocks to be drawn at the two ends of the dimension line, Off - Ordinary arrowheads or user-defined arrowhead block defined by the DIMBLK variable is used.

DIMSCALE   Real

The DIMSCALE variable controls the scale factor for all the size-related dimension variables such as those that affect text size, center mark size, arrow size, leader objects, etc. The DIMSCALE is not applied to the measured lengths, coordinates, angles, or tolerances. The default value for this variable is 1.0000; and in this case the dimensioning variables assume their preset values and the drawing is plotted at full scale. If the drawing is to be plotted at half the size, then the scale factor is the reciprocal of the drawing size. Hence the scale factor or the DIMSCALE value will be reciprocal of 1/2 which is 2/1 = 2.

0.0 - A default value based on the scaling between the current model space viewport and paper space is calculated. In case you are not using the paper space feature, then the scale factor is 1.0, >0 - A scale factor is computed that makes the text sizes, arrowhead sizes, and scaled distances to plot at their face value.

DIMSD1   Switch

The DIMSD1 (DIMension Suppress Dimension line 1) variable suppresses the drawing of the first dimension line when it is on. This variable is saved in the drawing and its initial value is off.

DIMSD2   Switch

The DIMSD1 (DIMension Suppress Dimension line 2) variable suppresses the drawing

of second dimension line when it is on. This variable is saved in the drawing and its initial value is off.

DIMSE1   Switch

The DIMSE1 variable is used to suppress drawing of the first extension line. When DIMSE1 (DIMension Suppress Extension line 1) is on, the first extension line is not drawn. This variable is saved in the drawing and its initial value is off.

DIMSE2   Switch

The DIMSE2 variable is used to suppress drawing of the second extension line. When DIMSE2 (DIMension Suppress Extension line 2) is on, the second extension line is not drawn. This variable is saved in the drawing and its initial value is off.

DIMSHO   Switch

DIMSHO variable governs the redefinition of dimension entities while dragging into some position. If DIMSHO (DIMension SHOw dragged dimensions) is on, associative dimensions will be computed dynamically as they are dragged. The DIMSHO value is saved in the drawing (not in a dimension style) and its initial value is on (1). Dynamic dragging reduces the speed of some computers and hence in such situations DIMSHO should be set off (0). However, when you are using the pointing device to specify the length of the leader in Radius and Diameter dimensioning, the DIMSHO setting is neglected and dynamic dragging is used.

DIMSOXD   Switch

If you want to place text inside the extension lines, you will have to set the DIMTIX variable on. And if you want to suppress the dimension lines and the arrow heads you will have to set the DIMDSOXD (DIMension Suppress Outside eXtension Dimension lines) variable on. DIMSOXD suppresses the drawing of dimension lines and the arrow heads when they are placed outside the extension lines. If DIMTIX is on and DIMSOXD is off and there is not enough space inside the extension lines for drawing the dimension lines, then dimension lines will be drawn outside the extension lines. In such a situation, if both DIMTIX and DIMSOXD are on, then the dimension line will be totally suppressed. DIMSOXD works only when DIMTIX is on. The DIMSOXD variable is saved in the drawing and its initial value is off.

DIMSTYLE   String

DIMSTYLE variable is used for displaying the name of the present dimension style. DIMSTYLE is a read-only variable and is saved in the drawing. You can change the dimension style using the DDIM or DIMSTYLE command.

DIMTAD   Integer

The DIMTAD (DIMension Text Above Dimension line) variable governs the vertical placement of the dimension text with respect to the dimension line. DIMTAD gets actuated when dimension text is drawn between the extension lines and is aligned with the dimension line, or when the dimension text is placed outside the extension lines. This variable is saved in the drawing and its initial value is 0.

0 - For this value the dimension text is placed at the center between the extension lines,
1 - The dimension text is placed above the dimension line and a single (unsplit) dimension line is drawn under it spanning between the extension lines. The exceptions to this arise when the dimension line is not horizontal and text inside the extension line is forced to be horizontal by making DIMTIH = 1. The space between the dimension line and the baseline of the lowest line of text is nothing but the prevailing DIMGAP value,
2 - The dimension text is placed on the side of the dimension line most remote from the defining points, 3 - The dimension text is

placed to tune to a JIS representation.

DIMTDEC   Integer

The DIMTDEC variable establishes the number of decimal places for the tolerance values for the primary units dimension. This variable is saved in the drawing and its initial value is 4.

DIMTFAC   Real

With the DIMTFAC (DIMension Tolerance scale FACtor) variable you can control the scaling factor of the text height of the tolerance values in relation to the dimension text height set by DIMTXT. Suppose DIMTFAC is set to 1.0 (the default value for DIMTFAC variable), then the text height of the tolerance text will be equal to the dimension text height. If DIMTFAC is set to a value of 0.50, the text height of the tolerance is half of the dimension text height. This variable is saved in the drawing and its initial value is 1.0000. It is important to remember that the scaling of tolerance text to any requirement is possible only when DIMTOL is on and DIMTM and DIMTP variable values are not identical, or when DIMLIM is on.

DIMTIH   Switch

The DIMTIH (DIMension Text Inside Horizontal) variable controls the placement of the dimension text inside the extension lines for Linear, Radius, Angular, and Diameter dimensioning. DIMTIH is effective only when the dimension text fits between the extension lines.

On - If DIMTIH is on (the default setting), it forces the dimension text inside the extension lines to be placed horizontally, rather than aligned, Off - In case DIMTIH is off, the dimension text is aligned with the dimension line.

DIMTIX   Switch

The DIMTIX variable draws the text between the extension lines. This variable is saved in the drawing and its initial value is

Off.

On - When DIMTIX is set to on, the dimension text is placed amidst the extension lines even if it would normally be placed outside the extension lines, Off - If DIMTIX is off, the placement of the dimension text depends on the type of dimension. For example, if the dimensions are Linear or Angular, the text will be placed inside the extension lines by AutoCAD if there is enough space available. While as for the Radius and Diameter dimensions, the text is placed outside the object being dimensioned.

DIMTM   Real

The DIMTM variable establishes the lower (minimum) tolerance limit for the dimension text. Tolerance is defined as the total amount by which a particular dimension is permitted to vary. The tolerance or limit values are drawn only if DIMTOL or DIMLIM variable is on. DIMTM (DImension Tolerance Minus) identifies the lower tolerance and DIMTP (DImension Tolerance Plus) identifies the upper tolerance. You can specify signed values for DIMTM and DIMTP variables. If DIMTOL is on and both DIMTM and DIMTP have same value, AutoCAD draws the "ñ" symbol followed by the tolerance value. If DIMTM and DIMTP hold different values, the upper tolerance is drawn above the lower tolerance. Also a positive (+) sign is appended to the DIMTP value if it is positive. For minus tolerance value (DIMTM), the negative of the value you enter (negative sign if you enter positive value and positive sign if you enter negative value) is displayed. Signs are not appended with zero. This variable is saved in the drawing and its initial value is 0.0000.

DIMTMOVE   Integer

The DIMTMOVE variable controls the dimension text movement rules. Its initial value is 0 and it is saved in drawing.

0 - Moves the dimension line with dimen-

sion text, 1 - Adds a leader when dimension text is moved, 2 - Allows text to be moved freely without a leader.

DIMTOFL   Switch

If DIMTOFL variable is turned on, a dimension line is drawn between the extension lines even if the text is located outside the extension lines. When DIMTOFL is off, for radius and diameter dimensions, the dimension line and the arrowheads are drawn inside the arc or circle, while the text and the leader are placed outside. This variable is saved in the drawing and its initial value is Off.

DIMTOH   Switch

The DIMTOH (DIMension Text Outside Horizontal) variable controls the orientation of the dimension text outside the extension lines. If DIMTOH is on, it forces the dimension text outside the extension lines to be placed horizontally, rather than aligned. In case DIMTOH is off, the dimension text is aligned with the dimension line. You must have noticed that the variable DIMTOH is same as DIMTIH variable except it controls text drawn outside the extension lines. This variable is saved in the drawing and its initial value is On.

DIMTOL   Switch

DIMTOL (DIMension with TOLerance) variable is used for controlling the appending of dimension tolerances to the dimension text. With DIMTM and DIMTP you can define the values of the lower and upper tolerances. If the DIMTOL variable is set on, the tolerances are appended to the default text. When DIMTOL is set on, DIMLIM variable is set off. This variable is saved in the drawing and its initial value is Off.

DIMTOLJ   Integer

DIMTOLJ variable establishes the vertical justification for the tolerance values with respect to the normal dimension text. This variable is saved in the drawing and its initial value is 1.

0 - Bottom, 1 - Middle, 2 - Top,.

DIMTP   Real

The DIMTP (DIMension Tolerance Plus) variable establishes the upper (maximum) tolerance limit for the dimension text. Tolerance is defined as the total amount by which a particular dimension is permitted to vary. The tolerance or limit values are drawn only if DIMTOL or DIMLIM variable is on. If DIMTOL is on and both DIMTM and DIMTP have same value, AutoCAD draws the "ñ" symbol followed by the tolerance value. If DIMTM and DIMTP hold different values, the upper tolerance is drawn above the lower tolerance. Also a positive (+) sign is appended to the DIMTP value if it is positive. This variable is saved in the drawing and its initial value is 0.0000.

DIMTSZ   Real

The DIMTSZ variable defines the size of oblique strokes (ticks) instead of arrowheads at the end of the dimension lines (just as in architectural drafting), for Linear, Radius, and Diameter dimensioning. This variable is saved in the drawing and its initial value is 0.0000.

0 - Arrows are drawn, >0 - Oblique strokes instead of arrows are drawn. The size of the ticks is computed as DIMTSZ*DIMSCALE. Hence if DIMSCALE factor is one then the size of the tick is equal to the DIMTSZ value. This variable is also used to determine whether dimension line and dimension text will get accommodated between the extension lines.

DIMTVP   Real

The DIMTVP (DIMension Text Vertical Position) variable, controls the vertical placement of the dimension text over or under the dimension line. In certain cases DIMTVP is used as DIMTAD to control the vertical position of the dimension text. DIMTVP value holds good only when DIMTAD is off. The vertical placing of the text is done by offsetting the dimension text. The amount of the vertical offset of dimension text is a product of text height and DIMTVP value. If the value of DIMTVP is 1.0, DIMTVP acts as DIMTAD. However if the value of the DIMTVP is less than 0.70, the dimension line is broken into two segments to accommodate the dimension text. This variable is saved in the drawing and its initial value is 0.0000.

DIMTXSTY   String

The DIMTXTSTY variable specifies the text style of the dimension. This variable is saved in the drawing and its initial value is "STANDARD".

DIMTXT   Real

The DIMTXT variable is used to control the height of the dimension text except if the current text style has a fixed height. This variable is saved in the drawing and its initial value is 0.1800.

DIMTZIN   Integer

With the DIMZIN variable you can control the suppression of the zeros for tolerance values. The variable is saved in the drawing and its initial value is 0.

0 - Suppresses zero feet and precisely zero inches, 1 - Includes zero feet and precisely zero inches, 2 - Includes zero feet and suppresses zero inches, 3 - Includes zero inches and suppresses zero feet. You can add 4 to the above values to omit the leading zeroes in all decimal dimensions. If you add 8, the trailing zeroes are omitted. If 12 (both 4 and 8) is added, the leading and the trailing zeroes are omitted.

DIMUNIT   Integer

The DIMUNIT variable establishes the linear units format for all dimension styles.

1 - Scientific units format, 2 - Decimal units

format, 3 - Engineering units format, 4 - Architectural (stacked) units format, 5 - Fractional (stacked) units format, 6 - Architectural, 7 - Fractional, 8 - Window Desktop.

DIMUPT   Switch

This variable governs the cursor functionality for User Positioned Text. This variable is saved in the drawing and its initial value is Off.

0 - The cursor controls the location of the dimension line only, 1 - The cursor controls the location of both the dimension text and the dimension line.

DIMZIN   Integer

The DIMZIN (DIMension Zero INch) controls the suppression of the inches part of a feet-inches dimension when the distance is integral number of feet or the suppression of the feet portion when the distance is less than one foot. This variable is saved in the drawing and its initial value is 0.

0 - Suppress zero feet and exactly zero inches, 1 - Include zero feet and, exactly zero inches, 2 - Include zero feet, suppress zero inches, 3 - Include zero inches, suppress zero feet. If the dimension has feet and a fractional inch part, the number of inches is included even if it is zero. This is independent of the DIMZIN setting. For example a dimension such as 1'-2/3" never exist. It will be in the form 1'-0 2/3". The integer values 0-3 of the DIMZIN variable control the feet and inch dimension only, while as you can add 4 to omit the leading zeroes in all decimal dimensions. For example 0.2600 becomes .2600. If you add 8, the trailing zeroes are omitted. For example, 4.9600 becomes 4.96. If 12 (both 4 and 8) is added, the leading and the trailing zeroes are omitted. For example, 0.2300 becomes .23.

DISPSILH   Integer

The DISPSILH variable governs the display of silhouette curves of body objects in a wireframe model. The variable is saved in

the drawing and its initial value is 0.

0 - Silhouette curves of body objects not displayed, 1 - Silhouette curves of body objects displayed.

DISTANCE   Real

The DISTANCE variable holds the distance value determined by the DIST command. This command is read-only.

DONUTID   Real

The DONUTID variable establishes the default inside diameter of a donut. The initial value for this variable is 0.5000.

DONUTOD   Real

The DONUTOD variable establishes the default outside diameter of a donut. It is important that the value of this variable be greater than zero. In case the value of DONUTID is greater than that of DONUTOD, then the two values are interchanged by the next command. The initial value for this variable is 1.0000.

DRAGMODE   Integer

The DRAGMODE variable establishes the Object Drag mode while carrying out editing operations.

0 - Dragging disabled, 1 - Dragging enabled if invoked, 2 - Auto. This variable is saved in the drawing and is initially set to 2.

DRAGP1   Integer

The DRAGP1 variable establishes the regen-drag input sampling rate. This variable is saved in registry and is initially set to a value of 10.

DRAGP2   Integer

The DRAGP2 variable establishes the fast-drag input sampling rate. This variable is saved in registry and is initially set to a value of 25.

DWGCHECK   Integer

The DWGCHECK variable determines

whether a drawing was last edited by a product other than AutoCAD. Its initial value is 0 and it is stored in registry.

0 - The dialog box display is suppressed,
1 - The dialog box will be displayed.

DWGCODEPAGE   String

The DWGCODEPAGE variable holds the drawing code page. When you create a new drawing, this variable is set to the system code page, Otherwise it is not maintained by AutoCAD. This variable describes the code page of the drawing. You can set this variable to any value by using the SYSCODEPAGE system variable or set it as undefined. It is a read-only variable and is saved in the drawing.

DWGNAME   String

The DWGNAME variable holds the name of the drawing as specified by the user. In case the drawing has not been assigned a name, the DWGNAME variable conveys that the drawing is unnamed. The drive and directory is also included if it was specified. This variable is a read-only variable.

DWGPREFIX   String

The DWGPREFIX variable holds the drive and directory prefix for the drawing. This variable is a read-only variable.

DWGTITLED   Integer

The DWGTITLED variable reflects whether the present drawing has been named.

0 - Indicates that the drawing has not been named, 1 - Indicates that the drawing has been named.

EDGEMODE   Integer

With the EDGEMODE variable you can control how the EXTEND and TRIM commands determine boundary and cutting edges.

0 - In this case the selected edge is used without an extension, 1 - The object is trimmed or extended to an imaginary extension of the cutting or boundary edges. This is the initial value for this variable.

ELEVATION   Real

The ELEVATION variable holds the current 3D elevation associated to the current UCS for the current space. This variable is saved in the drawing and has an initial value of 0.0000.

EXPERT   Integer

The issuance of some prompts is controlled with the EXPERT variable. The initial value for this variable is 0.

0 - All the prompts are issued, 1 - The "About to regen, proceed?" prompt and "Really want to turn the current layer off?" prompts are suppressed, 2 - The preceding prompts and "Block already defined. Redefine it?" (BLOCK command) and "A drawing with this name already exists. Overwrite it?" (SAVE or WBLOCK commands) are suppressed, 3 - The preceding prompts and the ones issued by LINETYPE if you try to load a linetype that is already loaded or create a new linetype in a file that already defines it are suppressed, 4 - The preceding prompts and the ones issued by UCS Save and VPORTS Save in case the name you provide already exists are suppressed, 5 - The preceding prompts and the ones issued by the DIMSTYLE Save option, and DIMOVERRIDE in case the dimension style name you provide already exists, are suppressed. Whenever the EXPERT command suppresses a prompt, the corresponding operation is carried out as if you have entered Y as the response to the prompt. The EXPERT command can influence menu macros, scripts, AutoLISP, and the command functions.

EXPLMODE   Integer

The EXPLMODE variable govern whether the EXPLODE command can explode nonuniformly scaled blocks. This variable is saved in the drawing and its initial value is 1.

0 - Nonuniformly scaled blocks cannot be exploded, 1 - Nonuniformly scaled blocks can be exploded.

EXTMAX   3D Point

The EXTMAX variable holds the upper-right point of the drawing extents and is saved in the drawing. The drawing extents increase outward when new objects are drawn and reduce only when ZOOM All or ZOOM Extents is used. The variable is reported in the World coordinates for the current space.

EXTMIN   3D Point

The EXTMIN variable holds the lower-left point of the drawing extents and is saved in the drawing. The drawing extents increase outward when new objects are drawn and reduce only when ZOOM All or ZOOM Extents is used. The variable is reported in the World coordinates for the current space.

EXTNAMES   Integer

The EXTNAMES variable controls the parameters for named object names (such as linetypes, lineweights and layers) stored in symbol tables. Its initial value is 1 and is saved in drawing.

0 - Uses Release 14 parameters, which limit names to 31 characters in length, 1 - Uses AutoCAD 2000 parameters. Names can be up to 255 characters in length.

FACETRATIO   Integer

The FACETRATIO variable sets the aspect ratio of faceting for cylindrical and conic ACIS solids. Its initial value is 0 and it is not saved.

0 - Creates an N by 1 mesh for cylindrical and conic ACIS solids, 1 - Creates an N by M mesh for cylindrical and conic ACIS solids.

FACETRES   Real

The FACETRES variable adjusts the

smoothness of shaded and objects whose hidden lines have been removed. This variable can be assigned values in the range of 0.010 to 10.0. The variable is saved in the drawing and has an initial value of 0.5.

FILEDIA   Integer

The FILEDIA variable suppresses the display of file dialog boxes. This variable is saved in registry and has an initial value of 1.

0 - The file dialog boxes are disabled. However you can make AutoCAD to display the file dialog box by entering a tilde (~) as the response to the prompt. This applied for AutoLISP and ADS functions also, 1 - The file dialog boxes are enabled except when a script or AutoLISP/ADS program is active in which case only a prompt appears.

FILLETRAD   Real

The FILLETRAD variable holds the current fillet radius and is saved in the drawing and its initial value is 0.5000.

FILLMODE   Integer

FILLMODE variable indicates whether objects drawn with SOLID command are filled in. This variable is saved in the drawing and its initial value is 1.

0 - Objects are not filled, 0 - Objects are filled.

FONTALT   String

The FONTALT variable specifies the alternate font to be used in case the specified font file cannot be found. In case you have not specified an alternate font, AutoCAD issues a warning. This variable is saved in registry and its initial value is "simplex.shx".

FONTMAP   String

The FONTMAP variable specifies the font mapping file to be used in case the specified font file cannot be found. This file holds one font mapping per line. The original font and the substitute font are separated by a

Appendix-D

semicolon (;). This variable is saved in registry and its initial value is "Acad.fmp".

FRONTZ   Real

The FRONTZ variable contains the front clipping plane offset (in current drawing units) from the target plane for the current viewport. You can determine the distance between the front clipping plane and the camera point by subtracting the FRONTZ value from the camera to target distance. This variable is saved in the drawing and is read-only.

FULLOPEN   Integer

The FULLOPEN variable checks whether the current drawing is partially open. Its value is not saved and it is a read-only variable.
0 - Indicates a partially open drawing,
1 - Indicates a fully open drawing

GRIDMODE   Integer

The GRIDMODE variable specifies whether the grid is turned on or off. This variable is saved in the drawing and its initial value is 0.
0 - The grid is turned off, 1 - The grid is turned on.

GRIDUNIT   2D point

The GRIDUNIT variable specifies the X and Y grid spacing for the current viewport. The changes made to the grid spacing are manifested only after using the REDRAW or REGEN command. This variable is saved in the drawing and its initial value is 0.5000,0.5000.

GRIPBLOCK   Integer

The GRIPBLOCK variable controls the assignment of grips in blocks. This variable is saved in registry and its initial value is 0.
0 - The grip is assigned only to the insertion point of the block, 1 - Grips are assigned to objects within the block.

GRIPCOLOR   Integer

The GRIPCOLOR variable controls the color of nonselected grips. It can take a value in the range of 1 to 255. This variable is saved in registry and its initial value is 5.

GRIPHOT   Integer

The GRIPHOT variable controls the color of selected grips. It can take a value in the range of 1 to 255. This variable is saved in registry and its initial value is 1.

GRIPS   Integer

With the GRIPS variable you can make use of selection set grips for the Stretch, Move, Rotate, Scale, and Mirror grip modes. This variable is saved in registry and its initial value is 1.
0 - Grips are disabled, 1 - Grips are enabled.

GRIPSIZE   Integer

The GRIPSIZE variable allows you to assign a size to the box drawn to show the grip. It sets its half height in pixels. This variable can be assigned a value in the range of 1 to 255. The variable is saved in registry and its initial value is 3.

HANDLES   Integer

The HANDLES variable is always on (1), which states that object handles are enabled and can be accessed by applications. This variable is saved in the drawing and is read-only.

HIDEPRECISION   Integer

The HIDEPRECISION variable sets the accuracy of hides and shades. Hides can be calculated in double precision or single precision.
0 - Single precision; uses less memory,
1 - Double precision; uses more memory

HIGHLIGHT   Integer

The HIGHLIGHT variable governs object highlighting. Objects selected with grips are not influenced.

0 - Object selection highlighting is disabled,
1 - Object selection highlighting is enabled.
This is the initial value for the variable.

HPANG   Real

The HPANG variable specifies the angle of
the hatch pattern. The initial value for this
variable is 0.

HPBOUND   Real

The HPBOUND variable governs the object type created by the BHATCH and
BOUNDARY commands. This variable is
saved in the drawing and its initial value is
1.
0 - A region is created, 1 - A polyline is created.

HPDOUBLE   Integer

The HPDOUBLE variable governs the
hatch pattern doubling for user-defined
patterns. The initial value of this variable is
0.
0 - Hatch pattern doubling disabled, 1 -
Hatch pattern doubling enabled.

HPNAME   String

The default hatch pattern name is established with HPNAME variable. The name
can be up to 34 characters and spaces are
not allowed. Empty string ("") is returned if
no default exists. To set no default enter a
period (.). The initial value of this variable
is "ANSI31".

HPSCALE   Real

The hatch pattern scale factor is specified
with HPSCALE variable. This variable cannot assume zero value. The initial value of
this variable is 1.0000.

HPSPACE   Real

The hatch pattern line spacing for user-defined simple patterns is specified by
HPSPACE variable. This variable cannot
assume zero value. The initial value of this
variable is 1.0000.

HYPERLINKBASE   String

The HYPERLINKBASE variable specifies
the path used for all relative hyperlinks in
the drawing. Its initial value is "" and is saved
in drawing.

IMAGEHLT   Integer

The IMAGEHLT variable Controls whether
the entire raster image or only the raster
image frame is highlighted. Its initial value
is 0 and is stored in registry.
0 - Highlights only the raster image frame,
1 - Highlights the entire raster image.

INDEXCTL   Integer

Controls whether layer and spatial indexes
are created and saved in drawing. Initial
value is 0.
0 - No indexes created, 1 - Layer index created, 2 - Spatial index is created, 3 - Layer
and spatial are created.

INETLOCATION   Real

Stores the Internet location used by
BROWSER. Initial value is "www
autodesk.com/acaduser".

INSBASE   3D point

The insertion base point established by the
BASE command is stored in this variable.
This point is defined in UCS coordinates
for the current space. The variable is saved
in the drawing and its initial value is
0.0000,0.0000,0.0000.

INSNAME   String

The INSNAME variable establishes the default block name for or INSERT commands.
To set no default enter a period (.). The initial value of this variable is "".

INSUNITS   Integer

The INSUNITS variable specifies a drawing units value, when you drag a block or
image from AutoCAD DesignCenter. Its initial value is 0 and it is stored in drawing.
0 - Unspecified (No units), 1 - Inches,

2 - Feet, 3 - Miles, 4 - Millimeters, 5- C e n - timeters, 6 - Meters, 7 - Kilometers, 8 - Microinches, 9 - Mils, 10 - Yards, 11 - Angstroms, 12 - Nanometers, 13 - Microns, 14 - Decimeters, 15 - Decameters, 16 - Hectometers, 17 - Gigameters, 18 - Astronomical Units, 19 - Light Years, 20 - Parsecs.

INSUNITSDEFSOURCE   Integer

The INSUNITSDEFSOURCE controls source content units value. Valid ranges between 0 to 20. Its initial value is 0 and it is stored in registry.

INSUNITSDEFTARGET   Integer

The INSUNITSDEFTARGET controls source content units value. Valid ranges between 0 to 20. Its initial value is 0 and it is stored in registry.

ISAVEBAK   Integer

Improves the speed of incremental saves, especially for large drawings on Windows. Initial value is 1.
0 - No BAK file is created, 1 - A BAK file is created.

ISAVEPERCENT   Integer

Determines the amount of wasted space tolerated in a drawing file. Initial value is 50.

ISOLINES   Integer

The ISOLINES variable specifies the number of isolines per surface on objects. The variable can accept a value in the range of 0 to 2047. This variable is saved in the drawing and its initial value is 4.

LASTANGLE   Real

The LASTANGLE variable holds the end angle of the last arc entered, with respect to the XY plane of the current UCS for the current space. This variable is a read-only variable.

LASTPOINT   3D point

The LASTPOINT variable holds the UCS coordinates for the current space of the most recently entered point. This variable is saved in the drawing and its initial value is 0.0000,0.0000,0.0000.

LASTPROMPT   String

The LASTPROMPT variable stores the last string echoed to the command line. Initial value is "".

LENSLENGTH   Real

The LENSLENGTH variable holds the length of the lens (in mm) used in perspective viewing for the current viewport. This variable is saved in the drawing.

LIMCHECK   Integer

This variable governs the drawing of objects outside the specified drawing limits. This variable is saved in the drawing and its initial value is 0.
0 - Object can be drawn outside the drawing limits, 1 - Object cannot be drawn outside the drawing limits.

LIMMAX   2D point

The upper-right drawing limits stated in World coordinates (for the current space) are held in the LIMMAX variable. This variable is saved in the drawing and its initial value is 12.0000,9.0000.

LIMMIN   2D point

The lower-left drawing limits stated in World coordinates (for the current space) are held in the LIMMIN variable. This variable is saved in the drawing and its initial value is 0.0000,0.0000.

LISPINIT   Integer

Specifies whether AutoLISP defined functions and variables are preserved when you open a new drawing. It is saved in registry and the initial value is 1.
0 - AutoLISP functions and variables are preserved, 1 - AutoLISP functions and variables are valid in current drawing only.

LOCALE   String
The LOCALE variable shows the ISO language code of the current AutoCAD version in use. The initial value of this variable is "enum" (varies by country)

LOGFILEMODE   Integer
Specifies whether the contents of the text window are written to a log file. Its initial value is 0.
0 - Log file is not maintained, 1 - Log file is maintained.

LOGFILENAME   String
Specifies the path for the log file. Initial value is "C:\ACADR14\acad.log".

LOGFILEPATH   String
The LOGFILEPATH variable specifies the path for the log files for all drawings in a session. The initial value varies depending on where you installed AutoCAD and is saved in registry.

LOGINNAME   String
The LOGINNAME variable shows the user's name as specified while configuring when AutoCAD is loaded.

LTSCALE   Real
The LTSCALE variable establishes global linetype scale factor. The variable is saved in the drawing and its initial value is 1.0000.

LUNITS   Integer
The LUNITS variable establishes the Linear Units mode. The variable is saved in the drawing and its initial value is 2.
1 - Scientific units mode, 2 - Decimal units mode, 3 - Engineering units mode, 4 - Architectural units mode, 5 - Fractional units mode.

LUPREC   Integer
The LUPREC variable establishes the linear units decimal places or denominator. The variable is saved in the drawing and its

initial value is 4.

LWDEFAULT   Enum
The LWDEFAULT variable controls the value for the default lineweight. The default lineweight can be set to any valid lineweight value in millimeters.

LWDISPLAY   Integer
The LWDISPLAY variable controls whether the lineweight is displayed on the Model or Layout tab. Its initial value 0 and is saved in drawing.
0 - Lineweight is not displayed,
1 - Lineweight is displayed.

LWUNITS   Integer
The LWUNITS variable controls whether lineweight units are displayed in inches or millimeters. Its initial value is 1 and is saved in registry.
0 - Inches, 1 - Millimeters.

MAXACTVP   Integer
The MAXACTVP variable specifies the maximum number of viewports to regenerate at one time. The initial value of this variable is 48.

MAXOBJMEM   Integer
Controls the object pager. Initial value is 0.

MAXSORT   Integer
The MAXSORT variable sets the maximum number of symbol names of file names that are to be sorted by listing commands. In case the total number of items is greater than this number, then no items are sorted. This variable is saved in registry and its initial value is 200.

MBUTTONPAN   Integer
The MBUTTONPAN variable controls the behavior of the third button or wheel on the pointing device. Its initial value is 1 and it is saved in registry.
0 - Supports the action defined in the

AutoCAD menu (.mnu) file, 1 - Supports panning by holding and dragging the button or wheel.

MEASUREINIT   Integer

The MEASUREMENT variable sets the drawing units as English or metric and controls which hatch pattern and linetype files an existing drawing uses when it's opened. It is saved in drawing and the initial value is 0.

0 - English(AutoCAD uses ANSI hatch patterns and linetypes0, 1 - Metric (AutoCAD uses ISOHatch and ISOLinetypes).

MEASUREMENT   Integer

The MEASUREMENT variable sets the drawing units as English or metric and controls which hatch pattern and linetype files for the current drawing only. It is saved in drawing and the initial value is 0.

0 - English(AutoCAD uses ANSI hatch patterns and linetypes0, 1 - Metric (AutoCAD uses ISOHatch and ISOLinetypes).

MENUCTL   Integer

The MENUCTL variable governs the page switching of the screen menu. This variable is saved in registry and its initial value is 1.

0 - For this value the screen menu does not switch pages in response to a keyboard command entry, 1 - For this value the screen menu switches pages in response to a keyboard command entry.

MENUECHO   Integer

The MENUECHO variable sets menu echo and prompt control bits. The initial value for this variable is 0. The variable is the addition of the following:

1 - The echo of menu items is suppressed, 2 - The display of system prompts during menu is suppressed, 4 - The ^P toggle of menu echoing is disabled, 8 - The input/output strings and debugging aid for DIESEL macros is displayed.

MENUNAME   String

The MENUNAME variable contains the MENUGROUP name. In case the prevailing primary menu has no MENUGROUP name, the menu file includes the path if the location of the file is not defined in the AutoCAD environment setting. This variable is saved in the application leader and is a read-only variable.

MIRRTEXT   Integer

The MIRRTEXT variable governs how the MIRROR command mirrors text. This variable is saved in the drawing and its initial value is 1.

0 - The text direction is retained, 1 - The text is mirrored.

MODEMACRO   String

The MODEMACRO variable shows a text string or DIESEL expression on the status line. This string reveals information like the name of the current drawing, time/date stamp, or special modes. The initial value of this variable is "".

MTEXTED   String

The MTEXTED variable sets the name of the program to be used for the editing of mtext objects. This variable is saved in registry and its initial value is "Internal".

NOMUTT   Short

The NOMUTT variable suppresses the message display (muttering) when it wouldn't normally be suppressed. Its initial value is 0 and it is not saved.

0 - Resumes normal muttering behavior, 1 - Suppresses muttering indefinitely.

OFFSETDIST   Real

This variable sets the default offset distance. If the value of this variable is less than zero then the offset distance can be specified with the through mode. If the value of this variable is greater than zero then the default offset distance is established. The initial value of this variable is 1.0000.

OFFSETGAPTYPE   Integer

The OFFSETGAPTYPE variable controls how to offset polylines when a gap is created as a result of offsetting the individual polyline segments. Its initial value is 0 and it is stored in registry.

0 - Extends the segments to fill the gap, 1 - Fills the gaps with a filleted arc segment where radius is the offset distance, 2 - Fills the gaps with a chamfered line segment.

OLEHIDE   Integer

Controls the display of OLE objects in AutoCAD. It is saved in registry and its initial value is 0.

0 - All OLE objects are visible, 1 - OLE objects are visible in paper space, 2 - OLE objects are visible in model space, 3 - No OLE objects are visible.

OLEQUALITY   Integer

The OLEQUALITY variable controls the default quality level for embedded OLE objects. Its initial value is 1 and it is stored in registry.

0 - Line art quality, such as an embedded spreadsheet, 1 - Text quality, such as an embedded Word document, 2 - Graphics quality, such as an embedded pie chart, 3 - Photograph quality, 4 - High quality photograph.

OLESTARTUP   Integer

The OLESTARTUP variable controls whether the source application of an embedded OLE object loads when plotting. Its initial value is 0 and it is saved in drawing.

0 - Does not load the OLE source application, 1 - Loads the OLE source application when plotting.

ORTHOMODE   Integer

The ORTHOMODE variable governs the orthogonal display of lines or polylines. This variable is saved in the drawing and its initial value is 0.

0 - The Ortho mode is turned off, 1 - The Ortho mode is turned on.

OSMODE   Integer

The OSMODE variable sets the running Object Snap modes using the following bit-codes:

0 - NONe object snap, 1 - ENDpoint object snap, 2 - MIDpoint object snap, 4 - CENter object snap, 8 - NODe object snap, 16 - QUAdrant object snap, 32 - INTersection object snap, 64 - INSertion object snap, 128 - PERpendicular object snap, 256 - TANgent object snap, 512 - NEArest object snap, 1024 - QUIck object snap, 2048 - APPint object snap. If you want to specify more than one object snap, enter the sum of their values. For example, if you want to specify the node and center object snaps, enter 4+8 = 12 as the value for the OSMODE variable. This variable is saved in the drawing and its initial value is 0.

OSNAPCOORD   Integer

Controls whether coordinates entered on the command line override running object snaps.

0 - Running object snap settings override keyboard entry, 1 - Keyboard entry overrides object snap settings, 2 - (Initial value) Keyboard entry overrides object snap setting except in scripts.

PAPERUPDATE   Integer

The PAPERUPDATE variable controls the display of a warning dialog when attempting to print a layout with a paper size different from the paper size specified by the default for the plotter configuration file. Its initial value is 0 and it is saved in registry.

0 - Displays a warning dialog box if the paper size specified in the layout is not supported by the plotter, 1 - Sets paper size to the configured paper size of the plotter configuration file.

PDMODE   Integer

The PDMODE variable sets Point Object Display mode. This variable is saved in the drawing and its initial value is 0.

PDSIZE   Real

The PDSIZE variable sets the display size of the point object. This variable is saved in the drawing and its initial value is 0.0000. 0 - For this value, point is created at 5 percent of the graphics height, >0 - In this case the value entered specifies the absolute size, <0 - In this case the value entered specifies the percentage of the viewport size.

PELLIPSE   Integer

The PELLIPSE controls the type of ellipse created with the ELLIPSE command. This variable is saved in the drawing and its initial value is 0. 0 - A true ellipse object is drawn, 1 - A polyline representation of an ellipse is drawn.

PERIMETER   Real

The PERIMETER variable holds the most recently perimeter value computed by AREA, DBLIST, or LIST commands. This variable is a read-only variable.

PFACEVMAX   Integer

The PFACEVMAX variable sets the maximum number of vertices per face. This variable is a read-only variable.

PICKADD   Integer

The PICKADD variable controls additive selection of objects. This variable is saved in registry and its initial value is 1. 0 - PICKADD variable is disabled, 1 - PICKADD variable is enabled. All the objects selected by any method are added to the selection set. If you want to remove objects from the selection set, hold down the Shift key and select the objects.

PICKAUTO   Integer

PICKAUTO variable controls the automatic windowing feature when the "Select object" prompt appears. This variable is saved in registry and its initial value is 1. 0 - PICKAUTO variable is disabled, 1 - PICKAUTO variable is enabled and a selection window is automatically drawn at the Select objects prompt.

PICKBOX   Integer

The PICKBOX variable sets the object selection target half height (in pixels). This variable is saved in registry and its initial value is 3.

PICKDRAG   Integer

The PICKDRAG variable governs the method of drawing a selection window: 0 - For this value the selection window is drawn by clicking the pointing device at one corner and then clicking again at the other corner of the window, 1 - For this value the selection window is drawn by clicking the pointing device at one corner, holding down the pick button, dragging the cursor, and finally releasing the pick button of the pointing device at the other corner of the window. This variable is saved in registry and its initial value is 0.

PICKFIRST   Integer

The PICKFIRST variable governs the method of object selection in such a manner that you can first select the object and then specify the desired edit or inquiry command. Its initial value is 1. 0 - PICKFIRST variable disabled, 1 - PICKFIRST variable enabled.

PICKSTYLE   Integer

The PICKSTYLE variable controls the associative hatch selection and group selection. This variable is saved in the drawing and its initial value is 1. 0 - Associative hatch selection and group selection not possible, 1 - Group selection possible, 2 - Associative hatch selection possible, 3 - Associative hatch selection and group selection possible.

PLATFORM   String
The PLATFORM variable specifies the platform of AutoCAD that is in use. This a read-only variable. Some of the platforms are: Microsoft Windows - Sun/SPARCstation, 386 DOS Extender - DECstation, Apple Macintosh - Silicon Graphics Iris Indigo

PLINEGEN   Integer
The PLINEGEN variable sets the linetype pattern generation around the vertices of a 2D polyline. This variable does not affect polylines with tapered segments. This variable is saved in the drawing and its initial value is 0.
0 - Polylines are generated with a dash at each vertex, 1 - Linetype is created in a continuous pattern around the vertices of the polyline.

PLINETYPE   Integer
Specifies whether AutoCAD uses optimized 2D polylines.
0 - PLINE creates old format polylines, 1 - PLINE creates optimized polylines, 2 - PLINE creates optimized polylines and the polylines in older drawings are converted on open.

PLINEWID   Real
The default polyline width is stored in this variable. This variable is saved in the drawing and its initial value is 0.0000

PLOTID   String
The PLOTID variable stores the current plotter's description. The plotter configuration can be changed by entering the plotter's full or partial description. This variable is saved in registry and its initial value is "".

PLOTROTMODE   Integer
The PLOTROTMODE variable controls the orientation of plots. This variable is saved in the drawing and its initial value is 1.
0 - The effective plotting area is rotated in order to align the corner with the Rotation icon with the paper at the lower-left for a rotation of 0, top-left for a rotation of 90, top-right for a rotation of 180, and lower left for a rotation of 270, 1 - The lower-left corner of the effective plotting area is aligned with the lower-left corner of the paper.

PLOTTER   Integer
The PLOTTER variable stores an integer number assigned for configured plotter. This integer number can be in the range of 0 to the number of configured plotters. You can change to some other configured plotter by entering the integer number assigned to the plotter. If you have 6 plotters, the valid numbers are 0, 1, 2, 3, 4, 5. This variable is saved in registry and its initial value is 0.

PLQUIET   Integer
The PLQUIET variable controls the display of optional dialog boxes and nonfatal errors for batch plotting and scripts. Its initial value is 0 and it is stored in registry.
0 - Displays plot dialog boxes and nonfatal errors, 1 - Logs nonfatal errors and doesn't display plot-related dialog boxes.

POLARADDANG   String
The POLARADDANG variable contains all user-defined polar angles. You can add up to 10 angles. Each angle can be up to 25 characters, separated with semicolons (;). its initial value is null and it is stored in registry.

POLARANG   Real
The POLARANG variable controls the polar angle increment. Values are 90, 45, 30, 22.5, 18, 15, 10, and 5. Its initial value is 90 and is saved in registry.

POLARDIST   Real

The POLARDIST variable controls the snap increment when the SNAPSTYL system variable is set to 1 (polar snap). Its initial value is 0.0000 and it is stored in registry.

POLARMODE   Integer

The POLARMODE variable Controls settings for polar and object snap tracking. Its initial value is 1 and it is saved in registry. The value is the sum of four bitcodes.
Polar angle measurements: 0 - Measure polar angles based on current UCS (absolute), 1 - Measure polar angles from selected objects (relative); Object snap tracking: 0 - Track orthogonally only, 2 - Use polar tracking settings in object snap tracking;Use additional polar tracking angles: 0 - No, 4 - Yes; Acquire object snap tracking points: 0 - Acquire automatically, 8 -Press SHIFT to acquire.

POLYSIDES   Integer

The POLYSIDES variable establishes the default number of sides for a polygon. This variable can take values in the range of 3 to 1024. The initial value of this variable is 4.

POPUPS   Integer

The POPUP variable shows the status of the presently configured display driver. This is a read-only variable.
0 - The dialog boxes, menu bar, pull-down menus, and icon menus are not supported, 1 - The dialog boxes, menu bar, pull-down menus, and icon menus are supported.

PROJECTNAME   String

Stores the current project name. Initial value is " ".

PROJMODE   Integer

The PROJMODE variable establishes the current Projection mode for Extend or Trim operations. Its initial value is 1.
0 - True 3D mode established (no projection), 1 - Projection to XY plane of the current UCS, 2 - Projection to current view plane.

PROXYGRAPHICS   Integer

Specifies whether images of proxy objects are saved in the drawing. The initial value is 1.
0 - Image is not saved, 1 - Image is saved.

PROXYNOTICE   Integer

Displays a notice when you open a drawing containing custom objects created by an application that is not present. The initial value is 1.
0 - No proxy warning displayed, 1 - Proxy warning displayed.

PROXYSHOW   Integer

Controls the display of proxy objects in a drawing. The initial value is 1.
0 - Proxy objects are not displayed, 1 - Graphic images are displayed for all proxy objects, 2 - Only bounding box is displayed for all proxy objects.

PSLTSCALE   Integer

The PSLTSCALE variable governs the paper space linetype scaling. This variable is saved in the drawing and its initial value is 1.
0 - Special linetype scaling not allowed. Linetype dash lengths depend on the drawing units of the space in which the objects were drawn, 1 - Linetype scaling governed by viewport scaling. In case TILEMODE is set to 0, dash lengths depend on the paper space drawing units, even if objects are in model space.

PSPROLOG   String

The PSPROLOG variable assigns a name for a prologue section which is to be read from the acad.psf file when PSOUT command is being used. This variable is saved in registry and its initial value is "".

PSQUALITY   Integer

The PSQUALITY variable governs the quality of rendering of PostScript images and also whether these images are drawn as filled objects or as outlines. This variable is saved in the drawing and its initial value is 75.

0 - The PostScript image generation is disabled, >0 - Any value greater than zero specifies the number of pixels per AutoCAD drawing unit for the PostScript resolution and fills outlines, <0 - Value less than zero specifies the number of pixels per AutoCAD drawing unit, but uses the absolute value. This causes the AutoCAD to display PostScript paths as non filled outlines.

PSTYLEMODE   Read only

The PSTYLEMODE variable indicates whether the current drawing is in a Color-Dependent or Named Plot Style mode. Its initial value is 0 and it is saved in drawing.

0 - Uses named plot style tables in the current drawing, 1 - Uses color-dependent plot style tables in the current drawing.

PSTYLEPOLICY   Integer

The PSTYLEPOLICY variable controls whether an object's color property is associated with its plot style. The new value you assign affects only newly created drawings. Its initial value is 1 and it is saved in registry.

0 - No association is made between color and plot style. 1 - An object's plot style is associated with its color.

PSVPSCALE   Real

The PSVPSCALE variable controls the view scale factor for all newly created viewports. The view scale factor is defined by comparing the ratio of units in paper space to the units in newly created model space viewports. A value of 0 means the scale factor is "Scaled to Fit". A scale must be a positive real value. Its initial value is 0 and it is not saved.

PUCSBASE   String

The PUCSBASE variable stores the name of the UCS that defines the origin and orientation of orthographic UCS settings in paper space only. Its initial value is "" and it is stored in drawing.

QTEXTMODE   Integer

The QTEXTMODE controls the Quick Text mode. This variable is saved in the drawing and its initial value is 0.

0 - The Quick Text mode is turned off and characters are displayed, 1 - The Quick Text mode is turned on and a box instead of text is displayed.

RASTERPREVIEW Integer

The RASTERPREVIEW variable determines whether the drawing preview images are saved with the drawing and in which format they will be saved. This variable is saved in the registry and its initial value is 1.

0 - No preview image created, 1 - Preview image is created.

REFEDITNAME   String

The REFEDITNAME variable indicates whether a drawing is in a reference-editing state; also, stores the reference file name. Its initial value is "" and it is not saved. This is a read-only variable.

REGENMODE   Integer

The REGENMODE variable controls the automatic regeneration of the drawing. This variable is saved in the drawing and its initial value is 1.

0 - REGENAUTO is turned off, 1 - REGENAUTO is turned on.

RE-INIT   Integer

The RE-INIT variable reinitializes the I/O ports, plotter, digitizer, display, and acad.pgp file. The following bit-codes are used for this process:

0 - Reinitialization not allowed (initial value), 1 - Reinitialization of digitizer port, 2 -

Reinitialization of plotter port, 4 - Reinitialization of digitizer, 8 - Reinitialization of display, 16 - Reinitialization of PGP file. You can specify more than one reinitialization by entering the sum of the values of the desired reinitializations.

RTDISPLAY   Integer

Controls the display of raster images during realtime zoom or pan.
0 - Displays raster image content, 1 - Displays raster image outline only (initial value).

SAVEFILE   String

The present auto-save filename is held in the SAVEFILE variable. This variable is a read-only variable. The initial value is "auto.sv$".

SAVEFILEPATH   String

The SAVEFILEPATH variable specifies the path to the directory for all automatic save files for the AutoCAD session. Its initial value is "C:\Temp\" and it is saved in registry.

SAVENAME   String

You can save the current drawing to a different name and this name is held in the SAVENAME variable. This variable is a read-only variable.

SAVETIME   Integer

AutoCAD has provided the facility of automatically saving your work at specific intervals. You can specify the automatic save time intervals (in minutes) with the SAVETIME variable. This variable is set to an initial value of 120.
0 - Automatic save facility is disabled, 0 - The drawing is saved according to the intervals specified. Once you make changes to the drawing, the SAVETIME timer starts. SAVE, SAVEAS, or QSAVE commands reset and restart this timer. AutoCAD saves the

drawing under the filename auto.sv$.

SCREENBOXES   Integer

The SCREENBOXES variable stores the number of boxes in the screen menu area of the graphics area. In case the screen menu is disabled, this variable is set to zero. The value of this variable is susceptible to change during an editing session on platforms that allow the AutoCAD graphics window to be resized or the screen menu to be reconfigured while you are in an editing session. This is a read-only variable.

SCREENMODE   Integer

The SCREENMODE variable holds a bit-code specifying the graphics/text state of the AutoCAD display. This is a read-only variable. Following are the bit values:
0 - Text screen is displayed, 1 - Graphics mode is displayed, 2 - Dual-screen display (text and graphics) is displayed.

SCREENSIZE   2D point

This variable holds the current viewport size in pixels. This is a read-only variable.

SDI   Integer

This variable controls whether AutoCAD runs in single- or multiple-document interface. Its initial value is 0 and it is saved in registry.
0 - Turns on multiple-drawing interface,
1 - Turns off multiple-drawing interface,
2 - (Read-only) Multiple-drawing interface is disabled, 3 - (Read-only) Multiple-drawing interface is disabled.

SHADEDGE   Integer

The SHADEDGE variable governs the shading of edges in rendering. This variable is saved in the drawing and its initial value is 3.
0 - Faces are shaded and edges are not highlighted, 1 - Faces are shaded and edges are drawn in background color, 2 - Faces are not filled and edges are in object color, 3 - Faces

are in object color and edges are drawn in background color.

SHADEDIF   Integer

The SHADEIF variable establishes the ratio of diffuse reflective light to ambient light (in percent of diffuse reflective light). This variable is saved in the drawing and its initial value is 70.

SHORTCUTMENU   Integer

This variable Controls whether Default, Edit, and Command mode shortcut menus are available in the drawing area. Its initial value is 11 and it is stored in registry. The following bitcodes are used by SHORTCUTMENU:

0 - Disables all Default, Edit, and Command mode shortcut menus, restoring R14 legacy behavior, 1 - Enables Default mode shortcut menus, 2 - Enables Edit mode shortcut menus, 4 - Enables Command mode shortcut menus. In this case, the Command mode shortcut menu is available whenever a command is active, 8 - Enables Command mode shortcut menus only when command options are currently available from the command line.

SHPNAME   String

The SHPNAME variable establishes the name of the default shape. The initial value for this variable is "". To set no default enter a period (.).

SKETCHINC   Real

The SKETCHINC specifies the record increment for the SKETCH command. This variable is saved in the drawing and its initial value is 0.1000.

SKPOLY   Integer

The SKPOLY variable decides whether SKETCH command generates lines or polylines. This variable is saved in the drawing and its initial value is 0.

0 - Lines are generated, 1 - polylines are generated.

SNAPANG   Real

The SNAPANG variable specifies the snap/grid rotation angle relative to the UCS for the current viewport. This variable is saved in the drawing and its initial value is 0. Changes to this variable are manifested only after a redraw is performed.

SNAPBASE   2D point

The SNAPBASE variable specifies the snap/grid origin point (in UCS X, Y coordinates) for the current viewport. This variable is saved in the drawing and its initial value is 0.0000,0.0000. Changes to this variable are manifested only after a redraw is performed.

SNAPISOPAIR   Integer

The SNAPISOPAIR variable controls the current isometric plane for the current viewport. This variable is saved in the drawing and its initial value is 0.

0 - Left, 1 - Top, 2 - Right.

SNAPMODE   Integer

The SNAPMODE variable controls the Snap mode. This variable is saved in the drawing and its initial value is 0.

0 - Snap disabled, 1 - Snap enabled for the current viewport.

SNAPSTYL   Integer

The SNAPSTYL variable establishes the snap style for the current viewport. This variable is saved in the drawing and its initial value is 0.

0 - Standard, 1 - Isometric.

SNAPTYPE   Integer

This variable the snap style for the current viewport. Its initial value is 0 and it is saved in registry.

0 - Grid, or standard snap, 1 - Polar snap.

SNAPUNIT   2D point

The SNAPUNIT variable specifies the X and

Y snap spacing for the current viewport. This variable is saved in the drawing and its initial value is 0.5000,0.5000. The changes to this variable are manifested only after a redraw is performed.

SOLIDCHECK   Integer

This variable turns the solid validation on and off for the current AutoCAD session. Its initial value is 1 and it is not saved.
0 - Turns off solid validation, 1 - Turns on solid validation.

SORTENTS   Integer

The SORTENTS variable governs the display of object sort order operations using the following values:
0 - SORTENTS is disabled, 1 - Sorts for object selection, 2 - Sorts for object snap, 4 - Sorts for redraw, 8 - Sorts for MSLIDE slide creation, 16 - Sorts for regens, 32 - Sorts for plotting, 64 - Sorts for PostScript output. More than one options can be selected by specifying the sum of the values of these options. Its initial value is 96. This value specifies sort operations for plotting and PostScript output.

SPLFRAME   Integer

The SPLFRAME variable governs the display of spline-fit polylines. This variable is saved in the drawing and its initial value is 0.
0 - The control polygon for spline fit polylines is not displayed. The fit surface of a polygon mesh is displayed while as the defining mesh is not displayed. Also invisible edges of 3D faces or polyface meshes are not displayed, 1 - The control polygon for spline fit polylines is displayed. The fit surface of a polygon mesh is not displayed while as the defining mesh is displayed. Also invisible edges of 3D faces or polyface meshes are displayed.

SPLINESEGS   Integer

The SPLINESEGS variable governs the number of line segments used to construct each spline. Hence with this variable you can control the smoothness of the curve. This variable is saved in the drawing and its initial value is 8. With this value a reasonably smooth curve is generated which does not need a much regeneration time. The greater the value of this variable, the smoother the curve and greater the regeneration time and the space occupied by the drawing file.

SPLINETYPE   Integer

The SPLINETYPE variable specifies the type of spline curve that will be generated by Spline option of PEDIT command. This variable is saved in the drawing and its initial value is 6.
5 - Quadratic B-spline is generated, 6 - Cubic B-spline is generated.

SURFTAB1   Integer

The SURFTAB1 variable governs the number of intervals (tabulated surfaces) to be generated for TABSURF and RULESURF commands along the path curve. This variable also defines the mesh density in the M direction for REVSURF and EDGESURF commands. This variable is saved in the drawing and its initial value is 6. In case the path curve is a line, arc, circle, spline-fit polyline, or an ellipse, the path curve is divided into intervals equal to the value of SURFTAB1 by the tabulation lines. Else, if the path curve is a polyline (not spline-fit), the tabulation lines are generated at the ends of the polyline segments and if there are any arc segments, each segment is divided into intervals equal to the value of SURFTAB1 by the tabulation lines.

SUTFTAB2   Integer

The SURFTAB2 variable defines the mesh density in the N direction for REVSURF and EDGESURF commands. This variable is saved in the drawing and its initial value is 6.

SURFTYPE   Integer

The SURFTYPE variable governs the type of surface-fitting to be performed by the Smooth option of PEDIT command. This variable is saved in the drawing and its initial value is 6.

5 - Quadratic B-spline surface, 6 - Cubic B-spline surface, 8 - Bezier surface.

SURFU   Integer

The SURFU variable specifies the surface density of polygon meshes in the M direction. This variable is saved in the drawing and its initial value is 6.

SURFV   Integer

The SURFV variable specifies the surface density of polygon meshes in the N direction. This variable is saved in the drawing and its initial value is 6.

SYSCODEPAGE   String

The SYSCODEPAGE variable expresses the system code page specified in the acad.xmf file. This variable is a read-only variable and is saved in the drawing.

TABMODE   Integer

The TABMODE variable governs the use of Tablet mode.

0 - Tablet mode disabled (initial value), 1 - Tablet mode enabled.

TARGET   3D point

The TARGET variable holds the position of the target point (in UCS coordinates) for the current viewport. This is a read-only variable and is saved in the drawing.

TDCREATE   Real

The TDCREATE variable holds the creation time and date of a drawing. This is a read-only variable and is saved in the drawing.

TDINDWG   Real

The TDINDWG variable holds the total editing time. This is a read-only variable and is saved in the drawing.

TDUCREATE   Real

This variable stores the universal time and date the drawing was created. This is a read-only variable and saved in drawing.

TDUPDATE   Real

The TDUPDATE variable holds the time and date of most recent update/save. This is a read-only variable and is saved in the drawing.

TDUSRTIMER   Real

The TDUSRTIMER variable stores the user elapsed timer. This is a read-only variable and is saved in the drawing.

TDUUPDATE   Real

This variable stores the universal time and date of the last update/save. This is a read-only variable and stored in drawing.

TEMPPREFIX   String

The TEMPPREFIX variable stores the directory name configured for the placement of temporary files. The path separator is included. This is a read-only variable.

TEXTEVAL   Integer

The TEXTEVAL variable determines the procedure of evaluation of text strings. The initial value for this variable is 0.

0 - All the responses to prompts for text strings and attributes are accepted as literals, 1 - In case the starting character of the text string is "(" or "!", it is treated as an AutoLISP expression. The TEXTEVAL setting does not affect the TEXT command. TEXT command accepts all input as literals.

TEXTFILL   Integer

The TEXTFILL variable governs the filling of TrueType fonts. This variable is saved in the registry and has an initial value of 1.

0 - The text is displayed as outlines, 1 - The

text is displayed as filled images.

TEXTQLTY   Real

The TEXTQLTY variable defines the resolution of TrueType, Bitstream, and Abode Type 1 fonts. The higher the value of this variable, the higher the resolution and lower the display and plotting speed. On the other hand the lower the value of this variable, the lower the resolution and higher the display and plotting speed. This variable is saved in the drawing and can take values in the range of 0 to 100.0, its initial value is 50.

TEXTSIZE   Real

The TEXTSIZE variable controls the text height of the text drawn with the current text style. But this is possible only if the style does not have a fixed height. This variable is saved in the drawing and its initial value is 0.2000.

TEXTSTYLE   String

The TEXTSTYLE variable stores the name of the current text style. This variable is saved in the drawing and its initial value is STANDARD.

THICKNESS   Real

The THICKNESS variable defines the current 3D thickness. This variable is saved in the drawing and its initial value is 0.0000.

TILEMODE   Integer

The TILEMODE variable governs entry into paper space and also how the AutoCAD viewports act. This variable is saved in the drawing and its initial value is 1.
0 - The paper space and viewport objects are enabled. The graphics area is cleared and you are prompted to use the MVIEW command to define viewports, 1 - Release 10 Compatibility mode is enabled. Automatically you are taken into Tiled Viewport mode and previously active viewport configuration is restored on the screen. Paper

space objects including viewport objects are not displayed. MSPACE, PSPACE, VPLAYER, and MVIEW commands are disabled.

TOOLTIPS   Integer

The TOOLTIPS variable is concerned with the Windows version of AutoCAD and determines the display of ToolTips. Its initial value is 1.
0 - The display of ToolTips is turned off, 1 - The display of ToolTips is turned on.

TRACEWID   Real

The TRACEWID variable establishes default value for the width of the trace. This variable is saved in the drawing and its initial value is 0.0500.

TRACKPATH   Integer

This variable controls the display of polar and object snap tracking alignment paths. Its initial value is 0 and it is saved in registry.
0 - Displays full screen object snap tracking path, 1 - Displays object snap tracking path only between the alignment point and From point to cursor location, 2 - Does not display polar tracking path, 3 - Does not display polar or object snap tracking paths.

TREEDEPTH   Integer

The TREEDEPTH variable specifies how many times the tree-structured spatial index may divide into branches. This variable is saved in the drawing and its initial value is 3020.
0 -The spatial index is totally suppressed. In this case the objects are processed in database order and hence it is not necessary to set the SORTENTS variable, >0 - TREEDEPTH variable is enabled. You can enter an integer of up to four digits. The first two digits indicate the depth of model space nodes and the second two digits indicate the depth of paper space nodes, <0 - If the value is negative then the model space

objects are treated as 2D objects. Negative values are relevant for 2D drawings. This way memory is more efficiently utilized and there is no trade-off with the performance.

TREEMAX   Integer

The TREEMAX variable sets the limit to the maximum number of nodes in the spatial index. This way the memory use during regeneration of drawing is limited. Its initial value is 10000000.

TRIMMODE   Integer

The TRIMMODE variable determines whether selected edges for chamfers and fillets will be trimmed.

0 - Selected edges are not trimmed after chamfering and filleting, 1 - Selected edges are trimmed after chamfering and filleting (initial value).

TSPACEFAC   Real

This variable controls the multiline text line spacing distance measured as a factor of text height. The valid values are 0.25 to 4.0. The initial value is 1 and it is not saved.

TSPACETYPE   Integer

Controls the type of line spacing used in multiline text. At Least adjusts line spacing based on tallest characters in a line. Exactly uses the specified line spacing, regardless of individual character sizes. Its initial value is 1 and it is not saved.

1 - At Least, 2 - Exactly.

TSTACKALIGN   Integer

This variable controls the vertical alignment of stacked text. Its initial value is 1 and it is saved in drawing.

0 - Bottom aligned, 1 - Center aligned, 2 - Top aligned.

TSTACKSIZE   Integer

This variable controls the percentage of stacked text fraction height relative to selected text's current height. The valid value

ranges from 1 to 127. Its initial value is 70 and it is saved in drawing.

UCSAXISANG   Integer

This variable sores the default angle when rotating the UCS around one of its axes using the X, Y, or Z options of the UCS command. The values must be entered as an angle in degrees (valid values are: 5, 10, 15, 18, 22.5, 30, 45, 90, 180). Its initial value is 90 and it is stored in registry.

UCSBASE   String

This variable stores the name of the UCS that defines the origin and orientation of orthographic UCS settings. The valid values include any named UCS. Its initial value is "World" and is stored in drawing.

UCSFOLLOW   Integer

The UCSFOLLOW variable controls the automatic displaying of a plan view when you switch from one UCS to another. All the viewports have the UCSFOLLOW facility and hence you need to specify the UCSFOLLOW setting separately for each viewport. This variable is saved in the drawing and its initial value is 0.

0 - Switch from one UCS to another, does not alter the view, 1 - Plan view of the new UCS is automatically displayed when you switch from one UCS to another.

UCSICON   Integer

The UCSICON variable displays the present UCS icon using bit-code for the current viewport. The value of this variable is the sum of the following:

1 - Icon display is enabled, 2 - The icon moves to the UCS origin if the icon display is enabled. In case more than one viewport is active, each of the viewport can have a different value for the UCSICON variable. If you are in paper space, the UCSICON variable will contain the setting for the UCS icon of the paper space. This variable is saved in the drawing and its initial value is

1.

UCSNAME   String

The UCSNAME variable contains the name of the current UCS. This is a read-only variable and is saved in the drawing. In case the current UCS is unnamed, then a null string is returned.

UCSORG   3D point

The coordinate value of the origin of the current UCS is held in the UCSORG variable. This is a read-only variable and is saved in the drawing.

UCSORTHO   Integer

This variable Determines whether the related orthographic UCS setting is restored automatically when an orthographic view is restored. Its initial value is 1 and it is stored in registry.

0 - Specifies that the UCS setting remains unchanged when an orthographic view is restored, 1 - Specifies that the related orthographic UCS setting is restored automatically when an orthographic view is restored.

UCSVIEW   Integer

This variable determines whether the current UCS is saved with a named view. Its initial value is 1 and is saved in registry.

0 - Does not save current UCS with a named view, 1 - Saves current UCS whenever a named view is created.

UCSVP   Integer

The UCSVP variable determines whether the UCS in active viewports remains fixed or changes to reflect the UCS of the currently active viewport. Its initial value is 1 and it is stored in drawing (Viewport specific).

0 - Unlocked; UCS reflects the UCS of the current viewport, 1 - Locked; UCS stored in viewport, and is independent of the UCS of the current viewport.

UCSXDIR   3D point

The X axis direction of the current UCS for the current space is held in UCSXDIR variable. This is a read-only variable and is saved in the drawing.

UCSYDIR   3D point

The Y axis direction of the current UCS for the current space is held in UCSYDIR variable. This is a read-only variable and is saved in the drawing.

UNDOCTL   Integer

The UNDOCTL variable holds a bit-code expressing the state of the UNDO command. This is a read-only variable. The value of this value is the addition of following values:

0 - UNDO command is disabled, 1 - UNDO command is enabled, 2 - Just one command can be undone, 4 - Auto-group mode is enabled, 8 - Some group is presently active.

UNDOMARKS   Integer

The UNDOMARKS variable contains the number of marks that have been put in the UNDO command's control stream by the Mark option. In case a group is presently active, the Mark and Back options cannot be accessed. This variable is a read-only variable.

UNITMODE   Integer

The UNITMODE variable governs the units display format. This variable is saved in the drawing and its initial value is 0.

0 - The fractional, feet and inches, and surveyor's angles are displayed as previously defined, 1 - The fractional, feet and inches, and surveyor's angles are displayed in the input format. This variable is saved in the drawing and its initial value is 0.

USERI1-5   Integer

Stores and retrieves integer values. Initial value is 0.

USERR1-5   Integer

Stores and retrieves real numbers. Initial value is 0.0000.

USERS1-5   Integer

Stores and retrieves text string data. Initial value is " ".

VIEWCTR   3D point

The VIEWCTR variable stores the center of view in the current viewport, defined in the UCS coordinates. This variable is a read-only variable and is saved in the drawing.

VIEWDIR   3D vector

The VIEWDIR variable contains the viewing direction in the current viewport expressed in the UCS coordinates. The camera position is expressed as a 3D offset from the target position. This variable is a read-only variable and is saved in the drawing.

VIEWMODE   Integer

The VIEWMODE variable governs Viewing mode for the current viewport using bit-code. The value for this variable is the addition of the following bit values:
0 - Viewing mode disabled, 1 - Perspective view active, 2 - Front clipping on, 4 - Back clipping on, 8 - UCS Follow mode on, 16 - Front clip not at eye. In case it is on, the front clipping plane is determined by the front clip distance stored in the FRONTZ variable. If it is off, the front clipping plane passes through the camera point and in this case FRONTZ variable is not taken into consideration. If the front clipping bit (2) is off then this flag is neglected. This variable is a read-only variable and is saved in the drawing.

VIEWSIZE   Real

The VIEWSIZE variable contains the view height in the current viewport and is defined in the drawing units. This variable is a read-only variable and is saved in the drawing.

VIEWTWIST   Real

The VIEWTWIST variable contains the view twist angle for the current viewport. This variable is a read-only variable and is saved in the drawing.

VISRETAIN   Integer

The VISRETAIN variable determines whether changes to the visibility of layers in xref are saved in the current drawing.
0 - Changes to On/Off, Freeze/Thaw, color, and linetype settings for the xref-dependent layers are not saved in the current drawing,
1 - Changes to the xref layer definitions in the current drawing are saved with the current drawing.

VSMAX   3D point

The VSMAX variable contains the upper-right corner of the virtual screen of the current viewport and is expressed in UCS coordinates. This variable is a read-only variable and is saved in the drawing.

VSMIN   3D point

The VSMIN variable contains the lower-left corner of the virtual screen of the current viewport and is expressed in UCS coordinates. This variable is a read-only variable and is saved in the drawing.

WHIPARC   Integer

This variable controls whether the display of circles and arcs is smooth. Its initial value is 0 and it is saved in registry.
0 - Circles and arcs are not smooth, but rather are displayed as a series of vectors,
1 - Circles and arcs are smooth, displayed as true circles and arcs.

WMFBKGND   Integer

The WMFBKGND variable controls whether the background display of AutoCAD objects is transparent in other applications when these objects are Output to a Windows metafile using the **WMFOUT** command, Copied to the Clipboard in AutoCAD and

Appendix-D

pasted as a Windows metafile or Dragged and dropped from AutoCAD as a Windows metafile. Its initial value is 1 and the values are not saved.

0 - The background is transparent, 1 - The background color is the same as the AutoCAD current background color.

WORLDUCS   Integer

The WORLDUCS variable expresses whether the UCS is the same as the WCS. This variable is a read-only variable.

0 - Current UCS and WCS are different, 1 - Current UCS and WCS are not different.

WORLDVIEW   Integer

The WORLDVIEW variable determines whether UCS changes to WCS during DVIEW or VPOINT commands. This variable is saved in the drawing and its initial value is 1.

0 - Current UCS is not changed, 1 - Current UCS is changed to WCS till the DVIEW or VPOINT command is in progress. The DVIEW and VPOINT command input is with respect to the current UCS.

WRITESTAT   Read-only

This variable indicates whether a drawing file is read-only or can be written to. For developers who need to determine write status through AutoLISP. Its initial value is 1 and values are not saved.

0 - Can't write to the drawing, 1 - Can write to the drawing.

XCLIPFRAME   Integer

This variable controls visibility of xref clipping boundaries and its initial value is 0.

0 - Clipping boundary is not visible, 1 - Clipping boundary is visible.

XEDIT   Integer

This variable controls whether the current drawing can be edited in-place when being referenced by another drawing. Its initial

value is 1 and it is saved in drawing.

0 - Can't use in-place reference editing, 1 - Can use in-place reference editing.

XFADECTL   Integer

This variable controls the fading intensity for references being edited in-place. Its initial value is 50 and it is stored in registry.

0 - 0 percent fading, minimum value, 90 - 0 percent fading, maximum value

XLOADCTL   Integer

Turns demand load on and off and controls whether it loads the original drawing or a copy. Initial value is 1.

0 - Turns off demand loading; entire drawing is loaded, 1 - Turns on demand loading; reference file is kept open, 2 - Turns on demand loading; a copy of reference file is opened.

XLOADPATH   String

Creates a path for storing temporary copies of demand-loaded xref files. Initial value is " ".

XREFCTL   Integer

The XREFCTL variable determines whether AutoCAD writes .xlg files (external reference log files). This variable is saved in registry and its initial value is 0.

0 - Xref log files are not written, 1 - Xref log files are written.

ZOOMFACTOR   Integer

This value accepts an integer between 3 and 100 as valid values. The higher the number, the more incremental the change applied by each mouse-wheel's forward/backward movement. Its initial value is 10 and it is stored in registry.

Index